Nanomaterials for Chemical Engineering

Nanomaterials for Chemical Engineering

Editor

Meiwen Cao

MDPI • Basel • Beijing • Wuhan • Barcelona • Belgrade • Manchester • Tokyo • Cluj • Tianjin

Editor
Meiwen Cao
College of Chemistry and
Chemical Engineering
China University of
Petroleum (Hua Dong)
Qingdao
China

Editorial Office
MDPI
St. Alban-Anlage 66
4052 Basel, Switzerland

This is a reprint of articles from the Special Issue published online in the open access journal *Nanomaterials* (ISSN 2079-4991) (available at: www.mdpi.com/journal/nanomaterials/special_issues/nano_chemical_engineering).

For citation purposes, cite each article independently as indicated on the article page online and as indicated below:

LastName, A.A.; LastName, B.B.; LastName, C.C. Article Title. *Journal Name* **Year**, *Volume Number*, Page Range.

ISBN 978-3-0365-7537-7 (Hbk)
ISBN 978-3-0365-7536-0 (PDF)

© 2023 by the authors. Articles in this book are Open Access and distributed under the Creative Commons Attribution (CC BY) license, which allows users to download, copy and build upon published articles, as long as the author and publisher are properly credited, which ensures maximum dissemination and a wider impact of our publications.

The book as a whole is distributed by MDPI under the terms and conditions of the Creative Commons license CC BY-NC-ND.

Contents

About the Editor vii

Meiwen Cao
Development of Functional Nanomaterials for Applications in Chemical Engineering
Reprinted from: *Nanomaterials* **2023**, *13*, 609, doi:10.3390/nano13030609 1

Tatyana Sergeevna Sazanova, Leonid Alexandrovich Mochalov, Alexander Alexandrovich Logunov, Mikhail Alexandrovich Kudryashov, Diana Georgievna Fukina and Maksim Anatolevich Vshivtsev et al.
Influence of Temperature Parameters on Morphological Characteristics of Plasma Deposited Zinc Oxide Nanoparticles
Reprinted from: *Nanomaterials* **2022**, *12*, 1838, doi:10.3390/nano12111838 5

Rama Gaur, Syed Shahabuddin, Irfan Ahmad and Nanthini Sridewi
Role of Alkylamines in Tuning the Morphology and Optical Properties of SnS_2 Nanoparticles Synthesized by via Facile Thermal Decomposition Approach
Reprinted from: *Nanomaterials* **2022**, *12*, 3950, doi:10.3390/nano12223950 17

Yabin Shao, Qing He, Lingling Xiang, Zibin Xu, Xiaoou Cai and Chen Chen
Strengthened Optical Nonlinearity of V_2C Hybrids Inlaid with Silver Nanoparticles
Reprinted from: *Nanomaterials* **2022**, *12*, 1647, doi:10.3390/nano12101647 29

Hongyuan Liu, Xunjun He, Jie Ren, Jiuxing Jiang, Yongtao Yao and Guangjun Lu
Terahertz Modulation and Ultrafast Characteristic of Two-Dimensional Lead Halide Perovskites
Reprinted from: *Nanomaterials* **2022**, *12*, 3559, doi:10.3390/nano12203559 39

Lintao Wang and Jing Yang
Zirconia-Doped Methylated Silica Membranes via Sol-Gel Process: Microstructure and Hydrogen Permselectivity
Reprinted from: *Nanomaterials* **2022**, *12*, 2159, doi:10.3390/nano12132159 51

Liaqat Ali, Abdul Manan and Bagh Ali
Maxwell Nanofluids: FEM Simulation of the Effects of Suction/Injection on the Dynamics of Rotatory Fluid Subjected to Bioconvection, Lorentz, and Coriolis Forces
Reprinted from: *Nanomaterials* **2022**, *12*, 3453, doi:10.3390/nano12193453 69

Andrey V. Shkolin, Evgeny M. Strizhenov, Sergey S. Chugaev, Ilya E. Men'shchikov, Viktoriia V. Gaidamavichute and Alexander E. Grinchenko et al.
Natural Gas Storage Filled with Peat-Derived Carbon Adsorbent: Influence of Nonisothermal Effects and Ethane Impurities on the Storage Cycle
Reprinted from: *Nanomaterials* **2022**, *12*, 4066, doi:10.3390/nano12224066 89

Chenxiaoyu Zhang, Shaobin Yang, Xu Zhang, Yingkai Xia and Jiarui Li
Extended Line Defect Graphene Modified by the Adsorption of Mn Atoms and Its Properties of Adsorbing CH_4
Reprinted from: *Nanomaterials* **2022**, *12*, 697, doi:10.3390/nano12040697 113

Veena Sodha, Syed Shahabuddin, Rama Gaur, Irfan Ahmad, Rajib Bandyopadhyay and Nanthini Sridewi
Comprehensive Review on Zeolite-Based Nanocomposites for Treatment of Effluents from Wastewater
Reprinted from: *Nanomaterials* **2022**, *12*, 3199, doi:10.3390/nano12183199 131

Qiuming Zhang, Xin Liao, Shaobo Liu, Hao Wang, Yin Zhang and Yongxiang Zhao
Tuning Particle Sizes and Active Sites of Ni/CeO$_2$ Catalysts and Their Influence on Maleic Anhydride Hydrogenation
Reprinted from: *Nanomaterials* **2022**, *12*, 2156, doi:10.3390/nano12132156 . . . . **161**

Shaobo Liu, Xin Liao, Qiuming Zhang, Yin Zhang, Hao Wang and Yongxiang Zhao
Crystal-Plane and Shape Influences of Nanoscale CeO$_2$ on the Activity of Ni/CeO$_2$ Catalysts for Maleic Anhydride Hydrogenation
Reprinted from: *Nanomaterials* **2022**, *12*, 762, doi:10.3390/nano12050762 . . . . **179**

Yihao Li, Hepan Zhao, Wei Xue, Fang Li and Zhimiao Wang
Transesterification of Glycerol to Glycerol Carbonate over Mg-Zr Composite Oxide Prepared by Hydrothermal Process
Reprinted from: *Nanomaterials* **2022**, *12*, 1972, doi:10.3390/nano12121972 . . . . **195**

Ana Luisa Farias Rocha, Ronald Zico de Aguiar Nunes, Robert Saraiva Matos, Henrique Duarte da Fonseca Filho, Jaqueline de Araújo Bezerra and Alessandra Ramos Lima et al.
Alternative Controlling Agent of *Theobroma grandiflorum* Pests: Nanoscale Surface and Fractal Analysis of Gelatin/PCL Loaded Particles Containing *Lippia origanoides* Essential Oil
Reprinted from: *Nanomaterials* **2022**, *12*, 2712, doi:10.3390/nano12152712 . . . . **211**

Yujie Yang, Zhen Liu, Hongchao Ma and Meiwen Cao
Application of Peptides in Construction of Nonviral Vectors for Gene Delivery
Reprinted from: *Nanomaterials* **2022**, *12*, 4076, doi:10.3390/nano12224076 . . . . **233**

Syed Sadiq Ali, Agus Arsad, Kenneth L. Roberts and Mohammad Asif
Effect of Inlet Flow Strategies on the Dynamics of Pulsed Fluidized Bed of Nanopowder
Reprinted from: *Nanomaterials* **2023**, *13*, 304, doi:10.3390/nano13020304 . . . . **255**

Silvia Sfameni, Tim Lawnick, Giulia Rando, Annamaria Visco, Torsten Textor and Maria Rosaria Plutino
Functional Silane-Based Nanohybrid Materials for the Development of Hydrophobic and Water-Based Stain Resistant Cotton Fabrics Coatings
Reprinted from: *Nanomaterials* **2022**, *12*, 3404, doi:10.3390/nano12193404 . . . . **271**

About the Editor

Meiwen Cao

Meiwen Cao received his Ph.D. degree in physical chemistry from the Institute of Chemistry, Chinese Academy of Sciences, in 2008 under the supervision of Prof. Yilin Wang. Then he joined the faculty of China University of Petroleum (East China) and works as an associate professor. Between 2017 and 2018, he worked as a senior visiting scholar in the University of Manchester with Prof. Jianren Lu. His research focuses on peptide-based supramolecular assembly for functional materials fabrication and application in drug delivery and gene therapy.

Editorial
Development of Functional Nanomaterials for Applications in Chemical Engineering

Meiwen Cao

State Key Laboratory of Heavy Oil Processing, Department of Biological and Energy Chemical Engineering, College of Chemistry and Chemical Engineering, China University of Petroleum (East China), Qingdao 266580, China; mwcao@upc.edu.cn

Nanomaterials are materials with particle sizes of less than 100 nm in at least one of their dimensions. The special structure of nanomaterials endows them unique characteristics different from those of bulk materials and individual atoms, which are known as surface effect, quantum size effect, and macroscopic quantum tunneling effect, etc. Therefore, nanomaterials usually exhibit novel optical, acoustic, electrical, magnetic, thermal, and catalytic properties, and have been widely applied in various fields. Nowadays, the production and application of nanomaterials in the chemical industry are very common, especially in the fields of new catalysts, selective adsorbents, corrosion prevention coatings, environmental protection, and other biopharmaceuticals and medical instruments, playing important roles in promoting the development of society and human beings.

This Special Issue, Nanomaterials for Chemical Engineering, focuses on the development of functional nanomaterials in the chemical engineering field. It collects 14 original research papers and 2 comprehensive review papers by the excellent scientists from relevant fields, covering the topics of development of novel nanomaterials and synthesis methods, experimental characterization, and computational modeling studies, as well as exploitation in devices and practical applications.

The synthesis, characterization, and property investigation of new nanomaterials are eternal prerequisites to meet the needs of nanotechnology development for chemical engineering applications. Sazanova et al. [1] reported the synthesis and characterization of zinc oxide (ZnO) nanoparticles by plasma-enhanced chemical vapor deposition, establishing that the synthesizing parameters of zinc source temperature and reactor temperature can effectively control the size and morphology of ZnO nanostructures. Gaur and coworkers [2] synthesized SnS_2 nanoparticles using a thermal decomposition approach and produced novel morphologies (e.g., nanoparticles, nanoplates, and flower-like morphologies assembled from flakes) by using different alkylamines as capping agents. Shao and coworkers [3] synthesized the Ag@MXene hybrids and studied their nonlinear optical characteristics, making great contributions to the development of ultrathin optoelectronic nanodevices and optical limiters. Liu et al. [4] fabricated 2D R-P type $(PEA)_2(MA)_2Pb_3I_{10}$ perovskite films on quartz substrates and studied their terahertz and ultrafast photoelectric response characteristics, demonstrating their potential applications in solar cells and photoelectric devices. Wang and coworkers [5] prepared methyl-modified ZrO_2-SiO_2 (ZrO_2-$MSiO_2$) membranes via the sol–gel method and characterized their physical–chemical properties. With excellent hydrothermal stability and regeneration capability, the ZrO_2-$MSiO_2$ membranes have significant potential in steam-stable hydrogen permselective applications. By incorporating Gyrotactic microbes to prevent the bioconvection of small particles and to improve consistency, Ali et al. [6] discussed the relevance of Lorentz and Coriolis forces on the kinetics of gyratory Maxwell nanofluids flowing against a continually stretched surface, contributing to the areas of elastomers, mineral productivity, paper-making, biosensors, and biofuels.

Many nanomaterials are highly porous in structure and have high specific surface area, which can be used for gas adsorption in the aims of purification and/or storage. Two research works in this Special Issue studied the adsorption and storage of natural gases on defined nanomaterials. [7,8] First, Shkolin et al. [7] synthesized an active carbon nanomaterial from cheap peat raw materials and investigated its structural–energetic and adsorption properties towards natural gas (ANG). The study provides insights into the solution for improving the safety and storage capacity of low-pressure gas storage systems. Second, Zhang et al. [8] reported the CH_4 adsorption properties on extended line defect (ELD) graphene according to the first principles of density functional theory (DFT). The Mn modification of ELD graphene was found to significantly affect the CH_4 adsorption. The specific molecular configurations and adsorption behaviors were discussed in detail in the paper. In another paper, Sodha et al. [9] presented a comprehensive review on zeolite-based nanocomposites for the treatment of effluents from wastewater. The review provides the basic knowledge about zeolites and highlights the types, synthesis, and removal mechanisms of zeolite-based materials for wastewater treatment along with the research gaps, being helpful for worldwide research on this topic.

Nanomaterials as applied as catalysts play important roles in many chemical engineering fields, being able to control reaction time, improve reaction speed and efficiency, save resources and energy, and improve economic benefits. Therefore, the study on catalytic nanomaterials attracts great research interests. The Special Issue presents two works focusing on fabrication of CeO_2-composited catalysts for maleic anhydride hydrogenation (MAH). Zhang et al. [10] prepared CeO_2-supported Ni catalysts with different Ni loadings and particle sizes by the impregnation method and investigated their hydrogenation performance. The work provides a theoretical and experimental basis for the preparation of high-activity catalysts for MAH. Liu et al. [11] synthesized CeO_2 supports with various shapes (e.g., nanocubes, nanorods, and nanoparticles) by using the hydrothermal technique, which were employed for supporting Ni species as catalysts for MAH. The study demonstrated morphology-dependent performances of the Ni/CeO_2 catalysts, which is helpful for developing novel catalysts for MAH. In another work, Li and coworkers [12] reported the synthesis of a series of Mg–Zr composite oxide catalysts by the hydrothermal method, aiming to catalyze the transesterification of glycerol with dimethyl carbonate to produce glycerol carbonate. The effects of the preparation method and Mg/Zr ratio on the catalytic performance and the deactivation of the catalysts were systematically investigated and discussed.

Nanomaterials have also found extensive applications in the fields of daily chemical products, cosmetics, pesticides, biomedicine etc., where the nanomaterials can be used as nanocarriers to load various active components for targeted delivery and controlled release. Rocha et al. [13] performed a systematic study on polymeric particles based on gelatin and poly-ε-caprolactone (PCL) containing essential oil from *Lippia origanoides*. The developed biocides have high physical stability and particle surface microtexture as well as pronounced bioactivity, which are efficient alternative controlling agents of *Conotrachelus humeropictus* and *Moniliophtora perniciosa*, the main pests of *Theobroma grandiflorum*. Yang et al. [14] presented a comprehensive review on peptide-based nanocarriers for gene delivery. The review puts forward discussion on the biological barriers for gene delivery, the peptide molecular design and assembly with DNA, the targeted delivery and controlled release of genome, the structure–function relationships of the delivery systems, and the current challenges and future perspectives in related fields, providing guidance towards the rational design and development of nonviral gene delivery systems.

Nanomaterials can also be engineered into the industrial production so as to improve the production process, elevate production efficiency, and optimize product performance. The work by Ali and coworkers [15] studied the effect of inlet flow strategies on the dynamics of pulsed fluidized bed of nanopowder. They changed the conventional single-drainage (SD) flow strategy to the modified double-drainage (MDD) flow strategy, which improved the production process by purging the primary flow during the non-flow period of the pulse to eliminate pressure buildup in the inlet flow line while providing a second drainage path to the residual gas. Sfameni et al. [16] reported the development of an efficient and eco-friendly procedure to form highly hydrophobic surfaces on cotton fabrics. By using a two-step treatment procedure, that is, first producing a hybrid silane film on cotton fabrics and then modifying with low-surface-energy components, the cotton fabrics were endowed with excellent water repellency. The work provides a new sustainable approach for fabric finishing and treatment.

In short, this Special Issue is expected to be interesting and enriching for readers by virtue of featuring all of the abovementioned high-quality original research works and comprehensive review papers. We give our sincere thanks to the excellent scholars that have made contributions. Currently, we are developing the second volume of the Special Issue, that is, "Nanomaterials for Chemical Engineering II". We welcome more excellent scholars to submit their excellent works in the area of synthesis and application of functional nanomaterials for chemical engineering.

Acknowledgments: This work was supported by the National Natural Science Foundation of China (22172194, 21872173).

Conflicts of Interest: The authors declare no conflict of interests.

References

1. Sazanova, T.S.; Mochalov, L.A.; Logunov, A.A.; Kudryashov, M.A.; Fukina, D.G.; Vshivtsev, M.A.; Prokhorov, I.O.; Yunin, P.A.; Smorodin, K.A.; Atlaskin, A.A.; et al. Influence of Temperature Parameters on Morphological Characteristics of Plasma Deposited Zinc Oxide Nanoparticles. *Nanomaterials* **2022**, *12*, 1838. [CrossRef] [PubMed]
2. Gaur, R.; Shahabuddin, S.; Ahmad, I.; Sridewi, N. Role of Alkylamines in Tuning the Morphology and Optical Properties of SnS2 Nanoparticles Synthesized by via Facile Thermal Decomposition Approach. *Nanomaterials* **2022**, *12*, 3950. [CrossRef] [PubMed]
3. Shao, Y.; He, Q.; Xiang, L.; Xu, Z.; Cai, X.; Chen, C. Strengthened Optical Nonlinearity of V2C Hybrids Inlaid with Silver Nanoparticles. *Nanomaterials* **2022**, *12*, 1647. [CrossRef] [PubMed]
4. Liu, H.; He, X.; Ren, J.; Jiang, J.; Yao, Y.; Lu, G. Terahertz Modulation and Ultrafast Characteristic of Two-Dimensional Lead Halide Perovskites. *Nanomaterials* **2022**, *12*, 3559. [CrossRef] [PubMed]
5. Wang, L.; Yang, J. Zirconia-Doped Methylated Silica Membranes via Sol-Gel Process: Microstructure and Hydrogen Permselectivity. *Nanomaterials* **2022**, *12*, 2159. [CrossRef] [PubMed]
6. Ali, L.; Manan, A.; Ali, B. Maxwell Nanofluids: FEM Simulation of the Effects of Suction/Injection on the Dynamics of Rotatory Fluid Subjected to Bioconvection, Lorentz, and Coriolis Forces. *Nanomaterials* **2022**, *12*, 3453. [CrossRef] [PubMed]
7. Shkolin, A.V.; Strizhenov, E.M.; Chugaev, S.S.; Men'shchikov, I.E.; Gaidamavichute, V.V.; Grinchenko, A.E.; Zherdev, A.A. Natural Gas Storage Filled with Peat-Derived Carbon Adsorbent: Influence of Nonisothermal Effects and Ethane Impurities on the Storage Cycle. *Nanomaterials* **2022**, *12*, 4066. [CrossRef] [PubMed]
8. Zhang, C.; Yang, S.; Zhang, X.; Xia, Y.; Li, J. Extended Line Defect Graphene Modified by the Adsorption of Mn Atoms and Its Properties of Adsorbing CH4. *Nanomaterials* **2022**, *12*, 697. [CrossRef] [PubMed]
9. Sodha, V.; Shahabuddin, S.; Gaur, R.; Ahmad, I.; Bandyopadhyay, R.; Sridewi, N. Comprehensive Review on Zeolite-Based Nanocomposites for Treatment of Effluents from Wastewater. *Nanomaterials* **2022**, *12*, 3199. [CrossRef]
10. Zhang, Q.; Liao, X.; Liu, S.; Wang, H.; Zhang, Y.; Zhao, Y. Tuning Particle Sizes and Active Sites of Ni/CeO$_2$ Catalysts and Their Influence on Maleic Anhydride Hydrogenation. *Nanomaterials* **2022**, *12*, 2156. [CrossRef] [PubMed]
11. Liu, S.; Liao, X.; Zhang, Q.; Zhang, Y.; Wang, H.; Zhao, Y. Crystal-Plane and Shape Influences of Nanoscale CeO$_2$ on the Activity of Ni/CeO$_2$ Catalysts for Maleic Anhydride Hydrogenation. *Nanomaterials* **2022**, *12*, 762. [CrossRef] [PubMed]
12. Li, Y.; Zhao, H.; Xue, W.; Li, F.; Wang, Z. Transesterification of Glycerol to Glycerol Carbonate over Mg-Zr Composite Oxide Prepared by Hydrothermal Process. *Nanomaterials* **2022**, *12*, 1972. [CrossRef] [PubMed]
13. Rocha, A.L.F.; de Aguiar Nunes, R.Z.; Matos, R.S.; da Fonseca Filho, H.D.; de Araújo Bezerra, J.; Lima, A.R.; Guimarães, F.E.G.; Pamplona, A.M.; Majolo, C.; de Souza, M.G.; et al. Alternative Controlling Agent of Theobroma grandiflorum Pests: Nanoscale Surface and Fractal Analysis of Gelatin/PCL Loaded Particles Containing *Lippia origanoides* Essential Oil. *Nanomaterials* **2022**, *12*, 2712. [CrossRef]
14. Yang, Y.; Liu, Z.; Ma, H.; Cao, M. Application of Peptides in Construction of Nonviral Vectors for Gene Delivery. *Nanomaterials* **2022**, *12*, 4076. [CrossRef] [PubMed]

15. Ali, S.S.; Arsad, A.; Roberts, K.L.; Asif, M. Effect of Inlet Flow Strategies on the Dynamics of Pulsed Fluidized Bed of Nanopowder. *Nanomaterials* **2023**, *13*, 304. [CrossRef]
16. Sfameni, S.; Lawnick, T.; Rando, G.; Visco, A.; Textor, T.; Plutino, M.R. Functional Silane-Based Nanohybrid Materials for the Development of Hydrophobic and Water-Based Stain Resistant Cotton Fabrics Coatings. *Nanomaterials* **2022**, *12*, 3404. [CrossRef] [PubMed]

Disclaimer/Publisher's Note: The statements, opinions and data contained in all publications are solely those of the individual author(s) and contributor(s) and not of MDPI and/or the editor(s). MDPI and/or the editor(s) disclaim responsibility for any injury to people or property resulting from any ideas, methods, instructions or products referred to in the content.

Article

Influence of Temperature Parameters on Morphological Characteristics of Plasma Deposited Zinc Oxide Nanoparticles

Tatyana Sergeevna Sazanova [1,*], Leonid Alexandrovich Mochalov [2], Alexander Alexandrovich Logunov [2], Mikhail Alexandrovich Kudryashov [2], Diana Georgievna Fukina [2], Maksim Anatolevich Vshivtsev [2], Igor Olegovich Prokhorov [2], Pavel Andreevich Yunin [3], Kirill Alexandrovich Smorodin [2], Artem Anatolevich Atlaskin [4] and Andrey Vladimirovich Vorotyntsev [1,2]

1. Laboratory of Membrane and Catalytic Processes, Nanotechnology and Biotechnology Department, Nizhny Novgorod State Technical University n.a. R.E. Alekseev, Minin Str. 24, 603950 Nizhny Novgorod, Russia; an.vorotyntsev@gmail.com
2. Chemical Engineering Laboratory, Research Institute for Chemistry, Lobachevsky State University of Nizhny Novgorod, Gagarin Ave. 23, 603022 Nizhny Novgorod, Russia; mochalovleo@gmail.com (L.A.M.); alchemlog@gmail.com (A.A.L.); mikhail.kudryashov1986@yandex.ru (M.A.K.); dianafuk@yandex.ru (D.G.F.); mvshivtcev@mail.ru (M.A.V.); igorprokhorov1998@yandex.ru (I.O.P.); smorodin.kirill.a@gmail.com (K.A.S.)
3. Department for Technology of Nanostructures and Devices, Institute for Physics of Microstructures of the Russian Academy of Science, Academic Str. 7, Afonino, 603087 Nizhny Novgorod, Russia; yunin@ipmras.ru
4. Laboratory of SMART Polymeric Materials and Technologies, Mendeleev University of Chemical Technology, Miusskaya Sq. 9, 125047 Moscow, Russia; atlaskin.a.a@muctr.ru
* Correspondence: yarymova.tatyana@yandex.ru

Abstract: Zinc oxide nanoparticles were obtained by plasma-enhanced chemical vapor deposition (PECVD) under optical emission spectrometry control from elemental high-purity zinc in a zinc–oxygen–hydrogen plasma-forming gas mixture with varying deposition parameters: a zinc source temperature, and a reactor temperature in a deposition zone. The size and morphological parameters of the zinc oxide nanopowders, structural properties, and homogeneity were studied. The study was carried out with use of methods such as scanning electron microscopy, X-ray structural analysis, and Raman spectroscopy, as well as statistical methods for processing and analyzing experimental data. It was established that to obtain zinc oxide nanoparticles with a given size and morphological characteristics using PECVD, it is necessary (1) to increase the zinc source temperature to synthesize more elongated structures in one direction (and vice versa), and (2) to decrease the reactor temperature in the deposition zone to reduce the transverse size of the deposited structures (and vice versa), taking into account that at relatively low temperatures instead of powder structures, films can form.

Keywords: zinc oxide; nanoparticles; PECVD; structure; morphology

1. Introduction

Currently, nanosized zinc oxide is one of the important materials in development for various fields of medicine and industry. Zinc oxide is a unique material because of its advantages such as cost efficiency and variable engineering properties, good biocompatibility and antibacterial properties, adjustable band-gap and particle size/shape, and many other features [1], making it applicable in a wide range of fields. In particular, zinc oxide nanoparticles can be used as: excellent antibacterial, antioxidant, antidiabetic and tissue regenerating agents [2]; innovative anticancer agents [3]; material for photocatalytic degradation of organic pollutant [4,5]; material for manufacture of electronic devices over flexible substrates [6]; an electron transport layer for quantum dot light-emitting diodes [7]; nanocomposite electrode material for supercapacitor [8]; and material for the production of gas sensors [9].

Nanosized zinc oxide is obtained in the form of rods (threads), ridges, honeycombs, rings, ribbons, springs (spirals), cells, tetrapods, as well as thin films and coatings. For the synthesis of all these numerous modifications, numerous preparation methods were proposed, of which two main groups can be distinguished: physical (sputtering, laser ablation, electrospraying, ball milling, electron beam evaporation, etc.) and chemical (microemulsion technique, sol–gel method, co-precipitation method, hydrothermal method, polyol method, chemical vapor deposition, etc.). The main advantage of the chemical methods is the possibility of obtaining particles with a given size, composition, and structure [10]. This is important because the shape and size of nanoparticles directly affects their properties. For example, Hammad T. and co-workers [11] reported changing a red shift from 3.62 to 3.33 eV when the average particle size of ZnO nanoparticles was increased from 11 to 87 nm, respectively. In another work, Mornani E. and co-workers [12] reported a change in the band gap from 4.45 to 4.08 eV with an increase in the average size of ZnO nanoparticles from 46 to 66 nm, respectively. Similarly, controlling nanoparticle sizes to determine their optimal range is significant for mechanical properties of nanoparticle assemblies [13], the rate of gas photofixation using powder catalysts based on nanoparticles [14], and others [15–19].

One of the most promising methods for producing zinc oxide nanoparticles of a given size and shape is plasma-enhanced chemical vapor deposition (PECVD). Plasma initiation makes it possible to significantly reduce the temperature of reactor walls and a deposition zone, as well as to eliminate the pollution possibility of the final product with equipment materials and to control the deposition zone temperature over a wider range, thereby setting the conditions for structure growth. Plasma initiation also makes it possible to achieve 100% conversion of initial substances due to establishing kinetic dependencies during plasma–chemical reactions. Thus, PECVD provides controllability, one stage, cost-effectiveness, high purity of resulting materials, as well as versatility and process scalability [20–22]. However, to widely use PECVD in the preparation of nanosized zinc oxide with given morphology, it is necessary to fundamentally study the influence of process parameters on the resulting product.

This work is devoted to studying zinc oxide nanoparticles obtained by the PECVD method from elemental high-purity zinc in a zinc–oxygen–hydrogen plasma-forming gas mixture with varying deposition parameters, namely a zinc source temperature and a reactor temperature in a deposition zone.

2. Experimental Section

2.1. Materials

Zinc of 5N purity (Changsha Rich Nonferrous Metals Co Ltd., Changsha, Hunan, China), high purity hydrogen (99.9999%) and oxygen 6.0 (99.9999%) (Horst Technologies Ltd., Dzerzhinsk, Nizhny Novgorod region, Russia) were used as components of a plasma-forming mixture. Hydrogen was used as a carrier gas, which also acted as a temperature stabilizer and regulator of the nanostructure growth.

2.2. Plasma–Chemical Synthesis

The scheme of a plasma–chemical installation is shown in Figure 1. The installation consisted of a gas supply system, a pumping system, a cuvette with initial zinc, and a pear-shaped plasma–chemical reactor made of high-purity quartz glass with a total volume of about 1200 cm^3. An external inductor, an RF generator (with an operating frequency of 40.68 MHz and a maximum power of 500 W), and a universal matching device were used to ignite an inductively coupled or mixed nonequilibrium plasma discharge.

A loading cuvette in the form of a boat with metal zinc was placed in a furnace with an external resistive heater and an internal thermocouple. A tank for collecting zinc oxide powder was placed into the reactor through a stainless-steel vacuum loading flange and was installed perpendicular to the carrier gas flow on a special movable holder. The loading cuvette, the powder tank, and the movable holder were made of high-purity quartz glass. To cool the reactor in the deposition zone, a circulation thermostat was used.

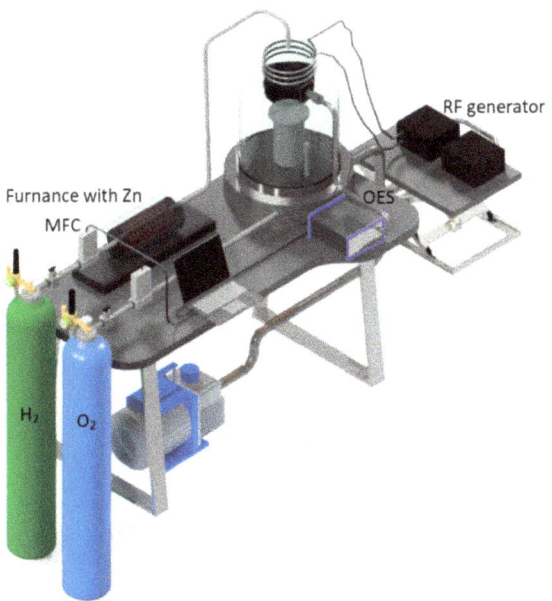

Figure 1. The scheme of a plasma–chemical installation with a pear-shaped reactor.

High-pure granulated Zn was loaded into the quartz furnace. High-pure H_2 was blown through the Zn source with the constant rate of 15 mL/min. The vapors of Zn were delivered by the career gas via the quartz lines heated up to 550 °C with the internal diameter of 6 mm directly into the plasma zone, where the formation of ZnO materials took place. The temperature of the lines and external surface of the plasma chamber was measured by a pyrometer. The temperature of the substrate holder was found by the internal thermocouple. High-pure O_2 was fed from below towards the main carrier gas with zinc vapor directly in the plasma zone with the constant rate of 15 mL/min. The total gas flow through the plasma–chemical reactor was set equal to 30 mL/min at a total pressure in the system of 0.1 Pa. The duration of each individual experiment (with the various deposition parameters) was one hour.

Before experiments, the installation was evacuated to a pressure of 1×10^{-3} Pa for several hours to remove traces of nitrogen and water from the walls of the reactor. Then, the powder tank was closed with a magnetic diaphragm of a special design.

The installation was also equipped with a high-resolution optical emission spectrometer HR4000CJ-UV-NIR (Avantes, The Netherlands) operating in the range of 180–1100 nm to control the excited intermediate species in the gas phase during the plasma–chemical process. The upper part of the plasma chamber was equipped with two plane-parallel windows made of special quartz glass of high transparency, maintained just after the inductor where the intensity of the lines was maximal.

According to the optical emission spectroscopy (OES) data (Figure 2), the composition of the plasma-forming mixture in all experiments included Zn (I), O (I), O (II), H (I), OH (I). Zn (II) emission lines were not observed.

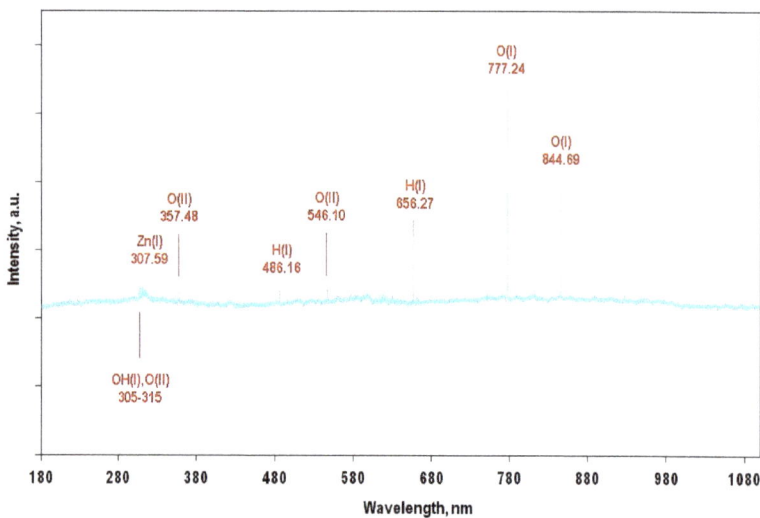

Figure 2. The optical emission spectra of the zinc–hydrogen–oxygen plasma (Zn:H$_2$:O$_2$ = 2:1:1).

2.3. Scanning Electron Microscopy and Energy-Dispersive X-ray Spectroscopy

The size-morphological characteristics of the zinc oxide samples were studied by scanning electron microscopy (SEM) using an electron microscope JSM-IT300LV (JEOL, Peabody, MA, USA) with an electron probe diameter of about 5 nm and a probe current of less than 0.5 nA (the operating voltage was 20 kV). SEM scanning was performed using low-energy secondary electrons and backscattered electrons under a low vacuum to eliminate the charge. As sample preparation for SEM, zinc oxide powders were applied onto carbon double-sided conductive tapes.

The size of the structures observed on SEM images was measured as the maximal diameter of their cross-section. For additional control of the measurement results, analysis of the images (determination of the average size of equivalent disk (D_{avg}) of structures' cross-section) was carried out using the method of watershed segmentation by a software SPMLab™ v5 (TopoMetrix, Santa Clara, CA, USA).

2.4. X-ray Structural Analysis

The structure of the zinc oxide samples was studied using X-ray diffraction (XRD) analysis on an X-ray diffractometer D8 Discover (Bruker, Germany) equipped with a sealed CuKα radiation source tube and a position-sensitive detector LynxEye. Diffraction patterns were obtained by θ/2θ scanning in the 2θ range of 10–66° with a step of 0.1°.

As sample preparation for the XRD analysis, a small amount of zinc oxide powders was placed in the center of a quartz disk. Then, about 3 drops of distilled water were added to the sample and it was spread to a thin layer with a glass rod. Next, the sample was placed in a desiccator to dry before the XRD analysis.

The results obtained were compared with the database PDF-2 Release 2011, namely PDF 01-071-6424 for ZnO and PDF 00-004-0831 for Zn.

2.5. Raman Spectroscopy

Raman spectra were studied on a spectroscopy complex NTEGRA Spectra Raman (NT-MDT, Moscow, Russia) using a laser with a wavelength of 473 nm. The radiation was focused by a 20× objective lens with an aperture of 0.45. The laser spot diameter was 5 μm. The power of the unfocused laser radiation was controlled by a silicon photodetector 11PD100-Si (Standa Ltd., Vilnius, Lithuania) and varied in the range from 1 mW to 1 μW. The Raman spectra analysis of the samples was carried out according to the scheme for reflection in the frequency range 80–800 cm^{-1} with a resolution of 0.7 cm^{-1}. As sample preparation for the Raman spectra analysis, zinc oxide powders were applied onto carbon double-sided conductive tapes. The measurements were carried out at room temperature.

2.6. Varying Temperature Parameters in Plasma–Chemical Deposition

To study the influence of a zinc source temperature on the morphology of ZnO nanoparticles, the temperature was varied from 370 to 470 °C. At the same time, the reactor temperature in the deposition zone was maintained at 250 °C. The plasma discharge power was 50 W.

To study the influence of a reactor temperature on the morphology of ZnO nanoparticles, the temperature was varied from 25 to 350 °C. At the same time, the zinc source temperature was maintained at 420 °C. The plasma discharge power was 50 W.

3. Results and Discussion

3.1. Zinc Source Temperature and Morphology of ZnO Nanoparticles

As a result of the plasma–chemical synthesis, ZnO nanoparticles with different shapes were obtained. According to the SEM images (Figure 3), the observed nanoparticles assumed a sphere-like, columnar, and rod-like shape depending on the zinc source temperature.

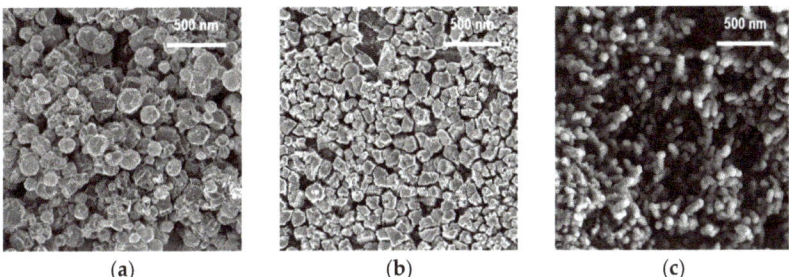

Figure 3. The SEM images of the ZnO nanopowders obtained by the PECVD method at the various zinc source temperatures: (**a**) 370 °C, (**b**) 420 °C, (**c**) 470 °C.

According to statistical processing of the SEM data (the sample size was 100 measurements), it was found that the average size (in the cross-section) of the deposited particles changed with an increase in the zinc source temperature, along with a variation in their shape (Figure 4).

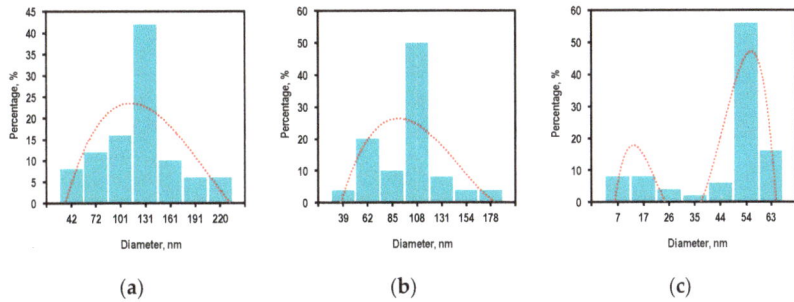

Figure 4. The transverse size distribution of the ZnO structures obtained by the PECVD method at the various zinc source temperatures: (**a**) 370 °C, (**b**) 420 °C, (**c**) 470 °C.

The average size of the ZnO structures decreased from 120 to 100 nm with a simultaneous decrease in the coefficient of variation from 46 to 44% caused by an increase in temperature from 370 to 420 °C. Next, the average size of the structures approached 45 nm with a coefficient of variation of 32% caused by a further increase in temperature to 470 °C. Such a change in the transverse size of the nanoparticles can be associated with the changeable mechanism of their growth and the corresponding redistribution of the Zn and O atoms in the spherical and elongated crystal structures.

The XRD analysis results for the ZnO structures obtained by the PECVD method at the various zinc source temperatures are shown in Figure 5. The diffraction peaks corresponding to (002), (101), (102), (103), (112), (201), (004), (202) planes are characteristic of a ZnO structure [16,23–29]. Such a diffraction pattern corresponds to a hexagonal structure of the wurtzite type. [23,24,29]. No other peaks associated with impurities were observed, indicating that high-purity ZnO nanoparticles were obtained. However, the diffraction peaks corresponding to (101) and (102) planes, which match with that of pure metallic Zn [30], were observed after increasing the zinc source temperature from 420 to 470 °C (Figure 5c). This means that there was an excess of Zn in this sample.

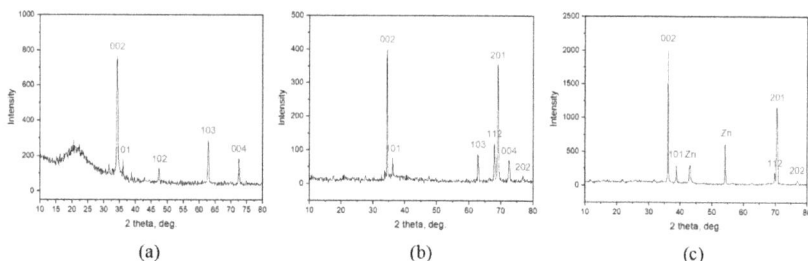

Figure 5. The XRD analysis results for the ZnO structures obtained by the PECVD method at the various zinc source temperatures: (**a**) 370 °C; (**b**) 420 °C; (**c**) 470 °C.

It should be noted that the XRD characterization revealed a strong preferred (002) orientation for all the obtained ZnO structures, indicating that the c-axis of the unit cell was aligned perpendicular to the horizontal plane of the deposition zone [28]. However, the (201) orientation became pronounced for the sample obtained at the source temperature of 420 °C (Figure 5b) and was slightly reduced with the increase in temperature from 420 to 470 °C (Figure 5c). Such a reorientation was probably related to the transition of the ZnO structures from spherical to elongated forms.

The Raman spectra for the ZnO structures obtained at the various zinc source temperatures are shown in Figure 6. The peak at about 435 cm^{-1} is typical for ZnO structures, while the peaks at about 275 cm^{-1} and 580 cm^{-1} do not correspond to ZnO normal modes. These peaks are additional vibrational modes and can be associated with defects and bond breaking, respectively [31–33].

Nevertheless, one must take into account the fact that the shape of Raman spectra depends on crystal orientation. In the case under consideration, based on the XRD data (Figure 5), the preferred orientation of the nanoparticles for all samples was (002). Moreover, for the sample obtained at the zinc source temperature of 420 °C (Figure 5b), the (201) orientation became pronounced, and the peak intensities in the Raman spectra (Figure 6) at about 275 cm^{-1} and 580 cm^{-1} had average values relative to other samples. The maximum intensities of these peaks were reached in another sample obtained at the source temperature of 470 °C, when the height of the (201) reflection on the XRD pattern (Figure 5c) was already reduced. Thus, the changes in the form of the Raman spectra can really be associated with defects and bond breaking in the case under consideration.

Analyzing the obtained Raman spectra, it can be concluded that the defectiveness of the ZnO structures (the peak at about 275 cm^{-1}) increases with the rise in the zinc source temperature. The degree of bond breaking (the peak at about 580 cm^{-1}) also has similar temperature dependence. Such a trend can be associated with increasing the Zn excess in the obtained ZnO structures, which can be led by bond breaking, and thereby causes the formation of defects similar to interstitial Zn.

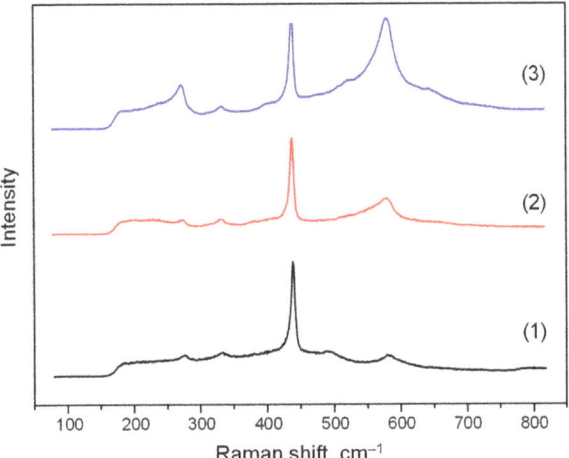

Figure 6. The Raman spectra of the ZnO structures obtained by the PECVD method at the various zinc source temperatures: (1) 370 °C, (2) 420 °C, (3) 470 °C.

3.2. Reactor Temperature and Size Distribution of ZnO Nanoparticles

As a result of the ZnO plasma–chemical synthesis at the reactor temperature in the deposition zone of 25 °C, a planar structure was formed instead of a nanoparticle one (Figure 7). At higher temperatures of the reactor deposition zone (250 and 350 °C), ZnO columnar structures were formed. Moreover, the average diameter (in cross-section) of the observed structures increased with a rise in the temperature.

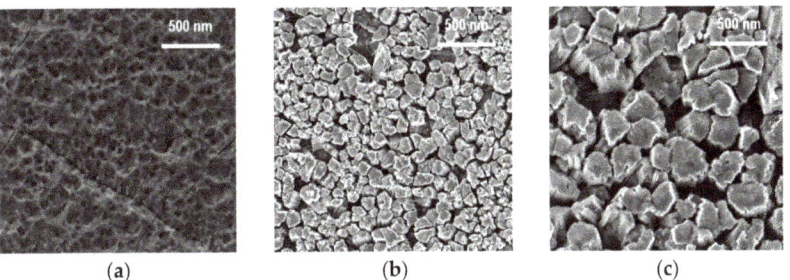

Figure 7. SEM images of ZnO nanopowders obtained by the PECVD method at various temperatures in the reactor deposition zone: (**a**) 25 °C, (**b**) 250 °C, (**c**) 350 °C.

According to statistical processing of the SEM data (the sample size was 100 measurements), it was established that the transverse diameter (at the widest part) of the ZnO columnar structures increased by three times with an increase in the reactor temperature in the deposition zone from 250 to 350 °C (Figure 8), and the coefficient of variation for the size range decreased from 44 to 27%.

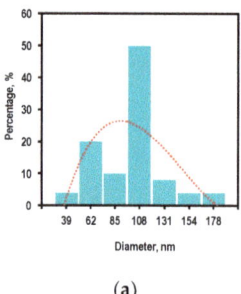 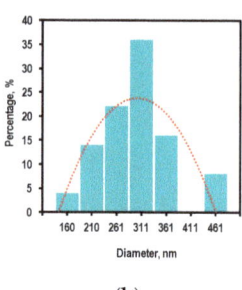

(a)   (b)

Figure 8. The transverse size distribution of the ZnO structures obtained by the PECVD method at the various temperatures in the reactor deposition zone: (**a**) 250 °C; (**b**) 350 °C.

The XRD analysis results for the ZnO structures obtained by the PECVD method at the various temperatures in the reactor deposition zone are shown in Figure 9. The diffraction peaks corresponding to (002), (101), (103), (112), (201), (004), and (202) planes are characteristic of a ZnO structure [16,23–29]. Such a diffraction pattern corresponds to a hexagonal structure of the wurtzite type [23,24,29]. No other peaks associated with impurities were observed, indicating that the high-purity ZnO nanoparticles were obtained.

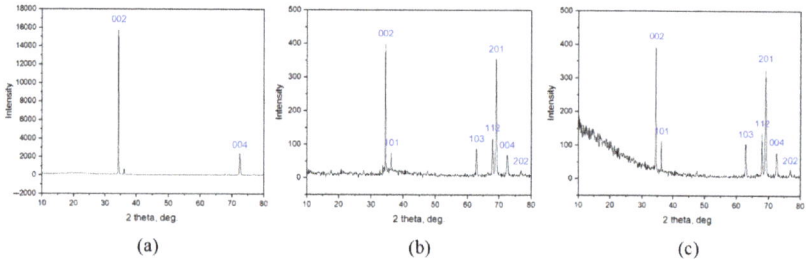

Figure 9. The XRD analysis results for the ZnO structures obtained by the PECVD method at the various temperatures in the reactor deposition zone: (**a**) 25 °C; (**b**) 250 °C; (**c**) 350 °C.

The XRD characterization revealed a strong preferred (002) orientation for all the obtained ZnO structures, indicating that the c-axis of the unit cell was aligned perpendicular to the horizontal plane of the deposition zone [28]. Notably, the diffraction pattern of the sample obtained at the temperature in the reactor deposition zone of 25 °C included only two pronounced peaks, namely (002) and (004). Such a pattern is characteristic of ZnO films [26]. Moreover, the (201) orientation became pronounced for the sample obtained at the temperature in the reactor deposition zone of 250 °C (Figure 9b) and was slightly reduced with the increase in temperature from 250 to 350 °C (Figure 9c). As already shown in Section 3.1, the presence of this peak is characteristic of the elongated ZnO structures.

The Raman spectra for the ZnO structures obtained at the various temperatures in the reactor deposition zone are shown in Figure 10. Similar to the case described in Section 3.1, three main regions can be distinguished: the typical ZnO peak at about 435 cm^{-1}, as well as the peaks at about 275 cm^{-1} and 580 cm^{-1}.

With a decrease in the reactor temperature from 350 to 250 °C, the intensity of all the observed peaks practically did not change; however, at a temperature of 25 °C, the intensity of the peak at about 435 cm^{-1} noticeably increased, and the peaks at about 275 cm^{-1} and 580 cm^{-1} were significantly smoothed out.

Comparing the Raman spectra for two experiments (Figures 6 and 10), it can be concluded that the size of the nanoparticles does not affect their defectiveness and the degree of bond breaking, in contrast with their shape.

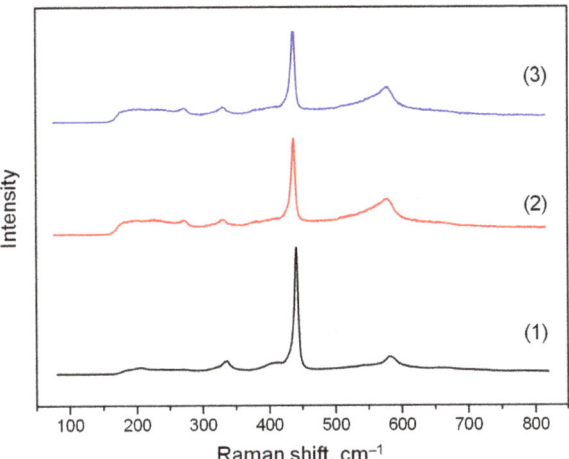

Figure 10. The Raman spectra of the ZnO structures obtained by the PECVD method at the various temperatures in the reactor deposition zone: (1) 25 °C; (2) 250 °C; (3) 350 °C.

3.3. Controlling the Size and Morphological Characteristics of ZnO Nanoparticles

The parameter controlling the morphology (shape) of the obtained nanoparticles is the zinc source temperature. Thus, the ZnO powders were formed with spherical, columnar, and rod-like particle shapes at the temperatures of 370, 420, and 470 °C, respectively; this was probably caused by a change in a mechanism of a plasma–chemical reaction. It was also found that the Zn excess in the deposited powder increased with the rise in the zinc source temperature. Based on the Raman spectra, it was shown that the Zn excess most likely led to intensification for the occurrence of structural defects and bond breaking.

The parameter controlling the dimensional characteristics of the obtained nanoparticles is the reactor temperature in the deposition zone. With a decrease in this parameter from 350 to 250 °C, the transverse size of the deposited ZnO particles was reduced by a factor of three. This effect could be associated with an increase in the specific input energy of the process due to the decrease in the temperature causing a fall in the concentration of the plasma-forming particles. Moreover, it is well known that the lattice parameters are temperature dependent; namely, increasing temperature can lead to lattice expansion with a subsequent increase in the resulting structures [23]. However, active cooling of the reactor in the deposition zone was not applicable in this case, since this parameter affected the relaxation rate of the particles at the surface of the powder tank. The lower the temperature, the more equilibrium the process would be, and therefore, the more perfect the structure to be deposited. When producing nanopowders, "perfect" does not mean "best", since such an equilibrium process led to the formation of dense monocrystalline layers, which was observed at the reactor temperature in the deposition region equal to 25 °C. It was also found that the size of the nanoparticles did not affect their defectiveness and the degree of bond breaking, in contrast with their shape.

It is also worth noting that an additional process parameter that controls the dimensional characteristics of the obtained nanoparticles can be the plasma discharge power as was shown in the previous work of the authors [34]. In that work, it was shown that the transverse diameter of the ZnO rod-like structures decreased by 30 times (from 900 to 30 nm) with an increase in the plasma discharge power from 30 to 70 W.

4. Conclusions

The direct one-stage synthesis of the ZnO nanoparticles was carried out by the PECVD method from elemental high-purity zinc in the zinc–oxygen–hydrogen plasma-forming mixture with the variable deposition parameters.

The dimensional and morphological parameters of the obtained ZnO powders were studied, as well as their structural properties and homogeneity. The study was carried out using methods such as SEM, XRD, and Raman spectroscopy, as well as statistical methods for processing and analyzing experimental data.

In order to determine the optimal parameters for the PECVD synthesis of the ZnO nanoparticles, a series of experiments were carried out. In each of them, one operating parameter was changed while the rest were constant.

It was found that the zinc source temperature was a parameter controlling the morphology (shape) of the obtained nanoparticles, and the reactor temperature in the deposition zone was a parameter controlling their dimensional characteristics. An additional parameter controlling the dimensional characteristics of the nanoparticles was the plasma discharge power.

Based on the analysis of the obtained experimental data, it can be concluded that, in order to obtain ZnO nanoparticles with given size and morphological characteristics in the PECVD process, it is necessary (1) to increase the zinc source temperature to obtain more elongated structures in one direction (and vice versa), (2) to increase the plasma discharge power for reducing the transverse size of the deposited structures (and vice versa), and (3) to lower the reactor temperature in the deposition zone to reduce the transverse size of the deposited structures (and vice versa). However, take into account that at relatively low temperatures instead of powder ones, film structures can form.

Author Contributions: Conceptualization, T.S.S.; data curation, L.A.M. and A.V.V.; funding acquisition, T.S.S.; investigation, T.S.S., L.A.M., A.A.L., M.A.K., D.G.F., M.A.V., I.O.P., P.A.Y., K.A.S., A.A.A. and A.V.V.; methodology, T.S.S. and L.A.M.; supervision, T.S.S.; visualization, M.A.V.; writing—original draft, T.S.S. All authors have read and agreed to the published version of the manuscript.

Funding: This work was supported by the Russian Science Foundation (grant no. 20-79-00138). Decoding the XRD spectra was supported by the Ministry of Science and Higher Education of the Russian Federation in the Framework of the Basic Part of the State Task [FSWE-2020-0008, project no. 0728-2020-0008].

Institutional Review Board Statement: "Not applicable" for studies not involving humans or animals.

Acknowledgments: The SEM studies were performed by the Collective Usage Center "New Materials and Resource-saving Technologies" (Lobachevsky State University of Nizhny Novgorod).

Conflicts of Interest: The authors declare no conflict of interest.

References

1. Borysiewicz, M.A. ZnO as a functional material, a review. *Crystals* **2019**, *9*, 505. [CrossRef]
2. Singh, T.A.; Sharma, A.; Tejwan, N.; Ghosh, N.; Das, J.; Sil, P.C. A state of the art review on the synthesis, antibacterial, antioxidant, antidiabetic and tissue regeneration activities of zinc oxide nanoparticles. *Adv. Colloid Interface Sci.* **2021**, *295*, 102495. [CrossRef] [PubMed]
3. Wiesmann, N.; Tremel, W.; Brieger, J. Zinc oxide nanoparticles for therapeutic purposes in cancer medicine. *J. Mater. Chem. B* **2020**, *8*, 4973–4989. [CrossRef] [PubMed]
4. Sultana, K.A.; Islam, M.T.; Silva, J.A.; Turley, R.S.; Hernandez-Viezcas, J.A.; Gardea-Torresdey, J.L.; Noveron, J.C. Sustainable synthesis of zinc oxide nanoparticles for photocatalytic degradation of organic pollutant and generation of hydroxyl radical. *J. Mol. Liq.* **2020**, *307*, 112931. [CrossRef]
5. Silva, J.M.P.; Andrade Neto, N.F.; Oliveira, M.C.; Ribeiro, R.A.P.; de Lazaro, S.R.; Gomes, Y.F.; Paskocimas, C.A.; Bomio, M.R.D.; Motta, F.V. Recent progress and approaches on the synthesis of Mn-doped zinc oxide nanoparticles: A theoretical and experimental investigation on the photocatalytic performance. *New J. Chem.* **2020**, *44*, 8805–8812. [CrossRef]
6. Tinoco, J.C.; Hernández, S.A.; Rodríguez-Bernal, O.; Vega-Poot, A.G.; Rodríguez-Gattorno, G.; Olvera, M.d.l.L.; Martinez-Lopez, A.G. Fabrication of Schottky barrier diodes based on ZnO for flexible electronics. *J. Mater. Sci. Mater. Electron.* **2020**, *31*, 7373–7377. [CrossRef]
7. Alexandrov, A.; Zvaigzne, M.; Lypenko, D.; Nabiev, I.; Samokhvalov, P. Al-, Ga-, Mg-, or Li-doped zinc oxide nanoparticles as electron transport layers for quantum dot light-emitting diodes. *Sci. Rep.* **2020**, *10*, 7496. [CrossRef]
8. Kumar, A. Sol gel synthesis of zinc oxide nanoparticles and their application as nano-composite electrode material for supercapacitor. *J. Mol. Struct.* **2020**, *1220*, 128654. [CrossRef]

9. Jaballah, S.; Benamara, M.; Dahman, H.; Lahem, D.; Debliquy, M.; El Mir, L. Formaldehyde sensing characteristics of calcium-doped zinc oxide nanoparticles-based gas sensor. *J. Mater. Sci. Mater. Electron.* **2020**, *31*, 8230–8239. [CrossRef]
10. Parashar, M.; Shukla, V.K.; Singh, R. Metal oxides nanoparticles via sol–gel method: A review on synthesis, characterization and applications. *J. Mater. Sci. Mater. Electron.* **2020**, *31*, 3729–3749. [CrossRef]
11. Hammad, T.M.; Salem, J.K.; Harrison, R.G. The influence of annealing temperature on the structure, morphologies and optical properties of ZnO nanoparticles. *Superlattices Microstruct.* **2010**, *47*, 335–340. [CrossRef]
12. Mornani, E.G.; Mosayebian, P.; Dorranian, D.; Behzad, K. Effect of calcination temperature on the size and optical properties of synthesized ZnO nanoparticles. *J. Ovonic Res.* **2016**, *12*, 75–80.
13. An, L.; Zhang, D.; Zhang, L.; Feng, G. Effect of nanoparticle size on the mechanical properties of nanoparticle assemblies. *Nanoscale* **2019**, *11*, 9563–9573. [CrossRef] [PubMed]
14. Li, J.; Li, Q.; Chen, Y.; Lv, S.; Liao, X.; Yao, Y. Size effects of Ag nanoparticle for N_2 photofixation over Ag/g-C3N4: Built-in electric fields determine photocatalytic performance. *Colloids Surfaces A Physicochem. Eng. Asp.* **2021**, *626*, 127053. [CrossRef]
15. Sorichetti, V.; Hugouvieux, V.; Kob, W. Structure and dynamics of a polymer–nanoparticle composite: Effect of nanoparticle size and volume fraction. *Macromolecules* **2018**, *51*, 5375–5391. [CrossRef]
16. Ismail, A.M.; Menazea, A.A.; Kabary, H.A.; El-Sherbiny, A.E.; Samy, A. The influence of calcination temperature on structural and antimicrobial characteristics of zinc oxide nanoparticles synthesized by Sol–Gel method. *J. Mol. Struct.* **2019**, *1196*, 332–337. [CrossRef]
17. Cai, B.; Pei, J.; Dong, J.; Zhuang, H.-L.; Gu, J.; Cao, Q.; Hu, H.; Lin, Z.; Li, J.-F. $(Bi,Sb)_2Te_3$/SiC nanocomposites with enhanced thermoelectric performance: Effect of SiC nanoparticle size and compositional modulation. *Sci. China Mater.* **2021**, *64*, 2551–2562. [CrossRef]
18. Ramya, M.; Nideep, T.K.; Nampoori, V.P.N.; Kailasnath, M. The impact of ZnO nanoparticle size on the performance of photoanodes in DSSC and QDSSC: A comparative study. *J. Mater. Sci. Mater. Electron.* **2021**, *32*, 3167–3179. [CrossRef]
19. Mahpudz, A.; Lim, S.L.; Inokawa, H.; Kusakabe, K.; Tomoshige, R. Cobalt nanoparticle supported on layered double hydroxide: Effect of nanoparticle size on catalytic hydrogen production by $NaBH_4$ hydrolysis. *Environ. Pollut.* **2021**, *290*, 117990. [CrossRef]
20. Mochalov, L.; Logunov, A.; Vorotyntsev, V. Structural and optical properties of As–Se–Te chalcogenide films prepared by plasma-enhanced chemical vapor deposition. *Mater. Res. Express* **2019**, *6*, 56407. [CrossRef]
21. Mochalov, L.; Logunov, A.; Markin, A.; Kitnis, A.; Vorotyntsev, V. Characteristics of the Te-based chalcogenide films dependently on the parameters of the PECVD process. *Opt. Quantum Electron.* **2020**, *52*, 197. [CrossRef]
22. Mochalov, L.; Logunov, A.; Prokhorov, I.; Sazanova, T.; Kudrin, A.; Yunin, P.; Zelentsov, S.; Letnianchik, A.; Starostin, N.; Boreman, G.; et al. Plasma-Chemical Synthesis of Lead Sulphide Thin Films for Near-IR Photodetectors. *Plasma Chem. Plasma Process.* **2021**, *41*, 493–506. [CrossRef]
23. Singh, P.; Kumar, A.; Kaushal, A.; Kaur, D.; Pandey, A.; Goyal, R.N. In situ high temperature XRD studies of ZnO nanopowder prepared via cost effective ultrasonic mist chemical vapour deposition. *Bull. Mater. Sci.* **2008**, *31*, 573–577. [CrossRef]
24. Purica, M. Optical and structural investigation of ZnO thin films prepared by chemical vapor deposition (CVD). *Thin Solid Film.* **2002**, *403–404*, 485–488. [CrossRef]
25. Kocyigit, A.; Orak, I.; Çaldıran, Z.; Turut, A. Current–voltage characteristics of Au/ZnO/n-Si device in a wide range temperature. *J. Mater. Sci. Mater. Electron.* **2017**, *28*, 17177–17184. [CrossRef]
26. Wu, J.-J.; Wen, H.-I.; Tseng, C.-H.; Liu, S.-C. Well-Aligned ZnO Nanorods via Hydrogen Treatment of ZnO Films. *Adv. Funct. Mater.* **2004**, *14*, 806–810. [CrossRef]
27. Savaloni, H.; Savari, R. Nano-structural variations of ZnO:N thin films as a function of deposition angle and annealing conditions: XRD, AFM, FESEM and EDS analyses. *Mater. Chem. Phys.* **2018**, *214*, 402–420. [CrossRef]
28. Franklin, J.B.; Zou, B.; Petrov, P.; McComb, D.W.; Ryan, M.P.; McLachlan, M.A. Optimised pulsed laser deposition of ZnO thin films on transparent conducting substrates. *J. Mater. Chem.* **2011**, *21*, 8178. [CrossRef]
29. Fan, H.; Jia, X. Selective detection of acetone and gasoline by temperature modulation in zinc oxide nanosheets sensors. *Solid State Ion.* **2011**, *192*, 688–692. [CrossRef]
30. Mondal, B.; Mandal, S.P.; Kundu, M.; Adhikari, U.; Roy, U.K. Synthesis and characterization of nano–zinc wire using a self designed unit galvanic cell in aqueous medium and its reactivity in propargylation of aldehydes. *Tetrahedron* **2019**, *75*, 4669–4675. [CrossRef]
31. Friedrich, F.; Nickel, N.H. Resonant Raman scattering in hydrogen and nitrogen doped ZnO. *Appl. Phys. Lett.* **2007**, *91*, 111903. [CrossRef]
32. Sann, J.; Stehr, J.; Hofstaetter, A.; Hofmann, D.M.; Neumann, A.; Lerch, M.; Haboeck, U.; Hoffmann, A.; Thomsen, C. Zn interstitial related donors in ammonia-treated ZnO powders. *Phys. Rev. B* **2007**, *76*, 195203. [CrossRef]
33. Montenegro, D.N.; Hortelano, V.; Martínez, O.; Martínez-Tomas, M.C.; Sallet, V.; Muñoz-Sanjosé, V.; Jiménez, J. Non-radiative recombination centres in catalyst-free ZnO nanorods grown by atmospheric-metal organic chemical vapour deposition. *J. Phys. D Appl. Phys.* **2013**, *46*, 235302. [CrossRef]
34. Sazanova, T.S.; Mochalov, L.A.; Logunov, A.A.; Fukina, D.G.; Vorotyntsev, I.V. Influence of plasma power on the size distribution of deposited zinc oxide nanorods. *IOP Conf. Ser. Mater. Sci. Eng.* **2021**, *1155*, 12093. [CrossRef]

Article

Role of Alkylamines in Tuning the Morphology and Optical Properties of SnS$_2$ Nanoparticles Synthesized by via Facile Thermal Decomposition Approach

Rama Gaur [1,*], Syed Shahabuddin [1,*], Irfan Ahmad [2] and Nanthini Sridewi [3,*]

1. Department of Chemistry, School of Technology, Pandit Deendayal Energy University, Knowledge Corridor, Raysan, Gandhinagar 382426, Gujarat, India
2. Department of Clinical Laboratory Sciences, College of Applied Medical Sciences, King Khalid University, Abha 61421, Saudi Arabia
3. Department of Maritime Science and Technology, Faculty of Defence Science and Technology, National Defence University of Malaysia, Kuala Lumpur 57000, Malaysia
* Correspondence: rama.gaur@sot.pdpu.ac.in (R.G.); syedshahab.hyd@gmail.com or syed.shahabuddin@sot.pdpu.ac.in (S.S.); nanthini@upnm.edu.my (N.S.); Tel.: +91-8585932338 (S.S.); +60-124-675-320 (N.S.)

Citation: Gaur, R.; Shahabuddin, S.; Ahmad, I.; Sridewi, N. Role of Alkylamines in Tuning the Morphology and Optical Properties of SnS$_2$ Nanoparticles Synthesized by via Facile Thermal Decomposition Approach. *Nanomaterials* **2022**, *12*, 3950. https://doi.org/10.3390/nano12223950

Academic Editor: Meiwen Cao

Received: 26 September 2022
Accepted: 3 November 2022
Published: 9 November 2022

Publisher's Note: MDPI stays neutral with regard to jurisdictional claims in published maps and institutional affiliations.

Copyright: © 2022 by the authors. Licensee MDPI, Basel, Switzerland. This article is an open access article distributed under the terms and conditions of the Creative Commons Attribution (CC BY) license (https://creativecommons.org/licenses/by/4.0/).

Abstract: The present study reported the synthesis of SnS$_2$ nanoparticles by using a thermal decomposition approach using tin chloride and thioacetamide in diphenyl ether at 200 °C over 60 min. SnS$_2$ nanoparticles with novel morphologies were prepared by the use of different alkylamines (namely, octylamine (OCA), dodecylamine (DDA), and oleylamine (OLA)), and their role during the synthesis was explored in detail. The synthesized SnS$_2$ nanostructures were characterized using an array of analytical techniques. The XRD results confirmed the formation of hexagonal SnS$_2$, and the crystallite size varied from 6.1 nm to 19.0 nm and from 2.5 to 8.8 nm for (100) and (011) reflections, respectively. The functional group and thermal analysis confirmed the presence of organics on the surface of nanoparticles. The FE-SEM results revealed nanoparticles, nanoplates, and flakes assembled into flower-like morphologies when dodecylamine, octylamine, and oleylamine were used as capping agents, respectively. The analysis of optical properties showed the variation in the bandgap and the concentration of surface defects on the SnS$_2$ nanoparticles. The role of alkylamine as a capping agent was explored and discussed in detail in this paper and the mechanism for the evolution of different morphologies of SnS$_2$ nanoparticles was also proposed.

Keywords: SnS$_2$ nanoparticles; thermal decomposition; nanoflakes; nanoflowers; capping agent; alkylamines

1. Introduction

The growing population around the world arouses energy concerns in all kinds of fields. It took us a long time to realize that "the ultimate source of energy sun is the ultimate source". The latest estimates by scientists prove that the amount of solar energy the sun gives to the earth in a single day is sufficient to meet the total energy needs of the world for 27 years at the current rate of consumption [1]. We only have to harness a very little fraction of it to meet all our energy needs for all time to come. Extensive research has been made over the years on solar cells to improve their efficiency and augment their commercialization. Hence, a continuous effort has been made in this area and is still being made to achieve maximum efficiency by modulating the materials used as the photoanode. Reports are available on the use of metal selenides as sensitizers and electron acceptors in dye-sensitized solar cells with an enhanced power conversion efficiency of up to 9.49% [2,3]. Owing to the negative traits such as toxicity and instability of selenides, we need to look for alternative inorganic materials as electron acceptors for future solar cells. The use of layered metal sulfides such as MoS$_2$, SnS$_2$, and WS$_2$ is considered a potential candidate for this purpose [4]. SnS$_2$ is a moderate band gap semiconductor (2–2.5 eV) with a layered structure. It has been reported to exhibit a CdI$_2$-type structure, where hexagonally ordered planes of Sn atoms are held between two hexagonally ordered planes of S atoms, and with adjacent sulfur layers [5,6]. It possesses excellent optical and electrical properties and is an important material for optoelectronic devices. Hence, SnS$_2$ is considered a potential candidate for various applications such as catalysis, solar cells, sensors, photodetectors, lithium and sodium ion batteries, light-emitting diodes, etc. [4–8].

Looking at the synthetic aspects and environmental concerns, SnS$_2$ is comparatively easy to synthesize and has no toxic effects on the environment. Various methods have been reported for the synthesis of SnS$_2$ nanoparticles, such as hydrothermal, sol–gel, laser ablation, chemical vapor deposition, etc. [4,9–11]. The reported synthetic approaches involve harsh conditions, high temperatures, long reaction times, and a lack of control over the shape and size of the materials. Alternate approaches such as solvent-assisted thermal decomposition allow the easy and facile synthesis of shape- and size-controlled nanomaterials.

Morphology-dependent studies on nanomaterials have revealed that the same material with different morphologies exhibits significantly different properties [12,13]. Research is being conducted in this direction to tune the morphology and surface characteristics by chemical methods. Researchers have reported different morphologies of SnS$_2$ such as flakes, nanosheets, worm-like shapes, nanorods, flower-like shapes, nanobelts, etc. [14–19].

Alkylamines have been reported to be an important class of stabilizers during the synthesis of colloidal semiconductor nanocrystals. A series of alkylamines have been reported as stabilizer/capping agents for the preparation of group II–VI and IV–VI nanomaterials [20]. The addition of these alkylamines during the synthesis of semiconductor nanocrystals aids nucleation and affects crystal growth [21]. The role of alkylamines in controlling morphological characteristics has always been debatable. A few reports suggest that the addition of alkylamine retards the kinetics by passivating the crystal surface, hence retarding crystal growth [22,23]. On the other hand, a few reports suggest their role as promoters by enhancing the nucleation kinetics and influencing crystal growth [24–26]. The above reports point towards the chemical interaction of alkylamines with precursor molecules, resulting in a pre-conditioned molecular precursor. Alkylamine-substituted precursors are decomposed to form nanocrystals with defined morphologies and unique optical and structural properties. The preferential adsorption of alkylamines on the surface of nanocrystals results in the accelerated growth of the other planes. The mechanistic insights into the role of alkylamines have been discussed by García-Rodríguez et al. (2014) in their reports [20]. The interaction of alkylamines with the growing crystal is a temperature-dependent phenomenon. The amine adsorption is minimum at low temperatures and is increased with an increase in the reaction temperature. Li et al. (2004) and Pradhan et al. (2007) suggested that alkylamines activate the precursors of ZnSe, ZnS, and CdSe nanocrystals at various reaction temperatures [25,27]. Similarly, Sun et al. reported that dodecylamine increases the rate of consumption of phosphine selenide precursor as well as the rate of CdSe nanocrystal growth [26]. In contrast, Guo et al. suggested that alkylamines decrease their reactivity instead [28]. Mourdikoudis et al. have reviewed oleyamine as a solvent, surfactant, and reducing agent, for the controlled preparation of a wide range of nanomaterials including metal oxides, metal sulfides, noble metals, and alloy nanocrystals [29]. These unique effects of alkylamines significantly improve the aspect of controlled morphology with desired properties.

From the above discussion, it is very clear that alkylamines play an important role during the synthesis by controlling crystal growth. The present paper aimed at exploring the role of alkylamines as a capping agent and their effect on the optical properties and the morphology of SnS$_2$ nanoparticles. In a continuation of previous work, the authors attempted the synthesis in the presence of different alkylamines (oleylamine, octylamine, and dodecylamine).

2. Experimental Section

2.1. Reagents

Tin(IV) chloride pentahydrate, thioacetamide (Sigma Aldrich® 99%, Ahemdabad, Gujarat, India), diphenyl ether (Sigma Aldrich® 99%, Ahemdabad, Gujarat, India), octylamine (Sigma Aldrich® 99%, Ahemdabad, Gujarat, India), oleylamine (Sigma Aldrich® 99%, Ahemdabad, Gujarat, India), dodecylamine (Sigma Aldrich® 99%, Ahemdabad, Gujarat, India), and Millipore® water were acquired. All the chemicals were used as received. Methanol used during the reaction was distilled before use.

2.2. Synthesis of SnS_2 Nanoparticles

The SnS_2 nanoparticles were synthesized using a simple thermal decomposition approach. In a typical synthesis, 1 mmol of $SnCl_4 \cdot 5H_2O$ and 1 mmol of CH_3CSNH_2 were added to 10 mL of diphenyl ether in a 50 mL round-bottom flask and were refluxed at 200 °C in the air for 1 h. After the completion of the reaction, a slurry was obtained and cooled to room temperature. A total of 30 mL of methanol was added to the slurry and the precipitate obtained was washed using an excess of methanol. The precipitate was dried overnight at 65 °C under a vacuum. For the preparation of nanoparticles in the presence of capping agents (1 mmol), oleylamine, octylamine, and dodecylamine were added to different reaction setups along with the Sn and S precursors to diphenyl ether in the initial step. The synthetic details and nomenclature of the SnS_2 samples, prepared in the present study, are given in Table 1.

Table 1. Synthetic details, nomenclature, and morphology of all SnS_2 samples.

Sample ID	Capping Agent	Color of the Product	Morphology	Crystallite Size		Sn:S Ratio (EDX)	Overall % Wt. Loss
				(100)	(011)		
S1	–	Dark Green	Flower-like	18.0	5.3	1:1.8	22.90
S2	Dodecylamine	Brown-green	Small nanoparticles	–	–	1:2.1	25.74
S3	Octylamine	Brown-green	Rosette-like morphology	31.5	8.8	1:2.2	21.6
S4	Oleylamine	Olive green	Nanoplates assembled into stacks	17.6	4.2	1:2.1	26.85

To investigate the role of alkylamines in detail, the synthesis was carried out by varying the amount of capping agent used. The results are discussed in Section 3.

2.3. Characterization

The as-prepared SnS_2 samples were characterized using an array of sophisticated characterization techniques. The SnS_2 nanostructure was analyzed for structural, compositional, and morphological characterization. The phase analysis was carried out using powder X-ray diffraction (Bruker AXS-D8 diffractometer, Cu-K$_\alpha$ radiation (λ = 1.5406 Å); 2θ range 5–90°; scan speed of 1° min^{-1}). The purity and stability of the as-prepared SnS_2 nanoparticles were analyzed by FT-IR spectroscopy (Perkin Elmer Spectrum 2, Mumbai, Maharashtra, India) and thermal gravimetric analysis (EXSTAR TG/DTA instrument (Hyderabad, Telangana, India); heating rate 10°/min, ambient air atmosphere). Optical properties were investigated using a diffuse reflectance spectrophotometer (DRS) (Perkin Elmer, (Ahemdabad, Gujarat, India)) and photoluminescence (PL) spectrophotometer (Perkin Elmer Model, (Ahemdabad, Gujarat, India)) in the wavelength range of 200 nm to 800 nm. The morphology of SnS_2 samples was analyzed using a field emission scanning electron microscope (Carl Zeiss, Bangalore, Karnataka, India) operating at 20 kV and equipped with an energy-dispersive X-ray analysis (EDXA, Bangalore, Karnataka, India) facility. For the FE-SEM analysis, the SnS_2 powders were sprinkled on clean aluminum stubs using conducting carbon tape and were gold coated for 30 s using a sputtering unit.

3. Results and Discussion

3.1. Structure and Phase Analysis

The XRD results for the SnS$_2$ nanoparticles prepared via the thermal decomposition approach using different capping agents confirmed the formation of hexagonal SnS$_2$ (JCPDS File No. 83-1705; Berndtite-2T phase) in all the samples (Figure 1). The XRD peaks at the 2θ values of 15.05°, 28.30°, 30.38°, 32.20°, 42.00°, 50.11°, and 52.63° were indexed to (001), (100), (002), (011), (012), (110), and (111) reflections of SnS$_2$, respectively. It was observed that the XRD pattern for samples S1 and S3 exhibited sharp and well-defined peaks indicating high crystallinity and morphology characteristics that are discussed later in Section 3.4. The presence of two sets of peaks (sharp and broad) in the XRD patterns of S1 and S3 implied that the growth of certain facets was restrained, and that a special morphology was formed [18].

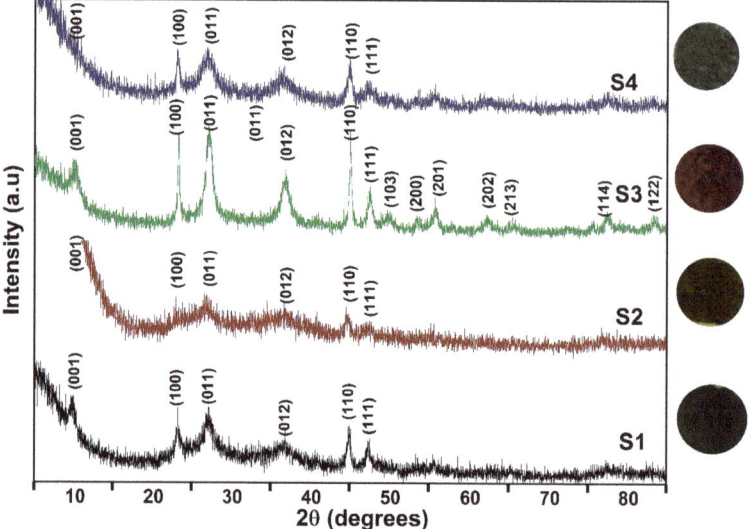

Figure 1. XRD plot of SnS$_2$ nanoparticles synthesized in the absence and presence of capping agent by thermal decomposition method.

On the other hand, the XRD patterns of samples S2 and S4 exhibited broad and poorly defined peaks, indicating a low crystallinity of the nanoparticles. The crystallite size of SnS$_2$ was calculated using the Debye–Scherrer formula. The crystallite size varied from 6.1 nm to 19 nm, as calculated using the (100) reflection, and 2.6 nm to 6.6 nm as calculated using the (011) reflection. The presence of the (001) peak in samples S1 and S3 indicated the formation of layered structures. The formation of highly crystalline nanoflakes assembled to form a flower-like structure for samples S1 and S2 was evident from the presence of sharp peaks in the XRD plot. In addition, the low intensity and poorly defined peaks in the XRD pattern pointed towards the formation of nanoparticles and nanoplates for samples S2 and S4.

3.2. Purity and Phase Stability

The purity and phase stability of the as-prepared SnS$_2$ samples were checked using FT-IR spectroscopy and thermogravimetric analysis (TGA), respectively. Figure 2a shows the FT-IR spectra of the SnS$_2$ nanoparticles synthesized using different capping agents by the thermal decomposition method. The IR spectra of the SnS$_2$ nanoparticles showed the presence of bands at around 3400 cm^{-1} and 1620 cm^{-1} attributed to stretching and bending vibrations of the hydroxyl group, indicating the presence of physisorbed moisture on the surface of the SnS$_2$ nanoparticles [18]. The band at about 3200 cm^{-1} was attributed to N–H stretching, and the IR bands at around 2920 cm^{-1}, 2840 cm^{-1}, and 1390 cm^{-1} were ascribed to asymmetric and symmetric stretching and bending of the C–H group, respectively [30]. The bands at around 1260 cm^{-1}, 870 cm^{-1}, and 660 cm^{-1}, were due to asymmetric and symmetric stretching and bending of the C–S group, respectively [31].

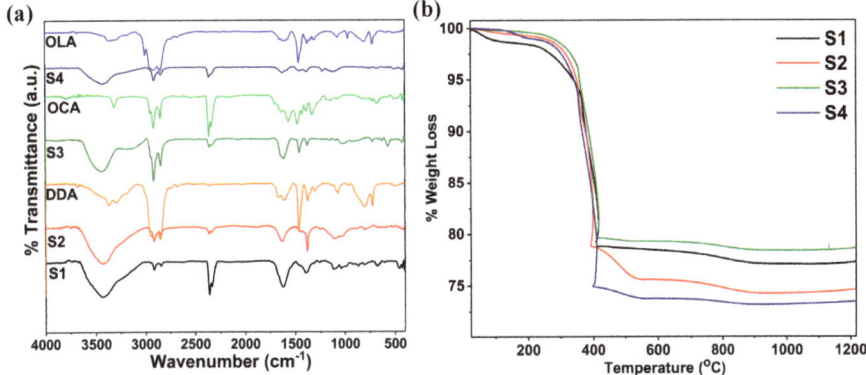

Figure 2. (a) IR and (b) TGA graph of SnS$_2$ nanoparticles synthesized in the absence and presence of capping agent by thermal decomposition method.

The IR spectra of all the SnS$_2$ nanoparticles exhibited a band at around 680 cm^{-1} due to Sn–S stretching. The IR band at around 1110 cm^{-1} was ascribed to C–N stretching [18]. The assignments of IR peaks for all the SnS$_2$ samples and capping agents used are listed in Table 2. The shift in the IR band positions confirmed the capping of nanoparticles with the different capping agents used (DDA, OCA, and OLA) on the surface of the SnS$_2$ nanoparticles.

Table 2. Assignments of IR peaks for the SnS$_2$ samples and the alkylamines used as capping agents.

S1	S2	(DDA)	S3	(OCA)	S4	(OLA)	Band Assignments
3425	3424	3366	3443	3314	3443	3373 and 3300	$\nu_{as}(NH_2)$ and $\nu_s(NH_2)$
3211	–	3292	–	2954	–	3005	δ (=C–H)
2920	2914	2923	2923	2923	2920	2950	ν_{as}C–H
2847	2847	2855	2849	2852	2844	2852	ν_sC–H
1619	1632	1611	1619	1620	1632	1611	δ (NH$_2$)
–	–	1630	–	1638	1592	1620	δ (C–C)
–	–	1461	1461	1477	–	1461	
1392	1384	1378	1380	1388	1384	1376	δ (CH$_2$)
1108	–	1303	–	1331	–	1307	
1097	1097	1073	1019	–	–	1070	δ (C–N)
–	–	966	–	–	–	966	δ (C–H) out of plane mode
–	–	801	–	–	–	801	
–	–	720	720	720	–	720	δ (C–C)
671	671	–	671	–	671	–	Sn–S

The TGA curves (Figure 2b) for all the SnS$_2$ samples showed a small weight loss of ~1% at around 100 °C due to the loss of physisorbed moisture. The single-step weight loss at around 390 °C was attributed to the phase transformation (oxidation) of SnS$_2$ to SnO$_2$ (theoretical weight loss = 17.6%) [32]. The observed weight loss values for S1, S2, S3, and S4 were 19.8%, 20.6%, 20.4% and 24.1%, respectively. The SnS$_2$ samples (S2, S3, and S4) prepared using capping agents exhibited marginally higher weight loss compared to those prepared in the absence of any capping agent (S1). This was attributed to the presence of more adsorbed organics on the surface of the SnS_2 nanoparticles (S2, S3, and S4), indicating the presence of capping agents on the SnS_2 nanoparticles. The TGA results were in agreement with the literature reports [18,32].

3.3. Optical Studies

The effect of the use of alkylamines on the optical properties of the SnS_2 nanoparticles was investigated using DRS and PL spectroscopy. Figure 3 shows the DRS and PL spectra for the SnS_2 nanoparticles. The DRS spectra exhibited band gap absorption for the SnS_2 nanoparticles in the range of 400 nm to 550 nm. The PL spectra exhibited an excitonic emission at around 550 nm and a defect emission at around 645 nm. The excitonic emission exhibited a blue shift for samples S1, S3, and S4 with respect to sample S2. The PL results were in agreement with the DRS results. The band gap was estimated from the Tauc plots shown in Figure 4. The band gap was found to vary from 2.31 eV to 3.50 eV. The variation in band gap was attributed to the difference in crystallite size, with sample S1 with the smallest crystallite size exhibiting a band gap of 3.50 eV.

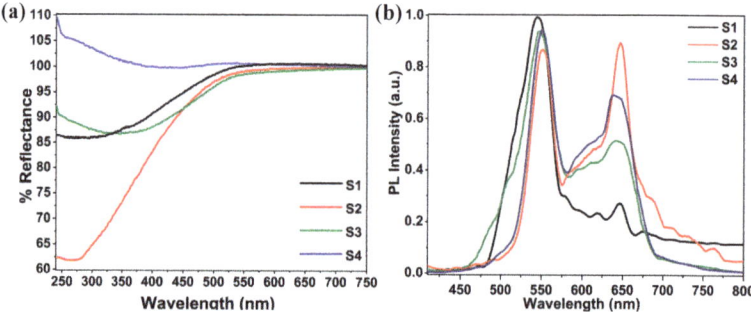

Figure 3. (**a**) DRS and (**b**) photoluminescence spectra of SnS_2 nanoparticles synthesized in the absence and presence of capping agent by thermal decomposition method.

Further investigation of the PL spectra indicated an increased concentration of surface defects for samples S2, S3, and S4 when the capping agent was used during the synthesis. The presence of excess surface defects imparted novel characteristics and modifications to the existing properties [33]. The $I_{exc}/I_{defects}$ ratio was found to vary from 2 to 7.3 (Table 3). Sample S2 was observed to exhibit the highest amount of surface defects ($I_{exc}/I_{defects}$ = 7.3) and the smallest crystallite size. Alkylamines are reported to play multiple roles as solvent surfactants and reducing agents during the synthesis of nanocrystals [29]. The growth of crystals in the presence of alkylamines results in twinning and stacking during crystal growth resulting in internal defects and influencing the final properties of nanocrystals [34,35].

Table 3. PL peak positions, FWHM, intensity ratio ($I_{excitonic}/I_{defects}$), band gap, and crystallite size of SnS_2 nanoparticles.

Sample ID	Peak Position (nm)		FWHM (nm)	($I_{excitonic}/I_{defects}$)	Band Gap (eV)	Crystallite Size (nm)		Sn:S Ratio
	Excitonic Emission	Defect Emission				(100)	(011)	
S1	544	645	37.5	2	2.86	18.0	5.3	1:1.8
S2	551	645	36.9	7.3	3.50	6.9	2.6	1:2
S3	548	647	53.4	3.6	2.33	7.3	3.0	1:2
S4	551	645	40.7	2.4	2.42	10.3	2.9	1:2

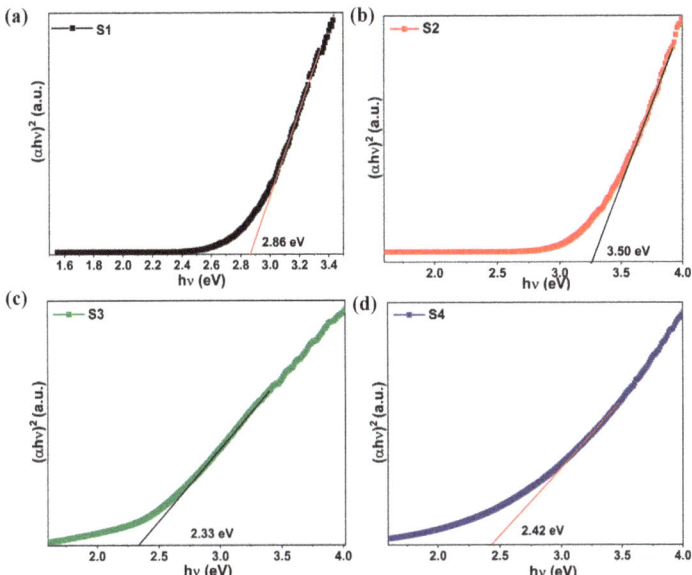

Figure 4. Tauc plot of SnS$_2$ nanoparticles synthesized in the absence and presence of capping agent by thermal decomposition method.

3.4. Morphological Analysis (FE-SEM and EDX Results)

The SnS$_2$ nanoparticles synthesized using different capping agents by the thermal decomposition approach were characterized using FE-SEM analysis. Figure 5 shows the FE-SEM images of the SnS$_2$ nanoparticles prepared in the absence and presence of capping agents such as octylamine, oleylamine, and dodecylamine. The presence of a capping agent played a vital role in influencing the morphology of the resulting nanoparticles. The FE-SEM analysis revealed the formation of a flower-like morphology for the pristine SnS$_2$ nanoparticles (S1) prepared by the thermal decomposition approach in the absence of a capping agent. A web-like microstructure of SnS$_2$ with a diameter of 800 nm was observed for Sample S1. The microstructures were formed by the assembly of flake-like nano-building units. The thickness of the flakes ranged from 10 nm to 15 nm.

On the other hand, the SnS$_2$ nanoparticles synthesized in the presence of capping agents (octylamine, oleylamine, and dodecylamine) exhibited particle-like, twisted-flower-like, and stacked-plate-like morphology, respectively. Sample S2, prepared in the presence of octylamine showed the formation of irregular particles. Sample S3, prepared in the presence of oleylamine, showed the formation of rosette-flower-like morphology. A closer look at the FE-SEM images revealed that the flowers were formed by the twisting and wrapping of linear structures, assembled to form rosette-like structures with a rosette-like morphology with a diameter of 1.2 microns, and the thickness of the linear structure was observed to be around 40 nm. Sample S4, prepared in the presence of dodecylamine, showed the formation of a stack of plate-like structures, or nanoplates assembled into stacks. The diameter of a typical nanoplate was around 250 nm and the thickness were around 30–35 nm.

The composition of all the SnS$_2$ samples (Sn:S) was analyzed using energy-dispersive X-ray analysis. The weight and atomic percent of Sn and S present in the SnS$_2$ samples synthesized by the thermal decomposition method are given in Table 1. The EDXA results indicated the presence of tin and sulfur in all the samples, and the Sn:S ratio varied from 1:1.8 to 1:2.2, which was close to the theoretical value (1:2).

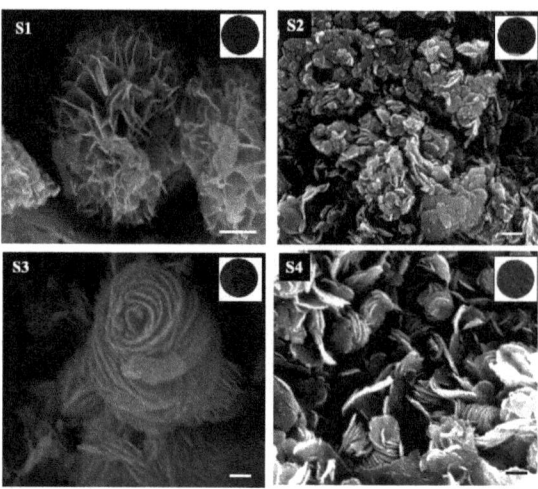

Figure 5. FE-SEM images of SnS$_2$ nanoparticles (S1–S4) prepared using the thermal decomposition approach in the absence and presence of capping agents. The inset shows the digital images of powder samples.

4. Mechanism for Morphology Evolution (Mechanism of Formation of SnS$_2$ Nanoparticles with Different Morphologies)

Scheme 1 depicts the proposed mechanism for the formation of SnS$_2$ nanoparticles with different morphologies using the thermal decomposition approach. The precursors (SnCl$_4$·5H$_2$O and CH$_3$CSNH$_2$ thioacetamide), when subjected to thermal decomposition, led to the formation of spherical SnS$_2$ seeds in the initial stage of the reaction. As the reaction proceeded, the nuclei grew to form flakes and strands. The flakes/strands were the primary building blocks for the hierarchical structures and the flakes/strands assembled to form flower-like structures [18]. The presence of capping agents during the thermal decomposition played a vital role in controlling the morphology of SnS$_2$ resulting in the formation of unique morphologies. Scheme 1 shows that the nuclei or basic building unit of SnS$_2$ are arranged differently in presence of the different capping agents. The physico-chemical properties of the capping different agents used during the synthesis decided the growth and assembly of the building blocks into nanostructures with a special morphology. The interaction of the surfactant molecule with the crystal seed drastically reduced the generation of nuclei. The reactant molecules then contributed to the characteristic growth of the nanocrystal.

The chain length of alkylamines also plays an important role in controlling the morphology of nanoparticles [21]. It is reported that higher activation energy and low reaction rate of amine with a longer carbon chain leads to the formation of smaller-sized quantum dots due to its higher capping capacity [36,37]. Hence, in this study, the growth was restricted, and this resulted in the formation of irregular particles or anisotropic crystals. Dodecylamine (C12) and oleylamine (C18), due to their longer alkyl(enyl) chains, resulted in the formation of irregular particles and stacks of nanoplates, respectively. On the other hand, octylamine (C8) resulted in the assembly of nuclei strands as a rosette-like morphology. Oleylamine (C18) acted as a surfactant and was adsorbed and subsequently passivated the surface, restricting further growth in planes, and resulting in the formation of nanoplates. The nanoplates were stacked together due to the interaction between the surfactant molecules adsorbed on the surface of the SnS$_2$ nanoplates. Octylamine (C8) on the other hand, due to its high polarity, led to the fusion of nanoplates in a helical manner, resulting in a flower-like morphology. Long aliphatic chains of dodecylamine

(C12) prevented the assembly and growth of nanoparticles and resulted in the formation of irregular particles.

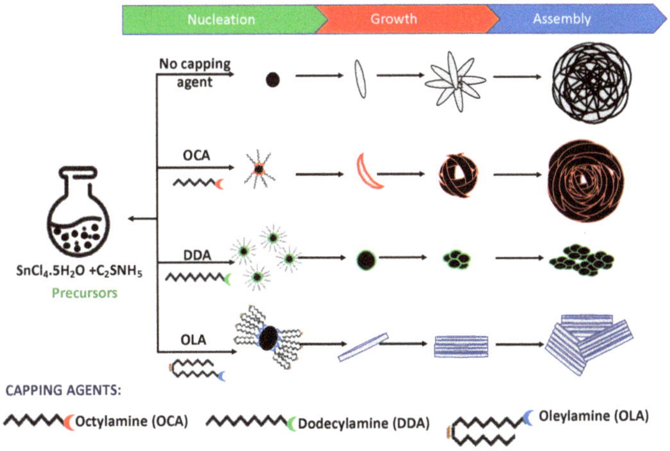

Scheme 1. Proposed mechanism for the evolution of different morphologies of SnS$_2$ prepared using thermal decomposition approach in the presence and absence of alkylamines.

5. Conclusions

SnS$_2$ nanoparticles with different morphologies were successfully synthesized by the thermal decomposition of SnCl$_4$·5H$_2$O and thioacetamide in the presence of different surfactants (dodecylamine, octylamine, and oleylamine). The use of alkylamine during the synthesis affected the nucleation and crystal growth and had a great influence on the morphology and optical properties of the SnS$_2$ nanoparticles. The SnS$_2$ nanoparticles showed the assembly of flakes into flower-like nanostructures, nanoparticles, and nanoplates stacked together. The formation of characteristic structures had an influence not only on the structural features but also on the optical properties of the SnS$_2$ nanoparticles. This approach was found to be beneficial for the surfactant-assisted synthesis of SnS$_2$ nanoparticles with unique morphologies. Alkylamines, due to their multifunctional characteristics, were an integral part of the nanocrystal synthesis. The use of alkylamines during the synthesis not only tailored the morphology but also influenced the properties of the nanomaterials. Thus, it is imperative to explore mechanistic insights into the role of alkylamines in nanocrystal synthesis. The careful optimization of the reaction parameters such as solvents, temperature, surfactants, etc., resulted in nanocrystals with a controlled shape and size for applications in solar cells, catalysis, environmental remediation, etc.

Author Contributions: Formal analysis, Methodology, Investigation, Writing—original draft, Writing—review & editing, R.G.; Funding acquisition, Writing—review & editing, I.A.; Writing—review & editing, S.S.; Funding acquisition, Writing—review & editing, N.S. All authors have read and agreed to the published version of the manuscript.

Funding: The authors would like to thank the Scientific Research Deanship at King Khalid University, Abha, Saudi Arabia through the Large Research Group Project under grant number 9RGP.02/219/43) and the Marine Pollution Special Interest Group, the National Defence University of Malaysia via SF0076-UPNM/2019/SF/ICT/6, for providing research facilities and funding.

Institutional Review Board Statement: Not applicable.

Informed Consent Statement: Not applicable.

Data Availability Statement: The data presented in this study are available on request from the corresponding author.

Acknowledgments: The authors would like to thank the Solar Research and Development Center (SRDC) for FE-SEM characterization and Pandit Deendayal Energy University for providing institutional fellowship, Scientific Research Deanship at King Khalid University, Abha, Saudi Arabia through the Large Research Group Project under grant number 9RGP.02/219/43) and the Marine Pollution Special Interest Group, the National Defence University of Malaysia via SF0076-UPNM/2019/SF/ICT/6, for providing research facilities and funding.

Conflicts of Interest: The authors declare no conflict of interest.

References

1. Shaikh, M.R.S. A review paper on electricity generation from Solar Energy. *Int. J. Res. Appl. Sci. Eng. Technol.* **2017**, *887*, 1884–1889. [CrossRef]
2. Guan, G.; Wu, J.; Huang, J.; Qian, X. Polynary metal selenide $CoSe_2/NiSe_2/MoSe_2$ porous nanospheres as efficient electrocatalytic materials for high-efficiency dye-sensitized solar cells. *J. Electroanal. Chem.* **2022**, *924*, 116888. [CrossRef]
3. Zhang, T.; Zhang, Q.; Li, Q.; Li, F.; Xu, L. Towards High-Performance Quantum Dot Sensitized Solar Cells: Enhanced Catalytic Activity and Stability of CuCo2Se4 Nanoparticles on Graphitic Carbon Nitride G-C3n4 Nanosheets. 2022. Available online: https://papers.ssrn.com/sol3/papers.cfm?abstract_id=4211162 (accessed on 1 September 2022).
4. Keshari, A.K.; Kumar, R. Nanostructured MoS_2-, SnS_2-, and WS_2-Based Anode Materials for High-Performance Sodium-Ion Batteries via Chemical Methods: A Review Article. *Energy Technol.* **2021**, *9*, 2100179. [CrossRef]
5. Huang, Y.; Sutter, E.; Sadowski, J.T.; Cotlt, M.; Monti, O.L.; Racke, D.A.; Sutter, P. Tin Disulfide: An Emerging Layered Metal Dichalcogenide Semiconductor: Materials Properties and Device Characteristics. *ACS Nano* **2014**, *8*, 10743–10755. [CrossRef]
6. Joseph, A.; Anjitha, C.R.; Aravind, A.; Aneesh, P.M. Structural, optical and magnetic properties of SnS_2 nanoparticles and photo response characteristics of p-Si/n-SnS_2 heterojunction diode. *Appl. Surf. Sci.* **2020**, *528*, 146977. [CrossRef]
7. Dong, W.; Lu, C.; Luo, M.; Liu, Y.; Han, T.; Ge, Y.; Xu, X. Enhanced UV–Vis photodetector performance by optimizing interfacial charge transportation in the heterostructure by SnS and $SnSe_2$. *J. Colloid Interface Sci.* **2022**, *621*, 374–384. [CrossRef]
8. Sun, Q.; Gong, Z.; Zhang, Y.; Hao, J.; Zheng, S.; Lu, W.; Wang, Y. Synergically engineering defect and interlayer in SnS_2 for enhanced room-temperature NO_2 sensing. *J. Hazard. Mater.* **2022**, *421*, 126816. [CrossRef]
9. Rajwar, B.K.; Sharma, S.K. Chemically synthesized Ti-doped SnS_2 thin films as intermediate band gap material for solar cell application. *Opt. Quantum Electron.* **2022**, *54*, 1–13. [CrossRef]
10. Xu, X.; Xu, F.; Zhang, X.; Qu, C.; Zhang, J.; Qiu, Y.; Wang, H. Laser-Derived Interfacial Confinement Enables Planar Growth of 2D SnS_2 on Graphene for High-Flux Electron/Ion Bridging in Sodium Storage. *Nano-Micro Lett.* **2022**, *14*, 1–16. [CrossRef] [PubMed]
11. Hayashi, K.; Kataoka, M.; Sato, S. Epitaxial Growth of SnS_2 Ribbons on a Au–Sn Alloy Seed Film Surface. *J. Phys. Chem. Lett.* **2022**, *13*, 6147–6152. [CrossRef] [PubMed]
12. Gaur, R. Morphology dependent activity of PbS nanostructures for electrochemical sensing of dopamine. *Mater. Lett.* **2020**, *264*, 127333. [CrossRef]
13. Gaur, R.; Jeevanandam, P. Synthesis and Characterization of Cd1-xZnxS (x = 0–1) Nanoparticles by Thermal Decomposition of Bis (thiourea) cadmium–zinc acetate Complexes. *ChemistrySelect* **2016**, *1*, 2687–2697. [CrossRef]
14. Yang, Y.B.; Dash, J.K.; Littlejohn, A.J.; Xiang, Y.; Wang, Y.; Shi, J.; Zhang, L.H.; Kisslinger, K.; Lu, T.M.; Wang, G.C. Large single crystal SnS_2 flakes synthesized from coevaporation of Sn and S. *Cryst. Growth Des.* **2016**, *16*, 961–973. [CrossRef]
15. Huang, L.; Cai, G.; Zeng, R.; Yu, Z.; Tang, D. Contactless Photoelectrochemical Biosensor Based on the Ultraviolet–Assisted Gas Sensing Interface of Three-Dimensional SnS_2 Nanosheets: From Mechanism Reveal to Practical Application. *Anal. Chem.* **2022**, *94*, 9487–9495. [CrossRef]
16. Zou, W.; Sun, L.H.; Cong, S.N.; Leng, R.X.; Zhang, Q.; Zhao, L.; Kang, S.Z. Preparation of worm-like SnS_2 nanoparticles and their photocatalytic activity. *J. Exp. Nanosci.* **2020**, *15*, 100–108. [CrossRef]
17. Sajjad, M.; Khan, Y.; Lu, W. One-pot synthesis of 2D SnS_2 nanorods with high energy density and long term stability for high-performance hybrid supercapacitor. *J. Energy Storage* **2021**, *35*, 102336. [CrossRef]
18. Gaur, R.; Jeevanandam, P. Synthesis of SnS_2 nanoparticles and their application as photocatalysts for the reduction of Cr (VI). *J. Nanosci. Nanotechnol.* **2018**, *18*, 165–177. [CrossRef]
19. Liu, J.; Wen, Y.; van Aken, P.A.; Maier, J.; Yu, Y. In situ reduction and coating of SnS_2 nanobelts for free-standing SnS@ polypyrrole-nanobelt/carbon-nanotube paper electrodes with superior Li-ion storage. *J. Mater. Chem. A* **2015**, *3*, 5259–5265. [CrossRef]
20. García-Rodríguez, R.; Liu, H. Mechanistic insights into the role of alkylamine in the synthesis of CdSe nanocrystals. *J. Am. Chem. Soc.* **2014**, *136*, 1968–1975. [CrossRef]
21. Rempel, J.Y.; Bawendi, M.G.; Jensen, K.F. Insights into the kinetics of semiconductor nanocrystal nucleation and growth. *J. Am. Chem. Soc.* **2009**, *131*, 4479–4489. [CrossRef]
22. Li, F.; Xie, Y.; Hu, Y.; Long, M.; Zhang, Y.; Xu, J.; Qin, M.; Lu, X.; Liu, M. Effects of alkyl chain length on crystal growth and oxidation process of two-dimensional tin halide perovskites. *ACS Energy Lett.* **2020**, *5*, 1422–1429. [CrossRef]

23. Talapin, D.V.; Rogach, A.L.; Kornowski, A.; Haase, M.; Weller, H. Highly luminescent monodisperse CdSe and CdSe/ZnS nanocrystals synthesized in a hexadecylamine-trioctylphosphine oxide-trioctylphospine mixture. *Nano Lett.* **2001**, *1*, 207–211. [CrossRef] [PubMed]
24. Jose, R.; Zhanpeisov, N.U.; Fukumura, H.; Baba, Y.; Ishikawa, M. Structure-property correlation of CdSe clusters using experimental results and first-principles DFT calculations. *J. Am. Chem. Soc.* **2006**, *128*, 629–636. [CrossRef]
25. Pradhan, N.; Reifsnyder, D.; Xie, R.; Aldana, J.; Peng, X. Surface ligand dynamics in growth of nanocrystals. *J. Am. Chem. Soc.* **2007**, *129*, 9500–9509. [CrossRef] [PubMed]
26. SSun, Z.H.; Oyanagi, H.; Nakamura, H.; Jiang, Y.; Zhang, L.; Uehara, M.; Yamashita, K.; Fukano, A.; Maeda, H. Ligand effects of amine on the initial nucleation and growth processes of CdSe nanocrystals. *J. Phys. Chem. C* **2010**, *114*, 10126–10131. [CrossRef]
27. Li, L.S.; Pradhan, N.; Wang, Y.; Peng, X. High quality ZnSe and ZnS nanocrystals formed by activating zinc carboxylate precursors. *Nano Lett.* **2004**, *4*, 2261–2264. [CrossRef]
28. Guo, Y.; Marchuk, K.; Sampat, S.; Abraham, R.; Fang, N.; Malko, A.V.; Vela, J. Unique challenges accompany thick-shell CdSe/nCdS (n > 10) nanocrystal synthesis. *J. Phys. Chem. C* **2012**, *116*, 2791–2800. [CrossRef]
29. Mourdikoudis, S.; Liz-Marzán, L.M. Oleylamine in nanoparticle synthesis. *Chem. Mater.* **2013**, *25*, 1465–1476. [CrossRef]
30. Queiroz, M.F.; Teodosio Melo, K.R.; Sabry, D.A.; Sassaki, G.L.; Rocha, H.A.O. Does the use of chitosan contribute to oxalate kidney stone formation? *Mar. Drugs* **2014**, *13*, 141–158. [CrossRef]
31. Rao, C.N.R.; Venkataraghavan, R.; Kasturi, T.R. Contribution to the infrared spectra of organosulphur compounds. *Can. J. Chem.* **1964**, *42*, 36–42. [CrossRef]
32. Sun, H.; Ahmad, M.; Luo, J.; Shi, Y.; Shen, W.; Zhu, J. SnS_2 nanoflakes decorated multiwalled carbon nanotubes as high performance anode materials for lithium-ion batteries. *Mater. Res. Bull.* **2014**, *49*, 319–324. [CrossRef]
33. Mozetič, M. Surface modification to improve properties of materials. *Materials* **2019**, *12*, 441. [CrossRef] [PubMed]
34. Villaverde-Cantizano, G.; Laurenti, M.; Rubio-Retama, J.; Contreras-Cáceres, R. Reducing Agents in Colloidal Nanoparticle Synthesis—An Introduction. In *Reducing Agents in Colloidal Nanoparticle Synthesis*; Mourdikoudis, S., Ed.; Royal Society of Chemistry: London, UK, 2021; pp. 1–27.
35. Rodrigues, T.S.; Zhao, M.; Yang, T.H.; Gilroy, K.D.; da Silva, A.G.; Camargo, P.H.; Xia, Y. Synthesis of colloidal metal nanocrystals: A comprehensive review on the reductants. *Chemistry* **2018**, *24*, 16944–16963. [CrossRef] [PubMed]
36. Hwang, S.; Choi, Y.; Jeong, S.; Jung, H.; Kim, C.G.; Chung, T.M.; Ryu, B.H. Low temperature synthesis of CdSe quantum dots with amine derivative and their chemical kinetics. *Jpn. J. Appl. Phys.* **2010**, *49*, 05EA03. [CrossRef]
37. Navazi, Z.R.; Nemati, A.; Akbari, H.; Davaran, S. The effect of fatty amine chain length on synthesis process of InP/ZnS quantum dots. *Orient. J. Chem.* **2016**, *32*, 2163–2169. [CrossRef]

Article

Strengthened Optical Nonlinearity of V$_2$C Hybrids Inlaid with Silver Nanoparticles

Yabin Shao [1], Qing He [2], Lingling Xiang [1], Zibin Xu [1], Xiaoou Cai [1] and Chen Chen [3,*]

[1] School of Jia Yang, Zhejiang Shuren University, Shaoxing 312028, China; shao_yabin@163.com (Y.S.); melody.xiang@vip.163.com (L.X.); xu_zibin@163.com (Z.X.); cai.xo@163.com (X.C.)
[2] Collaborative Innovation Center of Steel Technology, University of Science and Technology Beijing, Beijing 100083, China; 123simon@163.com
[3] College of Civil Engineering, East University of Heilongjiang, Harbin 150086, China
* Correspondence: chen_chen1600@163.com

Abstract: The investigation of nonlinear optical characteristics resulting from the light–matter interactions of two-dimensional (2D) nano materials has contributed to the extensive use of photonics. In this study, we synthesize a 2D MXene (V$_2$C) monolayer nanosheet by the selective etching of Al from V$_2$AlC at room temperature and use the nanosecond Z-scan technique with 532 nm to determine the nonlinear optical characters of the Ag@V$_2$C hybrid. The z-scan experiment reveals that Ag@V$_2$C hybrids usually exhibits saturable absorption owing to the bleaching of the ground state plasma, and the switch from saturable absorption to reverse saturable absorption takes place. The findings demonstrate that Ag@V$_2$C has optical nonlinear characters. The quantitative data of the nonlinear absorption of Ag@V$_2$C varies with the wavelength and the reverse saturable absorption results from the two-photon absorption, which proves that Ag@V$_2$C hybrids have great potential for future ultrathin optoelectronic devices.

Keywords: Ag@V$_2$C MXene; hybrids; Z-scan; nonlinear optical properties

1. Introduction

MXenes [1], 2D transition metal carbides and nitrides, exhibit the outstanding advantages of electrical conductivity, a beneficial elastic modulus and capacitance, an adaptable band gap and optical transparency [2–6]. Nonlinear optical (NLO) properties of MXene have drawn wide attention [7]. Among them, Ti$_3$C$_2$Tx is the first and most-studied MXene, and the research has been progressing vigorously regarding the linear optical properties [8–13].

Ti$_3$C$_2$Tx thin films was studied by Mochalin and Podila et al., and the findings exhibited a high modulation depth of 50% and high damage thresholds of 70 mJ cm^{-2} [11]. Ti$_3$C$_2$Tx was synthesized by Wen and Zhang et al. who revealed that the Ti$_3$C$_2$Tx maximum nonlinear absorption coefficient 10–21 m^2/V^2 had been increased tremendously compared with other 2D materials [12].

Thus far, the linear optical properties have been explored experimentally and theoretically in an extensive way, while little attention has been drawn to V$_2$CT$_X$'s optical nonlinear features as well as its related applications. The Z-scan technique proposed by Sheik-Bahae et al. is simple and accurate and is widely used to study the nonlinear properties of materials (particularly non-fluorescent) [13].

To achieve the application-oriented demands for MXenes, a large variety of modified techniques have been invented to obtain desired functionalities, such as the previous studies on colloidal solutions of nanoparticles (NPs) hybridized by graphene, TMDs, etc. [14].

Moreover, great breakthroughs have taken place in NLO applications via hybridizing NPs in the past few years in accordance with their defect states, surface control and excellent plasmonic properties [15–17]. In various experiments on its broadband and strong NLO response, V$_2$CT$_X$ nanosheets, a new member of the MXene family, has drawn extensive

attention due to its novel optical properties. Recent studies have shown that it has very good performance in hybrid mode-locking and good optical nonlinear properties, which indicates its good potential applications in acting as sensors and nonlinear components for lasers [18,19].

2. Materials and Methods

2.1. Preparation of Ag@V$_2$C Hybrids

First, 400 mesh V$_2$AlC (1 g) powder (Suzhou Beike Nano Technology Co., Ltd., Suzhou, China) is added slowly to 50% (v/v) HF solution, and it is stirred at room temperature for 90 h [16–18] to ensure the V-Al bonds are completely disconnected from V$_2$AlC with the V-C bonds intact. Second, to stratify the V$_2$CT$_X$, 30 mL tetraethylammonium hydroxide (TMAOH 25% in H$_2$O) is blended with 1 g multilayer V$_2$CT$_X$ and sonicated for 30 min at room temperature, and the precipitates are rinsed with deionized water to neutrality (pH $\geq$ 6). Then, the excess TMAOH is discarded from the product by repeated centrifugation at 3000 rpm. Finally, a 2D monolayer V$_2$CT$_X$ MXene material is obtained via freeze-drying.

The Ag@V$_2$C hybrids are made ready by mingling the V$_2$C hybrid dispersion and AgNO$_3$ solution. These are shown in Figure 1. First, 3 mL of original V$_2$C colloidal solution (1 mg/mL) in 30mL of AgNO$_3$ solution (1 mg/mL) is re-dispersed, and the mixture is sonicated for 30 min. Then, the obtained colloidal solution of hybrid nanocomposites of Ag@V$_2$C hybrids is centrifuged at 12,000 rpm for 20 min and re-dispersed in 30 mL deionized water.

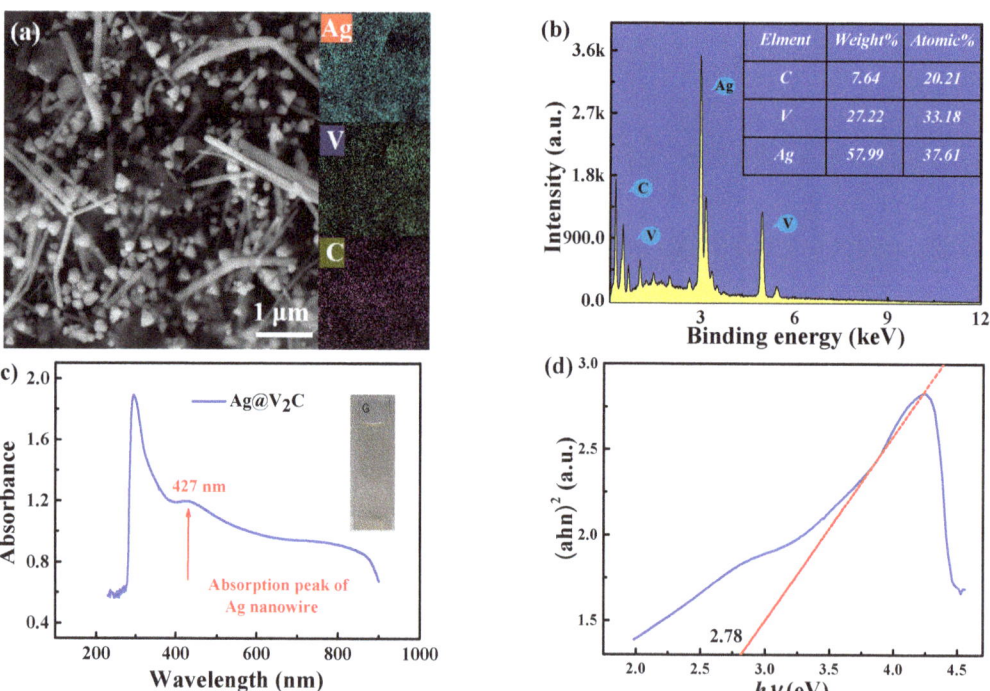

Figure 1. (**a**) SEM and EDS mappings of Ag V C elements. (**b**) EDS of Ag@V2C. The inset is atomic ratios for diverse elements. (**c**) Optical absorption spectra of Ag@V$_2$C. (**d**) Estimation of the band gap of Ag@V$_2$C.

2.2. Optical Experimental Setup

SEM on a ZEISS Sigma 300 (acceleration voltage: 15 kV, Dublin, CA, USA) was employed to perform the morphological study with a spectrometer (Ocean Optics 4000, CA, USA) to survey the UV–Vis–NIR spectra of monolayer V_2CT_X MXene. To evaluate the NLO features of V_2CT_X MXene flakes nanoparticles, we performed an experiment of a single beam open aperture (OA) Z-scan, with the help of a 6 ns 8 Hz Q-switched Nd:YAG nanosecond pulse laser (Surelite II, Continuum, San Jose, CA, USA) and an optical parametric oscillator (Continuum, APE OPO) to produce lasers beams of diverse wavelengths. In this experiment, the linear transmittance of the V_2CT_x hybrids solution was measured at 500 nm wavelength, and we found that it was 72%.

The laser beam with a waist diameter of 200 μm is focused through a lens with a focal length of 20 cm and projected onto a 2 mm diameter quartz cuvette filled with aqueous dispersion of V_2CT_X monolayer flake nanoparticles. The cuvette is installed on a computer-operated translation motion stage where the transmitted data for every z point is recorded by stored program. Their NLO properties are studied by Z-scan technology under 532 nm nanosecond laser pulses. The findings indicate that the Ag@ V_2CT_X has a large nonlinear absorption coefficient.

3. Results

Figure 1a illustrates the typically morphologic multilayer Ag@V_2C with coarse surface and the shape of accordion. The elements mapping data of SEM in combination with EDS illustrates that the Ag, V and C elements are evenly dispersed in the monolayer V_2C. In Figure 1b, the fundamental analysis of Ag@V_2C V_2C, the result of C: V: Ag ≈ 2:3:3 is obtained via the energy dispersive X-ray energy spectrum (EDS), while it is impracticable to measure the content of other elements by EDS due to the inadequate content.

In Figure 1c, the linear absorbance spectra of Ag@V_2C is investigated by using ultraviolet visible near infrared spectrophotometer to observe two absorption peaks—that is, 315 and 427 nm. These findings justify that the surface of Ag@V_2C is functionalized with Ag nanoparticles, and the positions of the peaks were predicted by theoretical calculations. In Figure 1d, the estimated band gap value of Ag@V_2C is 2.75~2.80 eV according to Kubelka-Munk's theory in previous data [19].

Hereinafter are the three different phases. In Figure 2a, the transmittance of Ag@V_2C hybrids remains flat before rising sharply when the Z position approaches zero, and a peak is observed, which reflects the SA property. In Figure 2b, the conversion from SA to RSA is observed. The transmittance goes up to the first peak to indicate SA property and then falls back to a valley to indicate the reverse saturable absorbtion RSA property when the Z position approaches zero. It then rises again to the second peak and falls back to the flat linear transmittance. In Figure 2c, the laser pulse energy mounts to 681 when a deep valley occurs, which indicates the RSA property. Figure 2a–c, in general, reveals a whole process of NLO properties, including SA, conversion and RSA.

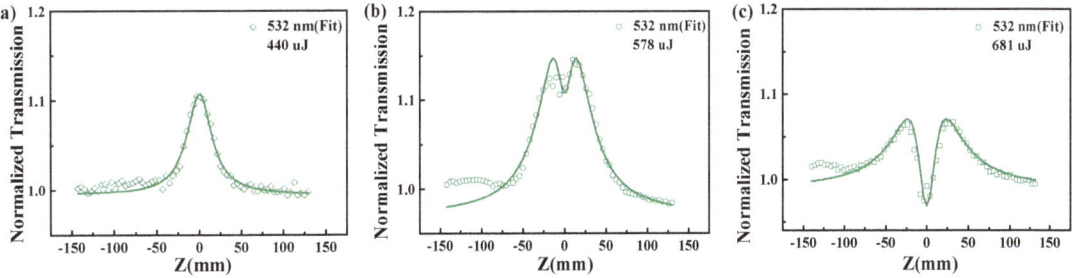

Figure 2. Open aperture Z-scan is conducted at the wavelength of 532 nm and the laser energies at (a) 440 μJ, (b) 578 μJ and (c) 681 μJ to obtain the normalized transmission of V_2CT_X hybrid colloid.

The physical mechanism of NLO properties can be expounded as follows in Figure 3. The above findings are related in the main to the changing energy of incident laser pulse and the nature property of MXene. It is generally believed that nonlinear susceptibility of the material results from intraband transition, interband transition, hot electron excitation and thermal effects. [20]. Specifically, when Ag@V$_2$C monolayer flakes are at low pulse energy, one photon absorption arises to indicate the SA property [21].

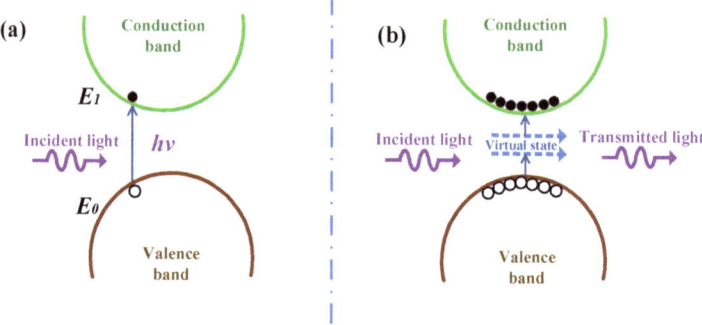

Figure 3. The physical mechanism of the NLO properties of Ag@V$_2$C monolayer flakes. (a) One photon absorption. (b) Two photon absorption.

Under laser irradiance, as the sample approaches the focal point, the laser energy increases abruptly, and the materials absorbs much incident light pulse when Ag@V$_2$C monolayer hybrids are pumped into an excited state, leaving a small amount at the ground state. Thus, more light penetrates the sample to make the transmittance higher and SA can be observed [22]. This phenomenon is termed ground-state bleaching of the plasmon band.

To remove the thermal effect's influence on the possibility of nonlinearity, a laser pulse was set with a short duration and low repetition rate (10 Hz and 6 ns). Moreover, as the incident pulse energy goes up to 1080 µJ, the conversion appears; as it reaches 1380 µJ, RSA occurs. The major causes of RSA may be due to the excited state absorption (ESA) and two-photon absorption (TPA) [23].

The quantitative findings of the Z-scan experiment are as follows. The quantitative relationship between laser energy intensity and optical path length can be expressed as follows [24]:

$$dI = -\alpha I dz \quad (1)$$

In Equation (1), I signifies the laser intensity, z signifies the optical path length, and α signifies the absorption coefficient. The absorption coefficient combining SA and RSA in Ag@V$_2$C monolayer hybrids is expressed in Equation (2) [25]:

$$\alpha(I) = \frac{\alpha_0}{1 + (I/I_s)} + \beta I \quad (2)$$

Here, α_0 signifies the linear absorption coefficient of Ag@V$_2$C monolayer hybrids, I signifies the laser intensity, I_s is the saturable intensity, and β is the positive nonlinear absorption coefficient. I in Equation (3) can also be expressed as follows in Equation (3):

$$I = \frac{I_0}{1 + z^2/z_0^2} \quad (3)$$

Therefore, Equation (2) can be reconsidered as:

$$\alpha(I_0) = \frac{\alpha_0}{1 + \frac{I_0}{(1+z^2/z_0^2)I_s}} + \frac{\beta I_0}{1 + z^2/z_0^2} \quad (4)$$

With the help of Equations (1)–(4), the data of normalized transmission in Figure 3 can be fitted. In this way, related parameters are obtained as shown in Table 1.

Table 1. The nonlinear optical parameters of Ag@V2C monolayer flake hybrids.

λ (nm)	I_0 (W/m^2)	I_s (W/m^2)	β (m/W)
532	1.1×10^{14}	0.61×10^6	-
	1.4×10^{14}	0.82×10^6	
	7.4×10^{13}	0.23×10^6	1.12×10^{-10}

To inspect the ultrafast carrier dynamics of Ag@V$_2$C hybrids, broadband transient absorption was studied. In Figure 4a, TA spectra of Ag@V$_2$C hybrids are illustrated by 2D map of TA signals that have been achieved temporally and spectrally. A constant pump fluence (6.4 × 10^3 mW/cm^2) and probe beam ranging from 450 to 600 nm are employed to measure broadband TA signals and obtain the 2D color coded maps, cut horizontally through each map five times to obtain five differential absorption spectra at different delay times (0, 3.2, 4.0, 6.6 and 11 ps).

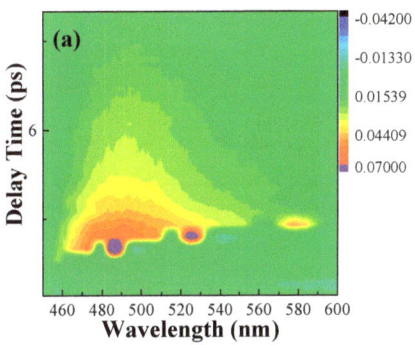
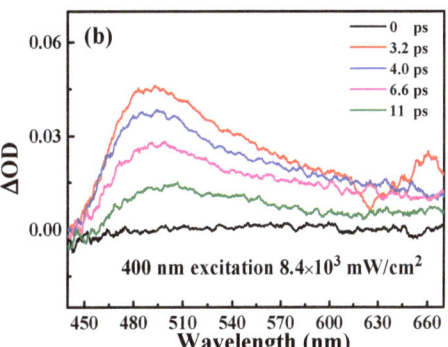

Figure 4. Carrier dynamics (at 400 nm pump) curves for Ag@V$_2$C. (a) Two-dimensional (2D) mapping of transient absorption spectra pumped at 400 nm with a fluence of 8.4 × 10^3 mW/cm^2, (b) Time and wavelength resolved transient absorption data of Ag@V$_2$C.

In Figure 4b, a positive absorption proves the generation of excited state absorption (ESA) in the spectral region as well as an ultrafast carrier relaxation process within the scale of ps. When the delay time increases, the amplitude of TA spectrum is reduced. The black curve, around the 0 ps delay time, signifies the unexcited Ag@V$_2$C hybrids. The peak value of Ag@V$_2$C hybrids, 485 nm, forms when the photo-induced absorption is brought about by the conversion of occupied–unoccupied states, which are found in the Z-scan experiment as RSA [26–28].

Then, the TA signal decays at ~13 ps and falls back to zero in the full waveband. The energy of the pump light markedly exceeds the energy bandgap of the Ag@V$_2$C nanosheet, the former 400 nm (~3.10 eV) laser, the latter being ~2.78 eV). Hence, with the incident laser pulse working up, electrons are excited and jump back and forth to the conduction band, while the holes remain in the valence band [29].

Soon afterwards is the conversion from electrons to hot electrons through photo-exciting with Fermi–Dirac distribution. The hot carriers will chill down shortly in sync with the decay process through e–e and e–ph diffusion on the conduction band, and with the carrier-phonon dispersing over a few relaxation processes, the hot carriers relax from the conduction band to the valence band and reunite with the holes.

In Figure 5a, the carrier dynamics are inspected at various wavelengths—that is, 480, 495, 515 and 540 nm, as shown in the figure. The optical transmission responses of Ag@V$_2$C hybrids are composed of two decay processes, a fast and a slow one. Coulomb-induced hot electrons are excited to be at the core state and be trapped by the surface state before releasing spare energy by dispersing the optical phonons (3.9 ps). The rest of the chilled electrons will experience nonradiative transition to fall back to ground state within 30.1 ps [30,31]. The above-mentioned biexponential decay function is formulated in Equation (3) [32].

$$\frac{\Delta T}{T} = A_1 \exp(-\frac{t}{\tau_1}) + A_2 \exp(-\frac{t}{\tau_2}) \quad (5)$$

where A_1 and A_2 signify the amplitudes of the fast, slow decay components, respectively; and τ_1 and τ_2 are the decay lifetimes of each component, correspondingly. The experimental data in Figure 5a justify the formulation of Equation (3) through which the fast and slow decay components, τ_1 and τ_2, are determined along with the rise of probe wavelengths owing to electrons at the lower energy states that are more apt to be detected and whose number is less likely to decrease than those at higher energy states. Similar properties can be seen in graphite [33].

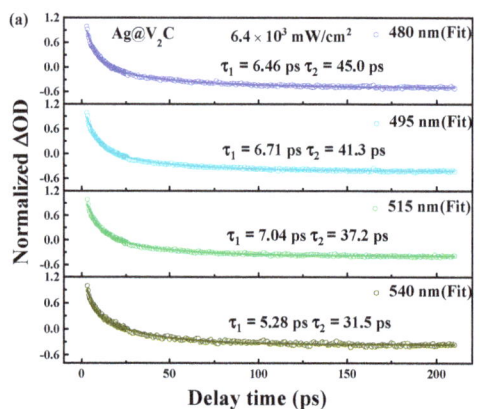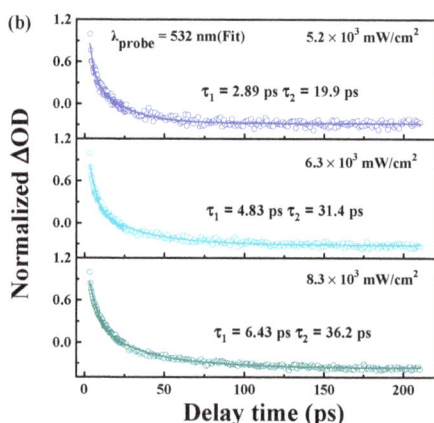

Figure 5. (a) Carrier dynamics (at 400 nm pump) curves for Ag@V$_2$C nanosheet at various probe wavelengths of 480, 495, 525, 515 and 540 nm, respectively. The scatters are experimental data while the solid lines are theoretical fit generated with pump fluence fixed at 6.4×10^3 mW/cm^2. (b) Carrier dynamics curves (at 400 nm pump) at different pump fluences 5.2×10^3, 6.3×10^3 and 8.3×10^3 mW/cm^2 with a probe wavelength fixed at 532 nm.

The pump fluence effect on carrier dynamics is inspected at the 500 nm probe wavelength. In Figure 5b, the findings of various pump-fluences (5.1×10^3, 6.4×10^3 and 8.1×10^3 mW/cm^2) accord with the results by applying Equation (3) and the corresponding parameters. With the rise of pump fluence, τ_1 goes up from 3.9 to 4.5 ps, and τ_2 goes from 11.1 to 13.2 ps, which is inseparable to the carrier density and its reliance on e–ph coupling [34]).

In brief, the faster decay part of 2D materials follow the principle of e–ph scattering, while the slower part that of ph–ph scattering [35], among which the necessary basis of the electrons' energy transfer is the electron–phonon interaction [36]. Ultimately, the improved efficiency of electron–phonon interaction promotes the cooling process, or in other words, it is high energy injection that speeds up electron decay. Similar properties can be seen in either quantum dots or 2D films [37–39].

The data shown in Figure 5 can be verified upon checking Equation (3) and please see in Table 2 regarding τ_1 and τ_2.

Table 2. Carrier dynamics parameters of the Ag@V$_2$C nanosheet.

	λ (nm)	τ$_1$ (ps)	τ$_2$ (ps)
Ag@V$_2$C	470	4.5	33.9
	485	4.6	36.5
	500	4.2	43.1
	520	3.9	45.8
V$_2$C nanosheet	470	3.8	18.0
	485	3.2	27.4
	500	4.6	22.5
	520	4.6	19.9

Resulting from the high density of Ag, those aroused electrons in the CB of Ag@V$_2$C hybrids tend to move to the d band of Ag atoms until the bleaching effect of VB disappears, which extends the aroused, composite electrons' lifetime and, accordingly, leads to more violent RSA properties [40]. Next, the electrons in the d band become excited functional groups to take in incident light before they move to a higher energy level and lead to an improved RSA effect. Then, the carriers that are excited and dressed with Ag nanoparticles will move from CB of Ti$_3$C to the sp band of the metal until they reach VB of Ag@V$_2$C hybrids. Compared with direct decay of pure Ti$_3$C$_2$ nanosheet, this takes a longer time scale (50 ps), and thus it enhances the RSA performance. Similar properties can be seen in other 2D materials [41].

Figure 6 exhibits the optical limiting response at 532 nm. With the increase of incident energy, the transmittance increases rapidly at first. When the incident energy continues to increase, the transmittance does not change drastically, which indicates that Ag@V$_2$C has superior optical limiting ability and can be used to manufacture optical limiting devices.

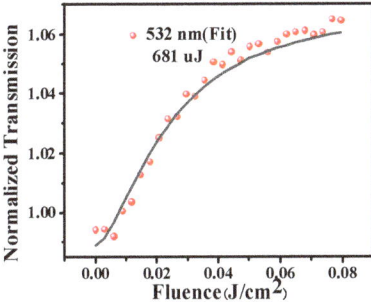

Figure 6. The optical limiting response at 532 nm.

4. Conclusions

Ag@V$_2$C ultrathin hybrids were synthesized via conventional etching, and remarkable NLO properties of Ag@V$_2$C hybrids were found through the OA z-scan technique at 532 nm. SA and RSA properties resulted mainly from GSB and TPA. Moreover, femtosecond transient absorption spectroscopy was adopted to inspect the ultrafast dynamics of the specimen, and we found that its decay contained a fast decay component (~4.5 ps) resulting from electron–phonon interactions and a slow component (~40 ps) from phonon–phonon interactions. Additionally, the two decay times increased with the pump fluence. The experiments confirmed that Ag@V$_2$C hybrids can be used in ultrafast optoelectronics and optical limiters.

Author Contributions: Data curation, C.C.; Funding acquisition, L.X., Z.X. and X.C.; Investigation, Y.S.; Resources, Q.H.; Writing—review and editing, L.X. All authors have read and agreed to the published version of the manuscript.

Funding: The article was supported by the following funds, Natural Science Foundation of Heilongjiang Province (Grant No. LH2021A019), East University of Heilongjiang Scientific Research Fund (Grant Nos. HDFHX210110, 210111, HDFKYTD202105) and Scientific research fund of Zhejiang Shuren University (Grant No. 2021R037).

Institutional Review Board Statement: Not applicable.

Informed Consent Statement: Not applicable.

Data Availability Statement: Data sharing is not applicable to this article.

Conflicts of Interest: The authors declare no conflict of interest.

References

1. Naguib, M.; Kurtoglu, M.; Presser, V.; Lu, J.; Niu, J.; Heon, M.; Hultman, L.; Gogotsi, Y.; Barsoum, M.W. Two-dimensional nanocrystals produced by exfoliation of Ti_3AlC_2. *Adv. Mater.* **2011**, *23*, 4248–4253. [CrossRef] [PubMed]
2. Keller, U. Recent developments in compact ultrafast lasers. *Nature* **2003**, *424*, 831–838. [CrossRef] [PubMed]
3. Mashtalir, O.; Naguib, M.; Mochalin, V.N.; Dall'Agnese, Y.; Heon, M.; Barsoum, M.W.; Gogotsi, Y. Intercalation and delamination of layered carbides and carbonitrides. *Nat. Commun.* **2013**, *4*, 1716. [CrossRef] [PubMed]
4. Lukatskaya, M.R.; Mashtalir, O.; Ren, C.E.; Dall'Agnese, Y.; Rozier, P.; Taberna, P.L.; Naguib, M.; Simon, P.; Barsoum, M.W.; Gogotsi, Y. Cation intercalation and high volumetric capacitance of two-dimensional titanium carbide. *Science* **2013**, *341*, 1502–1505. [CrossRef] [PubMed]
5. Shahzad, F.; Alhabeb, M.; Hatter, C.B.; Anasori, B.; Hong, S.M.; Koo, C.M.; Gogotsi, Y. Electromagnetic interference shielding with 2D transition metal carbides (MXenes). *Science* **2016**, *353*, 1137–1140. [CrossRef]
6. Peng, Q.; Guo, J.; Zhang, Q.; Xiang, J.; Liu, B.; Zhou, A.; Liu, R.; Tian, Y. Unique lead adsorption behavior of activated hydroxyl group in two-dimensional titanium carbide. *J. Am. Chem. Soc.* **2014**, *136*, 4113–4116. [CrossRef]
7. Jhon, Y.I.; Koo, J.; Anasori, B.; Seo, M.; Lee, J.H.; Gogotsi, Y.; Jhon, Y.M. Metallic MXene Saturable Absorber for Femtosecond Mode-Locked Lasers. *Adv. Mater.* **2017**, *29*, 1702496–1702503. [CrossRef]
8. Enyashin, A.; Ivanovskii, A. Two-dimensional titanium carbonitrides and their hydroxylated derivatives: Structural, electronic properties and stability of MXenes $Ti_3C_{2-x}N_x(OH)_2$ from DFTB calculations. *J. Solid State Chem.* **2013**, *207*, 42–48. [CrossRef]
9. Lashgari, H.; Abolhassani, M.; Boochani, A.; Elahi, S.; Khodadadi, J. Electronic and optical properties of 2D graphene-like compounds titanium carbides and nitrides: DFT calculations. *Solid State Commun.* **2014**, *195*, 61–69. [CrossRef]
10. Mauchamp, V.; Bugnet, M.; Bellido, E.P.; Botton, G.A.; Moreau, P.; Magne, D.; Naguib, M.; Cabioc'h, T.; Barsoum, M.W. Enhanced and tunable surface plasmons in two-dimensional Ti_3C_2 stacks: Electronic structure versus boundary effects. *Phys. Rev. B* **2014**, *89*, 235428. [CrossRef]
11. Berdiyorov, G. Optical properties of functionalized $Ti_3C_2T_2$ (T=F, O, OH) MXene: First-principles calculations. *AIP Adv.* **2016**, *6*, 055105. [CrossRef]
12. Hantanasirisakul, K.; Zhao, M.Q.; Urbankowski, P.; Halim, J.; Anasori, B.; Kota, S.; Ren, C.E.; Barsoum, M.W.; Gogotsi, Y. Fabrication of $Ti_3C_2T_x$ MXene transparent thin films with tunable optoelectronic properties. *Adv. Electron. Mater.* **2016**, *2*, 1600050. [CrossRef]
13. Dillon, A.D.; Ghidiu, M.J.; Krick, A.L.; Griggs, J.; May, S.J.; Gogotsi, Y.; Barsoum, M.W.; Fafarman, A.T. Highly conductive optical quality solution-processed films of 2D titanium carbide. *Adv. Funct. Mater.* **2016**, *26*, 4162–4168. [CrossRef]
14. Kim, I.Y.; Jo, Y.K.; Lee, J.M.; Wang, L.; Hwang, S.-J. Unique Advantages of Exfoliated 2D Nanosheets for Tailoring the Functionalities of Nanocomposites. *J. Phys. Chem. Lett.* **2014**, *5*, 4149–4161. [CrossRef] [PubMed]
15. Biswas, S.; Kole, A.K.; Tiwary, C.S.; Kumbhakar, P. Enhanced nonlinear optical properties of graphene oxide–silver nanocomposites measured by Z-scan technique. *RSC Adv.* **2016**, *6*, 10319–10325. [CrossRef]
16. Li, Z.; Dong, N.; Cheng, C.; Xu, L.; Chen, M.; Wang, J.; Chen, F. Enhanced nonlinear optical response of graphene by silver-based nanoparticle modification for pulsed lasing. *Opt. Mater. Express* **2018**, *8*, 1368–1377. [CrossRef]
17. Yan, X.; Wu, X.; Fang, Y.; Sun, W.; Yao, C.; Wang, Y.; Zhang, X.; Song, Y. Effect of silver doping on ultrafast broadband nonlinear optical responses in polycrystalline Ag-doped InSe nanofilms at near-infrared. *RSC Adv.* **2020**, *10*, 2959–2966. [CrossRef]
18. Ma, C.Y.; Huang, W.C.; Wang, Y.Z.; Adams, J.; Wang, Z.H.; Liu, J.; Song, Y.F.; Ge, Y.Q.; Guo, Z.Y.; Hu, L.P.; et al. MXene saturable absorber enabled hybrid mode-locking technology: A new routine of advancing femtosecond fiber lasers performance. *Nanophotonics* **2020**, *9*, 2451–2458. [CrossRef]
19. He, Q.; Hu, H.H.; Shao, Y.B.; Zhao, Z.Z. Switchable optical nonlinear properties of monolayer V_2CT_X MXene. *Optik* **2021**, *247*, 167629. [CrossRef]
20. Naguib, M.; Gogotsi, Y. Synthesis of Two-Dimensional Materials by Selective Extraction. *Acc. Chem. Res.* **2015**, *48*, 128–135. [CrossRef]
21. Jiang, X.; Liu, S.; Liang, W.; Luo, S.; He, Z.; Ge, Y.; Wang, H.; Cao, R.; Zhang, F.; Wen, Q. Broadband nonlinear photonics in few-layer MXene $Ti_3C_2T_x$ (T=F, O, or OH). *Laser Photonics Rev.* **2018**, *12*, 1700229. [CrossRef]
22. Dong, Y.; Chertopalov, S.; Maleski, K.; Anasori, B.; Hu, L.; Bhattacharya, S.; Rao, A.M.; Gogotsi, Y.; Mochalin, V.N.; Podila, R. Saturable absorption in 2D Ti_3C_2 MXene thin films for passive photonic diodes. *Adv. Mater.* **2018**, *30*, 1705714. [CrossRef] [PubMed]

23. Wang, J.; Chen, Y.; Li, R.; Dong, H.; Zhang, L.; Lotya, M.; Coleman, J.N.; Blau, W.J. Nonlinear optical properties of graphene and carbon nanotube composites. In *Carbon Nanotubes-Synthesis, Characterization, Applications*; Yellampalli, S., Ed.; IntechOpen: London, UK, 2011.
24. Sheik-Bahae, M.; Said, A.A.; Wei, T.; Hagan, D.J.; Stryland, E.W.V. Sensitive measurement of optical nonlinearities using a single beam. *IEEE J. Quantum Electron.* **1990**, *26*, 760–769. [CrossRef]
25. Wu, W.; Chai, Z.; Gao, Y.; Kong, D.; Wang, Y. Carrier dynamics and optical nonlinearity of alloyed CdSeTe quantum dots in glass matrix. *Opt. Mater. Express* **2017**, *7*, 1547–1556. [CrossRef]
26. Wang, J.; Ding, T.; Wu, K. Charge transfer from n-doped nanocrystals: Mimicking intermediate events in multielectron photocatalysis. *J. Am. Chem. Soc.* **2018**, *140*, 7791–7794. [CrossRef]
27. Lu, S.; Sui, L.; Liu, Y.; Yong, X.; Xiao, G.; Yuan, K.; Liu, Z.; Liu, B.; Zou, B.; Yang, B. White Photoluminescent Ti_3C_2 MXene Quantum Dots with Two-Photon Fluorescence. *Adv. Sci.* **2019**, *6*, 1801470. [CrossRef]
28. Wu, K.; Chen, J.; McBride, J.R.; Lian, T. Efficient hot-electron transfer by a plasmon-induced interfacial charge-transfer transition. *Science* **2015**, *349*, 632–635. [CrossRef]
29. Xie, Z.; Zhang, F.; Liang, Z.; Fan, T.; Li, Z.; Jiang, X.; Chen, H.; Li, J.; Zhang, H. Revealing of the ultrafast third-order nonlinear optical response and enabled photonic application in two-dimensional tin sulfide. *Photonics Res.* **2019**, *7*, 494–502. [CrossRef]
30. Zhang, H. Ultrathin Two-Dimensional Nanomaterials. *ACS Nano* **2015**, *10*, 9451–9469. [CrossRef]
31. Wibmer, L.; Lages, S.; Unruh, T.; Guldi, D.M. Excitons and Trions in One-Photon- and Two-Photon-Excited MoS2: A Study in Dispersions. *Adv. Mater.* **2018**, *30*, 1706702. [CrossRef]
32. Guo, J.; Shi, R.; Wang, R.; Wang, Y.; Zhang, F.; Wang, C.; Chen, H.; Ma, C.; Wang, Z.; Ge, Y.; et al. Graphdiyne-Polymer Nanocomposite as a Broadband and Robust Saturable Absorber for Ultrafast Photonics. *Laser Photonics Rev.* **2020**, *14*, 1900367–1900378. [CrossRef]
33. Breusing, M.; Ropers, C.; Elsaesser, T. Ultrafast Carrier Dynamics In Graphite. *Phys. Rev. Lett.* **2009**, *102*, 210–213. [CrossRef] [PubMed]
34. Xu, Y.; Jiang, X.-F.; Ge, Y.; Guo, Z.; Zeng, Z.; Xu, Q.-H.; Han, Z.; Yu, X.-F.; Fan, D. Size-dependent nonlinear optical properties of black phosphorus nanosheets and their applications in ultrafast photonics. *J. Mater. Chem. C* **2017**, *5*, 3007–3013. [CrossRef]
35. Gao, L.; Chen, H.; Zhang, F.; Mei, S.; Zhang, Y.; Bao, W.; Ma, C.; Yin, P.; Guo, J.; Jiang, X.; et al. Ultrafast Relaxation Dynamics and Nonlinear Response of Few-Layer Niobium Carbide MXene. *Small Methods* **2020**, *4*, 2000250. [CrossRef]
36. Brongersma, M.L.; Halas, N.J.; Nordlander, P. Plasmon-induced hot carrier science and technology. *Nat. Nanotechnol.* **2015**, *10*, 25–34. [CrossRef]
37. Urayama, J.; Norris, T.B.; Singh, J.; Bhattacharya, P. Observation of Phonon Bottleneck in Quantum Dot Electronic Relaxation. *Phys. Rev. Lett.* **2001**, *86*, 4930–4933. [CrossRef]
38. Wang, J.; Wu, C.; Dai, Y.; Zhao, Z.; Wang, A.; Zhang, T.; Wang, Z.L. Achieving ultrahigh triboelectric charge density for efficient energy harvesting. *Nat. Commun.* **2017**, *8*, 88. [CrossRef]
39. Kameyama, T.; Sugiura, K.; Kuwabata, S.; Okuhata, T.; Tamai, N.; Torimoto, T. Hot electron transfer in Zn–Ag–In–Te nanocrystal–methyl viologen complexes enhanced with higher-energy photon excitation. *RSC Adv.* **2020**, *10*, 16361–16365. [CrossRef]
40. Kalanoor, B.S.; Bisht, P.B.; Akbar, A.S.; Baby, T.T. Optical nonlinearity of silver-decorated graphene. *J. Opt. Soc. Am. B* **2012**, *29*, 669–675. [CrossRef]
41. Yu, Y.; Si, J.; Yan, L.; Li, M.; Hou, X. Enhanced nonlinear absorption and ultrafast carrier dynamics in graphene/gold nanoparticles nanocomposites. *Carbon* **2019**, *148*, 72–79. [CrossRef]

Article

Terahertz Modulation and Ultrafast Characteristic of Two-Dimensional Lead Halide Perovskites

Hongyuan Liu [1], Xunjun He [2,*], Jie Ren [2], Jiuxing Jiang [3], Yongtao Yao [4] and Guangjun Lu [5,*]

1. School of Computer Science, Harbin University of Science and Technology, Harbin 150080, China
2. Key Laboratory of Engineering Dielectric and Applications (Ministry of Education), School of Electrical and Electronic Engineering, Harbin University of Science and Technology, Harbin 150080, China
3. College of Science, Harbin University of Science and Technology, Harbin 150080, China
4. National Key Laboratory of Science and Technology on Advanced Composites in Special Environments, Harbin Institute of Technology, Harbin 150080, China
5. School of Electronic and Information Engineering/School of Integrated Circuits, Guangxi Normal University, Guilin 541004, China
* Correspondence: hexunjun@hrbust.edu.cn (X.H.); lv-guangjun@163.com (G.L.)

Abstract: In recent years, two-dimensional (2D) halide perovskites have been widely used in solar cells and photoelectric devices due to their excellent photoelectric properties and high environmental stability. However, the terahertz (THz) and ultrafast responses of the 2D halide perovskites are seldom studied, limiting the developments and applications of tunable terahertz devices based on 2D perovskites. Here, 2D R-P type (PEA)$_2$(MA)$_2$Pb$_3$I$_{10}$ perovskite films are fabricated on quartz substrates by a one-step spin-coating process to study their THz and ultrafast characteristics. Based on our homemade ultrafast optical pump–THz probe (OPTP) system, the 2D perovskite film shows an intensity modulation depth of about 10% and an ultrafast relaxation time of about 3 ps at a pump power of 100 mW due to the quantum confinement effect. To further analyze the recombination mechanisms of the photogenerated carriers, a three-exponential function is used to fit the carrier decay processes, obtaining three different decay channels, originating from free carrier recombination, exciton recombination, and trap-assisted recombination, respectively. In addition, the photoconductor changes ($\Delta\sigma$) at different pump–probe delay times are also investigated using the Drude-Smith model, and a maximum difference of 600 S/m is obtained at $\tau_p = 0$ ps for a pump power of 100 mW. Therefore, these results show that the 2D (PEA)$_2$(MA)$_2$Pb$_3$I$_{10}$ film has potential applications in high-performance tunable and ultrafast THz devices.

Keywords: 2D R-P type perovskite; OPTP; terahertz modulation; ultrafast characteristics; Drude-Smith model

1. Introduction

In recent years, terahertz (THz) waves have received great attention from various researchers because of their special properties and promising applications in many fields. Currently, THz generators and detectors have been reported and consistently demonstrated, while THz functional devices have faced great challenges due to a lack of the appropriate nature materials [1–3]. THz modulators, as key components of THz communication fields, can modulate the amplitude, phase, and polarization of THz waves by exciting active materials to implement different functions [4–6]. Traditional semiconductors can achieve a high modulation speed among these active materials with ultrafast optical pumping. To achieve a high modulation depth, however, a high power is required to produce photogenerated carriers, thus significantly restricting their application fields [7]. Recently, three-dimensional (3D) organic–inorganic hybrid halide perovskites have achieved unprecedented and rapid developments in the field of photoelectric devices due to their high absorption efficiency, adjustable energy band, high defect tolerance, and good carrier transport performance [8,9]. However, these conventional 3D halide perovskites are sensitive and unstable to water, light, and heat, meaning that there are high requirements for their preparation and preservation environments, thereby limiting their wide commercial applications [10,11]. For the previously studied $CH_3NH_3PbI_3$ perovskite, for example, the methylamino cation $CH_3NH_3^+$ (MA) is extremely soluble in water, and thus readily produces PbI_2 in the perovskite, causing irreversible chemical damages [11].

To solve the above issues, a commonly used method is to replace the organic cation in 3D perovskites with a hydrophobic long-chain organic cation to construct the 2D perovskites, thus enhancing the resistance to water molecules and environmental factors (such as ultraviolet rays and heat), and as a result, greatly improving the stability of the organic–inorganic hybrid perovskites [12–17]. Currently, 2D organic–inorganic halide perovskites have been rapidly developed in the field of optoelectronics due to their high stability, tunable photoelectric property, and high quantum efficiency [18–21]. For example, Karunadasa et al. first prepared a 2D organic–inorganic hybrid perovskite solar cell with an active layer of $(PEA)_2(MA)_2Pb_3I_{10}$, which can still maintain a high efficiency when placed in an environment with a relative humidity of 52% for 46 days [18]. After that, Sargent et al. designed $(PEA)_2(MA)_{n-1}Pb_nI_{n+1}$ (n > 40) perovskite systems, which can realize a short-circuit current of up to 19.12 mA/cm^{-2} and conversion efficiency of 15.3% [19]. In 2016, Jinwoo et al. used the $(PEA)_2(MA)_{n-1}Pb_nBr_{n+1}$ (n = 1~4) perovskite as a light-emitting layer to prepare high-efficiency quasi-2D light-emitting diodes (LEDs) with a current efficiency of 4.90 cd/A [20]. In 2021, Xu et al. fabricated highly efficient quasi-2D perovskite light-emitting diodes with a maximum brightness of 35,000 cd/m^2 and maximum external quantum efficiency (EQE) of 12.4% [21]. Although 2D perovskites have been extensively studied in photoelectric devices, the THz and ultrafast characteristics of the 2D perovskites are seldom reported, limiting the developments and applications of the tunable and ultrafast THz devices based on 2D perovskites [22].

In this paper, we investigated the THz and ultrafast responses of a 2D $(PEA)_2(MA)_2Pb_3I_{10}$ perovskite film to reveal the decay mechanisms of the photogenerated carriers. Firstly, $(PEA)_2(MA)_2Pb_3I_{10}$ perovskite films with n = 3 (PEA content of 50%) were prepared on the quartz substrates by a one-step spin-coating process. Then, the structural and photoelectric properties of the prepared films were characterized by the scanning electron microscope (SEM), X-ray diffraction (XRD), ultraviolet–visible (UV–vis) spectroscopy, and photoluminescence (PL) technologies, respectively. Next, the THz and ultrafast responses were measured using our homemade ultrafast optical pump–THz probe (OPTP) system, obtaining an intensity modulation depth of about 10% and an ultrafast relaxation time of about 3 ps for the pump power of 100 mW. To further discover the recombination mechanisms of the photogenerated carriers, finally, the decay process of the carriers was fitted using a three-exponential formula. Therefore, the prepared 2D $(PEA)_2(MA)_2Pb_3I_{10}$ perovskite has broad application prospects in the design of high-performance tunable and ultrafast THz devices.

2. Structure of 2D Perovskites

Generally, 2D organic–inorganic hybrid perovskites are expressed as $(RNH_3)_2A_{n-1}M_nX_{3n+1}$ (n = 1, 2, 3, 4 . . . . . .), where R is the organic group, and n is the number of the stacked diagonal octahedral layers, namely the number of organic layers. When n is equal to 1 or a finite integer, the $(RNH_3)_2A_{n-1}M_nX_{3n+1}$ can be considered as a pure-2D or quasi-2D structure, while becoming a 3D structure as n = ∞, as shown in Figure 1. Thus, 2D organic–inorganic perovskites can be constructed by replacing the A in 3D structures with the hydrophobic organic macromolecules [20]. Figure 1a shows the structural schematics of the $(PEA)_2(MA)_{n-1}Pb_nI_{3n+1}$ perovskites with different PEA percentages, in which the organic and inorganic layers are interbedded with each other, resulting in the formation of a natural multi-quantum well structure, as shown in Figure 1b. Therefore, the 2D organic–inorganic perovskites not only have excellent environmental stability due to the existence of hydrophobic molecules resisting moisture, light, and heat, but they also have a large exciton binding energy and excellent photoelectric properties due to the presence of the quantum well structure [18,23].

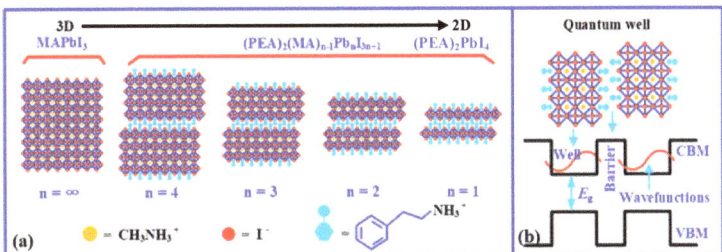

Figure 1. Structural schematic of the perovskites: (**a**) evolution process of perovskites from 3D to 2D and (**b**) multi-quantum well structure.

3. Fabrications and Characterization of 2D Perovskite Films

To obtain high-quality 2D perovskite films, it is very important to follow the fabrication procedures to ensure the growth of high-quality grains. Here, a one-step spin-coating process is applied to fabricate 2D perovskite films. Figure 2 shows the fabrication process of the 2D $(PEA)_2(MA)_2Pb_3I_{10}$ perovskite film. The detailed fabrication steps are as follows: firstly, the MAI and PEAI powders are dissolved in a DMF solution at a ratio of 1:1 to form a precursor solution. Then, the precursor solution is deposited on the surface of a 1 cm × 1 cm quartz substrate by the one-step spin-coating process with two consecutive stages (the first stage at 1000 r/min for 10 s, and the second with 3000 r/m for 50 s). Moreover, in the 30 s of the second spin-coating stage, the antisolvent (chlorobenzene) is continuously dripped onto the substrate to quickly form a uniform film. After spin-coating, the fabricated samples are transferred to a hot plate and annealed at 100 °C for 10 min to remove the residual solvents and transit the intermediate solvate phase into the perovskite, finally forming homogeneous 2D $(PEA)_2(MA)_2Pb_3I_{10}$ perovskite films. All of the fabrication steps are implemented inside a nitrogen-filled glove box at room temperature.

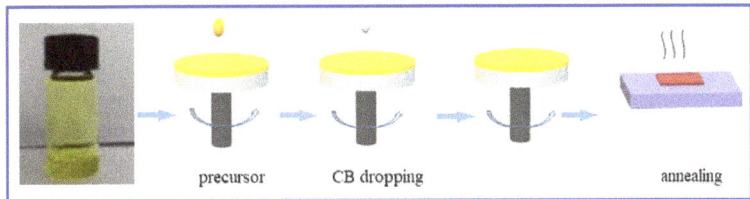

Figure 2. Fabrication process for $(PEA)_2(MA)_2Pb_3I_{10}$ perovskite film.

To examine the structural morphologies and optical properties of the as-fabricated perovskite films, next, we caried out different measurements and characterizations using SEM, XRD, UV–vis, and PL technologies, as shown in Figure 3. Figure 3a shows an optical microscope image of the 2D $(PEA)_2(MA)_2Pb_3I_{10}$ perovskite film, which indicates that the perovskite film is very flat and dense owing to the uniform distributions of the precursor solution on the substrate. Figure 3b displays a top-view SEM micrograph of the 2D $(PEA)_2(MA)_2Pb_3I_{10}$ perovskite film, in which the perovskite appears to have nanorod-like crystalline features at the length scale of hundreds of nanometers and a few pinholes due to tight contact between the grain boundaries. Figure 3c shows an XRD pattern of the perovskite film. In the XRD pattern, there are two diffraction peaks at 14.2° and 28.72°, corresponding to the (111) and (222) crystal planes, respectively, which is consistent with the previously reported crystal plane positions [24]. Moreover, no characteristic peaks associated with PbI_2 or other redundant phases are observed, suggesting that the fabricated films were fully generated with a high crystallinity. Figure 3d presents the UV–vis absorption and PL emission spectra of the as−grown perovskite film. The PL emission spectrum shows only a PL peak with a full width at half maximum (FWHM) of 47 nm centered near 710 nm, while the absorption spectrum shows that the perovskite film has a broad absorption spectrum with three absorption peaks between 600 and 750 nm, indicating a bandgap of ~1.96 eV, as shown in Figure 3e. Moreover, these exciton absorption peaks mainly arise from the low n−member perovskite compounds, demonstrating the presence of 2D perovskites as well as multiphase [25]. In addition, non-zero values of absorbance below the absorption onset are also observed, which is consistent with the fact that the rougher film morphology of perovskites results in a large amount of scattering. Therefore, the above measurement results demonstrate that the as-grown perovskite is a 2D layer structure.

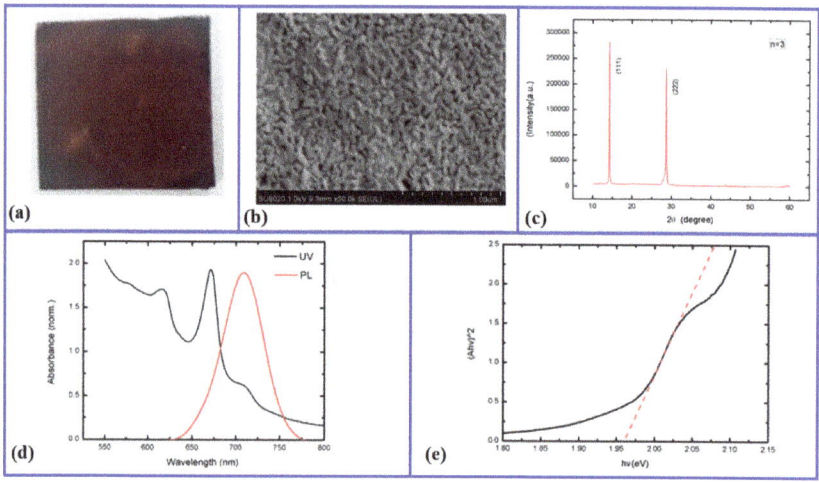

Figure 3. Structural morphologies and optical properties of $(PEA)_2(MA)_2Pb_3I_{10}$ perovskite: (**a**) optical microscope image, (**b**) SEM image, (**c**) XRD pattern, (**d**) absorption profile and PL spectrum, and (**e**) bandgap extraction.

4. Terahertz and Ultrafast Responses of 2D Perovskite Films

As is well-known, THz waves are very sensitive to changes in external circumstances. When the semiconductor films are pumped by external light, for example, the THz waves passing through them would be modulated due to the existence of photogenerated carriers. Moreover, the more photogenerated carriers, the greater the modulation depth of the device [26]. Thus, the application prospects of the materials can be further developed according to their modulation depths and speeds. A 2D $(PEA)_2(MA)_2Pb_3I_{10}$ perovskite is a direct bandgap semiconductor and can absorb the photons with energy larger than its bandgap width when pumped by the external light, producing great photogenerated carriers due to the transition of the electrons as a result. To evaluate the THz modulation depth and ultrafast characteristics of the fabricated 2D perovskite films, we set up an OPTP system mainly consisting of the ZnTe crystal-based THz generation and detection beams and an optical pump beam photoexciting the samples, as shown in Figure 4.

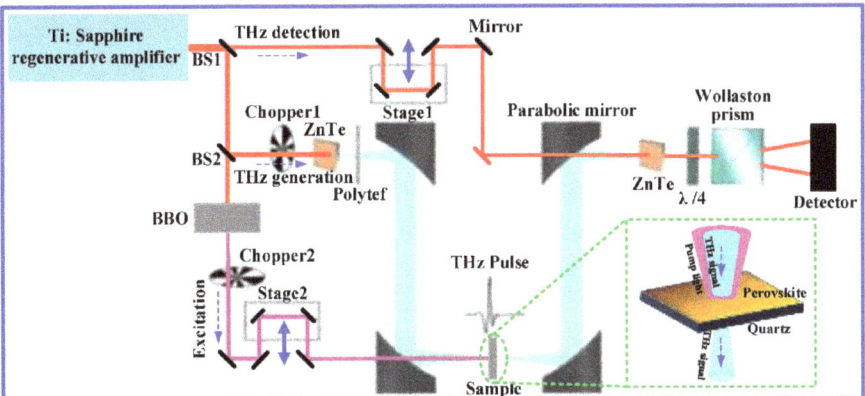

Figure 4. Schematic diagram of the homemade OPTP measurement system.

In this homemade system, an amplified Ti–sapphire laser with a pulse duration of 90 ps, wavelength of 800 nm, spectral width of 28 nm, and repetition rate of 75 MHz is used as the optical source for the generation and detection of the THz signal and the photoexcitation of the samples [27]. The laser output beam is split into three beams, where one beam is employed to excite the ZnTe crystal to generate a THz pulse, the second beam is used to detect the THz pulse via free-space electro–optic sampling in a ZnTe crystal, and the third part is used to generate a frequency-doubled 400 nm pump pulse using a barium borate (BBO) crystal to excite the 2D perovskite sample. Moreover, the 400 nm pump beam has an energy of 3.1 eV higher than the bandgap of 2D perovskites (1.96 eV), which can photoinduced free carriers and excitons. In addition, the diameter of the pump beam is 5 mm, which is larger than the diameter (2 mm) of the focused THz beam to ensure uniform photoexcitation. Thus, the ultrafast response measurements in the OPTP system are carried out by varying the delay time (τ_p) between pump and detection beams using a translational delay stage, while terahertz time-domain spectrum measurements are performed by fixing the pump pulse at the desired position and sampling the THz pulse using another translational delay stage. As a result, the frequency-dependent terahertz spectroscopy can be obtained after the Fourier transform [28].

To examine the THz modulation ability of the fabricated perovskite films, next, the THz time–domain spectra across the sample, fabricated onto a quartz substrate with a thickness of 2 mm, are measured using our OPTP system at different pump powers. The measurement results are shown in Figure 5a, where the gray dotted line is the reference value of the quartz substrate without the perovskite film. There is a significant time delay between the reference and the sample, demonstrating that the perovskite film were fabricated on the substrate. Figure 5b shows the normalized THz transmission spectra of the fabricated sample for different laser excitations, clearly observing a gradual reduction in the THz transmission with the increase in pump power. The change in the transmission spectra can be attributed to the generation of free carriers in the perovskites. For example, in the absence of a pump beam, the carriers in the perovskite are in the thermal balance state, and there is no observed split in the energy level of the perovskite. In this case, the perovskite has a few carriers that can freely move, and thus obtains a transmission intensity of about 90% at 1 THz. Once the perovskite film is pumped with different pump powers. However, the carriers are generated in the perovskite, breaking the thermal equilibrium state of the perovskite. By further increasing the pump power, the numbers of electrons and holes in the conductivity and valence bands of the perovskite can be gradually increased, leading to a reduction in terahertz transmission intensity, achieving a THz intensity change of nearly 5% at 1THz for a pump power of 100 mW.

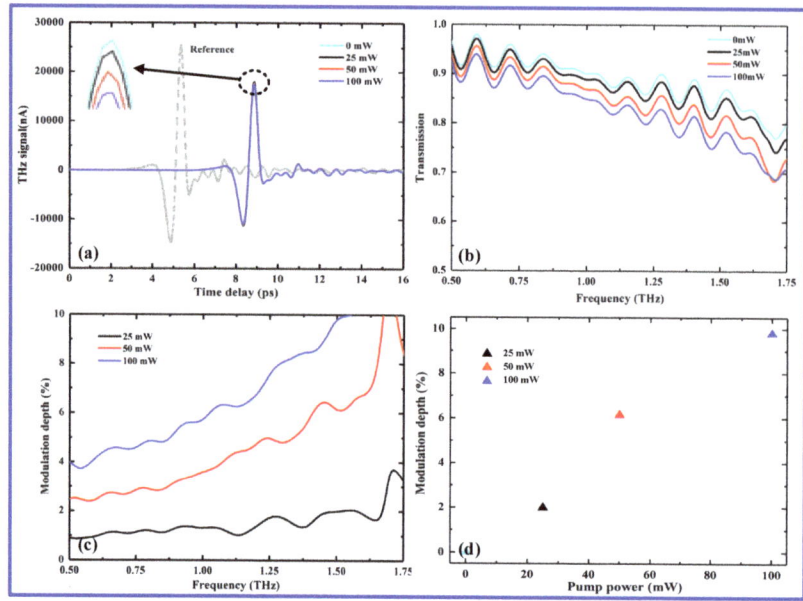

Figure 5. Terahertz performances of the fabricated $(PEA)_2(MA)_2Pb_3I_{10}$ perovskite films at different irradiation powers of 25 mW, 50 mW, and 100 mW: (**a**) terahertz time–domain transmission spectra, in which the inset shows the magnification values at the peak, and the dashed line represents the time-domain transmission spectrum of the reference substrate without perovskite film; (**b**) terahertz frequency-domain transmission spectra; (**c**) modulation depth over the broadband range of 0.5–1.75 THz; and (**d**) modulation depth of different pump powers at 1.5 THz.

To further assess the modulation performance of the $(PEA)_2(MA)_2Pb_3I_{10}$ perovskite film, a modulation depth (*MD*) was introduced to quantify the modulation ability of the perovskite film at different irradiation powers, which can be expressed as

$$MD = \frac{\int P_{laser-off}(\omega)d\omega - \int P_{laser-on}(\omega)d\omega}{P_{laser-off}(\omega)} \tag{1}$$

where $P_{laser\text{-}on}(\omega)$ and $P_{laser\text{-}off}(\omega)$ represent the THz amplitudes as the laser pump beam is turned on and off, respectively [29]. As shown in Figure 5c,d, the intensity modulation depth of the fabricated 2D $(PEA)_2(MA)_2Pb_3I_{10}$ perovskite films is gradually enhanced with the increase in pump power, showing a linear increase for different pump powers. The maximum intensity MD is found to be about 10% at 1.5 THz for the pump power of 100 mW. Moreover, the lower MD can be further improved by increasing the pump power. Therefore, the $(PEA)_2(MA)_2Pb_3I_{10}$ perovskite is demonstrated as a promising material that can implement a highly efficient THz modulation.

Next, to explore the ultrafast relaxation response of the photogenerated carriers, the homemade OPTP system is used to monitor the dynamic decay process of the photogenerated carriers by varying the relative delay time (τ_p) between the pump and detection beams. In this experiment, the sample is excited by a femtosecond laser beam with a 400 nm wavelength at the normal incidence, and the probing THz electric field vector is parallel to the plane of the surface of the sample [30]. Following photoexcitation, generally, the relative change in the THz electric field is proportional to the photoinduced conductivity of the pumped material due to the presence of free charges. Thus, the dynamics of charge carriers are manifested as the photoinduced THz transmission changes ($-\Delta T/T_0$) in the samples at the peak of the THz pulse as a function of pump–probe delay. Figure 6 shows the transient THz transmission dynamics following the photoexcitation of the prepared 2D (PEA)$_2$(MA)$_2$Pb$_3$I$_{10}$ perovskite film for a range of pump powers. In these experimental results, it is noted that the nonequilibrium carriers relax at ultrafast speeds, fully recovering the equilibrium state within a dozen picosecond time scale for the pump power of 25 mW. Moreover, the fast relaxation becomes increasingly significant with the increase in the pump power, indicating that the 2D (PEA)$_2$(MA)$_2$Pb$_3$I$_{10}$ perovskite has potential application prospects in ultrafast THz devices. This ultrafast phenomenon can be attributed to the inherent multi-quantum well structure in the 2D perovskites.

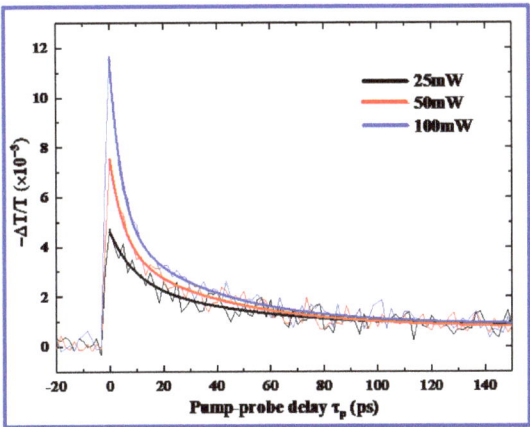

Figure 6. Carrier transient dynamics of 2D (PEA)$_2$(MA)$_2$Pb$_3$I$_{10}$ perovskite film at different pump powers (The thin and thick solid curves correspond to the measured and fitted results).

To further understand the ultrafast relaxation dynamics of the 2D (PEA)$_2$(MA)$_2$Pb$_3$I$_{10}$ perovskite, a triexponential decay function is used to fit the measured THz transient dynamics at different pump powers, extracting the carry lifetimes of the ultrafast processes to discover the recombination channels of the photoexcited free carriers and excitons. Thus, the triexponential decay function is given by the following express [31]

$$f(t) = A_1 e^{-\frac{t}{\tau_1}} + A_2 e^{-\frac{t}{\tau_2}} + A_3 e^{-\frac{t}{\tau_3}} \quad (2)$$

where τ_1, τ_2, and τ_3 are the lifetimes of different relaxation processes, respectively. A_1, A_2, and A_3 are the corresponding coefficients of each lifetime component, which determine the weights of the decaying and nondecaying components separately. By fitting the measured THz transient changes obtained using our OPTP system, the lifetimes are extracted for different pump powers, as summarized in Table 1. As observed in Table 1, the lifetimes of the three components are $\tau_1 \sim 10$ ps, $\tau_2 \sim 33$ ps, and $\tau_3 \sim 2$ ns for a pump power of 25 mW, respectively. With the increase in the pump power, the initial fast relaxation process becomes faster, while the slow process becomes slower, obtaining $\tau_1 \sim 3$ ps, $\tau_2 \sim 18$ ps, and $\tau_3 \sim 6$ ns for the pump power of 100 mW as a result. These results indicate that such a decay process usually involves three recombination pathways: monomolecular recombination, bimolecular recombination, and Auger recombination [32]. At a lower pump power, the photogenerated carrier relaxations are dominated by the monomolecular decay (τ_3), corresponding to the slow process, whereas at a higher pump power, the recombination channels are dominated by the bimolecular decay (τ_2) and Auger decay (τ_1), corresponding to the fast process. Thus, the monomolecular decay component observed at a lower pump power arises most likely from trap-assisted recombinations, depending on the trap cross-section, energetic depth, density, and distribution. For higher pump power, the bimolecular decay originates from the overlaps of electron and hole wavefunctions, while the Auger process results from the exciton–exciton scatterings, where the excitons are localized inside the QW structures, providing an additional channel for the fast relaxation of free carriers [30].

Table 1. Extracted lifetimes of different pump powers by the triexponential fitting.

Power Lifetime	25 mW	50 mW	100 mW
τ_1 (ps)	10 ± 0.2	6 ± 0.2	3 ± 0.1
τ_2 (ps)	32 ± 0.8	30 ± 0.9	18 ± 1.2
τ_3 (ps)	1830 ± 30	3341 ± 50	5854 ± 50

To gain further insights into the ultrafast relaxation behaviors with three exponential decay components (τ_1, τ_2, and τ_3), the spectral dispersions of the THz photoinduced conductivity ($\Delta\sigma$) at different pump–probe delay times (τ_p) were derived from the measured THz transmission transient dynamics using the following expression [33]:

$$\Delta\sigma(\omega, \tau_p) \approx -\frac{n+1}{Z_0} \frac{\Delta T(\omega, \tau_p)}{T(\omega, \tau_p)} / d \text{ [S/m]} \quad (3)$$

where n is the refractive index of the quartz substrate, whose value is 1.95 at the terahertz range; $Z_0 = 377\ \Omega$ is the impedance of free space; and d is the thickness of the perovskite film.

Figure 7a shows a typical trace of THz transmission for the pump power of 100 mW, which reveals the transient dynamics of free carriers and excitons in the 2D perovskite film, as discussed above. Figure 7b displays the THz electric field changes at different pump–probe delay times, as shown in Figure 7a (blue, red, and black solid curves correspond to τ_p of 0, 5, and 113 ps, respectively, in which the ΔE is enlarged by over ten times for clarity), while Figure 7c displays the changes in the THz photoconductivity extracted using the corresponding THz electric field changes, which is shown by scattered points. It is noted that the variations of the THz intensity and photoconductivity are increasingly weakened with the increase in the pump–probe delay time. For example, at $\tau_p = 0$ ps, the photogenerated carriers start to decay and relax quickly, and THz intensity and conductivity exhibit maximal changes due to the existence of abundant photogenerated carriers. As τ_p is increased from 0 to 5.0 ps, the change in THz intensity and conductivity is gradually decreased due to the recombination of the photogenerated carriers. At $\tau_p = 113$ ps, however, the system is almost restored to the initial balanced state, and the photogenerated carriers have been fully recombined, leading to the minimal change of the THz intensity and conductivity. Such a change trend can be attributed to the change in the photoinduced free carry density [34]. In addition, the Drude–Smith model is used to further fit the extracted photoinduced THz conductivities (solid curves of Figure 7c), and the corresponding deviations are shown in Figure 7d. It is noticed that in the early process, the extracted photoinduced THz conductivities show a considerable disparity from the Drude–Smith model due to the complicated THz responses, as shown in the top row of Figure 7d. Such a remarked deviation can result from the contributions of both the charge carrier transport and exciton–phonon scattering [35]. For the slow decay process ($\tau_p = 113$ ps), however, the extracted values agree well with the fitting values (see the bottom row of Figure 7d), indicating a primary contribution from the defect trapping process [36].

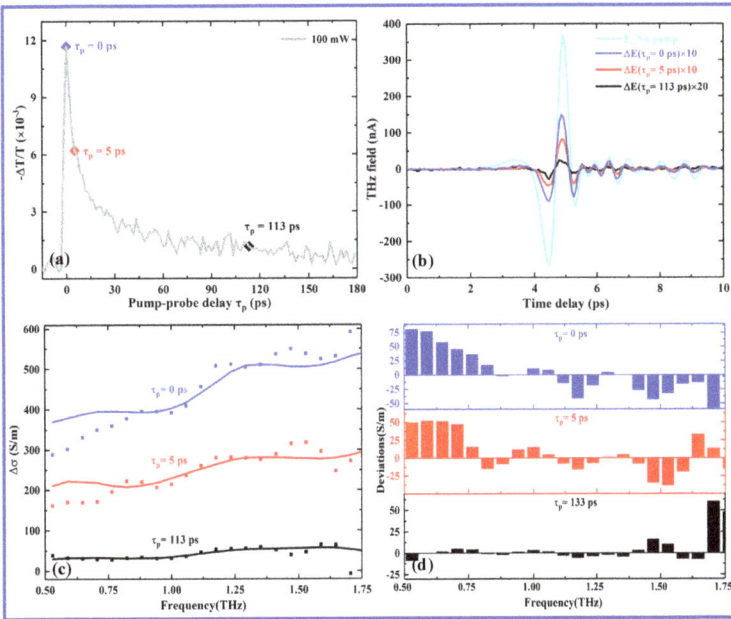

Figure 7. Changes in the THz transmission field, and photoinduced conductivity of the $(PEA)_2(MA)_2Pb_3I_{10}$ film at a pump power of 100 mW: (**a**) THz transmission change at different pump−probe delay times, (**b**) THz field change at τ_p = 0, 5 and 113.3 ps, (**c**) THz photoinduced conductivity changes at τ_p = 0, 5 and 113.3 ps, and (**d**) deviations between the extracted and fitted photoinduced conductivities.

5. Conclusions

In summary, we prepared the $(PEA)_2(MA)_2Pb_3I_{10}$ perovskite films by a one-step spin-coating process and characterized them by different measuring methods. The SEM, XRD, UV–vis, and PL measurements demonstrate that the as-grown $(PEA)_2(MA)_2Pb_3I_{10}$ perovskite films are a 2D layer structure. The OPTP measurements show that the 2D $(PEA)_2(MA)_2Pb_3I_{10}$ perovskite film can achieve an MD of up to 10% at 1.5 THz and an ultrafast relaxation time of about 3ps at an illumination power of 100 mW. Moreover, the fitting results obtained by a three-exponential function reveal that the decay mechanism involves the monomolecular, bimolecular, and Auger recombination processes, corresponding to the free carrier relaxation, exciton recombination, and trap-assisted recombination, respectively. In addition, the changes in the photogenerated conductivity at different pump–probe delay times were extracted and fitted using the measured THz transient dynamics and the Drude-Smith model, respectively, obtaining a maximum change of 600 S/m at τ_p = 0 ps. Therefore, these results show that the 2D $(PEA)_2(MA)_2Pb_3I_{10}$ film has potential applications in high-performance tunable and ultrafast THz devices.

Author Contributions: Conceptualization, X.H., J.R. and H.L.; methodology, H.L. and J.R.; validation, H.L. and G.L.; resources, X.H. and G.L.; data curation, J.J. and Y.Y.; writing—original draft preparation, H.L., J.R. and G.L.; writing—review and editing, J.R., X.H. and G.L.; visualization, Y.Y.; supervision, X.H.; project administration, J.J.; funding acquisition, X.H. and G.L. All authors have read and agreed to the published version of the manuscript.

Funding: The work is supported by the National Natural Science Foundation of China (Nos. 62075052), the Science Foundation of the National Key Laboratory of Science and Technology on Advanced Composites in Special Environments (JCKYS2020603C009 and 6142905212711), and the Natural Science Foundation of Heilongjiang Province (LH2019F022).

Institutional Review Board Statement: Not applicable.

Informed Consent Statement: Not applicable.

Data Availability Statement: Data is contained within the article.

Conflicts of Interest: The authors declare no conflict of interest.

References

1. Wang, L.; Zhang, Y.; Guo, X.; Chen, T.; Liang, H.; Hao, X.; Hou, X.; Kou, W.; Zhao, Y.; Zhou, T.; et al. A review of THz modulators with dynamically tunable metasurfaces. *Nanomaterials* **2019**, *9*, 965. [CrossRef] [PubMed]
2. Edmund, L. Terahertz applications: A source of fresh hope. *Nat. Photonics* **2007**, *1*, 257–258.
3. Chen, H.T.; Padilla, W.J.; Zide, J.M.O.; Gossard, A.C.; Taylor, A.J.; Averitt, R.D. Active terahertz metamaterial devices. *Nature* **2006**, *444*, 597–600. [CrossRef] [PubMed]
4. Rahm, M.; Li, J.S.; Padilla, W.J. THz wave modulators: A brief review on different modulation techniques. *J. Infrared Millim. Terahertz Waves* **2013**, *34*, 1–27. [CrossRef]
5. Kleine-Ostmann, T.; Pierz, K.; Hein, G.; Dawson, P.; Koch, M. Audio signal transmission over THz communication channel using semiconductor modulator. *Electron. Lett.* **2004**, *40*, 124–126. [CrossRef]
6. Qiao, J.; Wang, S.; Wang, Z.; He, C.; Zhao, S.; Xiong, X.; Zhang, X.; Tao, X.; Wang, S.L. Ultrasensitive and broadband all-optically controlled THz modulator based on MoTe$_2$/Si van der Waals heterostructure. *Adv. Opt. Mater.* **2020**, *8*, 2000160. [CrossRef]
7. Ono, M.; Hata, M.; Tsunekawa, M.; Nozaki, K.; Sumikura, H.; Chiba, H.; Notomi, M. Ultrafast and energy-efficient all-optical switching with graphene-loaded deep-subwavelength plasmonic waveguides. *Nat. Photonics* **2020**, *14*, 37–43. [CrossRef]
8. Solanki, A.; Yadav, P.; Turner-Cruz, S.H.; Lim, S.S.; Saliba, M.; Sum, T.C. Cation influence on carrier dynamics in perovskite solar cells. *Nano Energy* **2019**, *58*, 604–611. [CrossRef]
9. Kojima, A.; Teshima, K.; Shirai, Y.; Miyasaka, T. Organometal halide perovskites as visible-light sensitizers for photovoltaic cells. *J. Am. Chem. Soc.* **2009**, *131*, 6050–6051. [CrossRef]
10. Rong, Y.; Liu, L.; Mei, A.; Li, X.; Han, H. Beyond efficiency: The challenge of stability in mesoscopic perovskite solar cells. *Adv. Energy Mater.* **2015**, *5*, 1501066. [CrossRef]
11. Wang, Z.; Shi, Z.; Li, T.; Chen, Y.; Huang, W. Stability of perovskite solar cells: A prospective on the substitution of the A cation and X anion. *Angew. Chem. Int. Ed.* **2017**, *56*, 1190–1212. [CrossRef]
12. Lai, H.; Kan, B.; Liu, T.; Zheng, N.; Xie, Z.; Zhou, T.; Wan, X.; Zhang, X.D.; Liu, Y.S.; Chen, Y. Two-dimensional Ruddlesden–Popper perovskite with nanorod-like morphology for solar cells with efficiency exceeding 15%. *J. Am. Chem. Soc.* **2018**, *140*, 11639–11646. [CrossRef]
13. Tsai, H.; Nie, W.; Blanco, J.C.; Stoumpos, C.C.; Asadpour, R.; Harutyunyan, B.; Neukirch, A.J.; Verduzco, R.; Kanatzidis, M.G.; Mohite, A.D.; et al. High-efficiency two-dimensional Ruddlesden–Popper perovskite solar cells. *Nature* **2016**, *536*, 312–316. [CrossRef]
14. Chen, Y.; Sun, Y.; Peng, J.; Tang, J.; Zheng, K.; Liang, Z. 2D Ruddlesden–Popper perovskites for optoelectronics. *Adv. Mater.* **2018**, *30*, 1703487. [CrossRef]
15. Ren, H.; Yu, S.; Chao, L.; Xia, Y.; Sun, Y.; Zuo, S.; Li, F.; Niu, T.; Zhang, L.; Chen, Y.; et al. Efficient and stable Ruddlesden–Popper perovskite solar cell with tailored interlayer molecular interaction. *Nat. Photonics* **2020**, *14*, 154–163. [CrossRef]
16. Zhao, T.; Chueh, C.C.; Chen, Q.; Rajagopal, A.; Jen, A.K.-Y. Defect passivation of organic-inorganic hybrid perovskites by diammonium iodide toward high-performance photovoltaic devices. *ACS Energy Lett.* **2016**, *1*, 757–763. [CrossRef]
17. Niu, T.; Ren, H.; Wu, B.; Xia, Y.; Xie, X.; Yang, Y.; Gao, X.; Chen, Y.; Huang, W. Reduced-dimensional perovskite enabled by organic diamine for efficient photovoltaics. *J. Phys. Chem. Lett.* **2019**, *10*, 2349–2356. [CrossRef]
18. Smith, I.C.; Hoke, E.T.; Solis-Ibarra, D.; McGehee, M.D.; Karunadasa, H.I. A layered hybrid perovskite solar-cell absorber with enhanced moisture stability. *Angew. Chem. Int. Ed.* **2014**, *53*, 11232–11235. [CrossRef]
19. Quan, L.N.; Yuan, M.; Comin, R.; Voznyy, O.; Beauregard, E.M.; Hoogland, S.; Buin, A.; Kirmani, A.R.; Zhao, K.; Amassian, A.; et al. Ligand-stabilized reduced-dimensionality perovskites. *J. Am. Chem. Soc.* **2016**, *138*, 2649–2655. [CrossRef]
20. Byun, J.; Cho, H.; Wolf, C.; Jang, M.; Sadhanala, A.; Friend, R.H.; Yang, H.; Lee, T. Efficient visible quasi-2D perovskite light-emitting diodes. *Adv. Mater.* **2016**, *28*, 7515–7520. [CrossRef]
21. Xu, Q.; Wang, R.; Jia, Y.L.; He, X.; Deng, Y.; Yu, F.; Zhang, Y.; Ma, X.; Chen, P.; Zhang, Y.; et al. Highly efficient quasi-two dimensional perovskite light-emitting diodes by phase tuning. *Organ. Electron.* **2021**, *98*, 106295. [CrossRef]
22. Stoumpos, C.C.; Cao, D.H.; Clark, D.J.; Young, J.; Rondinelli, J.M.; Jang, J.I.; Hupp, J.T.; Kanatzidis, M.G. Ruddlesden–Popper hybrid lead iodide perovskite 2D homologous semiconductors. *Chem. Mater.* **2016**, *28*, 2852–2867. [CrossRef]
23. Milot, R.L.; Sutton, R.J.; Eperon, G.E.; Haghighirad, A.A.; Hardigree, J.M.; Miranda, L.; Snaith, H.J.; Johnston, M.B.; Herz, L.M. Charge-carrier dynamics in 2D hybrid metal-halide perovskites. *Nano Lett.* **2016**, *16*, 7001–7007. [CrossRef]
24. Shang, Q.; Wang, Y.; Zhong, Y.; Mi, Y.; Qin, L.; Zhao, Y.; Qiu, X.; Liu, X.; Zhang, Q. Unveiling structurally engineered carrier dynamics in hybrid quasi-two-dimensional perovskite thin films toward controllable emission. *J. Phys. Chem. Lett.* **2017**, *8*, 4431–4438. [CrossRef]

25. Zheng, Y.; Deng, H.; Jing, W.; Li, C. Bandgap energy tuning and photoelectrical properties of self-assembly quantum well structure in organic-inorganic hybrid perovskites. *Acta Phys. Sin.* **2011**, *60*, 2433. [CrossRef]
26. Yan, H.J.; Ku, Z.L.; Hu, X.F.; Zhao, W.; Zhong, M.; Zhu, Q.; Lin, X.; Jin, Z.; Ma, G. Ultrafast terahertz probes of charge transfer and recombination pathway of $CH_3NH_3PbI_3$ perovskites. *Chin. Phys. Lett.* **2018**, *35*, 028401. [CrossRef]
27. Molis, G.; Adomavicius, R.; Krotkus, A.; Bertulis, K.; Giniunas, L.; Pocius, J.; Danielius, R. Terahertz time-domain spectroscopy system based on femtosecond Yb: KGW laser. *Electron. Lett.* **2007**, *43*, 1. [CrossRef]
28. Nuss, M.C.; Orenstein, J. Terahertz time-domain spectroscopy. In *Millimeter and Submillimeter Wave Spectroscopy of Solids, Topics in Applied Physics*; Springer: Berlin/Heidelberg, Germany, 1998; Volume 74, pp. 7–50.
29. Lee, K.S.; Kang, R.; Son, B.; Kim, D.; Yu, N.; Ko, D.K. Characterization of optically-controlled terahertz modulation based on a hybrid device of perovskite and silicon. In Proceedings of the Conference on Lasers and Electro-Optics/Pacific Rim, Singapore, 31 July–4 August 2017; p. s1712.
30. Wehrenfennig, C.; Eperon, G.E.; Johnston, M.B.; Snaith, H.J.; Herz, L.M. High charge carrier mobilities and lifetimes in organolead trihalide perovskites. *Adv. Mater.* **2014**, *26*, 1584–1589. [CrossRef] [PubMed]
31. Qin, Z.; Zhang, C.; Chen, L.; Wang, X.; Xiao, M. Charge carrier dynamics in sn-doped two-dimensional lead halide perovskites studied by terahertz spectroscopy. *Front. Energy Res.* **2021**, *9*, 658270. [CrossRef]
32. Johnston, M.B.; Herz, L.M. Hybrid perovskites for photovoltaics: Charge-carrier recombination, diffusion, and radiative efficiencies. *Acc. Chem. Res.* **2016**, *49*, 146–154. [CrossRef] [PubMed]
33. Chanana, A.; Liu, X.; Zhang, C.; Vardeny, Z.V.; Nahata, A. Ultrafast frequency-agile terahertz devices using methylammonium lead halide perovskites. *Sci. Adv.* **2018**, *4*, eaar7353. [CrossRef]
34. Lin, D.; Ma, L.; Ni, W.; Wang, C.; Zhang, F.; Dong, H.; Gurzadyan, G.G.; Nie, Z. Unveiling hot carrier relaxation and carrier transport mechanisms in quasi-two-dimensional layered perovskites. *J. Mater. Chem. A* **2020**, *8*, 25402–25410. [CrossRef]
35. Blancon, J.C.; Tsai, H.; Nie, W.; Stoumpos, C.C.; Pedesseau, L. Extremely efficient internal exciton dissociation through edge states in layered 2D perovskites. *Science* **2017**, *355*, 1288–1292. [CrossRef]
36. Delport, G.; Chehade, G.; Lédée, F.; Diab, H.; Milesi-Brault, C.; Trippé-Allard, G.; Even, J.; Lauret, J.; Deleporte, E.; Garrot, D. Exciton–exciton annihilation in two-dimensional halide perovskites at room temperature. *J. Phys. Chem. Lett.* **2019**, *10*, 5153–5159. [CrossRef]

Article

Zirconia-Doped Methylated Silica Membranes via Sol-Gel Process: Microstructure and Hydrogen Permselectivity

Lintao Wang and Jing Yang *

School of Urban Planning and Municipal Engineering, Xi'an Polytechnic University, Xi'an 710048, China; wanglt912@163.com
* Correspondence: jingy76@163.com; Tel.: +86-29-62779357

Abstract: In order to obtain a steam-stable hydrogen permselectivity membrane, with tetraethylorthosilicate (TEOS) as the silicon source, zirconium nitrate pentahydrate ($Zr(NO_3)_4 \cdot 5H_2O$) as the zirconium source, and methyltriethoxysilane (MTES) as the hydrophobic modifier, the methylmodified ZrO_2-SiO_2 (ZrO_2-$MSiO_2$) membranes were prepared via the sol-gel method. The microstructure and gas permeance of the ZrO_2-$MSiO_2$ membranes were studied. The physical-chemical properties of the membranes were characterized by Fourier transform infrared spectroscopy (FTIR), X-ray photoelectron spectroscopy (XPS), X-ray diffraction (XRD), transmission electron microscopy (TEM), scanning electron microscope (SEM), and N_2 adsorption–desorption analysis. The hydrogen permselectivity of ZrO_2-$MSiO_2$ membranes was evaluated with Zr content, temperature, pressure difference, drying control chemical additive (glycerol) content, and hydrothermal stability as the inferred factors. XRD and pore structure analysis revealed that, as n_{Zr} increased, the $MSiO_2$ peak gradually shifted to a higher 2θ value, and the intensity gradually decreased. The study found that the permeation mechanism of H_2 and other gases is mainly based on the activation–diffusion mechanism. The separation of H_2 is facilitated by an increase in temperature. The ZrO_2-$MSiO_2$ membrane with n_{Zr} = 0.15 has a better pore structure and a suitable ratio of micropores to mesopores, which improved the gas permselectivities. At 200 °C, the H_2 permeance of $MSiO_2$ and ZrO_2-$MSiO_2$ membranes was 3.66×10^{-6} and 6.46×10^{-6} mol·m^{-2}·s^{-1}·Pa^{-1}, respectively. Compared with the $MSiO_2$ membrane, the H_2/CO_2 and H_2/N_2 permselectivities of the ZrO_2-$MSiO_2$ membrane were improved by 79.18% and 26.75%, respectively. The added amount of glycerol as the drying control chemical additive increased from 20% to 30%, the permeance of H_2 decreased by 11.55%, and the permselectivities of H_2/CO_2 and H_2/N_2 rose by 2.14% and 0.28%, respectively. The final results demonstrate that the ZrO_2-$MSiO_2$ membrane possesses excellent hydrothermal stability and regeneration capability.

Keywords: microporous membrane; H_2 permselectivity; zirconia-doped SiO_2 membrane; hydrothermal stability; regeneration

1. Introduction

It is well-known that hydrogen is a clean energy source [1]. At present, there are many ways to obtain H_2, but the biggest problem preventing its commercialization is the purification and separation of H_2. The purification of H_2 can be achieved in three main ways: pressure swing adsorption, cryogenic distillation, and membrane separation [2,3]. Although pressure swing adsorption and cryogenic distillation can be operated commercially, the economic benefits are low. The main commercial application of membranes in gas separation is the separation of hydrogen from nitrogen, methane, and argon in an ammonia sweep gas stream. In the past few years, hundreds of new polymer materials have been reported, and only eight or nine polymer materials have been used to make gas separation membrane bases. Surprisingly few of them were used to make industrial membranes [4]. Membrane separation technology is also one of the most promising hydrogen purification

methods. At high temperatures, the H_2 separation membrane has attracted much attention in the application of membrane reactors, for example, in the steam reforming of natural gas. The characteristics of lower energy consumption and investment cost, as well as simple operation, have made the membrane separation method widely concerned. Compared to other techniques of hydrogen purification, the membrane separation method offers more energy efficiency and environmental friendliness. The quality of the separation membrane directly affects the separation performance. Therefore, it is very necessary to choose suitable materials to prepare efficient and stable membrane materials [5].

In recent years, research on hydrogen separation membranes has mainly focused on molecular sieves, alloys, and microporous silica [6–9]. Prominently, silica membranes have received extensive attention for H_2 separation due to their advantages such as high permselectivity and considerable thermal stability [10]. Amorphous silica membranes derived from sol-gel and chemical vapor deposition (CVD) methods received an enormous amount of attention [11,12]. They have a stable chemical structure and molecular sieve mechanism, which can separate hydrogen across a broad temperature range [13]. However, silica membranes have demonstrated poor steam and thermal stability. Due to the hydrophilic nature of silica membranes, if they are frequently exposed to humid, low-temperature atmospheres, the flux and permselectivity of H_2 will largely be reduced [14,15]. The study found that the addition of hydrophobic groups can reduce the affinity of silica for water and improve gas permselectivity [15]. At this stage, methyl, vinyl, perfluorodecalin, etc., are often used in the hydrophobic modification of gas permeation separation in silica membrane, and the effect of increasing the hydrothermal stability is obvious [16]. Wei et al. [17] prepared perfluorodecyl hydrophobically modified silica membranes. The results demonstrated that the addition of perfluorodecyl made the modified silica membrane change from hydrophilic to hydrophobic. The membrane exhibited excellent hydrothermal stability at 250 °C and a water vapor molar ratio of 5%. Debarati et al. [18] hydrophobically modified the surface of ceramic membranes with polydimethylsiloxane to achieve a contact angle of 141°. It shows that the surface of the ceramic membrane with the methyl group is highly hydrophobic. Somayeh et al. [19] also demonstrated that vinyl-modified silica particles can improve the hydrophobic properties of the membrane.

It is well-known that the addition of metal oxides (such as TiO_2 [20,21], Al_2O_3 [22,23], Fe_2O_3 [24], CoO [25], and ZrO_2 [26,27]) can not only enhance the hydrothermal stability of the membrane, but also further improve the antifouling ability and performance of the membrane. This indicates that the incorporation of metal oxides can form mixed oxide network structures that are more stable than amorphous silica materials [28,29]. In particular, ZrO_2 is an excellent transition metal oxide, which is often utilized in studies of gas membrane separation. Li et al. [30] prepared a zirconia membrane via the polymeric sol-gel method, which possessed H_2 permeance of about 5×10^{-8} mol·m^{-2}·s^{-1}·Pa^{-1}, H_2/CO_2 permselectivity of 14, and outstanding hydrothermal stability under a steam pressure of 100 KPa. Gu et al. [31] used the sol-gel method to prepare the microporous zirconia membrane. After being treated with 0.50 mol·L^{-1} of H_2SO_4, the hydrogen permeance was (2.8 to 3.0) $\times 10^{-6}$ mol·Pa^{-1}·m^{-2}·s^{-1}. The H_2 permselectivities of the equimolar binary system were 6 and 9, respectively. Doping ZrO_2 in the SiO_2 matrix can improved the hydrophobicity and hydrothermal stability of the membrane matrix, and further enhance the gas permeability. Numerous studies have been carried out on ZrO_2-doped silica materials/membranes. According to mesoporous stabilized zirconia intermediate layers, Gestel et al. [32] revealed a much better membrane setup. For various CO_2/N_2 combinations, the as-prepared membrane demonstrated permselectivities of 20–30 and CO_2 permeances of 1.5 to 4 m^3/(m^2·h·bar). Ahn et al. [33] prepared a silica-zirconia membrane with hydrogen permselectivity on a porous alumina support using tetraethyl orthosilicate (TEOS) and tert-butanol zirconium (IV) under 923 K conditions by chemical vapor deposition. The H_2 permeance of the obtained membrane was 3.8×10^{-7} mol·m^{-2}·s^{-1}·Pa^{-1}, and the permselectivities for CO_2 and N_2 were 1100 and 1400, respectively. Hove et al. [34] compared the gas permeation properties of a hybrid silica (BTESE) membrane, Zr-doped BTESE membrane, and silica membrane before and after hydrothermal treatment under the same circumstances. At 100 °C, the hybrid silica membrane and Zr-doped BTESE membrane maintained good hydrothermal stability, while the silica membrane lost selectivity for all the studied gases. After hydrothermal treatment at 200 or 300 °C, the CO_2 permeance of the Zr-doped BTESE membrane decreased significantly, and the H_2/CO_2 permselectivity increased significantly, by 65.71%. So far, many scholars have demonstrated the effect of different conditions during preparation on the properties of zirconia-doped silica materials/membranes. The influence of the Zr/Si molar ratio on the microstructure of the membrane and the permeability of the gas is crucial. Unfortunately, there are few reports in this regard. Furthermore, the effects of methyl modification on the microstructure and steam stability of ZrO_2-SiO_2 membranes were rarely described in papers. Some scholars have found that adding a drying control chemical additive (DCCA) in the process of preparing the membrane via the sol-gel method can effectively reduce the uneven shrinkage of the membrane during the heating process and during the calcining process [35], and improve the gas permselectivity of the membrane.

In this paper, methyl-modified ZrO_2-SiO_2 (ZrO_2-$MSiO_2$) materials/membranes with various Zr/Si molar ratios (n_{Zr}) were fabricated. Glycerol was chosen to be the DCCA. The impact of n_{Zr} on the microstructures and H_2 permselectivities of ZrO_2-$MSiO_2$ membranes was thoroughly addressed. The water vapor stability of ZrO_2-$MSiO_2$ membranes was investigated further by comparing the gas permeability characteristics of the ZrO_2-$MSiO_2$ membranes before and after steam treatment. The heat regeneration performance of ZrO_2-$MSiO_2$ membranes was also investigated.

2. Materials and Methods

2.1. Preparation of $MSiO_2$ Sols

The $MSiO_2$ sols were prepared by tetraethylorthosilicate (TEOS, purchased from Xi'an chemical reagent Co., Ltd., Xi'an, China) as a silica source, methyltriethoxysilane (MTES, purchased from Hangzhou Guibao Chemical Co., Ltd., Hangzhou, China) as a hydrophobic modified agent, anhydrous ethanol (EtOH, purchased from Tianjin Branch Micro-Europe Chemical Reagent Co., Ltd., Tianjin, China) as a solvent, and nitric acid (HNO_3, purchased from Sichuan Xilong Reagent Co., Ltd., Chengdu, China) as a catalyst. To begin, TEOS, MTES, and EtOH were completely combined in a three-necked flask using a magnetic stirrer. The flask was correctly immersed in an ice-water combination. The solution was then agitated for 50 min using a magnetic stirrer to ensure thorough mixing. The H_2O and HNO_3 combination was then dropped into the mixture while it was still being stirred. The reaction mixture was then agitated in a three-necked flask at a constant temperature of 60 °C for 3 h to yield the $MSiO_2$ sol.

2.2. Preparation of ZrO_2 Sols

In a three-necked flask, 0.6 M zirconium nitrate pentahydrate ($Zr(NO_3)_4 \cdot 5H_2O$, purchased from Tianjin Fuchen Chemical Reagent Co., Ltd., Tianjin, China) and 0.2 M oxalic acid ($C_2H_2O_4 \cdot 2H_2O$, purchased from Tianjin HedongHongyan Chemical Reagent Co., Ltd., Tianjin, China) solutions were combined at a molar ratio of 4.5:1.0 and agitated. The aforementioned mixture was then treated with 35% (v/v) glycerol (GL, purchased from Tianjin Kemiou Chemical Reagent Co., Ltd., Tianjin, China), and stirring was maintained in a water bath at 50 °C for 3 h to yield the ZrO_2 sols.

2.3. Preparation of ZrO_2-$MSiO_2$ Sols

The ZrO_2 sols were aged for 12 h at 25 °C. The ZrO_2 sols and EtOH were then added to the $MSiO_2$ sols and stirred for 60 min to create the required ZrO_2-$MSiO_2$ sols. The n_{Zr} ratio was 0, 0.08, 0.15, 0.3, and 0.05. The ZrO_2-$MSiO_2$ sols were diluted three times with ethanol after 12 h. GL was used as a drying control agent at 0%, 10%, 20%, and 30% (DCCA). After 60 min of stirring, ZrO_2-$MSiO_2$ sols with varied GL contents were obtained.

2.4. Preparation of ZrO$_2$-MSiO$_2$ Materials

The ZrO$_2$-MSiO$_2$ sols were then placed individually in petri plates for gelation at 30 °C. The gel materials were ground and pulverized with a mortar, and then calcined at a heating rate of 0.5 °C·min^{-1} at 400 °C for 2 h under nitrogen atmosphere protection, and then cooled down naturally. The ZrO$_2$-MSiO$_2$ materials with different n$_{Zr}$ were prepared. The ZrO$_2$-MSiO$_2$ materials with n$_{Zr}$ = 0 are also referred to as "MSiO$_2$" materials.

2.5. Preparation of ZrO$_2$-MSiO$_2$ Membranes

The ZrO$_2$-MSiO$_2$ membranes were coated on top of composite interlayers supported by porous α-alumina discs. The discs are 5 mm-thick and 30 mm in diameter, with a porosity of 40% and an average pore size of 100 nm. ZrO$_2$-MSiO$_2$ membranes were effectively prepared by dip-coating the substrates in three-fold ethanol-diluted silica sol for 7 s, then drying and calcining them. Each sample was dried at 30 °C for 3 h before being calcined at 400 °C in a temperature-controlled furnace in a N$_2$ environment with a ramping rate of 0.5 °C·min^{-1} and a dwell period of 2 h. The dip-coating-drying-calcining process was repeated three times. Figure 1 demonstrates the preparation process of the ZrO$_2$-MSiO$_2$ materials/membranes. The ZrO$_2$-MSiO$_2$ membranes with n$_{Zr}$ = 0 are also referred to as "MSiO$_2$" membranes.

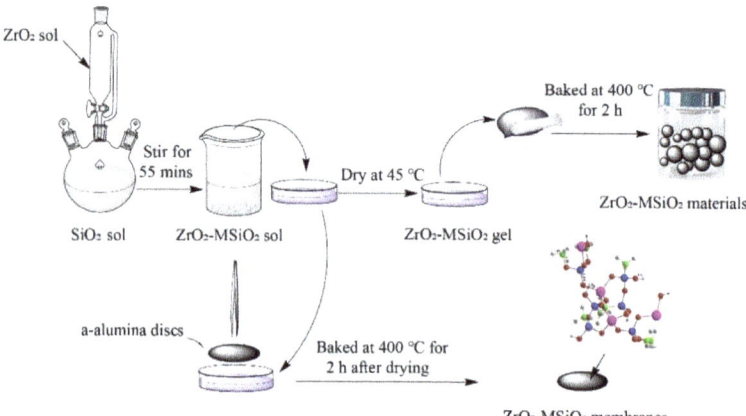

Figure 1. Schematic diagram of the preparation of ZrO$_2$-MSiO$_2$ materials/membranes.

2.6. Steam Treatment and Regeneration of ZrO$_2$-MSiO$_2$ Membranes

The ZrO$_2$-MSiO$_2$ membranes were subjected to a 7-day steam stability test in which they were placed into saturated steam at 25 °C. After steam treatment, for thermal regeneration of ZrO$_2$-MSiO$_2$ membranes, they were processed at a calcination temperature of 350 °C, with the same calcination technique as before. The gas permeances of ZrO$_2$-MSiO$_2$ membranes were investigated after steam treatment and regeneration, respectively.

2.7. Characterizations

Using Fourier transform infrared spectroscopy, the functional groups of ZrO$_2$-MSiO$_2$ materials were characterized (FTIR, Spotlight 400 and Frontier, PerkinElmer Corporation, Waltham, MA, USA), and the wavelength measuring range was 400 to 4000 cm^{-1} using the KBr compression technique. Using a Rigaku D/max-2550pc X-ray diffractometer (XRD, Rigaku D/max-2550pc, Hitachi, Tokyo, Japan) with CuKα radiation at 40 kV and 40 mA, the ZrO$_2$-MSiO$_2$ materials' phase structure was found. The X-ray photoelectron spectra (XPS) were acquired on a K-Alpha X-ray photoelectron spectroscope from Thermo Fisher Scientific with AlKα excitation and were calibrated regarding the signal of adventitious carbon (XPS, ESCALAB250xi, Thermo Scientific, Waltham, MA, USA). The binding energy estimates were derived using the C (1s) line at 284.6 eV as the reference point. Transmission electron microscopy (TEM, JEM 2100F, JEOL, Tokyo, Japan) was utilized to investigate the ZrO_2-$MSiO_2$ powders' crystallization. Operating at 5 kV, scanning electron microscopy (SEM, JEOL JSM-6300, Hitachi, Tokyo, Japan) was utilized to study the surface morphologies of the ZrO_2-$MSiO_2$ membranes. N_2 adsorption–desorption measurements were conducted using an automated Micromeritics, ASAP2020 analyzer (ASAP 2020, Micromeritics, Norcross, GA, USA). The ZrO_2-$MSiO_2$ materials' BET surface area, pore volume, and pore size distribution were determined.

Figure 2 is a schematic of the experimental setup used to evaluate the performance of single gas permeation. Prior to the experiment, the pressure and temperature were set to the desired values for thirty minutes to allow the gas permeation to stabilize. The permeation properties of $MSiO_2$ and ZrO_2-$MSiO_2$ membranes were evaluated using H_2, CO_2, and N_2. The gas permeability was determined based on the outlet gas flow. The gas permselectivity values (ideal permselectivities) were calculated by the permeance ratio between two gases.

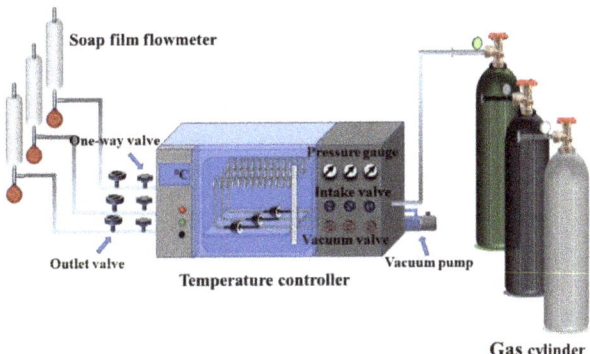

Figure 2. Single gas permeation experiment device diagram.

3. Results

3.1. Chemical Structure Analysis

FTIR spectra were used to investigate the functional groups of ZrO_2-$MSiO_2$ materials. The FTIR spectra of ZrO_2-$MSiO_2$ materials containing various n_{Zr} contents are displayed in Figure 3. The absorption peak at around 3448 cm^{-1} was assigned to the stretching and bending vibration of the -OH group from the absorbed water. The absorption peak at 1630 cm^{-1} corresponds to Si-OH and Zr-OH on the surface of ZrO_2-$MSiO_2$ materials [36]. The antisymmetric stretching vibration absorption peak -CH_3 at 2985 cm^{-1} was mainly from unhydrolyzed TEOS and MTES. The absorption peak at 1278 cm^{-1} was attributed to the Si-CH_3 group. It is also the main hydrophobic functional group of the membrane. The absorption peak observed at 1050 cm^{-1} was attributed to the Si-O-Si bond [37]. Compared with the materials with n_{Zr} = 0, the materials with n_{Zr} = 0.08–0.5 all showed a new absorption peak at the wavenumber of 448 cm^{-1}. This was related to the formation of Zr-O bonds [38]. Meanwhile, with the increase of n_{Zr}, the peak at 1050 cm^{-1} shifted to around 1100 cm^{-1}. This may be ascribed to the fact that partial substitution of Zr atoms for Si atoms in the Si-O-Si network to form Zr-O-Si bonds occurred [39], breaking the symmetry of SiO_2 and leading to the shift of peak positions. However, there was no obvious Zr-O-Si bond in the FTIR spectrum of ZrO_2-$MSiO_2$ materials due to the overlap of the Zr-O-Si bond with Si-O-Si [40]. Furthermore, the decrease in the intensity of the silanol band at 779 and 835 cm^{-1} with increasing n_{Zr} could be attributed to the substitution of Si-OH bonds by Zr-O-Si bonds [41]. It demonstrates the formation of Zr-O-Si bonds in the produced materials.

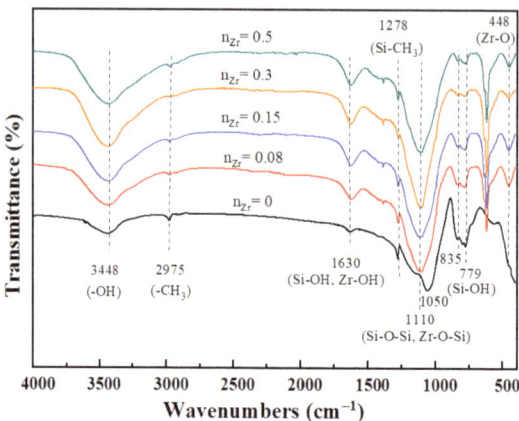

Figure 3. FTIR spectra curves of ZrO_2-$MSiO_2$ materials with various n_{Zr}.

3.2. Phase Structure Analysis

The XRD patterns of the ZrO_2-$MSiO_2$ materials with varied n_{Zr} are presented in Figure 4. The peaks of amorphous SiO_2 were concentrated at $2\theta = 23.1°$ [42]. The SiO_2 peak moved progressively towards higher 2θ values as n_{Zr} rose, and it slowly dropped in intensity. This is attributable to the replacement of the portion of silicon atoms by the inserted Zr atoms, producing Zr-O-Si bonds, resulting in a drop in the SiO_2 concentration. The peaks corresponding to a crystalline tetragonal structure of zirconia are clearly apparent in the ZrO_2-$MSiO_2$ materials with $n_{Zr} = 0.15$–0.5. The (101), (112), and (202) reflection planes of the body-centered ZrO_2 (t-ZrO_2) tetragonal phase were ascribed to the large diffraction peaks occurring at 60.2°, 50.7°, and 30.2°, respectively (JCPDS No. 79-1771). XRD analysis demonstrated that with the growth of the Zr concentration, the peak intensity corresponding to t-ZrO_2 progressively increased. In other words, the content of t-ZrO_2 increased with the growth in Zr content. Combined with the FTIR analysis, the Zr element in ZrO_2-$MSiO_2$ materials may exist in the form of Zr-O-Si bonds and t-ZrO_2.

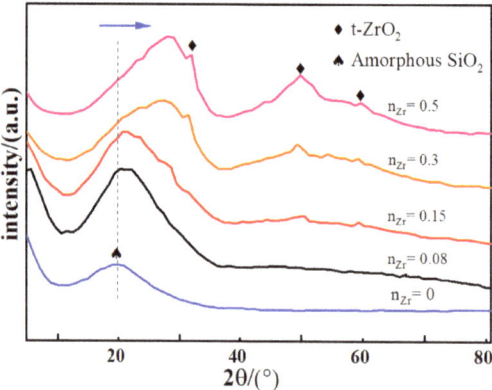

Figure 4. XRD patterns of ZrO_2-$MSiO_2$ materials with various n_{Zr}.

To further investigate the presence of Zr and Si species in the ZrO_2-$MSiO_2$ materials, the XPS measurement was conducted. The survey XPS spectrum of ZrO_2-$MSiO_2$ material with $n_{Zr} = 0.15$ is shown in Figure 5. Figure 5 demonstrates that C, O, Si, and Zr elements are present in the ZrO_2-$MSiO_2$ material, which indicates the successful incorporation of Zr into the silica frameworks. Figure 6 presents the Si 2p and Zr 3d XPS spectra of the

ZrO$_2$-MSiO$_2$ sample with n$_{Zr}$ = 0.15. In Figure 6a, the peaks at the binding energies of 102.8 and 104.7 eV correspond to Si-C and Si-O bonds, respectively. In Figure 6b, the peaks at 186.6 and 183.3 eV correspond to the Zr-O 3d$_{3/2}$ and Zr-O 3d$_{5/2}$ peaks, respectively.

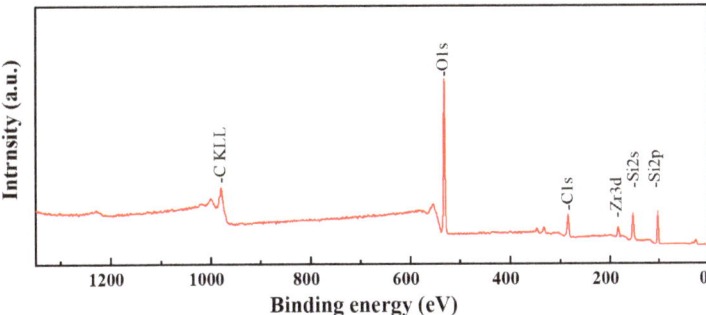

Figure 5. The survey XPS spectrum of ZrO$_2$-MSiO$_2$ materials.

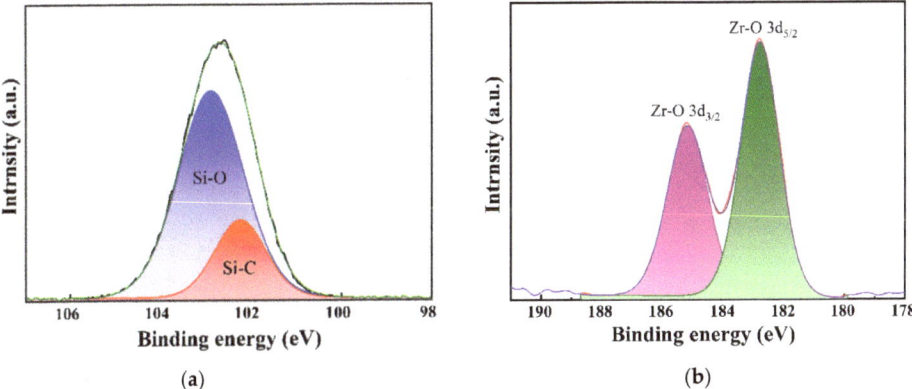

Figure 6. XPS peak decomposition for the (**a**) Si 2p and (**b**) Zr 3d photoelectron peaks of the ZrO$_2$-MSiO$_2$ materials.

3.3. TEM Analysis

The TEM micrographs of the ZrO$_2$-MSiO$_2$ material with n$_{Zr}$ = 0 and 0.15 at 400 °C under nitrogen atmosphere are illustrated in Figure 7. Figure 7a depicts that the silica particles in MSiO$_2$ materials are amorphous, while in Figure 7b, a small amount of particles with darker color appear and are mixed in the silica skeleton, which may be due to the presence of t-ZrO$_2$. Overall, the ZrO$_2$-MSiO$_2$ materials with n$_{Zr}$ = 0.15 still maintained the amorphous state.

3.4. Pore Structure Analysis

The N$_2$ adsorption-desorption isotherm of the ZrO$_2$-MSiO$_2$ materials with various n$_{Zr}$ are shown in Figure 8a. According to the Brunauer-Deming-Deming-Teller (BDDT) classification, the ZrO$_2$-MSiO$_2$ materials showed a type I adsorption isotherm, while n$_{Zr}$ = 0 indicated the formation of microporous structures. The isotherms for the four samples (n$_{Zr}$ = 0.08–0.5) all showed a similar trend, which could be categorized as type IV isotherms. However, the shapes of the hysteresis loops for the four samples were different, implying the variation of pore structures. A significant proportion of adsorption occurred in the range of low relative pressure, $P/P_0 < 0.1$, indicating that the materials contain a large quantity of micropores. The shape of the hysteresis loop of the ZrO$_2$-MSiO$_2$ materials with n_{Zr} = 0.5 was altered, indicating the presence of larger mesopores or macropores. In addition, the distributions of pore size for all samples are depicted in Figure 8b. It is found that the samples with n_{Zr} = 0.08–0.5 showed a broader pore size distribution and a larger mean pore size than the samples with n_{Zr} = 0. The conclusion was also confirmed by the pore structure parameters of ZrO_2-$MSiO_2$ materials with various n_{Zr} in Table 1. The average pore size, BET specific surface area, and total pore volume of the ZrO_2-$MSiO_2$ materials gradually increased with n_{Zr} = 0.08 and 0.15, and the pore size distribution became wider. However, the total pore volume of the ZrO_2-$MSiO_2$ materials with n_{Zr} = 0.3 and 0.5 decreased instead. The fact is that the bond length of Zr-O (1.78 Å) was slightly longer than that of Si-O (1.64 Å) [43]. Figure 9 shows the molecular structure models of $MSiO_2$, ZrO_2-$MSiO_2$, and t-ZrO_2 crystallites, respectively. Hence, the formation of Zr-O-Si bonds contributes to the formation of pores. With the increase of Zr content, more and more t-ZrO_2 crystallites were formed and distributed in the framework of $MSiO_2$ materials, and the internal pore structure of the ZrO_2-$MSiO_2$ materials was hindered from shrinking and pore collapse, resulting in the decrease of the BET surface area and pore volume. From Table 1, it can be seen that the ZrO_2-$MSiO_2$ materials with n_{Zr} = 0.15 had the maximal total pore volume (0.43 $cm^3 \cdot g^{-1}$), BET surface area (616.77 $m^2 \cdot g^{-1}$), and the minimum mean pore size (2.19 nm).

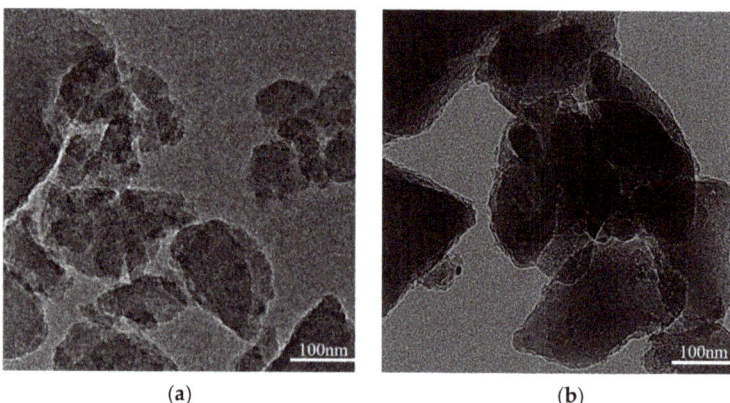

Figure 7. TEM images of ZrO_2-$MSiO_2$ materials with n_{Zr} = (**a**) 0 and (**b**) 0.15.

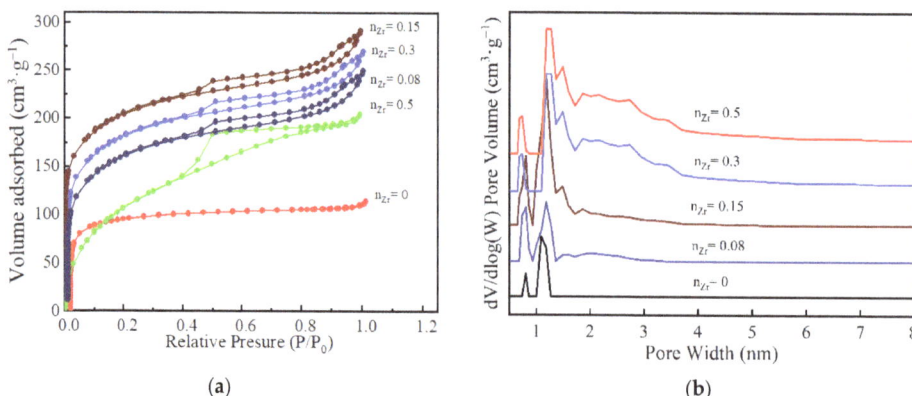

Figure 8. (**a**) The nitrogen adsorption-desorption isotherms and (**b**) the corresponding pore size distribution curves for the ZrO_2-$MSiO_2$ materials with various n_{Zr}.

Table 1. Pore structure parameters of the ZrO$_2$-MSiO$_2$ materials with various n$_{Zr}$.

n$_{Zr}$	BET Surface Area (m$^2 \cdot$g^{-1})	Average Pore Size (nm)	V$_{total}$ (STP) (cm$^3 \cdot$g^{-1})	V$_{micro}$ (STP) (cm$^3 \cdot$g^{-1})	V$_{micro}$/V$_{total}$ (%)
0	389.38	1.75	0.23	0.15	65.22
0.08	579.96	2.08	0.37	0.13	35.14
0.15	616.77	2.19	0.43	0.12	27.91
0.3	606.35	2.35	0.41	0.09	21.95
0.5	545.32	3.58	0.38	0.08	21.05

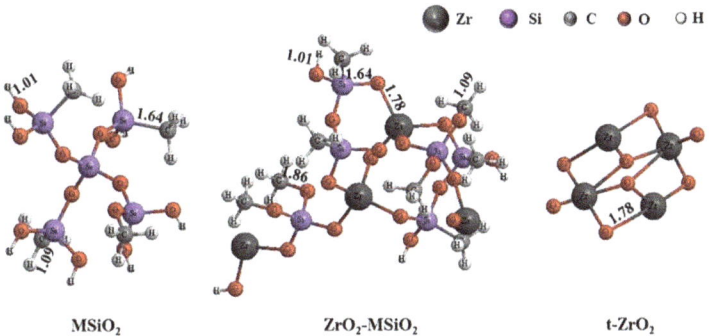

Figure 9. Molecular structural models of MSiO$_2$, ZrO$_2$-MSiO$_2$, and t-ZrO$_2$.

3.5. Gas Permselectivity Analysis

3.5.1. The Influence of n$_{Zr}$

Figure 10 depicts the influence of n$_{Zr}$ on the gas permeabilities and H$_2$ permselectivities of ZrO$_2$-MSiO$_2$ membranes with varying n$_{Zr}$ and 0% DCCA addition at 25 °C and 0.1 MPa. In Figure 10a, with the increase of n$_{Zr}$, the H$_2$, CO$_2$, and N$_2$ permeances of the samples increased until n$_{Zr}$ = 0.15, and then decreased. Compared with the MSiO$_2$ membranes (n$_{Zr}$ = 0), the H$_2$, CO$_2$, and N$_2$ permeance of the ZrO$_2$-MSiO$_2$ membranes with n$_{Zr}$ = 0.15 increased by 50.95%, 26.74%, and 36.36%, respectively. From the pore structure analysis (Table 1), the overall pore volumes of the ZrO$_2$-MSiO$_2$ membranes grew somewhat with increasing n$_{Zr}$ until n$_{Zr}$ = 0.15, and then decreased, which can explain why the ZrO$_2$-MSiO$_2$ membranes with n$_{Zr}$ = 0.15 had the highest permeance to each gas. For the same membrane, the order of gas molecular permeance is H$_2$ > CO$_2$ > N$_2$. Gas permeance decreased with increasing d$_k$ (0.289, 0.33, and 0.364 nm, respectively), indicating that all membranes exhibited molecular sieve properties. However, when n$_{Zr}$ ≥ 0.15, the permeance of CO$_2$ decreased more closely to that of N$_2$. This behavior was related to the fact that following heat treatment at 400 °C, the Zr-O-Si bonds and t-ZrO$_2$ crystallites generated in the ZrO$_2$-SiO$_2$ membranes will generate a significant number of Brønsted acid sites. High acidity leads to a reduction in the affinity of the membranes for CO$_2$, hence lowering the CO$_2$ permeance [44].

It can be observed from Figure 10b that compared with MSiO$_2$ membranes, the H$_2$/CO$_2$ and H$_2$/N$_2$ permselectivities of ZrO$_2$-MSiO$_2$ membranes with n$_{Zr}$ = 0.15 increased by 22.93% and 33.04%, respectively. Combined with the previous characterization test, it was found that the ZrO$_2$-MSiO$_2$ membranes with n$_{Zr}$ = 0.15 had a good pore structure, which is beneficial to improve the permselectivity of gas. In addition, the acidic sites formed by the ZrO$_2$-MSiO$_2$ membranes reduced the affinity of the membranes for CO$_2$ and helped to separate it from H$_2$. However, the permselectivities of ZrO$_2$-MSiO$_2$ membranes after n$_{Zr}$ = 0.15 showed a decreasing trend. Compared with the ZrO$_2$-MSiO$_2$ membranes with n$_{Zr}$ = 0.15, the H$_2$/CO$_2$ and H$_2$/N$_2$ permselectivities of the membranes with n$_{Zr}$ = 0.5 decreased by 9.35% and 20.15%, respectively. As a result, just because the n$_{Zr}$ concentration is larger, it does not indicate that the separation effect is better. Since the Zr-O and Si-O bonds in zirconium-substituted siloxane rings are longer than in pure siloxane rings, for the ZrO_2-$MSiO_2$ membranes with n_{Zr} = 0.5, the number of siloxane rings containing Zr increased, and the pore size of the membranes became larger. Meanwhile, a large number of t-ZrO_2 crystals were produced, which led to the shrinkage of the pore structure inside the membranes and the collapse of the pores, resulting in the decrease of the permselectivities of the membranes.

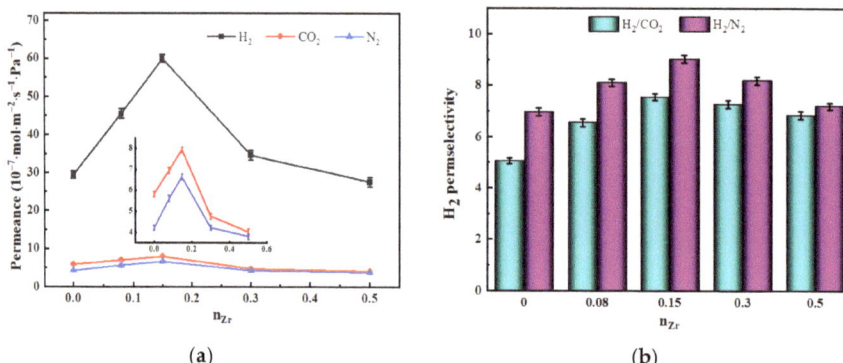

Figure 10. (a) Gas permeance and (b) H_2 permselectivities of ZrO_2-$MSiO_2$ membranes with various n_{Zr} and 0% DCCA addition at a pressure difference of 0.1 MPa and 25 °C.

SEM images of surface topography for $MSiO_2$ and ZrO_2-$MSiO_2$ (n_{Zr} = 0.15) membranes calcined at 400 °C are shown in Figure 11. Compared to the $MSiO_2$ membranes, the particle size and distribution of the ZrO_2-$MSiO_2$ membranes were more uniform, the membranes' surfaces had no obvious defects, and the surface was uniform and smooth. The particle size of $MSiO_2$ membranes was between 1.1 and 5.8 nm, while the particle size of ZrO_2-$MSiO_2$ membranes was between 1.3 and 8.9 nm. The formed ZrO_2-$MSiO_2$ membranes with a smooth surface and uniform membrane pores were more conducive to the gas permselectivitiy.

Figure 11. SEM images of surface topography for (a) $MSiO_2$ and (b) ZrO_2-$MSiO_2$ (n_{Zr} = 0.15) membranes with 0% DCCA addition calcined at 400 °C.

3.5.2. The Influence of Temperature

The permeances and permselectivities of the $MSiO_2$ and ZrO_2-$MSiO_2$ (n_{Zr} = 0.15) membranes with 0% DCCA addition at a pressure difference of 0.1 MPa and temperature changing from 25 to 200 °C are shown in Figure 12. Obviously, with the increasing temperature, the permeance of H_2 of $MSiO_2$ and ZrO_2-$MSiO_2$ membranes increased gradually, as seen in Figure 12a. From 25 to 200 °C, the H_2 permeance of $MSiO_2$ and ZrO_2-$MSiO_2$ membranes rose by 19.66% and 7.12%, respectively, demonstrating that the H_2 permeation behavior in the two membranes followed the activated diffusion transport mechanism. In contrast, the permeabilities of CO_2 and N_2 were similar to the Knudsen diffusion trend, whereby both slightly decreased. The CO_2 permeance of $MSiO_2$ and ZrO_2-$MSiO_2$ membranes decreased by 31.46% and 30.20% from 25 to 200 °C, respectively, and the N_2 permeance decreased by 18.60% and 29.98%, respectively. The major explanations for the decrease in CO_2 and N_2 permeance were the violent movement of molecules and the rise in the mean free path as temperature increased.

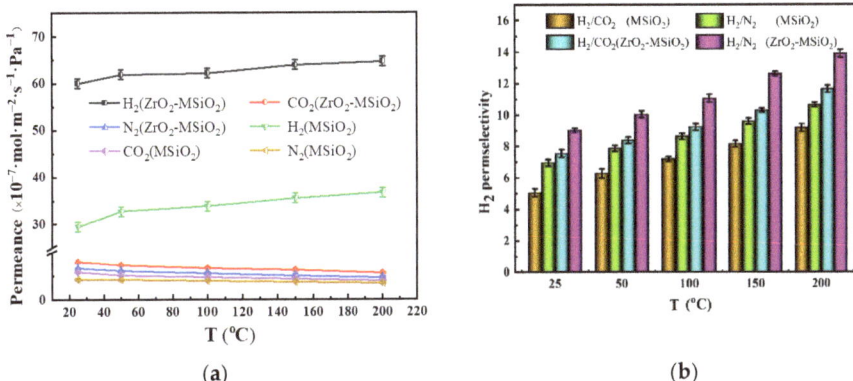

Figure 12. Influence of temperature on the (**a**) gas permeances and (**b**) H_2 permselectivities of $MSiO_2$ and ZrO_2-$MSiO_2$ (n_{Zr} = 0.15) membranes with 0% DCCA addition at a pressure difference of 0.1 MPa.

In Figure 12b, it can be seen that the permselectivities in the membranes gradually increased with the increase of temperature. At 200 °C, compared with the $MSiO_2$ membranes, the H_2/CO_2 and H_2/N_2 permselectivities of the ZrO_2-$MSiO_2$ membranes increased by 21.11% and 23.34%, respectively. The above results show that the ZrO_2-$MSiO_2$ membranes had better permselectivity and permeance of H_2 than the $MSiO_2$ membranes under the same conditions.

Combined with the preceding studies, it was determined that a rise in temperature facilitated the separation of H_2 and that the separation process of H_2 from CO_2 and N_2 is dominated by activation diffusion, which is described by the Arrhenius equation [45]:

$$P = P_0 \exp\left(-\frac{E_a}{RT}\right) \quad (1)$$

In Formula (1), P is the permeation rate, E_a is the apparent activation energy, and P_0 is a constant, which depends on the pore wall–gas molecule interaction, gas selective layer thickness, and pore shape and tortuosity [46]. For linear fitting, $1000/RT$ was used as the abscissa and $\ln P$ as the ordinate, and the slope of the fitting equation might be used to obtain the apparent activation energy. Figure 13 depicts the Arrhenius fitting diagrams for the three gases.

Figure 13 demonstrates that the apparent activation energy of H_2 is positive, while that of various other gases is negative. This is related to the gas-activated transport behavior, whereby there are two parallel transmission channels for gas through the membrane: one is through selective micropores, with gas transport processed by a thermally activated surface diffusion mechanism in the micropore state [47], and the other is through larger pores [48]. The activation energies for CO_2 and N_2 are negative, indicating that there are permeation pathways large enough in these types of membranes to allow the diffusion of these larger gas. It is generally believed that E_a is composed of two parts [46], the adsorption heat, Q_{st}, of gas on the surface of the membrane and the activation energy, E_m, of gas flowing through the solid surface. The larger the E_m is, the harder it is for the gas to diffuse, E_a

$= E_m - Q_{st}$. Arrhenius equation parameter values are shown in Table 2. Table 2 shows that the E_m values of the gases in the ZrO_2-$MSiO_2$ membrane are all less than those in the $MSiO_2$ membrane. It shows that the structure of ZrO_2-$MSiO_2$ membranes is not as dense as that of $MSiO_2$ membranes. This is in good accordance with the N_2 adsorption–desorption results. This finding also shows that ZrO_2 doping successfully diminishes the densification of the SiO_2 network. The higher porosity of the ZrO_2-$MSiO_2$ membranes allows the gas to cross the membrane pore barrier using their kinetic energy. Therefore, the gas (H_2, CO_2, and N_2) permeance of ZrO_2-$MSiO_2$ membranes is higher than that of $MSiO_2$ membranes.

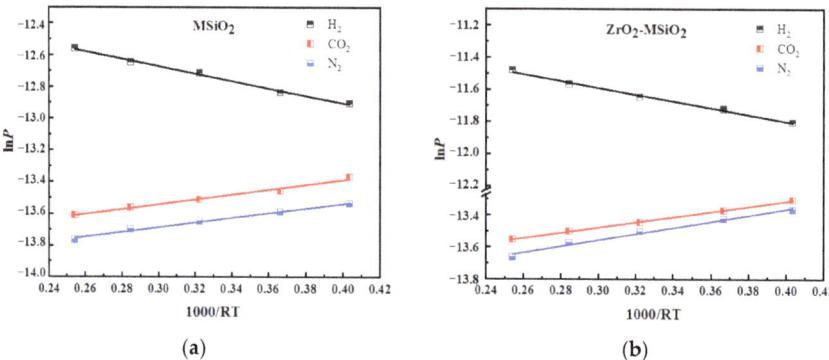

Figure 13. Arrhenius plots of different gases of (**a**) $MSiO_2$ and (**b**) ZrO_2-$MSiO_2$ (n_{Zr} = 0.15) membranes.

Table 2. Arrhenius equation parameter values of $MSiO_2$ and ZrO_2-$MSiO_2$ (n_{Zr} = 0.15) membranes.

Membrane	Gases	E_a (kJ·mol^{-1})	Q_{st} (kJ·mol^{-1})	E_m (kJ·mol^{-1})
$MSiO_2$	H_2	2.32	6.00	8.32
	CO_2	−1.53	24.00	22.47
	N_2	−1.48	18.00	16.52
ZrO_2-$MSiO_2$	H_2	2.10	6.00	8.10
	CO_2	−1.64	24.00	22.36
	N_2	−1.94	18.00	16.06

Table 3 displays the E_a of H_2, pore diameter, H_2 permeance, and H_2 permselectivities for several membranes/films prepared by other researchers. It is challenging to concurrently enhance the membranes' permselectivity and gas permeability, as seen in Table 3. Generally, the larger the average pore size of the membrane, the higher the permeance to H_2, which is accompanied by a smaller E_a. Meanwhile, the E_a of H_2 is related to the interaction of H_2 with the membrane pore wall. It can be seen from Table 3 that the as-prepared ZrO_2-$MSiO_2$ membrane has a large H_2 permselectivity compared to other membranes.

Table 3. E_a of H_2, pore diameter, H_2 permeance, and H_2 permselectivities for various membranes/films prepared by other researchers.

Type	Temperature/Pressure	E_a of H_2 (kJ·mol^{-1})	Pore Diameter (nm)	H_2 Permeance (mol·m^{-2}·s^{-1}·Pa^{-1})	H_2 Permselectivities	
					H_2/CO_2	H_2/N_2
SiO_2 [49]	200 °C, 2 bar	-	0.30–0.54	4.62×10^{-7}	3.7	10.5
ZIF-7-SiO_2 [50]	200 °C	-	5	8×10^{-7}	8.78	11.8
Pd-SiO_2 [51]	200 °C, 0.3 MPa	-	0.57	7.26×10^{-7}	4.3	14
ZrO_2 [52]	350 °C	-	4.95	5.3×10^{-8}	14.3	3.1
ZrO_2-SiO_2 [27]	550 °C	7.0	0.165	1.8×10^{-7}	-	-
BTDA-DDS polyimide [53]	30 °C, 5 MPa	-	-	2.52×10^{-9}	5.16	193.21
Cellulose acetate [54]	25 °C, 1 bar	-	-	3.55×10^{-9}	-	30.3
ZrO_2-$MSiO_2$ *	200 °C, 0.1 MPa	2.10	2.19	6.46×10^{-6}	11.64	13.88

* In this work.

3.5.3. The Influence of Pressure Difference

Figure 14 illustrates the effect of pressure difference on the gas permeances and H_2 permselectivities of $MSiO_2$ and ZrO_2-$MSiO_2$ (n_{Zr} = 0.15) membranes at 200 °C with 0% DCCA addition. In Figure 14a, it can be observed that the H_2 permeance of the ZrO_2-$MSiO_2$ membranes with n_{Zr} = 0.15 improved with the increase of the pressure difference, and the pressure dependence increased. However, the H_2 permeance of $MSiO_2$ membranes remained basically unchanged. The $MSiO_2$ and ZrO_2-$MSiO_2$ membranes increased their H_2 permeance by 2.16% and 19.96%, respectively, when the pressure was increased from 0.10 to 0.40 MPa. In Figure 14b, the H_2 permselectivities of $MSiO_2$ membranes did not change significantly, and the H_2/CO_2 and H_2/N_2 permselectivities decreased by 5.27% and 1.12%, respectively. However, the H_2 permselectivities of the ZrO_2-$MSiO_2$ membranes with n_{Zr} = 0.15 changed greatly, where the H_2/CO_2 and H_2/N_2 permselectivities decreased by 22.04% and 21.12%, respectively. Clearly, it can be seen that the permselectivity of the ZrO_2-$MSiO_2$ membranes was reduced more than that of the $MSiO_2$ membranes, that is, the pressure had a relatively small effect on the gas permeation of the $MSiO_2$ membranes. This phenomenon is attributed to the relatively dense $MSiO_2$ membranes with micropores as the main component. With the increase of the pressure difference between the two sides of the membranes, the power of the gas passing through the ZrO_2-$MSiO_2$ membranes increased, making it easier for the gas to pass through the mesopores or even the macropores, which has a greater impact on the permselectivity of the membranes. At the same time, it is shown that the H_2 diffusion mechanism of ZrO_2-$MSiO_2$ membranes was different from that of $MSiO_2$ membranes due to the influence of doping ZrO_2. The gas transport in the ZrO_2-$MSiO_2$ membranes follows the surface diffusion mechanism. In addition, when the pressure difference increased to 0.4 MPa, the permselectivities of H_2/CO_2 and H_2/N_2 were still higher than their respective Knudsen diffusion (4.69 and 3.74, respectively), indicating that they still have good gas permeation performance under high pressure.

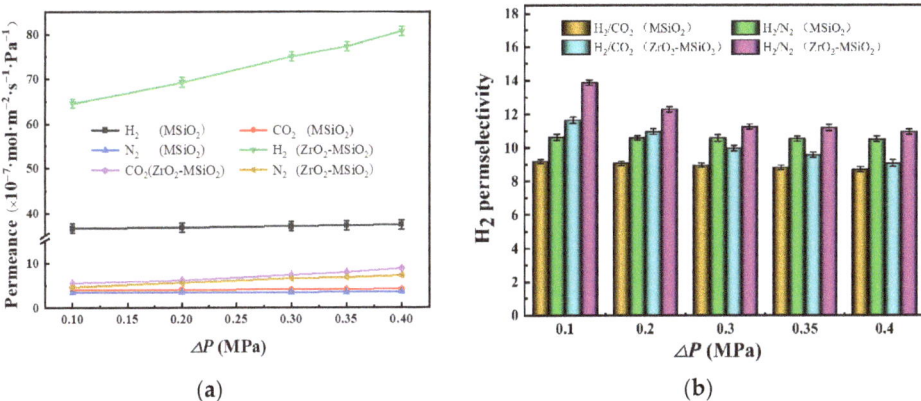

Figure 14. Influence of pressure difference on the (**a**) gas permeances and (**b**) H_2 permselectivities of $MSiO_2$ and ZrO_2-$MSiO_2$ (n_{Zr} = 0.15) membranes with 0% DCCA addition at 200 °C.

3.5.4. The Influence of the DCCA

Comparing and analyzing the previous results, it was found that the addition of DCCA (glycerol) by the sol-gel method can effectively reduce the uneven shrinkage of the membrane when it is heated during firing. The gas permeances and H_2 permselectivities of ZrO_2-$MSiO_2$ membranes (n_{Zr} = 0.15) with various DCCA additions at 200 °C and 0.1 MPa are shown in Figure 15.

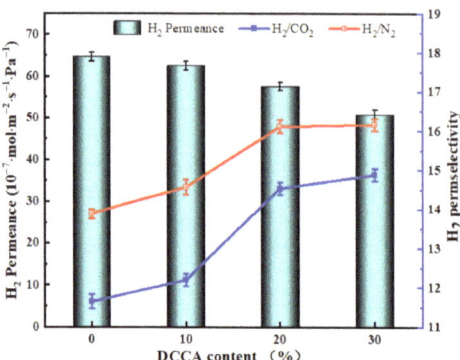

Figure 15. H_2 permeances and permselectivities of ZrO_2-$MSiO_2$ membranes (n_{Zr} = 0.15) with various DCCA additions at 200 °C and 0.1 MPa.

Figure 15 demonstrates that the H_2 permeance of the ZrO_2-$MSiO_2$ membranes reduced by 21.30% as the DCCA addition increased from 0 to 30%. The permselectivities of H_2/CO_2 and H_2/N_2 increased by 21.77% and 14.07%, respectively. However, compared with the 20% membranes, the H_2/CO_2 and H_2/N_2 permselectivities of the membranes with the addition of 30% only increased by 2.14% and 0.28%, respectively. Figure 16 shows the relationship between DCCA (GL) contents and $F_{H2} \times \alpha_{H2}$ (F_{H2} is the H_2 permeance, α_{H2} is the permselective of H_2). It was clearly observed that the addition of DCCA from 0 to 20% enhanced the $F_{H2} \times \alpha_{H2}$ value, and after more than 20%, the $F_{H2} \times \alpha_{H2}$ value showed a lower level. This is attributed to the fact that the addition of glycerol will gradually surround the sol particles, reduce the agglomeration caused by the collision of the colloid particles, accelerate the creation of the sol-gel network, and improve the stability of the sol structure [52]. In addition, the membrane layer collapsed and cracked easily during the drying and calcination processes, and the addition of glycerol can effectively reduce the liquid–gas surface tension to a certain extent, thereby protecting the gel skeleton from deformation [55]. However, the particles of the sol were surrounded by steric effects when too much GL was added, making the sol sticky and difficult to dry to form a membrane. The extension of the drying time makes the cross-linking of the sol more thorough and eventually leads to the densification of the membrane, which is not conducive to gas separation. From the above point of view, the membranes with 20% DCCA addition were the more worthy choice.

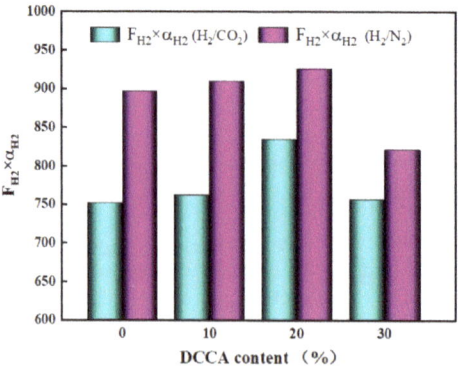

Figure 16. The relationship between DCCA addition and $F_{H2} \times \alpha_{H2}$.

3.5.5. Steam Treatment and Regeneration Analysis

Figure 17 shows the effects of steam treatment and thermal regeneration on the gas permeances (H_2, CO_2, and N_2) and H_2 permselectivities of $MSiO_2$ and ZrO_2-$MSiO_2$ (n_{Zr} = 0.15) membranes with 0% DCCA addition at a pressure difference of 0.1 MPa and 25 °C. The permeances of H_2, CO_2, and N_2 for $MSiO_2$ and ZrO_2-$MSiO_2$ membranes appear to have reduced after steam treatment. After steam aging for 7 days, the permeance of H_2 for $MSiO_2$ and ZrO_2-$MSiO_2$ membranes dropped by 20.63% and 3.70%, respectively, as compared to untreated fresh samples, and the permselectivities of H_2/CO_2 and H_2/N_2 for $MSiO_2$ membranes decreased by 1.59% and 1.04%, respectively, whereas those of ZrO_2-$MSiO_2$ membranes increased by 0.09% and 0.43%, respectively.

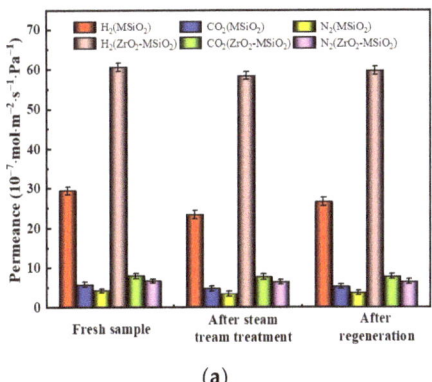 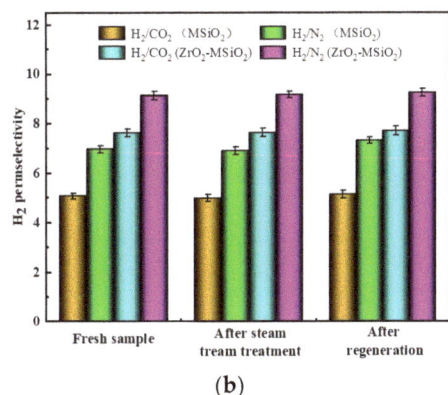

Figure 17. Effect of hydrothermal conditions on the (**a**) gas permeances and (**b**) H_2 permselectivities of $MSiO_2$ and ZrO_2-$MSiO_2$ (n_{Zr} = 0.15) membranes with 0% DCCA addition at a pressure difference of 0.1 MPa and 25 °C.

The gas permeances (H_2, CO_2, and N_2), as well as the permselectivities of H_2/CO_2 and H_2/N_2 for two membranes, all exhibit an increased trend after regeneration by calcination at 350 °C. However, as compared to untreated fresh samples, the H_2 permeances of $MSiO_2$ and ZrO_2-$MSiO_2$ membranes after regeneration dropped by 9.96% and 1.65%, respectively, whereas the permselectivities of H_2/CO_2 and H_2/N_2 for $MSiO_2$ membranes improved by 1.12% and 4.71%, respectively, and those for ZrO_2-$MSiO_2$ membranes increased by 0.08% and 1.21%, respectively. The decrease in gas permeances in both membranes suggests that membrane pore shrinking occurs after calcination at 350 °C. Lower permeance and greater permselectivities are produced as a result of the smaller pores. This is attributed to the partial Zr atoms replacing the Si atoms in Si-O-Si to form more stable Zr-O-Si bonds, which further improves the hydrothermal stability of the membrane material. Therefore, the above results indicated that the ZrO_2-$MSiO_2$ membranes had a better hydrothermal stability and reproducibility than $MSiO_2$ membranes.

4. Conclusions

The ZrO_2-$MSiO_2$ membranes were manufactured to enhance the steam stability and H_2 permselectivity of SiO_2 membranes. It was found that with the increase of ZrO_2 content, the pore size distribution of the materials became wider and the average pore size increased, indicating that the doping of ZrO_2 had the effect of expanding the pores. The ZrO_2-$MSiO_2$ membranes with n_{Zr} = 0.15 had a good pore structure and suitable micropore/mesoporous ratio, which is beneficial to improve the permeance of natural gas. At 200 °C, the H_2/CO_2 and H_2/N_2 permselectivities of ZrO_2-$MSiO_2$ membranes were 79.18% and 26.75% greater than those of $MSiO_2$ membranes, respectively. Furthermore, when the pressure was increased to 0.4 MPa, the permselectivities were still higher than their respective Knudsen diffusion, indicating that they still had good gas permeance at high pressure. With the addition of DCCA from 20% to 30%, the H_2/CO_2 and H_2/N_2 permselectivities of ZrO_2-$MSiO_2$ membranes only increased by 2.14% and 0.28%, respectively, and the $F_{H2} \times \alpha_{H2}$ value with 20% addition was the highest. In conclusion, it is worthwhile to choose 20% GL as the DCCA addition for ZrO_2-$MSiO_2$ membranes. Compared with the untreated fresh sample, after 7 days of water vapor aging, the permeance of ZrO_2-$MSiO_2$ membranes to H_2 decreased by only 3.70%, and the permselectivities of H_2/CO_2 and H_2/N_2 increased by only 0.09% and 0.43%, respectively. After regeneration at 350 °C, the H_2 permeance of the ZrO_2-$MSiO_2$ membranes decreased by 1.65%, and the permselectivities of H_2/CO_2 and H_2/N_2 increased by 0.08% and 1.21%. It is enough to show that the prepared ZrO_2-$MSiO_2$ membranes had a good hydrothermal stability and certain regeneration performance. In the future, the influence of high temperature (for example, $\geq$300 °C) and mixed gases on the gas permeances and permselectivities of ZrO_2-$MSiO_2$ membranes should be explored, which is important to the practical engineering applications.

Author Contributions: Conceptualization, L.W. and J.Y.; writing—original draft preparation, L.W. and J.Y.; funding acquisition, J.Y. All authors have read and agreed to the published version of the manuscript.

Funding: This work was supported by the Scientific Research Project of Shaanxi province of China (2022SF-287) and the Scientific Research Project of Shaanxi Education Department, China (19JC017).

Institutional Review Board Statement: Not applicable.

Informed Consent Statement: Not applicable.

Data Availability Statement: Not applicable.

Conflicts of Interest: The authors declare no conflict of interest.

References

1. Nazir, H.; Louis, C.; Jose, S.; Prakash, J.; Muthuswamy, N.; Buan, M.E.; Flox, C.; Chavan, S.; Shi, X.; Kauranen, P.; et al. Is the H_2 economy realizable in the foreseeable future? Part I: H_2 production methods. *Int. J. Hydrogen Energy* **2020**, *45*, 13777–13788. [CrossRef]
2. Nordio, M.; Wassie, S.A.; Van Sint Annaland, M.; Tanaka, D.A.P.; Sole, J.L.V.; Gallucci, F. Techno-economic evaluation on a hybrid technology for low hydrogen concentration separation and purification from natural gas grid. *Int. J. Hydrogen Energy* **2020**, *46*, 23417–23435. [CrossRef]
3. Cai, L.; Cao, Z.; Zhu, X.; Yang, W. Improved hydrogen separation performance of asymmetric oxygen transport membranes by grooving in the porous support layer. *Green Chem. Eng.* **2020**, *2*, 96–103. [CrossRef]
4. Lv, B.; Luo, Z.; Deng, X.; Chen, J.; Fang, C.; Zhu, X. Study on dry separation technology of a continuous gas-solid separation fluidized bed with a moving scraper ( )–Separation performance. *Powder Technol.* **2021**, *377*, 565–574. [CrossRef]
5. Koutsonikolas, D.E.; Pantoleontos, G.; Karagiannakis, G.; Konstandopoulos, A.G. Development of H_2 selective silica membranes: Performance evaluation through single gas permeation and gas separation tests. *Sep. Purif. Technol.* **2021**, *264*, 118432. [CrossRef]
6. Wu, R.; Yue, W.; Li, Y.; Huang, A. Ultra-thin and high hydrogen permeable carbon molecular sieve membrane prepared by using polydopamine as carbon precursor. *Mater. Lett.* **2021**, *295*, 129863. [CrossRef]
7. Singla, S.; Shetti, N.P.; Basu, S.; Mondal, K.; Aminabhavi, T.M. Hydrogen production technologies—Membrane based separation, storage and challenges. *J. Environ. Manag.* **2022**, *302*, 113963. [CrossRef]
8. Yan, E.; Huang, H.; Sun, S.; Zou, Y.; Chu, H.; Sun, L. Development of Nb-Ti-Co alloy for high-performance hydrogen separating membrane. *J. Membr. Sci.* **2018**, *565*, 411–424. [CrossRef]
9. Farina, L.; Santucci, A.; Tosti, S. Plasma Enhancement Gases separation via ceramic porous membranes for plasma exhaust processing system of DEMO. *Fusion Eng. Des.* **2021**, *169*, 112484. [CrossRef]
10. Zhang, D.; Zhang, X.; Zhou, X.; Song, Y.; Jiang, Y.; Lin, B. Phase stability and hydrogen permeation performance of BaCo0·4Fe0·4Zr0·1Y0·1O3-δ ceramic membranes. *Ceram. Int.* **2021**, *48*, 9946–9954. [CrossRef]
11. Bakoglidis, K.D.; Palisaitis, J.; Dos Santos, R.B.; Rivelino, R.; Persson, P.O.; Gueorguiev, G.K.; Hultman, L. Self-Healing in Carbon Nitride Evidenced As Material Inflation and Superlubric Behavior. *ACS Appl. Mater. Interfaces* **2018**, *10*, 16238–16243. [CrossRef]
12. Kakanakova-Georgieva, A.; Gueorguiev, G.; Sangiovanni, D.G.; Suwannaharn, N.; Ivanov, I.G.; Cora, I.; Pécz, B.; Nicotra, G.; Giannazzo, F. Nanoscale phenomena ruling deposition and intercalation of AlN at the graphene/SiC interface. *Nanoscale* **2020**, *12*, 19470–19476. [CrossRef] [PubMed]

13. Thirumal, V.; Yuvakkumar, R.; Kumar, P.S.; Ravi, G.; Keerthana, S.P.; Velauthapillai, D. Facile single-step synthesis of [email-protected] hybrid nanocomposite by CVD method to remove hazardous pollutants. *Chemosphere* **2021**, *286*, 131733. [CrossRef] [PubMed]
14. Kanezashi, M.; Asaeda, M. Hydrogen permeation characteristics and stability of Ni-doped silica membranes in steam at high temperature. *J. Membr. Sci.* **2006**, *271*, 86–93. [CrossRef]
15. Rosli, A.; Ahmad, A.L.; Low, S.C. Anti-wetting polyvinylidene fluoride membrane incorporated with hydrophobic polyethylene-functionalized-silica to improve CO_2 removal in membrane gas absorption. *Sep. Purif. Technol.* **2019**, *221*, 275–285. [CrossRef]
16. Yang, J.; Chen, J. Hydrophobic modification and silver doping of silica membranes for H_2/CO_2 separation. *J. CO2 Util.* **2013**, *3–4*, 21–29. [CrossRef]
17. Wei, Q.; Ding, Y.L.; Nie, Z.R.; Liu, X.G.; Li, Q.Y. Wettability, pore structure and performance of perfluorodecyl-modified silica membranes. *J. Membr. Sci.* **2014**, *466*, 114–122. [CrossRef]
18. Mukherjee, D.; Kar, S.; Mandal, A.; Ghosh, S.; Majumdar, S. Immobilization of tannery industrial sludge in ceramic membrane preparation and hydrophobic surface modification for application in atrazine remediation from water. *J. Eur. Ceram. Soc.* **2019**, *39*, 3235–3246. [CrossRef]
19. Karimiab, S.; Mortazavia, Y.; Khodadadia, A.A.; Holmgrenb, A.; Korelskiyc, D.; Hedlund, J. Functionalization of silica membranes for CO_2 separation. *Sep. Purif. Technol.* **2020**, *235*, 116207. [CrossRef]
20. Khan, A.A.; Maitlo, H.A.; Khan, I.A.; Lim, D.; Zhang, M.; Kim, K.-H.; Lee, J.; Kim, J.-O. Metal oxide and carbon nanomaterial based membranes for reverse osmosis and membrane distillation: A comparative review. *Environ. Res.* **2021**, *202*, 111716. [CrossRef]
21. Kurt, T.; Topuz, B. Sol-gel Control on Mixed Network Silica Membranes for Gas Separation. *Sep. Purif. Technol.* **2020**, *255*, 117654. [CrossRef]
22. Zhang, Y.; Huang, B.; Mardkhe, M.K.; Woodfield, B.F. Thermal and hydrothermal stability of pure and silica-doped mesoporous aluminas. *Microporous Mesoporous Mater.* **2019**, *284*, 60–68. [CrossRef]
23. Gu, Y.; Hacarlioglu, P.; Oyama, S.T. Hydrothermally stable silica–alumina composite membranes for hydrogen separation. *J. Membr. Sci.* **2008**, *310*, 28–37. [CrossRef]
24. Teo, H.T.; Siah, W.R.; Yuliati, L. Enhanced adsorption of acetylsalicylic acid over hydrothermally synthesized iron oxide-mesoporous silica MCM-41 composites. *J. Taiwan Inst. Chem. Eng.* **2016**, *65*, 591–598. [CrossRef]
25. Uhlmann, D.; Smart, S.; da Costa, J.C.D. H_2S stability and separation performance of cobalt oxide silica membranes. *J. Membr. Sci.* **2011**, *380*, 48–54. [CrossRef]
26. Chang, C.H.; Gopalan, R.; Lin, Y.S. A comparative study on thermal and hydrothermal stability of alumina, titania and zirconia membranes. *J. Membr. Sci.* **1994**, *91*, 27–45. [CrossRef]
27. Yoshida, K.; Hirano, Y.; Fujii, H.; Tsuru, T.; Asaeda, M. Hydrothermal stability and performance of silica-zirconia membranes for hydrogen separation in hydrothermal conditions. *J. Chem. Eng. Jpn.* **2001**, *34*, 523–530. [CrossRef]
28. Díez, B.; Roldán, N.; Martín, A.; Sotto, A.; Perdigón-Melón, J.; Arsuaga, J.; Rosal, R. Fouling and biofouling resistance of metal-doped mesostructured silica/polyethersulfone ultrafiltration membranes. *J. Membr. Sci.* **2017**, *526*, 252–263. [CrossRef]
29. Wang, L.; Yang, J.; Mu, R.; Guo, Y.; Hou, H. Sol-Gel Processed Cobalt-Doped Methylated Silica Membranes Calcined under N_2 Atmosphere: Microstructure and Hydrogen Perm-Selectivity. *Materials* **2021**, *14*, 4188. [CrossRef]
30. Li, L.; Hong, Q. Gas separation using sol–gel derived microporous zirconia membranes with high hydrothermal stability. *Chin. J. Chem. Eng.* **2015**, *23*, 1300–1306. [CrossRef]
31. Gu, Y.; Kusakabe, K.; Morooka, S. Sulfuric acid-modified zirconia membrane for use in hydrogen separation. *Sep. Purif. Technol.* **2001**, *24*, 489–495. [CrossRef]
32. Van Gestel, T.; Velterop, F.; Meulenberg, W.A. Zirconia-supported hybrid organosilica microporous membranes for CO_2 separation and pervaporation. *Sep. Purif. Technol.* **2020**, *259*, 118114. [CrossRef]
33. Ahn, S.J.; Takagaki, A.; Sugawara, T.; Kikuchi, R.; Oyama, S.T. Permeation properties of silica-zirconia composite membranes supported on porous alumina substrates. *J. Membr. Sci.* **2017**, *526*, 409–416. [CrossRef]
34. Hove, M.T.; Luiten-Olieman, M.; Huiskes, C.; Nijmeijer, A.; Winnubst, L. Hydrothermal stability of silica, hybrid silica and Zr-doped hybrid silica membranes. *Purif. Technol.* **2017**, *189*, 48–53. [CrossRef]
35. Goswami, K.P.; Pugazhenthi, G. Effect of binder concentration on properties of low-cost fly ash-based tubular ceramic membrane and its application in separation of glycerol from biodiesel. *J. Clean. Prod.* **2021**, *319*, 128679. [CrossRef]
36. Pan, G.S.; Gu, Z.H.; Yan, Z.; Li, T.; Hua, G.; Yan, L. Preparation of silane modified SiO_2 abrasive particles and their Chemical Mechanical Polishing (CMP) performances. *Wear* **2011**, *273*, 100–104. [CrossRef]
37. Azmiyawati, C.; Niami, S.S.; Darmawan, A. Synthesis of silica gel from glass waste for adsorption of Mg^{2+}, Cu^{2+}, and Ag^+ metal ions. *IOP Conf. Ser. Mater. Sci. Eng.* **2019**, *509*, 012028. [CrossRef]
38. Sumanjit; Rani, S.; Mahajan, R.K. Equilibrium, kinetics and thermodynamic parameters for adsorptive removal of dye Basic Blue 9 by ground nut shells and Eichhornia. *Arab. J. Chem.* **2016**, *9*, S1464–S1477. [CrossRef]
39. Xiong, R.; Li, X.; Ji, S.; Sun, X.; He, J. Thermal stability of ZrO_2–SiO_2 aerogel modified by Fe(III) ion. *J. Sol-Gel Sci. Technol.* **2014**, *72*, 496–501. [CrossRef]
40. Del Monte, F.; Larsen, W.; Mackenzie, J.D. Chemical Interactions Promoting the ZrO_2 Tetragonal Stabilization in ZrO_2–SiO_2 Binary Oxides. *J. Am. Ceram. Soc.* **2010**, *83*, 1506–1512. [CrossRef]

41. Wang, W.; Zhou, J.; Wei, D.; Wan, H.; Zheng, S.; Xu, Z.; Zhu, D. ZrO_2-functionalized magnetic mesoporous SiO_2 as effective phosphate adsorbent. *J. Colloid Interface Sci.* **2013**, *407*, 442–449. [CrossRef] [PubMed]
42. Musić, S.; Filipović-Vinceković, N.; Sekovanić, L. Precipitation of amorphous SiO_2 particles and their properties. *Braz. J. Chem. Eng.* **2011**, *28*, 89–94. [CrossRef]
43. Zheng, W.; Bowen, K.H.; Li, J.; Dabkowska, I.; Gutowski, M. Electronic structure differences in ZrO_2 vs. HfO_2. *J. Phys. Chem. A* **2005**, *109*, 11521–11525. [CrossRef] [PubMed]
44. Kazunari, K.; Jyunichi, I.; Hideaki, M.; Satoshi, F. Evaluation of hydrogen permeation rate through zirconium pipe. *Nucl. Mater. Energy* **2018**, *16*, 12–18.
45. Lin, R.B.; Xiang, S.; Xing, H.; Zhou, W.; Chen, B. Exploration of porous metal–organic frameworks for gas separation and purification. *Coord. Chem. Rev.* **2017**, *378*, 87–103. [CrossRef]
46. Hong, Q.; Chen, H.; Li, L.; Zhu, G.; Xu, N. Effect of Nb content on hydrothermal stability of a novel ethylene-bridged silsesquioxane molecular sieving membrane for H_2/CO_2 separation. *J. Membr. Sci.* **2012**, *S421–S422*, 190–200.
47. Boffa, V.; Blank, D.; Elshof, J. Hydrothermal stability of microporous silica and niobia–silica membranes. *J. Membr. Sci.* **2008**, *319*, 256–263. [CrossRef]
48. Liu, L.; Wang, D.K.; Martens, D.L.; Smart, S.; da Costa, J.C.D. Binary gas mixture and hydrothermal stability investigation of cobalt silica membranes. *J. Membr. Sci.* **2015**, *493*, 470–477. [CrossRef]
49. Qureshi, H.F.; Nijmeijer, A.; Winnubst, L. Influence of sol–gel process parameters on the micro-structure and performance of hybrid silica membranes. *J. Membr. Sci.* **2013**, *446*, 19–25. [CrossRef]
50. He, D.; Zhang, H.; Ren, Y.; Qi, H. Fabrication of a novel microporous membrane based on ZIF-7 doped 1,2-bis(triethoxysilyl)ethane for H_2/CO_2 separation. *Microporous Mesoporous Mater.* **2022**, *331*, 111674. [CrossRef]
51. Song, H.; Zhao, S.; Lei, J.; Wang, C.; Qi, H. Pd-doped organosilica membrane with enhanced gas permeability and hydrothermal stability for gas separation. *J. Mater. Sci.* **2016**, *51*, 6275–6286. [CrossRef]
52. Beyler, A.P.; Boye, D.M.; Hoffman, K.R.; Silversmith, A.J. Fluorescence enhancement in rare earth doped sol-gel glass by N,N dimethylformamide as a drying control chemical additive. *Phys. Procedia* **2011**, *13*, 4–8. [CrossRef]
53. García, M.G.; Marchese, J.; Ochoa, N.A. Aliphatic–aromatic polyimide blends for H_2 separation. *Int. J. Hydrogen Energy* **2010**, *35*, 8983–8992. [CrossRef]
54. Nikolaeva, D.; Azcune, I.; Tanczyk, M.; Warmuzinski, K.; Jaschik, M.; Sandru, M.; Dahl, P.I.; Genua, A.; Lois, S.; Sheridan, E.; et al. The performance of affordable and stable cellulose-based poly-ionic membranes in CO_2/N_2 and CO_2/CH_4 gas separation. *J. Membr. Sci.* **2018**, *564*, 552–561. [CrossRef]
55. Chen, F.; Ji, Z.; Qi, Q. Effect of liquid surface tension on the filtration performance of coalescing filters. *Sep. Purif. Technol.* **2019**, *209*, 881–891. [CrossRef]

Article

Maxwell Nanofluids: FEM Simulation of the Effects of Suction/Injection on the Dynamics of Rotatory Fluid Subjected to Bioconvection, Lorentz, and Coriolis Forces

Liaqat Ali [1,*], Abdul Manan [2] and Bagh Ali [3]

1 School of Sciences, Xi'an Technological University, Xi'an 710021, China
2 Department of Physics and Mathematics, Faculty of Sciences, Superior University, Lahore 54000, Pakistan
3 Department of Computer Science and Information Technology, Faculty of Sciences, Superior University, Lahore 54000, Pakistan
* Correspondence: liaqat@xatu.edu.cn

Abstract: In this study, the relevance of Lorentz and Coriolis forces on the kinetics of gyratory Maxwell nanofluids flowing against a continually stretched surface is discussed. Gyrotactic microbes are incorporated to prevent the bioconvection of small particles and to improve consistency. The nanoparticles are considered due to their valuable properties and ability to enhance thermal dissipation, which is important in heating systems, advanced technology, microelectronics, and other areas. The main objective of the analysis is to enhance the rate of heat transfer. An adequate similarity transformation is used to convert the primary partial differential equations into non-linear dimensionless ordinary differential equations. The resulting system of equations is solved using the finite element method (FEM). The increasing effects of the Lorentz and Coriolis forces induce the velocities to moderate, whereas the concentration and temperature profiles exhibit the contrary tendency. It is observed that the size and thickness of the fluid layers in the axial position increase as the time factor increases, while the viscidity of the momentum fluid layers in the transverse path decreases as the time factor decreases. The intensity, temperature, and velocity variances for the suction scenario are more prominent than those for the injection scenario, but there is an opposite pattern for the physical quantities. The research findings are of value in areas such as elastomers, mineral productivity, paper-making, biosensors, and biofuels.

Keywords: Maxwell nanofluid; finite element analysis; suction/injection; grid independence analysis; Coriolis force

Citation: Ali, L.; Manan, A.; Ali, B. Maxwell Nanofluids: FEM Simulation of the Effects of Suction/ Injection on the Dynamics of Rotatory Fluid Subjected to Bioconvection, Lorentz, and Coriolis Forces. *Nanomaterials* 2022, *12*, 3453. https://doi.org/10.3390/ nano12193453

Academic Editor: Henrich Frielinghaus

Received: 27 August 2022
Accepted: 20 September 2022
Published: 2 October 2022

Publisher's Note: MDPI stays neutral with regard to jurisdictional claims in published maps and institutional affiliations.

Copyright: © 2022 by the authors. Licensee MDPI, Basel, Switzerland. This article is an open access article distributed under the terms and conditions of the Creative Commons Attribution (CC BY) license (https:// creativecommons.org/licenses/by/ 4.0/).

1. Introduction

The heat and mass transfer analysis of the non-Newtonian hydrodynamic boundary layer flow phenomenon has attracted the interest of many researchers due to the enormous number of potential applications in engineering and industry. The well-known Newtonian liquids (liquids with a sequential strain-stress correlation) basic theory is incapable of elucidating the fluids' internal microstructure. Non-Newtonian liquids (liquids with a sequential strain-stress correlation) include quince paste, animal blood, cement sludges, esoteric lubricating oils, effluent slurry, and liquids containing synthetic polymer additives. One such rate-type non-Newtonian fluid model is called the Maxwell nanofluid model which predicts the stress relaxation time. The extensive choice of methodological and engineering applications associated with Maxwell nanofluids, such as biochemical, gasoline, polymer, and nutrition release, has motivated many investigators to scrutinize the features of Maxwell nanofluids with respect to numerous geometrical and substantial limitations. The convective Maxwell hybrid nanofluid stream in a sturdy channel was studied using the Laplace transform strategy [1], whereby the authors developed a solution to dynamical problems involving Maxwell fluid fractionally. The Caputo fractional differential function was used for the energy dissipation assessment of hydromagnetic Maxwell nanofluid flow over an elongating penetrable surface with Dufour and Soret ramifications. Jawad et al. [2] used HAM to procure estimated analytical results. Jamshed [3] exploited the Keller box technique (KBT) to investigate the fluidity of an mhd Maxwell nanofluid over a non-linearly elongating sheet in terms of viscous dissipation and entropy propagation. Ali et al. [4] investigated buoyant, induced transitory bio-convective Maxwell nanoliquid spinning three-dimensional flows over the Riga surface for chemically reactive and activating energy using a finite element stratagem. Dulal et al. [5] presented results of a study on mhd radiative heat transfer of nanofluids induced by a plate through a porous medium with chemical reaction. Very recently, various authors have explored boundary layer Maxwell nanofluid flow past a different geometric environment. These include Ahmed et al. [6], who reported on mixed convective 3D flow over a vertical stretching cylinder with a shooting technique, Gopinath et al. [7] who explored convective-radiative boundary layer flow of nanofluids with viscous-Ohmic dissipation, Bilal et al. [8] who presented the significance of the Coriolis force on the dynamics of the Carreau–Yasuda rotating nanofluid subject to gyrotactic microorganisms, Ahmed [9] who investigated the effect of a heat source on the stagnation point fluid flow via an elongating revolving plate using a numerical approach, M. Bilal [10] who used the HAM technique to investigate chemically reactive impacts on magnetised nanofluid flow over a rotary pinecone, Amirsom et al. [11] who estimated the influence of bioconvection on three-dimensional nanofluid flow induced by a bi-axial stretching sheet, Prabhavathi et al. [12] who used FEM to investigate CNT nanofluid flow through a cone with thermal slip scenarios, Zohra et al. [13] who used mhd micropolar fluid bio-nanoconvective Naiver slip flow in a stretchable horizontal channel, and Gopinath et al. [14] who reported on diffusive mhd nanofluid flow past a non-linear stretching/shrinking sheet with viscous-Ohmic dissipation and thermal radiation.

Nanofluids are fluids that incorporate an appropriate distribution of metal and metallic nanoparticles at the nano size and are engineered to perform specific functions [15–18]. The literature suggests that the presence of nanoparticles in a base fluid has a significant impact on the thermophysical properties of the fluid, particularly those fluids with inadequate permittivity characteristics based on theoretical and experimental investigations [19–21]. Applications in virtually every field of engineering and science relating to convective nanofluid heat transfer flow have stimulated the interest of many scientists and engineers. These include the use of diamond and silica nanoparticles to enhance the electrical characteristics of lubricants, the use of liquids containing nanoparticles to absorb sunlight in solar panels, and exploitation of the antimicrobial properties of zinc and titanium oxide particles for biomedical engineering applications, such as drug delivery and pharmacological treatment [22–25].

The bioconvection phenomenon occurs as a result of the existence of a density gradient in the flow field. Consequently, the movement of particles at the macroscopic level enhances the density stratification of the base liquid in one direction. The presence of gyrotactic microorganisms in nanofluid flow has attracted the interest of many researchers due to their potential application in relation to enzyme function, bio-sensors, biotechnology, drug delivery, and biofuels. These applications have motivated researchers to undertake numerical studies on bioconvective nanofluid flow with microorganisms in different flow field geometries. Chu et al. [26] investigated bioconvective Maxwell nanoliquid flow using a reversible, regularly pivoting sheet in the presence of non-linear radiative and heat emitter influences using a homotopy analysis method. Sreedevi et al. [27] investigated the influence of Brownian motion and thermophoresis on Maxwell three-dimensional nanofluid flow over a stretching sheet with thermal radiation. Rao et al. [28] explored bioconvection in conventional reactive nanoliquid flow over a vertical cone with gyrotactic microorganisms embedded in a permeable medium. Awais Ali et al. [29], using an Adams–Bash strategy (ABS), statistically explored the Lie group, to investigate bio-convective nanoliquid supporting and opposing flow with motile microorganisms. To determine the Arrhenius activation energy of bio-convective nanoliquid flow through a stretchable surface, Paluru [30] undertook a heat and mass transfer analysis of MWCNT-kerosene nanofluid flow over a wedge with thermal radiation. Transient bio-convective Carreau nanofluid flow with gyrotactic microorganisms past a horizontal slender stretching sheet was considered by Elayarani et al. [31] to investigate heat and mass transfer effects in the presence of thermal radiation, multi-slip conditions, and magnetic fields by employing the ANFIS (adaptive neuro-fuzzy inference system) model. Bagh et al. [32] reported on the g-jitter impact on magnetohydrodynamic non-Newtonian fluid over an inclined surface by applying a finite element simulation. Umar et al. [33] investigated the optimized Cattaneo–Christov heat and mass transference flow of bio-convective Carreau nanofluid with microorganisms, influenced by a longitudinal straining cartridge with convective limitations. Al-Hussain [34] developed an analytical model based on the Cattaneo–Christov transit law for a bio-convective magnetic nanofluid stream via a whirling cone immersed in an asymmetric penetrable surface in the context of cross-diffusion, Navier-slip, and Stefan blowing effects.

The careful review of the literature detailed above shows that little attention has been paid to the self-motile denitrifying microbes contained in Maxwell nanofluid spinning flows through an elongating sheet under an externally applied magnetic field. To the best of our knowledge, none of the studies cited has considered the interpreted problem. Consequently, the main objective of this study was to explore the mass and heat transfer impacts on transitory hydromagnetic Maxwell spinning nanofluid 3D radiative flow comprising microbes and suction/injection processes. Many authors [35–37] have examined mhd nanofluid flow using different numerical techniques. Here, the flow-governing associated non-linear PDEs are computed using a finite volume technique [38,39] by adopting a weighted residual strategy. The varied flow field properties for a variety of substantial factors are explained and illustrated graphically. The computing results generated using Matlab source code were validated against previous studies and determined to show acceptable consistency. The values of the friction factor, Nusselt, and Sherwood numbers are simulated and addressed in tabulated form. The computational evaluation can be used for gasoline, polymers, precise nutrition release, engine lubricants, paint rheology, biosensors, medicine delivery, and biofuels.

Research Queries

The following research questions are addressed in this study:

1. What effects do relaxation of the Deborah number, the Coriolis effect, and an applied magnetic field force have on the hydrodynamics of heat flux, fluid viscosity, and concentration level variances using injection/suction?
2. What impact do Brownian motion and thermophoresis have on heat and mass transfer rates and the skin friction factor for suction/injection?
3. How do Brownian motion, the relaxation Deborah number, and time-dependent factors impact on the temperature profile?
4. What is the bioconvection impact on the motile dispersal function with suction/injection?

2. Mathematical Geometry

The transitory magneto-hydrodynamic 3D rotational flow of Maxwell nanofluid over a bidirectional elongating surface is investigated. Figure 1 depicts the fluid dynamic layout and coordinate structure of the articulated problem, with the flow, constrained to $z \geq 0$. With a rotational consistent velocity Ω, the nanofluid flow rotates around the z-axis. When $z = t = 0.0$, the sheet is extended along the x-axis having $u_w = \tilde{a}x$ velocity. In the axial direction, a static and uniform magnetic field of magnitude B_0 is implemented. An induced magnetic Reynolds number leads to a reduced magnetic field, which results in minimal Hall current and Ohmic inefficiency [40]. To avoid causing sedimentation, gyrotactic microbes are utilized to maintain convectional stability. The external temperature and intensity are signified by T_∞, and C_∞, N_∞, respectively, while the temperature and intensity at the surface are represented by T_w, and C_w, N_w, respectively. For the current elaborated problem, $V = (u_1(x,y,z), u_2(x,y,z), u_3(x,y,z))$ is assumed to be the velocity field. The equations of mass conservation, linear moments, temperature, and concentrations are formulated as a result of the preceding assertions [41–43]:

$$\partial_x u_1 + \partial_y u_2 + \partial_z u_3 = 0 \tag{1}$$

$$\rho_{nf}(\partial_t u_1 + u_1 \partial_x u_1 + u_2 \partial_y u_1 + u_3 \partial_z u_1 - 2\Omega u_2 + \lambda_1 \varrho_{u_1}) = -\partial_x p + \mu_{nf} \partial_{zz} u_1 - \sigma_{nf} B_0^2 u_1 \tag{2}$$

$$\rho_{nf}(\partial_t u_2 + u_1 \partial_x u_2 + u_2 \partial_y u_2 + u_3 \partial_z u_2 - 2\Omega u_1 + \lambda_1 \varrho_{u_2}) = -\partial_y p + \mu_{nf} \partial_{zz} u_2 - \sigma_{nf} B_0^2 u_2 \tag{3}$$

$$\rho_{nf}(\partial_t u_3 + u_1 \partial_x u_3 + u_2 \partial_y u_3 + u_3 \partial_z u_3) = -\partial_z p + \mu_{nf} \partial_{zz} u_3 \tag{4}$$

$$\partial_t T + u_1 \partial_x T + u_2 \partial_y T + u_3 \partial_z T = \alpha_{nf} \partial_{zz} T + \tau^* \{ D_b \partial_z C \partial_z T + \frac{D_T}{T_\infty}(\partial_z T)^2 \} \tag{5}$$

$$\partial_t C + u_1 \partial_x C + u_2 \partial_y C + u_3 \partial_z C = D_b \partial_{zz} C + \frac{D_T}{T_\infty} \partial_{zz} T \tag{6}$$

$$\partial_t N + u_1 \partial_x N + u_2 \partial_y N + u_3 \partial_z N + \frac{bWc}{(C_w - C_\infty)} [\partial_z (N \partial_z C)] = D_m \partial_{zz} N \tag{7}$$

where,

$$\varrho_{u_1} = \Big\{ u_1^2 \partial_{xx} u_1 + u_2^2 \partial_{yy} u_1 + u_3^2 \partial_{zz} u_1 + 2u_1 u_2 \partial_{xy} u_1 + 2u_2 u_3 \partial_{yz} u_1 + 2u_1 u_3 \partial_{xz} u_1$$
$$- 2\Omega(u_1 \partial_x u_2 + u_2 \partial_y u_2 + u_3 \partial_z u_2) + 2\Omega(u_2 \partial_x u_1 - u_1 \partial_y u_1) \Big\},$$

$$\varrho_{u_2} = \Big\{ u_1^2 \partial_{xx} u_2 + u_2^2 \partial_{yy} u_2 + u_3^2 \partial_{zz} u_2 + 2u_1 u_2 \partial_{xy} u_2 + 2u_2 u_3 \partial_{yz} u_2 + 2u_1 u_3 \partial_{xz} u_2$$
$$- 2\Omega(u_1 \partial_x u_1 + u_2 \partial_y u_1 + u_3 \partial_z u_1) + 2\Omega(u_2 \partial_x u_2 - u_1 \partial_y u_2) \Big\}$$

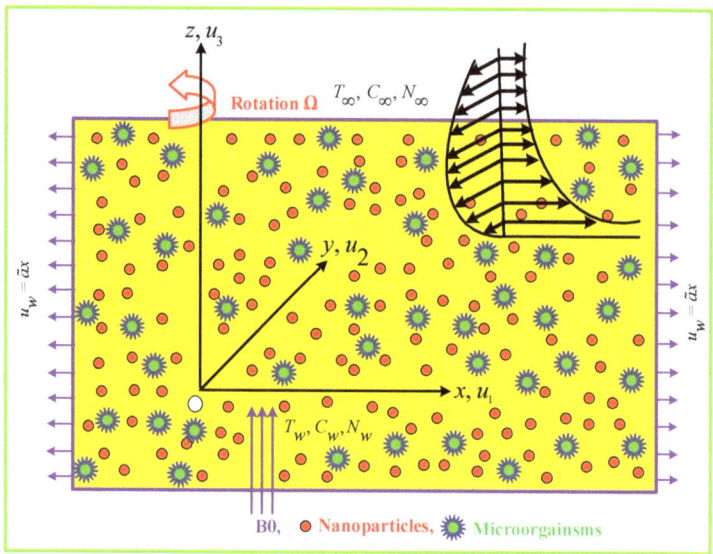

Figure 1. Flow Sketch.

Here, (C, N, T) represent the nanoparticle density, micro-organism concentration, and fluid temperature, (D_b, D_T, D_m) are the Brownian motion, thermophoresis and microorganism diffusion; $(\lambda_1, \rho_{n_f}, \mu_{n_f}, \alpha_{n_f})$ are, respectively, the relaxation time, density, dynamic viscosity, and thermal diffusivity of the nanofluid. The boundary conditions are [44,45]:

$$t < 0 : u_1 = 0, \ u_2 = 0, \ u_3 = 0, \ C = (C_\infty), \ N = (N_\infty), \ T = (T_\infty) \tag{8}$$

$$t \geq 0 : u_1 = a(x), \ u_3 = -w_0, \ u_2 = 0, \ C = (C_w), \ N = (N_w), \ T = (T_w), \text{ when } z = 0 \tag{9}$$

$$t \geq 0 : u_1 \to 0, \ u_2 \to 0, \ C \to C_\infty, \ N \to N_\infty, \ T \to T_\infty, \text{ when } z \to \infty. \tag{10}$$

The following similarity transforms are used to alleviate the complexity of the articulated problem as [41,44]:

$$\left.\begin{array}{l}\Gamma = \tilde{a}t, \ u_1 = \tilde{a}x\dfrac{\partial F(\zeta,\eta)}{\partial \eta}, \ u_2 = \tilde{a}xG(\zeta,\eta), \ u_3 = -\sqrt{\tilde{a}\nu\zeta}F(\zeta,\eta), \ \zeta = 1 - e^{-\Gamma}, \eta = \sqrt{\dfrac{\tilde{a}xz^2}{\zeta\nu}} \\ C = (C_w - C_\infty)\Phi(\zeta,\eta) + C_\infty, \ N = (N_w - N_\infty)\chi(\zeta,\eta) + N_\infty, \ T = (T_w - T_\infty)\theta(\zeta,\eta) + T_\infty\end{array}\right\} \tag{11}$$

In the context of Equation (11), Equation (1) is justified, and Equations (2)–(10) are transmuted into the non-linear PDEs illustrated below in (ζ, η) form:

$$F''' + \dfrac{\eta}{2}F'' - \dfrac{\zeta\eta}{2}F'' + \zeta\{FF'' - F'^2 - M^2F' + 2\lambda G + \beta\varsigma_{u_1}\} - \zeta(1-\zeta)\dfrac{\partial F'}{\partial \zeta} = 0 \tag{12}$$

$$G'' + \dfrac{\eta}{2}G' - \dfrac{\zeta\eta}{2}G' + \zeta\{FG' - 2\lambda F' - M^2G - F'G + \beta\varsigma_{u_2}\} - \zeta(1-\zeta)\dfrac{\partial G}{\partial \zeta} = 0 \tag{13}$$

$$\theta'' - \dfrac{\eta}{2}(\zeta - 1)Pr\theta' + \zeta Pr F\theta' + N_b Pr\theta\Phi + N_t Pr\theta'^2 - \zeta(1-\zeta)Pr\dfrac{\partial \theta}{\partial \zeta} = 0 \tag{14}$$

$$\Phi'' + 0.5\eta Le(1-\zeta)\Phi' + Le\zeta F\Phi' + N_t N_b^{-1}\theta'' - \zeta(1-\zeta)Le\dfrac{\partial \Phi}{\partial \zeta} = 0 \tag{15}$$

$$\chi'' + \dfrac{Lb}{2}(1-\zeta)Lb\chi' + \zeta LbF\chi' - Pe\Phi''(\delta_1 + \chi) + Pe\chi'\Phi' = Lb\zeta(1-\zeta)\dfrac{\partial \chi}{\partial \zeta} \tag{16}$$

$$\left.\begin{array}{l}F(\zeta,\eta) = \Gamma, \ F'(\zeta,\eta) = \theta(\zeta,\eta) = 1, \ G(\zeta,\eta) = 0, \Phi(\zeta,\eta) = \chi(\zeta,\eta) = 1, \ \zeta \geq 0, \text{ when } \eta = 0 \\ F'(\zeta,\eta) \to 0, \ \theta(\zeta,\eta) \to 0, \ G(\zeta,\eta) \to 0, \ \Phi(\zeta,\eta) \to 0, \ \chi(\zeta,\eta) \to 0, \ \zeta \geq 0, \text{ when } \eta \to \infty\end{array}\right\} \tag{17}$$

where $\varsigma_{u_1} = 2FF'F'' - F^2F''' - 2\lambda FG'$, $\varsigma_{u_2} = 2FF'G' - F^2G'' - 2\lambda F'^2 + 2\lambda FF'' - 2\lambda G^2$, and primes $(', '', ''')$ denote the derivatives w.r.t (η). Here rotation, magnetized, Prandtl and Lewis numbers, Brownian factor, bioconvection Lewis number, thermophoresis, Peclet number, relaxation Deborah number, microorganism concentration difference, and suction/injection are, respectively, λ, M, Pr, Le, N_b, Lb, N_t, Pe, β, δ_1, and Γ factors are described as:

$$\lambda = \dfrac{\Omega}{a}, \ M = \sqrt{\dfrac{\sigma_{n_f}B_0^2}{\rho_f\tilde{a}}}, \ Pr = \dfrac{\nu}{\alpha_{n_f}}, \ Le = \dfrac{\nu}{D_B}, \ Lb = \dfrac{\nu}{D_m}, \ N_b = \tau\nu^{-1}D_B(C_w - C_\infty),$$

$$N_t = \dfrac{D_T(\tau T_w - \tau T_\infty)}{\nu T_\infty}, \ \beta = \lambda_1 a, \ Pe = \dfrac{bW_c}{D_m}, \ \delta_1 = \dfrac{N_\infty}{N_w - T_\infty}, \ \Gamma = \dfrac{w_0}{\sqrt{\tilde{a}\nu\zeta}}.$$

The physical quantities (Sherwood, Nusselt) numbers, and the coefficient of skin friction are expressed here as:

$$Nu_x = \dfrac{xq_w}{\kappa(T_w - T_\infty)}, \ Shr = \dfrac{xq_m}{D_B(C_w - C_\infty)}, \ C_{fx} = \dfrac{\tau_w^x}{\rho u_1^2}, \ C_{fy} = \dfrac{\tau_w^y}{\rho u_1^2}. \tag{18}$$

here, the skin friction tensors at the wall are represented as $\tau_w^x = \mu \left(\frac{\partial u_1}{\partial z}\right)_{z=0}$ (along the x-axis) and $\tau_w^y = \mu \left(\frac{\partial u_2}{\partial z}\right)_{z=0}$ (along the y-axis), the heat flux, and the mass at the surface is $qm = -D_B \left(\frac{\partial C}{\partial z}\right)_{z=0}$, and $qw = -\kappa \left(\frac{\partial T}{\partial z}\right)_{z=0}$. Taking Equation (11), we get:

$$\begin{cases} C_{f_x} Re_x^{1/2} = \frac{F''(0)}{\sqrt{\zeta}}, C_{f_y} Re_x^{1/2} = \frac{G'(0)}{\sqrt{\zeta}}, \\ Nu_x Re_x^{1/2} = -\frac{[\theta'(0)]}{\sqrt{\zeta}}, Shr_x Re_x^{1/2} = -\frac{[\Phi'(0)]}{\sqrt{\zeta}}. \end{cases} \quad (19)$$

3. Computational Procedure

The finite element analysis (FEA) is a computational approach for discovering numerical approximations to ODEs and PDEs with complicated boundary conditions. This is an efficient approach for resolving technological problems, especially those involving fluid diversities. This methodology represents an excellent numerical strategy for solving a variety of real-world problems, particularly for heat transfer via fluids and biomaterials [46]. Reddy [47,48] presents a layout of the Galerkin finite element methodology (GFEM), summarizing the main elements of this methodology. This methodology is an unparalleled computational methodology in the field of engineering, is valuable for evaluating integral governing equations incorporating fluid diversities, and is an extremely effective methodology for resolving numerous non-linear problems [49–51]. To evaluate the set of Equations (12)–(16) along the boundary condition (17), firstly, we assume:

$$F' = P \quad (20)$$

The system of Equations (12)–(17) reduced as:

$$P'' - \frac{\eta}{2}(\zeta - 1)P' + \zeta(FP' - P^2 + 2\lambda G - M^2 P + \beta(2FPP' - F^2 P'' - 2\beta FG')) = \zeta \frac{\partial P}{\partial \zeta} - \zeta^2 \frac{\partial P}{\partial \zeta} \quad (21)$$

$$G'' + \frac{1}{2}(1-\zeta)\eta G' + \zeta(FG' + \beta(2FPG' - F^2 G'' - 2\lambda P^2 + 2\beta FP' - 2\beta G^2))$$
$$-PG - 2\beta P = -\zeta^2 \frac{\partial(G)}{\partial \zeta} + \zeta \frac{\partial(G)}{\partial \zeta} \quad (22)$$

$$\theta'' - \frac{\eta}{2}(\zeta-1)Pr\theta' + Pr\zeta F\theta' + Pr\theta'(N_b \Phi' + N_t \theta') = Pr\zeta(1-\zeta)\frac{\partial \theta}{\partial \zeta} \quad (23)$$

$$\Phi'' + \frac{\eta}{2}Le(1-\zeta)\Phi' + Le\zeta F\Phi' + N_t N_b^{-1}\theta''^2 = \zeta(1-\zeta)Le\frac{\partial \Phi}{\partial \zeta} \quad (24)$$

$$\chi'' + \frac{Lb}{2}(1-\zeta)\eta\chi' + \zeta LbF\chi' - Pe\Phi''(\delta_1 + \chi) + Pe\chi'\Phi' = Lb\zeta(1-\zeta)\frac{\partial \chi}{\partial \zeta} \quad (25)$$

$$\left.\begin{aligned} F(\zeta,\eta) = \Gamma, \ G(\zeta,\eta) = 0, \ P(\zeta,0) = \theta(\zeta,\eta) = \Phi(\zeta,\eta) = \chi(\zeta,\eta) = 1, \ \zeta \geq = 0, \ at \ \eta = 0 \\ P(\zeta,\eta) \to 0, \ G(\zeta,\eta) \to 0, \ \theta(\zeta,\eta) \to 0, \ \Phi(\zeta,\eta) \to 0, \ \chi(\zeta,\eta) \to 0, \ \zeta \geq 0, \ as \ \eta \to \infty \end{aligned}\right\} \quad (26)$$

For numerical calculation, the plate length has been specified as $\zeta = 1.0$ and the thickness as $\eta = 5.0$. The Equations (20)–(25) have a variational form that can be represented as:

$$\int_{\Omega_e} w_{f_1}\{F' - P\}d\Omega_e = 0 \tag{27}$$

$$\int_{\Omega_e} w_{f_2}\left\{P'' + \frac{1}{2}(1-\zeta)\eta P' + \zeta(FP' - P^2 + 2\lambda P - M^2 P + \beta(2FPP' - F^2 P'' - 2\lambda FG'))\right.$$

$$\left. -\zeta(1-\zeta)\frac{\partial P}{\partial \zeta}\right\}d\Omega_e = 0 \tag{28}$$

$$\int_{\Omega_e} w_{f_3}\left\{G'' + \frac{1}{2}(1-\zeta)\eta G' + \zeta(FG' + \beta(2FPG' - F^2 G'' - 2\lambda P^2 + 2\lambda FP' - 2\lambda G^2))\right.$$

$$\left. -PG - 2\lambda P + (\zeta - 1)\zeta\frac{\partial(G)}{\partial \zeta}\right\}d\Omega_e = 0 \tag{29}$$

$$\int_{\Omega_e} w_{f_4}\left\{\theta'' + \frac{Pr}{2}(1-\zeta)\eta\theta' + Pr\zeta F\theta' + N_b Pr\theta'\Phi' + N_t Pr(\theta')^2 - Pr\zeta(1-\zeta)\frac{\partial \theta}{\partial \zeta}\right\}d\Omega_e = 0 \tag{30}$$

$$\int_{\Omega_e} w_{f_5}\left\{\Phi'' - \frac{\eta}{2}Le(\zeta-1)\Phi' + Le(\zeta F\Phi' + \frac{N_t}{LeN_b}(\theta'')^2 + (\zeta-1)\zeta\frac{\partial \Phi}{\partial \zeta})\right\}d\Omega_e = 0 \tag{31}$$

$$\int_{\Omega_e} w_{f_6}\left\{\chi'' + \frac{Lb}{2}(1-\zeta)\eta\chi' + \zeta LbF\chi' - Pe\left(\Phi''(\delta_1 + \chi) + \chi'\Phi'\right) - \zeta(1-\zeta)Lb\frac{\partial \chi}{\partial \zeta}\right\}d\Omega_e = 0. \tag{32}$$

Here $w_{f_s}(s = 1, 2, 3, 4, 5, 6)$ stand for trial functions. Let the domain (Ω_e) be divided into 4−nodded elements. The associated approximations of the finite element are:

$$F = \sum_{j=1}^{t}[\dot{F}_j Y_j(\zeta,\eta)], \ P = \sum_{j=1}^{t}[\dot{P}_j Y_j(\zeta,\eta)], \ G = \sum_{j=1}^{t}[\dot{G}_j Y_j(\zeta,\eta)], \ \theta = \sum_{j=1}^{t}[\dot{\theta}_j Y_j(\zeta,\eta)], \ \Phi = \sum_{j=1}^{t}[\dot{\Phi}_j Y_j(\zeta,\eta)]. \tag{33}$$

here, Y_j (j = 1,2,3,4) and $t = 4$ For Ω_e, the linear-interpolation key functions are defined as follows:.

$$\left.\begin{array}{l} Y_1 = \dfrac{(\zeta_{e+1} - \zeta)(\eta_{e+1} - \eta)}{(\zeta_{e+1} - \zeta_e)(\eta_{e+1} - \eta_e)}, \ Y_2 = \dfrac{(\zeta - \zeta_e)(\eta_{e+1} - \eta)}{(\zeta_{e+1} - \zeta_e)(\eta_{e+1} - \eta_e)} \\ Y_3 = \dfrac{(\zeta - \zeta_e)(\eta - \eta_e)}{(\zeta_{e+1} - \zeta_e)(\eta_{e+1} - \eta_e)}, \ Y_4 = \dfrac{(\zeta_{e+1} - \zeta)(\eta - \eta_e)}{(\zeta_{e+1} - \zeta_e)(\eta_{e+1} - \eta_e)} \end{array}\right\} \tag{34}$$

Therefore, the stiffness element matrix, matrix of unknowns and the force vector/matrix for the finite element model are followed as:

$$\begin{bmatrix} [L^{11}] & [L^{12}] & [L^{13}] & [L^{14}] & [L^{15}] & [L^{16}] \\ [L^{21}] & [L^{22}] & [L^{23}] & [L^{24}] & [L^{25}] & [L^{26}] \\ [L^{31}] & [L^{32}] & [L^{33}] & [L^{34}] & [L^{35}] & [L^{36}] \\ [L^{41}] & [L^{42}] & [L^{43}] & [L^{44}] & [L^{45}] & [L^{46}] \\ [L^{51}] & [L^{52}] & [L^{53}] & [L^{54}] & [L^{55}] & [L^{56}] \\ [L^{61}] & [L^{62}] & [L^{63}] & [L^{64}] & [L^{65}] & [L^{66}] \end{bmatrix} \begin{bmatrix} \{F\} \\ \{P\} \\ \{G\} \\ \{\theta\} \\ \{\Phi\} \\ \{\chi\} \end{bmatrix} = \begin{bmatrix} \{R_1\} \\ \{R_2\} \\ \{R_3\} \\ \{R_4\} \\ \{R_5\} \\ \{R_6\} \end{bmatrix} \tag{35}$$

where $[L_{mn}]$ and $[R_m]$ $(m, n = 1, 2, 3, 4, 5, 6)$ are expressed as:

$$L_{ij}^{11} = \int_{\Omega_e} Y_i \frac{dY_j}{d\eta} d\Omega_e, L_{ij}^{12} = -\int_{\Omega_e} Y_i Y_j d\Omega_e, L_{ij}^{13} = L_{ij}^{14} = L_{ij}^{15} = L_{ij}^{21} = L_{ij}^{24} = L_{ij}^{25} = L_{ij}^{26} = 0,$$

$$L_{ij}^{22} = -\int_{\Omega_e} \frac{dY_i}{d\eta} \frac{dY_j}{d\eta} d\Omega_e + \frac{1}{2}(1-\zeta)\eta \int_{\Omega_e} Y_i \frac{dY_j}{d\eta} d\Omega_e + \zeta \int_{\Omega_e} \bar{F} Y_i \frac{dY_j}{d\eta} d\Omega_e - \zeta \int_{\Omega_e} \bar{P} Y_i Y_j d\Omega_e - \zeta(1-\zeta) \int_{\Omega_e} Y_i \frac{dY_j}{d\zeta} d\Omega_e$$

$$+ 2\beta\zeta \int_{\Omega_e} \bar{F}\bar{P} Y_i \frac{dY_j}{d\eta} d\Omega_e + \beta\zeta \int_{\Omega_e} \bar{F}^2 \frac{dY_i}{d\eta} \frac{dY_j}{d\eta} d\Omega_e - M^2\zeta \int_{\Omega_e} Y_i Y_j d\Omega_e, L_{ij}^{23} = 2\lambda\zeta \int_{\Omega_e} Y_i Y_j d\Omega_e$$

$$- 2\beta\lambda \int_{\Omega_e} \zeta \bar{F} Y_i \frac{dY_j}{d\eta} d\Omega_e, L_{ij}^{31} = L_{ij}^{34} = L_{ij}^{35} = L_{ij}^{36} = 0, L_{ij}^{32} = 2\lambda\zeta \int_{\Omega_e} Y_i Y_j d\Omega_e - 2\lambda\beta \int_{\Omega_e} \zeta \bar{F} Y_i \frac{dY_j}{d\eta} d\Omega_e,$$

$$L_{ij}^{33} = -\int_{\Omega_e} \frac{dY_i}{d\eta} \frac{dY_j}{d\eta} d\Omega_e + \frac{1}{2}(1-\zeta)\eta \int_{\Omega_e} Y_i \frac{dY_j}{d\eta} d\Omega_e + \zeta \int_{\Omega_e} \bar{F} Y_i \frac{dY_j}{d\eta} d\Omega_e - \zeta \int_{\Omega_e} \bar{P} Y_i Y_j d\Omega_e - \zeta(1-\zeta) \int_{\Omega_e} Y_i \frac{dY_j}{d\zeta} d\Omega_e$$

$$+ 2\beta\zeta \int_{\Omega_e} \bar{F}\bar{P} Y_i \frac{dY_j}{d\eta} d\Omega_e + \beta\zeta \int_{\Omega_e} \bar{F}^2 \frac{dY_i}{d\eta} \frac{dY_j}{d\eta} d\Omega_e - 2\lambda\beta \int_{\Omega_e} \zeta \bar{G} Y_i Y_j d\Omega_e, L_{ij}^{41} = L_{ij}^{42} = L_{ij}^{43} = 0,$$

$$L_{ij}^{44} = -\int_{\Omega_e} \frac{dY_i}{d\eta} \frac{dY_j}{d\eta} d\Omega_e + \frac{Pr}{2}(1-\zeta)\eta \int_{\Omega_e} Y_i \frac{dY_j}{d\eta} d\Omega_e + Pr\zeta \int_{\Omega_e} \bar{F} Y_i \frac{dY_j}{d\eta} d\Omega_e + PrN_b \int_{\Omega_e} \bar{\Phi}' Y_i \frac{dY_j}{d\eta} d\Omega_e$$

$$+ PrN_t \int_{\Omega_e} \bar{\theta}' Y_i \frac{dY_j}{d\eta} d\Omega_e - Pr\zeta(1-\zeta) \int_{\Omega_e} Y_i \frac{dY_j}{d\zeta} d\Omega_e, L_{ij}^{45} = L_{ij}^{46} = L_{ij}^{51} = L_{ij}^{52} = L_{ij}^{53} = L_{ij}^{56} = 0,$$

$$L_{ij}^{54} = -\frac{N_t}{N_b} \int_{\Omega_e} \frac{dY_i}{d\eta} \frac{dY_j}{d\eta} d\Omega_e, L_{ij}^{55} = -\int_{\Omega_e} \frac{dY_i}{d\eta} \frac{dY_j}{d\eta} d\Omega_e + \frac{Le}{2}(1-\zeta)\eta \int_{\Omega_e} Y_i \frac{dY_j}{d\eta} d\Omega_e + Le\zeta \int_{\Omega_e} \bar{F} Y_i \frac{dY_j}{d\eta} d\Omega_e$$

$$- Le\zeta(1-\zeta) \int_{\Omega_e} Y_i \frac{dY_j}{d\zeta} d\Omega_e, L_{ij}^{61} = L_{ij}^{62} = L_{ij}^{63} = L_{ij}^{64} = 0,$$

$$L_{ij}^{65} = -Pe\delta_1 \int_{\Omega_e} \frac{dY_i}{d\eta} \frac{dY_j}{d\eta} d\Omega_e, L_{ij}^{66} = -\int_{\Omega_e} \frac{dY_i}{d\eta} \frac{dY_j}{d\eta} d\Omega_e + \frac{Lb}{2}(1-\zeta)\eta \int_{\Omega_e} Y_i \frac{dY_j}{d\eta} d\Omega_e + Lb\zeta \int_{\Omega_e} \bar{F} Y_i \frac{dY_j}{d\eta} d\Omega_e$$

$$- Pe \int_{\Omega_e} \bar{\Phi}' Y_i \frac{dY_j}{d\eta} d\Omega_e - Pe \int_{\Omega_e} \bar{\Phi}'' Y_i d\varphi_j d\Omega_e - Lb\zeta(1-\zeta) \int_{\Omega_e} Y_i \frac{dY_j}{d\zeta} d\Omega_e,$$

and

$$R_i^1 = \Gamma, \quad R_i^2 = -\oint_{\Gamma_e} Y_i n_\eta \frac{\partial P}{\partial \eta} ds, \quad R_i^3 = -\oint_{\Gamma_e} Y_i n_\eta \frac{\partial G}{\partial \eta} ds, \quad R_i^4 = -\oint_{\Gamma_e} Y_i n_\eta \frac{\partial \theta}{\partial \eta} ds,$$

$$R_i^5 = -\oint_{\Gamma_e} Y_i n_\eta \frac{\partial \Phi}{\partial \eta} ds - \frac{Nt}{Nb} \oint_{\Gamma_e} Y_i n_\eta \frac{\partial \theta}{\partial \eta} ds, \quad R_i^6 = -\oint_{\Gamma_e} Y_i n_\eta \frac{\partial \chi}{\partial \eta} ds. \tag{36}$$

Here, $\bar{F} = \sum_{j=1}^{t}(\bar{F}_j Y_j)$, $\bar{G} = \sum_{j=1}^{t}(\bar{G}_j Y_j)$, $\bar{P} = \sum_{j=1}^{t}(\bar{P}_j Y_j)$, $\bar{\theta}' = \sum_{j=1}^{t}(\bar{\theta}'_j Y_j)$, and $\bar{\Phi}' = \sum_{j=1}^{t}(\bar{\Phi}'_j Y_j)$ are key values that are probably supposed to be renowned. In order to linearize the acquired 61,206 equations with the 10^{-5} needed precision, we perform six function evaluations at each node.

4. Results and Discussion

This section describes through FE analysis how suction/injection impacts the mechanisms of a Maxwell spinning fluid when it is impacted by the Coriolis effect, magnetohydrodynamic effects, and micro-organisms. Three different patterns of arcs are mapped on fluctuating values of the intravenous injection/suction (Γ) factor for every figure for these significant quantities, as follows: $(\Gamma = -0.2)$ (suction), $(\Gamma = 0.0)$ (static), and $(\Gamma = 0.2)$ (injection). The following are the predefined values for the parameters involved: $\beta = 0.1$, $\lambda = 1.0 = M$, $N_b = 0.2 = N_t$, $Le = 10$, $Pr = Lb = 5.0$, $Pe = 0.5$, $\delta_1 = 0.2$. An analysis of mesh separation is executed to show that the finite element simulations are accurate. The entire zone is split into various grid concentrations of mesh sizes, and there is no further modification after (100×100) has been observed, so all simulations are based on this mesh size (Table 1). For distinctive scenarios, comparisons with previous research are provided in Tables 2 and 3 to determine the remedy methodology's accuracy. In certain restrictive instances, it is observed that the existing mathematical evaluations correlate very well with the current investigation. The friction coefficient, as well as the axial and transverse indications $-F''(0), -G(0)$, are calculated using finite element analysis and are summarized in Table 2 for various values of the rotatory factor $(\lambda) = 0.0, 1.0, 2.0, 5.0$ when $(\zeta) = 1$. The table shows that the computational findings achieved are consistent with the results reported by [52,53]. Furthermore, in Table 3, the Nusselt quantity $-\theta(0)$ outputs are consistent with those reported by Bagh et al. [54] and Mustafa et al. [55], who present FEA findings for a variety of values β, λ, Pr, and determine that they are satisfactorily correlated. As a result, certainty in statistical computing is increased, and it is confirmed that the finite element evaluations obtained using the Matlab program show a strong rate of convergence.

Table 1. Meshing analysis for various mesh dimensions when $\zeta = 1.0$.

Grid Size	$-F''(\zeta,0)$	$-G'(\zeta,0)$	$-\theta'(\zeta,0)$	$-\Phi'(\zeta,0)$	$-\chi'(\zeta,0)$
15 × 15	1.7050	0.6876	0.6954	3.7399	4.5771
40 × 40	1.6946	0.6764	0.7399	3.4025	4.8073
70 × 70	1.6935	0.6742	0.7538	3.3371	4.7750
100 × 100	1.6932	0.6736	0.7556	3.3265	4.7565
120 × 120	1.6931	0.6736	0.7558	3.3264	4.7562

Table 2. Assessment of $-F''(0)$ and $-G'(0)$ for different values of λ when $\zeta = 1$ and other parameters are fixed at zero.

λ	Ali [52]		Wang [53]		Present Results	
	$-F''(0)$	$-G'(0)$	$-F''(0)$	$-G'(0)$	$-F''(0)$	$-G'(0)$
0	01.00000	00.00000	01.0000	00.0000	01.00000	00.00000
1	01.32501	00.83715	01.3250	00.8371	01.32501	00.83715
2	01.65232	01.28732	01.6523	01.2873	01.65232	01.28732
5	02.39026	02.15024	–	–	02.39026	02.15024

Table 3. Assessment of $\{-\theta'(0)\}$ at $\zeta = 1$ at various values of Pr, λ and other parameters are fixed at zero.

Pr	β	λ	Ali [54]	Shafique [55]	Present Results (FEM)
1.0	0.20	0.2	00.546683	00.54670	00.5466828
–	0.40	–	00.528090	00.52809	00.5280903
–	0.60	–	–	00.51009	00.5100870
–	0.80	–	00.492547	00.49255	00.4925468

4.1. Variations of Velocity Profiles

Figures 2–5 illustrate the primary and secondary velocity dispersion for various values of the magnetism factor, rotating factor, unsteady factor, and meditation Deborah quantity. Figure 2a,b illustrates the effect of various values of the magnetism factor on the velocity profiles $G'(\zeta, \eta)$ and $F'(\zeta, \eta)$. The presence of frictional factors in the context of a Lorentz effect is caused by the incorporation of a stimulating external magnetization and results in the transverse momentum declines shown in Figure 2a, whereas the axial momentum exhibits an inverse relation, as shown in Figure 2b. The axial $F'(\zeta, \eta)$ and transverse $G'(\zeta, \eta)$ for various rotating parametric inputs are shown in Figure 3a,b. Figure 3a shows that the Coriolis force causes the transverse momentum to decrease for increasing values of the rotation factor, whereas Figure 3b demonstrates the reverse effect. Figure 4a,b show that the size and thickness of the momentum fluid layers in the axial position increase as the time factor increases, while the viscidity of the momentum fluid layers in the transverse path decreases as the time factor decreases. As a result, it is clear that the unsteadiness factor is crucial for influencing the transverse momentum. Physically, a reduced quantity of fluids is pinched axially with the enhanced viscoelastic effects and fluid is pushed away in a radial direction. Figure 5a,b shows that the Deborah quantity is (β) over the velocity profiles for various values of the tranquility factor. The presence of thermoelastic impacts in the context of delivering the best results in a deflation of the building of transverse momentum is shown in Figure 5a, whereas the tangential momentum exhibits an inverse correlation, as shown in Figure 5b. The increasing relative strength of the rheological effect is associated with a higher meditation quantity, resulting in a decrease in velocity. Additionally, these graphs demonstrate that the $F'(\zeta, \eta)$ profile decreases with increase in $\Gamma = 0.2$ (injection), but is significantly increased when $\Gamma = -0.2$ (suction).

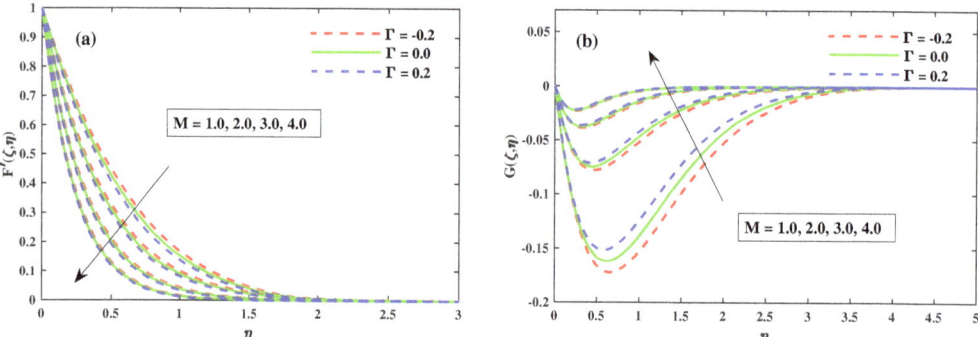

Figure 2. Influence of M on G along y-direction in (**b**), and F' along x-direction in (**a**) when $\zeta = 1$.

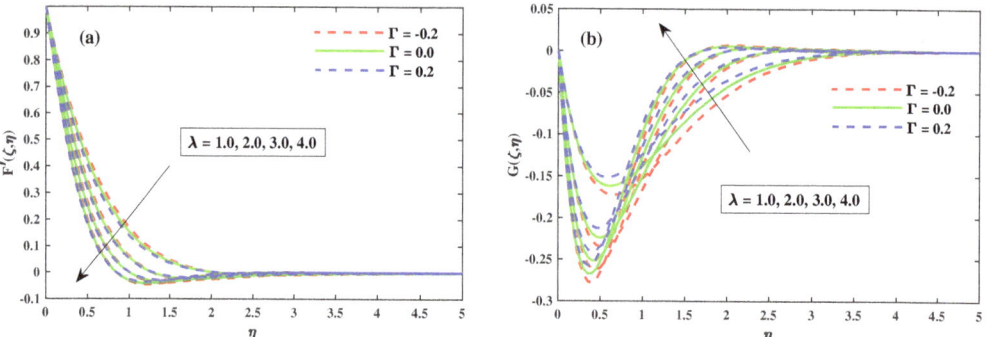

Figure 3. Influence of λ on G along y-direction in (**b**) and F' along x-direction in (**a**) when $\zeta = 1$.

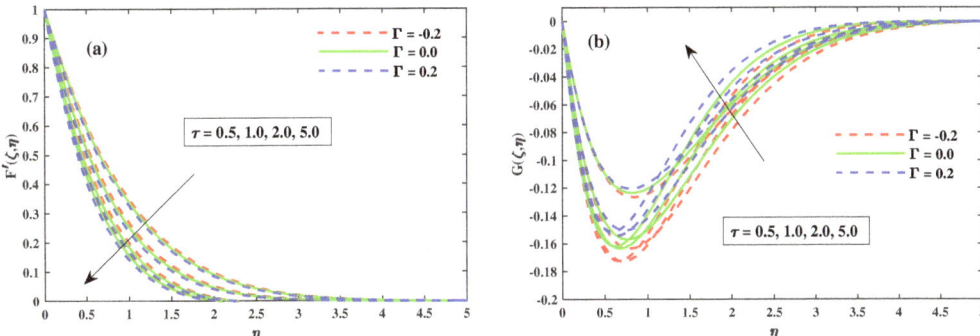

Figure 4. Influence of τ on G along y-direction in (**b**) and F' along x-direction in (**a**) when $\zeta = 1$.

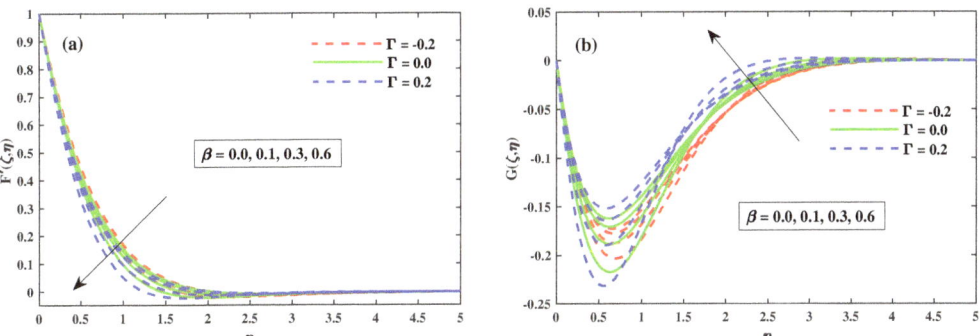

Figure 5. Influence of β on G along y-direction in (**b**) and F' along x-direction in (**a**) when $\zeta = 1$.

Figure 6a,b shows the graphics of $C_{f_x}\sqrt{Re_x}$ (friction factor) across the transverse and $C_{f_y}\sqrt{Re_x}$ (axial direction) closer to the surface for the $\zeta(0:0.2:1)$ spectrum and for $M(1:1:5)$. As shown in Figure 6a, increasing $\zeta(0 \to 1)$ gradually increases the spread of $(C_{f_x}\sqrt{Re_x})$ until no significant difference is observed. In contrast, the $(C_{f_x}\sqrt{Re_x})$ value adjacent to the plate substrate decreases significantly when M is increased. Figure 6b shows that when boosting $\zeta(0 \to 1)$, the spread of $(C_{f_x}\sqrt{Re_x})$ changes steadily up to a consistent rate, and then there is no significant variation, whereas M continues to increase. A large discrepancy in values adjacent to the surface of the sheet, $(C_{f_y}\sqrt{Re_x})$ can be observed. Physically, the application of a magnetic field normal to the direction of fluid flow gives rise to a force known as Lorentz force. Figure 7a,b shows that the dispersion of $(C_{f_x}\sqrt{Re_x})$ tends to increase at a consistent rate up to a certain point, after which no significant variation for enhancing $\zeta(0 \to 1)$ occurs. When λ increases, however, there is a substantial decrease in $(C_{f_x}\sqrt{Re_x})$. When $\zeta(0 \to 1)$ is enhanced, the dissemination of $(C_{f_x}\sqrt{Re_x})$ is substantially decreased until no significant change is detected, as shown in Figure 7b, while λ is increased. Furthermore, it is apparent from these infographics that the basic values of $(C_{f_x}\sqrt{Re_x})$ and $(C_{f_y}\sqrt{Re_x})$ for the scenario of $\Gamma = 0.2$ (injection) are smaller than those for the scenario of $\Gamma = -0.2$ (suction).

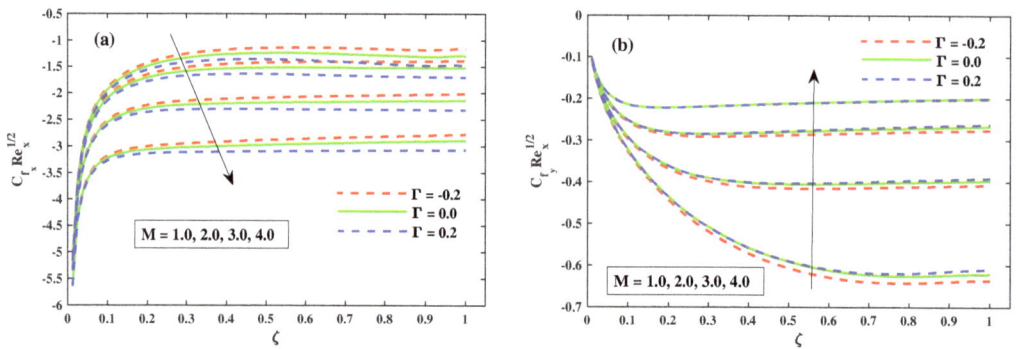

Figure 6. Influence of M on $Cf_x Re_x^{1/2}$ along x-direction in (**a**), and $Cf_y Re_y^{1/2}$ along y-direction in (**b**).

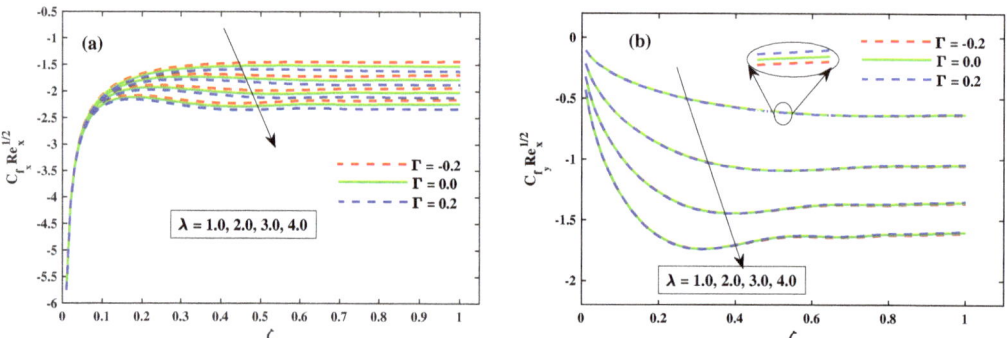

Figure 7. Impact of λ of $Cf_x Re_x^{1/2}$ along x-direction in (**a**), and $Cf_y Re_y^{1/2}$ along y-direction in (**b**).

4.2. Temperature Profiles

Figures 8–10 illustrate the $\theta(\zeta, \eta)$ dissemination when varying the factors involved. The thermal configurations in Figure 8 are enhanced by the magnetic field factor. The cumulative induced resultant force, also known as the resistor Lorentz force, governs the flow momentum between the externally applied magnetic effect and the inner electromagnetic force, as shown in Figure 8a, whereas the wall thickness of the heat transfer performance increases with increasing λ, as shown in Figure 8b. Figure 8a,b show how the thermophoretic factor (N_t) and the Brownian motion factor (N_b) affect the temperature profile. The dispersion of the temperature profile appears to grow as N_b and N_t inclines. Physically, Nt apply a force on the neighbour particles, the force moving the particles from a hot region to a cold region. Figure 10a,b show the impact of (β) and the time-dependent (τ) on the temperature profile. The tranquility of Deborah's number and the unsteady factor are enhanced, as are the $\theta(\zeta, \eta)$ profiles. Furthermore, it can be seen from these graphs that the temperature decreases with the intensity of $\Gamma = 0.2$ (injection), while increasing with the $\Gamma = -0.2$ (suction) factor. Illustrations of the Nusselt quantity $(Nu_x Re_x^{1/2})$ at $(0.1:0.1:0.3) Nt\&Nb$ for $M\&\lambda$ are shown in Figure 11a,b. The dispersion of $(Nu_x\sqrt{Re_x})$ decreases subsequently as M and λ are accelerated. For increasing values of $Nt\&Nb$, a substantial deterioration in $(Nu_x\sqrt{Re_x})$ occurs close to the panel substrate. Additionally, the figure indicates that for $\Gamma = 0.2$ (injection) there is a relatively large quantity of $(Nu_x\sqrt{Re_x})$.

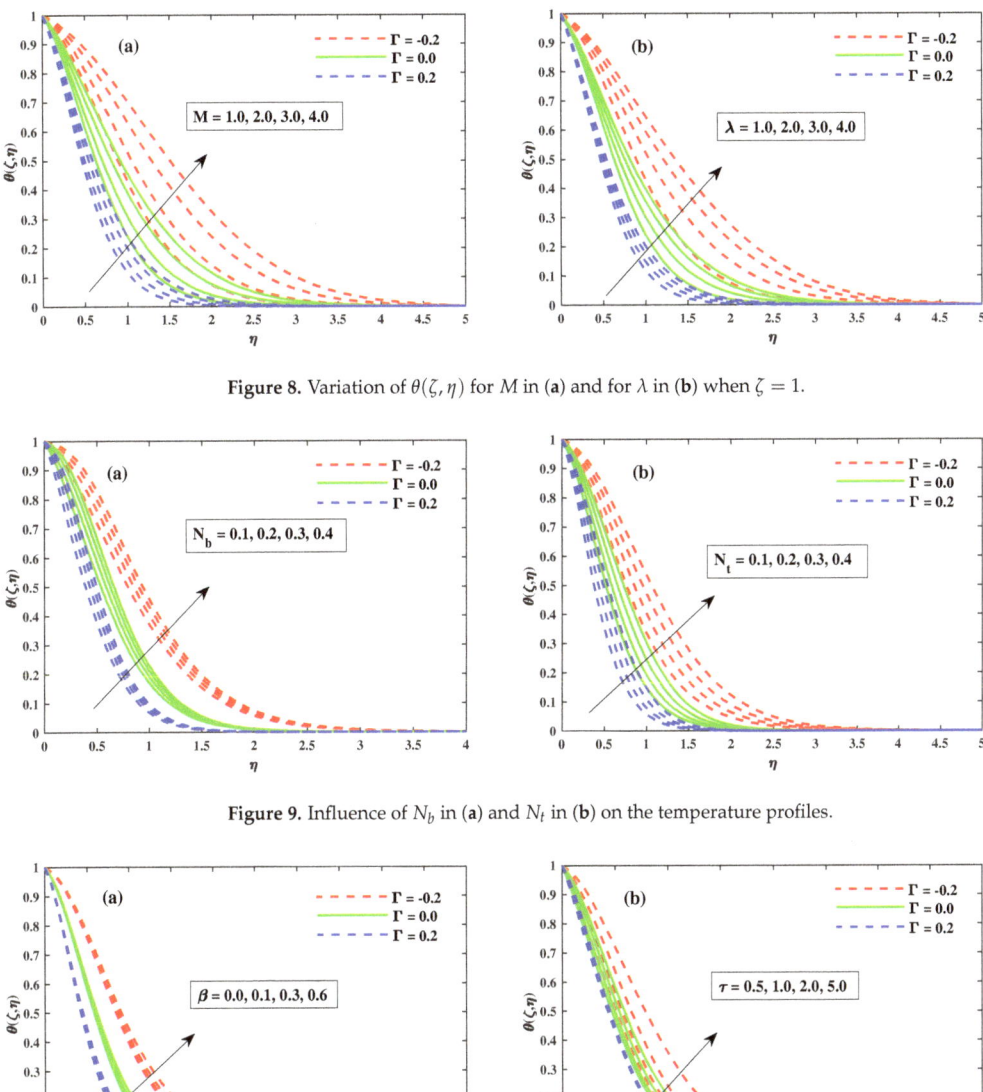

Figure 8. Variation of $\theta(\zeta, \eta)$ for M in (**a**) and for λ in (**b**) when $\zeta = 1$.

Figure 9. Influence of N_b in (**a**) and N_t in (**b**) on the temperature profiles.

Figure 10. The variation of temperature against β in (**a**) and for τ in (**b**).

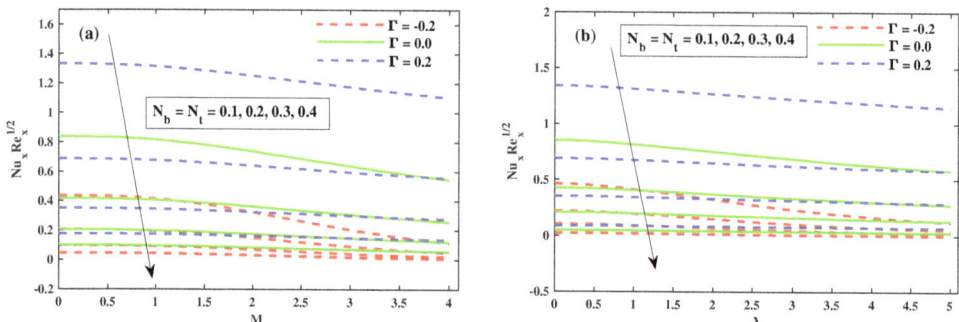

Figure 11. The variation of Nusselt number with N_b, N_t against M in (**a**) and against λ in (**b**).

4.3. Concentration Distributions

Figure 12a,b illustrates that $\Phi(\zeta,\eta)$ varies with the magnetic M, the rotating factor λ, the Lewis Le, and the Deborah number (β). The concentration profiles are augmented as the magnetic field, rotating field, and relaxation Deborah parameters are enhanced, as illustrated in Figures 12a,b and 13b, respectively. Furthermore, Figure 13a demonstrates that deterioration in the organism's density increases the Lewis number Le. Physically, a high Lewis number corresponds to a low mass diffusivity, causing the species concentration in the nanofluid to decrease. Figure 14a,b show the progressive behavioural patterns of the local Sherwood number $(Shr_x\sqrt{Re_x})$ at $(0.1:0.1:0.3)Nt\&Nb$ for $M(0:1:5)\&\lambda(0:1:5)$. The dispersion of $(Shr_x\sqrt{Re_x})$ is reduced as M and λ are increased. In the case of enhancing $Nt\&Nb$, however, a conflicting pattern is observed, and the $\Gamma = 0.2$ (injection) scenario is higher $(Shr_x\sqrt{Re_x})$ than the $\Gamma = -0.2$ (suction) specific case. Figures 15 and 16a,b show $(\chi(\zeta,\eta))$ for variation in M, the rotating parameter λ, the bioconvection Lewis number Lb, and the Peclet number (Pe). The microbe dispersion profile is intensified as the magnetic factor M and rotation factor λ inputs increase, and it notably tumbles in the context of the bioconvection Lewis number Lb and the Peclet number (Pe) (see Figure 16a,b). Furthermore, it can be seen in the infographics that the microbe dispersion profile $\chi(\zeta,\eta)$ decreases when the $\Gamma = 0.2$ (injection) parameter is used, but it is fractionally increased when the $\Gamma = -0.2$ (suction) factor is used. Figure 17a,b shows the trend in the microbe concentration quantity $Re_x^{1/2}N_x$ for $M(0:1:5)\&\lambda(0:1:5)$ at $Nt\&Nb(0.1:0.1:0.3)$, respectively. The $Re_x^{1/2}N_x$ decreases as M and λ increase, whereas the $Re_x^{1/2}N_x$ increases as $Nt\&Nb$ increase. It is also observed that for $\Gamma = 0.2$ (injection), $Re_x^{1/2}N_x$ is higher than for $\Gamma = -0.2$ (suction).

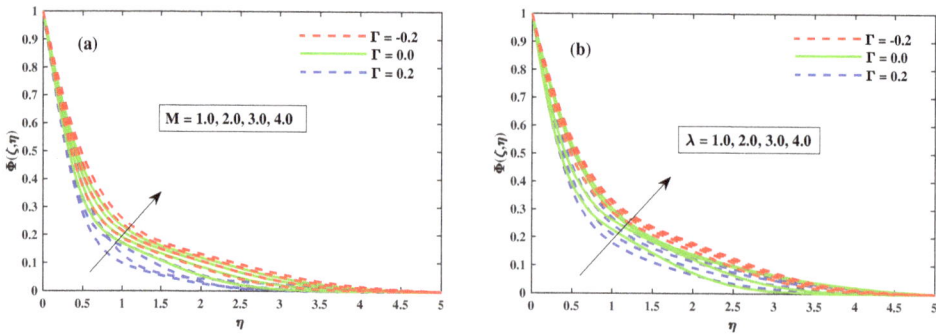

Figure 12. Variation in $\Phi(\zeta,\eta)$ for M in (**a**) and for λ in (**b**) when $\zeta = 1$.

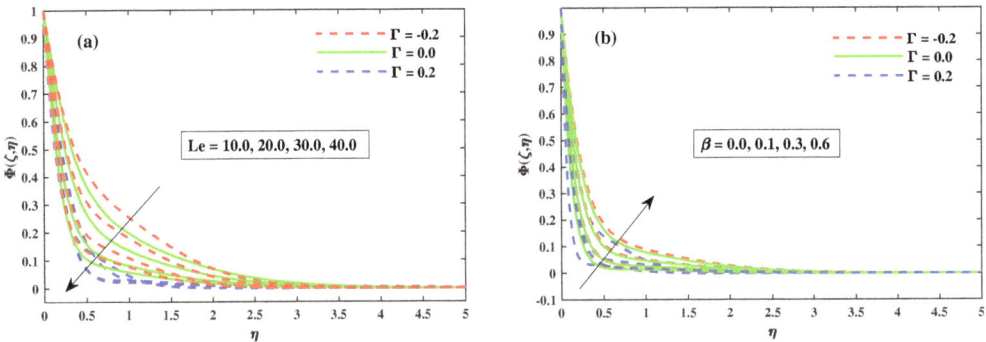

Figure 13. The variation in $\Phi(\zeta, \eta)$ against Le in (**a**) and for β in (**b**).

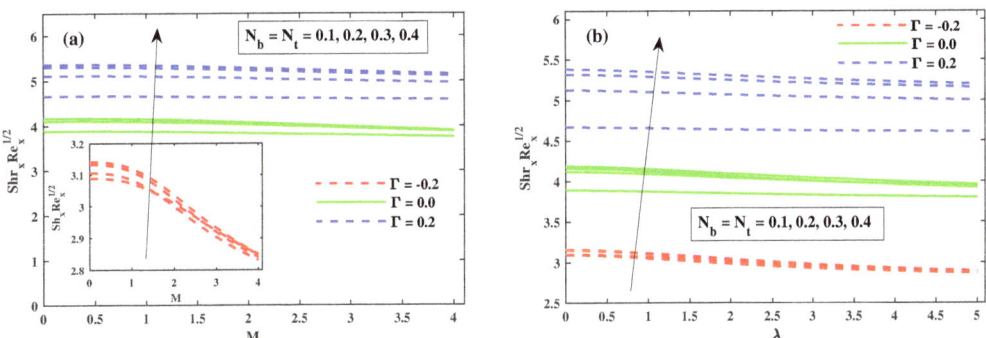

Figure 14. The effect of N_b, N_t for M in (**a**) and for λ in (**b**) on Sherwood number.

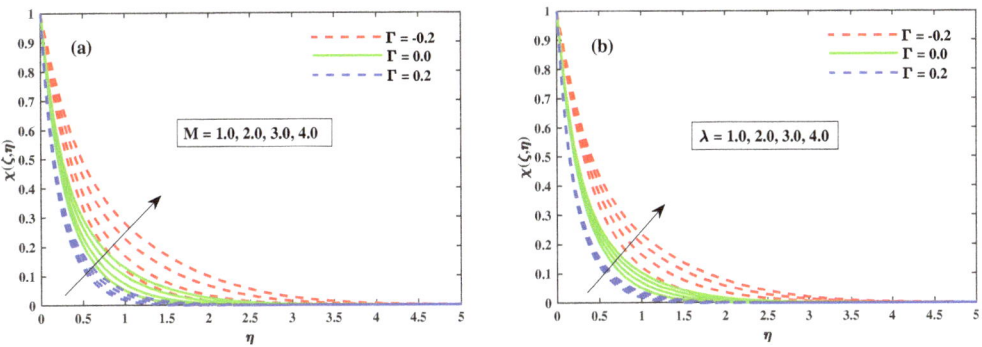

Figure 15. Influence of M in (**a**) and λ in (**b**) on $\chi(\zeta, \eta)$ when $\zeta = 1$.

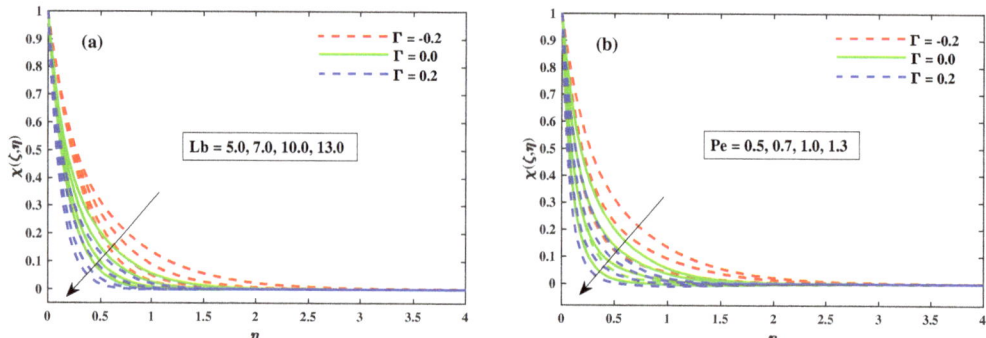

Figure 16. Impact of Lb in (**a**) and Pe in (**b**) on $\chi(\zeta, \eta)$ at $\zeta = 1$.

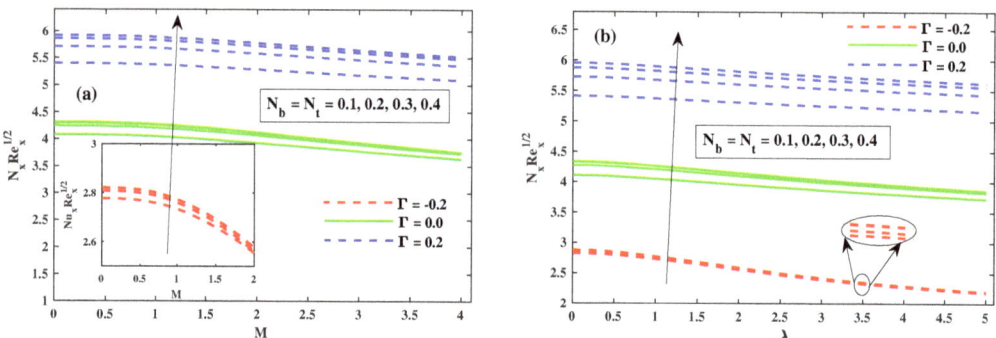

Figure 17. Fluctuation in $N_x Re_x^1/2$ for N_b, N_t, along M in (**a**) and along λ in (**b**).

5. Concluding Remarks

In this article, a finite element simulation was exploited to investigate Maxwell nanofluid flows over a bidirectional elongating surface with bio-convection, suction/injection, Coriolis, and Lorentz forces for three-dimensional spinning flow. Based on the results, the following inferences can be made:

1. Increase in the Coriolis and Lorentz's forces has a decreasing impact on the velocity magnitude, and
   - has a significant influence on the temperature dispersion and concentration.
   - intensifies the impact of $Cf_x Re_x^{1/2}$.
   - the Coriolis force causes the transverse momentum to decrease for increasing values of the rotation factor.
   - with the infusion capability, the velocity, temperature, and concentration components are reduced.

2. It is becoming increasingly evident that the simultaneous enhancement of Brownian and thermophoresis factors has a negative effect on the distribution of temperature, and
   - a declining impact on $Nu_x Re_x^{1/2}$, and positive effects on $Shr_x Re_x^{1/2}$.
   - injection is associated with a larger amount in $Nu_x Re_x^{1/2}$.
   - the injection case has a larger $Shr_x Re_x^{1/2}$ and $Re_x^{1/2} N_x$ compared to the suction case.

- the wall thickness of the heat transfer performance increases with increasing rotating parameter.

3. Higher input to the relaxation Deborah number and the unsteady parameter has a negative impact on the magnitude of the primary and secondary velocity, but
   - has substantial consequences for temperature dispersion.
   - for tiny particles, the volume fraction shows rising effects with higher relaxation Deborah number.

4. Motile microorganism viscosity reduces in the context of augmented bioconvection Peclet and Lewis numbers

This study has involved an analysis of the parameters that their impact on dynamic of fluid flow problems and can be extended in future to include Blasius and Sakiadis flow, and Darcy–Forchheimer and thermoelastic Jeffrey nanofluids.

Author Contributions: L.A. writing—original draft preparation, resources, formal analysis, validation, project administration; A.M. review and editing, investigation and B.A. thoroughly checked the mathematical modeling, corrections, validation. All authors finalized the manuscript after its internal evaluation. All authors have read and agreed to the published version of the manuscript.

Funding: This work is supported by the Xi'an Technological University, with scientific research start-up fund (Grant No. 0853-302020678) in support of this research project.

Data Availability Statement: Not applicable.

Acknowledgments: This work is supported by Xi'an Technological University. The authors would like to acknowledge and express their gratitude to the School of Sciences for their continuous support to work at Xi'an Technological University, Xi'an, China.

Conflicts of Interest: The authors declare no conflict of interest.

Nomenclature

The following abbreviations are used in this manuscript:

T	Non-dimensional temperature, K	T_w	Temperature at surface, K
T_∞	Temperature away from the surface, K	σ	Conductivity of fluid
D_T	Thermophoretic dispersion	U_w	Velocity, s^{-1}
Cf_x	Skin friction at x-direction	C_p	Specific heat
Cf_y	Skin friction at y-direction	B_0	Magnetic field strength
Nu_x	Nusselt number	ρ_f	Density of fluid, Kgm^{-3}
v_f	Kinematic viscosity of fluid, $m^{-2}s^{-1}$	M	Magnetic parameter
Cp	Specific heat at constant pressure, $JKg^{-1}K^{-1}$	Ω	Rotating parameter
Pe	Peclet number	W_c	Optimum cell swimming
Re_x	Local Reynold number	f	Base fluid
β	Deborah number	Γ	Suction/injection
D_B	Brownian diffusion coefficient	Pr	Prandtl number
Nb	Brownian motion parameter	q_w	Surface heat flux, Wm^{-2}
N	Intensity of microbes	Le	Lewis number
ρ_p	Nanoparticle density	Nt	Thermophoresis parameter
τ_w	External heat transfer factor	Le	Lewis number

References

1. Ali, R.; Asjad, M.I.; Aldalbahi, A.; Rahimi-Gorji, M.; Rahaman, M. Convective flow of a Maxwell hybrid nanofluid due to pressure gradient in a channel. *J. Therm. Anal. Calorim.* **2020**, *43*, 1319–1329. [CrossRef]
2. Jawad, M.; Saeed, A.; Gul, T. Entropy Generation for MHD Maxwell Nanofluid Flow Past a Porous and Stretching Surface with Dufour and Soret Effects. *Braz. J. Phys.* **2021**, *13*, 469–480. [CrossRef]
3. Jamshed, W. Numerical investigation of MHD impact on maxwell nanofluid. *Int. Commun. Heat Mass Transf.* **2021**, *120*, 104973. [CrossRef]
4. Ali, L.; Ali, B.; Ghori, M.B. Melting effect on Cattaneo–Christov and thermal radiation features for aligned MHD nanofluid flow comprising microorganisms to leading edge: FEM approach. *Comput. Math. Appl.* **2022**, *109*, 260–269. [CrossRef]

5. Pal, D.; Mandal, G. Magnetohydrodynamic nonlinear thermal radiative heat transfer of nanofluids over a flat plate in a porous medium in existence of variable thermal conductivity and chemical reaction. *Int. J. Ambient. Energy* **2021**, *42*, 1167–1177. [CrossRef]
6. Ahmed, A.; Khan, M.; Ahmed, J. Mixed convective flow of Maxwell nanofluid induced by vertically rotating cylinder. *Appl. Nanosci.* **2020**, *10*, 5179–5190. [CrossRef]
7. Pal, D.; Mandal, G. Hydromagnetic convective–radiative boundary layer flow of nanofluids induced by a non-linear vertical stretching/shrinking sheet with viscous–Ohmic dissipation. *Powder Technol.* **2015**, *279*, 61–74. [CrossRef]
8. Ahmad, B.; Ahmad, M.O.; Ali, L.; Ali, B.; Hussein, A.K.; Shah, N.A.; Chung, J.D. Significance of the Coriolis force on the dynamics of Carreau–Yasuda rotating nanofluid subject to Darcy–forchheimer and gyrotactic microorganisms. *Mathematics* **2022**, *10*, 2855. [CrossRef]
9. Ahmed, J.; Khan, M.; Ahmad, L. Stagnation point flow of Maxwell nanofluid over a permeable rotating disk with heat source/sink. *J. Mol. Liq.* **2019**, *287*, 110853. [CrossRef]
10. Bilal, S.; Ur Rehman, K.; Mustafa, Z.; Malik, M. Maxwell nanofluid flow individualities by way of rotating cone. *J. Nanofluids* **2019**, *8*, 596–603. [CrossRef]
11. Amirsom, N.A.; Uddin, M.; Basir, M.F.M.; Ismail, A.; Beg, O.A.; Kadir, A. Three-dimensional bioconvection nanofluid flow from a bi-axial stretching sheet with anisotropic slip. *Sains Malays.* **2019**, *48*, 1137–1149. [CrossRef]
12. Prabhavathi, B.; Reddy, P.S.; Vijaya, R.B. Heat and mass transfer enhancement of SWCNTs and MWCNTs based Maxwell nanofluid flow over a vertical cone with slip effects. *Powder Technol.* **2018**, *340*, 253–263. [CrossRef]
13. Zohra, F.T.; Uddin, M.J.; Ismail, A.I. Magnetohydrodynamic bio-nanoconvective Naiver slip flow of micropolar fluid in a stretchable horizontal channel. *Heat Transf. Res.* **2019**, *48*, 3636–3656. [CrossRef]
14. Pal, D.; Mandal, G. Double diffusive magnetohydrodynamic heat and mass transfer of nanofluids over a nonlinear stretching/shrinking sheet with viscous-Ohmic dissipation and thermal radiation. *Propuls. Power Res.* **2017**, *6*, 58–69. [CrossRef]
15. Ali, B.; Ali, L.; Abdal, S.; Asjad, M.I. Significance of Brownian motion and thermophoresis influence on dynamics of Reiner-Rivlin fluid over a disk with non-Fourier heat flux theory and gyrotactic microorganisms: A Numerical approach. *Phys. Scr.* **2021**, *96*, 94001. [CrossRef]
16. Mandal, G.; Pal, D. Entropy generation analysis of radiated magnetohydrodynamic flow of carbon nanotubes nanofluids with variable conductivity and diffusivity subjected to chemical reaction. *J. Nanofluids* **2021**, *10*, 491–505. [CrossRef]
17. Ali, L.; Liu, X.; Ali, B.; Din, A.; Al Mdallal, Q. The function of nanoparticle's diameter and Darcy-Forchheimer flow over a cylinder with effect of magnetic field and thermal radiation. *Case Stud. Therm. Eng.* **2021**, *28*, 101392. [CrossRef]
18. Mandal, G. Convective-radiative heat transfer of micropolar nanofluid over a vertical non-linear stretching sheet. *J. Nanofluids* **2016**, *5*, 852–860. [CrossRef]
19. Sreedevi, P.; Reddy, P.S.; Chamkha, A.J. Magneto-hydrodynamics heat and mass transfer analysis of single and multi-wall carbon nanotubes over vertical cone with convective boundary condition. *Int. J. Mech. Sci.* **2018**, *135*, 646–655. [CrossRef]
20. Ali, L.; Liu, X.; Ali, B. Finite Element Analysis of Variable Viscosity Impact on MHD Flow and Heat Transfer of Nanofluid Using the Cattaneo–Christov Model. *Coatings* **2020**, *10*, 395. [CrossRef]
21. Uddin, M.; Kabir, M.; Bég, O.A.; Alginahi, Y. Chebyshev collocation computation of magneto-bioconvection nanofluid flow over a wedge with multiple slips and magnetic induction. *Proc. Inst. Mech. Eng. Part N J. Nanomater. Nanoeng. Nanosyst.* **2018**, *232*, 109–122. [CrossRef]
22. Rout, B.; Mishra, S.; Thumma, T. Effect of viscous dissipation on Cu-water and Cu-kerosene nanofluids of axisymmetric radiative squeezing flow. *Heat Transf. Res.* **2019**, *48*, 3039–3054. [CrossRef]
23. Sheri, S.R.; Thumma, T. Numerical study of heat transfer enhancement in MHD free convection flow over vertical plate utilizing nanofluids. *Ain Shams Eng. J.* **2018**, *9*, 1169–1180. [CrossRef]
24. Thumma, T.; Mishra, S. Effect of viscous dissipation and Joule heating on magnetohydrodynamic Jeffery nanofluid flow with and without multi slip boundary conditions. *J. Nanofluids* **2018**, *7*, 516–526. [CrossRef]
25. Bég, O.A.; Kabir, M.N.; Uddin, M.J.; Izani Md Ismail, A.; Alginahi, Y.M. Numerical investigation of Von Karman swirling bioconvective nanofluid transport from a rotating disk in a porous medium with Stefan blowing and anisotropic slip effects. *Proc. Inst. Mech. Eng. Part C J. Mech. Eng. Sci.* **2021**, *235*, 3933–3951. [CrossRef]
26. Chu, Y.M.; Aziz, S.; Khan, M.I.; Khan, S.U.; Nazeer, M.; Ahmad, I.; Tlili, I. Nonlinear radiative bioconvection flow of Maxwell nanofluid configured by bidirectional oscillatory moving surface with heat generation phenomenon. *Phys. Scr.* **2020**, *95*, 105007. [CrossRef]
27. Sreedevi, P.; Reddy, P.S. Combined influence of Brownian motion and thermophoresis on Maxwell three-dimensional nanofluid flow over stretching sheet with chemical reaction and thermal radiation. *J. Porous Media* **2020**, *4*, 23. [CrossRef]
28. Rao, M.V.S.; Gangadhar, K.; Chamkha, A.J.; Surekha, P. Bioconvection in a Convectional Nanofluid Flow Containing Gyrotactic Microorganisms over an Isothermal Vertical Cone Embedded in a Porous Surface with Chemical Reactive Species. *Arab. J. Sci. Eng.* **2021**, *46*, 2493–2503. [CrossRef]
29. Awais, M.; Awan, S.E.; Raja, M.A.Z.; Parveen, N.; Khan, W.U.; Malik, M.Y.; He, Y. Effects of Variable Transport Properties on Heat and Mass Transfer in MHD Bioconvective Nanofluid Rheology with Gyrotactic Microorganisms: Numerical Approach. *Coatings* **2021**, *11*, 231. [CrossRef]

30. Sreedevi, P.; Sudarsana Reddy, P. Heat and mass transfer analysis of MWCNT-kerosene nanofluid flow over a wedge with thermal radiation. *Heat Transf.* **2021**, *50*, 10–33. [CrossRef]
31. Elayarani, M.; Shanmugapriya, M.; Kumar, P.S. Intensification of heat and mass transfer process in MHD carreau nanofluid flow containing gyrotactic microorganisms. *Chem. Eng. Process. Process. Intensif.* **2021**, *160*, 108299. [CrossRef]
32. Ali, B.; Raju, C.; Ali, L.; Hussain, S.; Kamran, T. G-Jitter impact on magnetohydrodynamic non-Newtonian fluid over an inclined surface: Finite element simulation. *Chin. J. Phys.* **2021**, *71*, 479–491. [CrossRef]
33. Farooq, U.; Waqas, H.; Khan, M.I.; Khan, S.U.; Chu, Y.M.; Kadry, S. Thermally radioactive bioconvection flow of Carreau nanofluid with modified Cattaneo-Christov expressions and exponential space-based heat source. *Alex. Eng. J.* **2021**, *60*, 3073–3086. [CrossRef]
34. Alhussain, Z.A.; Renuka, A.; Muthtamilselvan, M. A magneto-bioconvective and thermal conductivity enhancement in nanofluid flow containing gyrotactic microorganism. *Case Stud. Therm. Eng.* **2021**, *23*, 100809. [CrossRef]
35. Sreedevi, P.; Reddy, P.S. Effect of SWCNTs and MWCNTs Maxwell MHD nanofluid flow between two stretchable rotating disks under convective boundary conditions. *Heat Transf. Res.* **2019**, *48*, 4105–4132. [CrossRef]
36. Ali, L.; Wu, Y.J.; Ali, B.; Abdal, S.; Hussain, S. The crucial features of aggregation in TiO_2-water nanofluid aligned of chemically comprising microorganisms: A FEM approach. *Comput. Math. Appl.* **2022**, *123*, 241–251. [CrossRef]
37. Kumar, P.; Poonia, H.; Ali, L.; Areekara, S. The numerical simulation of nanoparticle size and thermal radiation with the magnetic field effect based on tangent hyperbolic nanofluid flow. *Case Stud. Therm. Eng.* **2022**, *37*, 102247. [CrossRef]
38. Sheri, S.R.; Thumma, T. Heat and mass transfer effects on natural convection flow in the presence of volume fraction for copper-water nanofluid. *J. Nanofluids* **2016**, *5*, 220–230. [CrossRef]
39. Ali, L.; Liu, X.; Ali, B.; Mujeed, S.; Abdal, S.; Mutahir, A. The Impact of Nanoparticles Due to Applied Magnetic Dipole in Micropolar Fluid Flow Using the Finite Element Method. *Symmetry* **2020**, *12*, 520. [CrossRef]
40. Ali, B.; Rasool, G.; Hussain, S.; Baleanu, D.; Bano, S. Finite element study of magnetohydrodynamics (MHD) and activation energy in Darcy–Forchheimer rotating flow of Casson Carreau nanofluid. *Processes* **2020**, *8*, 1185. [CrossRef]
41. Abbas, Z.; Javed, T.; Sajid, M.; Ali, N. Unsteady MHD flow and heat transfer on a stretching sheet in a rotating fluid. *J. Taiwan Inst. Chem. Eng.* **2010**, *41*, 644–650. [CrossRef]
42. Babu, M.J.; Sandeep, N. 3D MHD slip flow of a nanofluid over a slendering stretching sheet with thermophoresis and Brownian motion effects. *J. Mol. Liq.* **2016**, *222*, 1003–1009. [CrossRef]
43. Hayat, T.; Muhammad, T.; Shehzad, S.; Alsaedi, A. Three dimensional rotating flow of Maxwell nanofluid. *J. Mol. Liq.* **2017**, *229*, 495–500. [CrossRef]
44. Ali, B.; Hussain, S.; Nie, Y.; Hussein, A.K.; Habib, D. Finite element investigation of Dufour and Soret impacts on MHD rotating flow of Oldroyd-B nanofluid over a stretching sheet with double diffusion Cattaneo Christov heat flux model. *Powder Technol.* **2021**, *377*, 439–452. [CrossRef]
45. Rosali, H.; Ishak, A.; Nazar, R.; Pop, I. Rotating flow over an exponentially shrinking sheet with suction. *J. Mol. Liq.* **2015**, *211*, 965–969. [CrossRef]
46. Ali, B.; Hussain, S.; Nie, Y.; Ali, L.; Hassan, S.U. Finite element simulation of bioconvection and cattaneo-Christov effects on micropolar based nanofluid flow over a vertically stretching sheet. *Chin. J. Phys.* **2020**, *68*, 654–670. [CrossRef]
47. Reddy, J.N. *Solutions Manual for an Introduction to the Finite Element Method*; McGraw-Hill: New York, NY, USA, 1993; p. 41.
48. Jyothi, K.; Reddy, P.S.; Reddy, M.S. Carreau nanofluid heat and mass transfer flow through wedge with slip conditions and nonlinear thermal radiation. *J. Braz. Soc. Mech. Sci. Eng.* **2019**, *41*, 415. [CrossRef]
49. Ali, B.; Nie, Y.; Khan, S.A.; Sadiq, M.T.; Tariq, M. Finite Element Simulation of Multiple Slip Effects on MHD Unsteady Maxwell Nanofluid Flow over a Permeable Stretching Sheet with Radiation and Thermo-Diffusion in the Presence of Chemical Reaction. *Processes* **2019**, *7*, 628. [CrossRef]
50. Ali, L.; Ali, B.; Liu, X.; Ahmed, S.; Shah, M.A. Analysis of bio-convective MHD Blasius and Sakiadis flow with Cattaneo-Christov heat flux model and chemical reaction. *Chin. J. Phys.* **2021**, *12*, 554–571. [CrossRef]
51. Ibrahim, W.; Gadisa, G. Finite Element Method Solution of Boundary Layer Flow of Powell-Eyring Nanofluid over a Nonlinear Stretching Surface. *J. Appl. Math.* **2019**, *2019*. [CrossRef]
52. Ali, B.; Naqvi, R.A.; Ali, L.; Abdal, S.; Hussain, S. A Comparative Description on Time-Dependent Rotating Magnetic Transport of a Water Base Liquid With Hybrid Nano-materials Over an Extending Sheet Using Buongiorno Model: Finite Element Approach. *Chin. J. Phys.* **2021**, *70*, 125–139. [CrossRef]
53. Wang, C. Stretching a surface in a rotating fluid. *Z. Angew. Math. Phys. ZAMP* **1988**, *39*, 177–185. [CrossRef]
54. Ali, B.; Nie, Y.; Hussain, S.; Manan, A.; Sadiq, M.T. Unsteady magneto-hydrodynamic transport of rotating Maxwell nanofluid flow on a stretching sheet with Cattaneo–Christov double diffusion and activation energy. *Therm. Sci. Eng. Prog.* **2020**, *20*, 100720. [CrossRef]
55. Shafique, Z.; Mustafa, M.; Mushtaq, A. Boundary layer flow of Maxwell fluid in rotating frame with binary chemical reaction and activation energy. *Results Phys.* **2016**, *6*, 627–633. [CrossRef]

Article

Natural Gas Storage Filled with Peat-Derived Carbon Adsorbent: Influence of Nonisothermal Effects and Ethane Impurities on the Storage Cycle

Andrey V. Shkolin [1,2,*], Evgeny M. Strizhenov [1,2], Sergey S. Chugaev [1,2], Ilya E. Men'shchikov [2], Viktoriia V. Gaidamavichute [2], Alexander E. Grinchenko [2] and Anatoly A. Zherdev [1]

[1] Research Institute of Power Engineering, Bauman Moscow State Technical University, Baumanskaya 2-ya str. 5, 105005 Moscow, Russia
[2] Frumkin Institute of Physical Chemistry and Electrochemistry, Russian Academy of Sciences, Leninskii Prospect, 31, Build. 4, 119071 Moscow, Russia
* Correspondence: shkolin@phyche.ac.ru

Citation: Shkolin, A.V.; Strizhenov, E.M.; Chugaev, S.S.; Men'shchikov, I.E.; Gaidamavichute, V.V.; Grinchenko, A.E.; Zherdev, A.A. Natural Gas Storage Filled with Peat-Derived Carbon Adsorbent: Influence of Nonisothermal Effects and Ethane Impurities on the Storage Cycle. *Nanomaterials* 2022, 12, 4066. https://doi.org/10.3390/nano12224066

Academic Editor: Meiwen Cao

Received: 24 October 2022
Accepted: 16 November 2022
Published: 18 November 2022

Publisher's Note: MDPI stays neutral with regard to jurisdictional claims in published maps and institutional affiliations.

Copyright: © 2022 by the authors. Licensee MDPI, Basel, Switzerland. This article is an open access article distributed under the terms and conditions of the Creative Commons Attribution (CC BY) license (https://creativecommons.org/licenses/by/4.0/).

Abstract: Adsorbed natural gas (ANG) is a promising solution for improving the safety and storage capacity of low-pressure gas storage systems. The structural–energetic and adsorption properties of active carbon ACPK, synthesized from cheap peat raw materials, are presented. Calculations of the methane–ethane mixture adsorption on ACPK were performed using the experimental adsorption isotherms of pure components. It is shown that the accumulation of ethane can significantly increase the energy capacity of the ANG storage. Numerical molecular modeling of the methane–ethane mixture adsorption in slit-like model micropores has been carried out. The molecular effects associated with the displacement of ethane by methane molecules and the formation of a molecule layered structure are shown. The integral molecular adsorption isotherm of the mixture according to the molecular modeling adequately corresponds to the ideal adsorbed solution theory (IAST). The cyclic processes of gas charging and discharging from the ANG storage based on the ACPK are simulated in three modes: adiabatic, isothermal, and thermocontrolled. The adiabatic mode leads to a loss of 27–33% of energy capacity at 3.5 MPa compared to the isothermal mode, which has a 9.4–19.5% lower energy capacity compared to the thermocontrolled mode, with more efficient desorption of both methane and ethane.

Keywords: mixture adsorption; nanoporous carbon; natural gas storage; methane; ethane; heat of adsorption; cyclic adsorption; ideal adsorbed solution theory; numerical molecular simulation; molecular dynamic

1. Introduction

Natural gas as one of the most common, cheap, and environmentally friendly fuels may in the near future become the main energy source for countries adhering to the principles of sustainable development. However, based on the experience of countries using natural gas as the main energy source for domestic consumption, it can be noted that among the main difficulties on the way to a large-scale energy transition to natural gas is the issue of gas delivery and back-up storage close to the consumer. The energy crisis in Europe in 2021–2022 has highlighted the importance of building an appropriate gas infrastructure with back-up storage to mitigate problems with gas supplies and uneven consumption. In addition, the events of 26 September 2022 on the Nord Stream gas pipelines also exposed the security problem of the existing gas infrastructure.

In recent years, the development of adsorbed natural gas (ANG) systems has been considered the most promising solution to improve the safety [1,2] and capacity of a low-pressure gas storage system [3]. However, there are several well-known difficulties that prevent ANG large-scale implementation: high requirements for the adsorbent, stability, uniformity and reproducibility of its adsorption properties; the exothermic nature of adsorption, leading to predominantly negative thermal effects, which should be compensated for during charging and discharging, and to an increase in the duration of these processes [4]; and a decrease in storage efficiency due to the accumulation of C_{2+} hydrocarbon impurities during cyclic operation [5].

The study of the impurities' influence on the efficiency of adsorption gas storage is presented in the scientific literature much less often than studies on the thermal effects. This is mainly due to the complexity and laboriousness of conducting a cyclic experiment. A study [6] showed that when using natural gas with a methane content of about 92.18% mol., after 700 natural gas charging cycles, corresponding to about 250,000 km of vehicle mileage, the efficiency of the storage system drops by 50% due to the accumulation of impurity hydrocarbons C_{2+}. The ethane accumulation ended around cycle 30, but the efficiency continued to decline through 700 cycles due to the accumulation of heavier hydrocarbons. In [6], an adsorption layer with high thermal conductivity due to expanded natural graphite (ENG) was used, which together with a water thermostat significantly reduced the thermal effects in the charging and discharging processes. The authors of [7] supplied gas in the radial direction from the central collector (pipes with 40 channels) to reduce the heterogeneity of thermal effects. The work used natural gas with a methane content of 90.68% mol. After 10 cycles (the study included 30 cycles in total), the charging mass of gas decreased by 20%, and the discharging mass of gas by only 2% compared with the first cycle: according to the authors, the loss in discharging due to the accumulation of impurities fully appeared already from the first cycle and subsequently weakly changed. In [8], the cyclic process of charging and discharging gas from an ANG tank at room temperature on commercial activated carbon Maxsorb MSC-30 was studied. When charging natural gas with 85.45% mol. methane, the gravimetric excess adsorption decreased to 33% after 100 cycles and continued to slowly decrease until it reached 25% by the 1000th cycle. Volumetric storage capacity decreased to 50% after the first 100 cycles and remained constant thereafter. The authors showed that periodic regeneration by degassing at 400 °C for 2 h makes it possible to remove impurities and reactivate active carbon. These works were not aimed to investigate the combined effect of the impurity accumulation and the thermal control of the adsorber or the thermal effects of adsorption. In [9], the method of mathematical modeling was used to study the cyclic processes of charging and discharging natural gas containing 88% mol. methane for two limit cases: isothermal and adiabatic. The authors noted that non-isothermal effects in the adiabatic process can be more negative to the performance of the gas charging–discharging cycle than the presence of impurities in natural gas. In the case of the adiabatic process using natural gas with impurities, the accumulation efficiency decreased by 37–40%. However, the authors noted that they limited themselves to only C_1–C_4 hydrocarbons, excluding the accumulation of heavier hydrocarbons from the model. The authors of [10] studied the effect of impurities in natural gas (methane content 90.58% mol.) in a full-size system with and without thermal control (heating during discharging). It was found that for 20 cycles without thermal control, the useful volumetric storage capacity decreased by 16%, and the amount of gas discharged by 10%. However, with the use of thermal management, efficiency losses have been reduced to 7% and 5%, respectively, due to more efficient ethane removal. However, thermal control showed no significant improvement in reducing the accumulation of hydrocarbons heavier than C_2H_6, which continued to displace methane and ethane from the system.

In this paper, we study the properties of active carbon synthesized from cheap peat raw materials, its properties for adsorption of methane and ethane, the two main components of natural gas, and their mixture adsorption by the methods of the ideal adsorbed solution theory (IAST) and molecular modeling. The cyclic mode of charging and discharging of a model scaled natural gas storage equipped with a synthesized adsorbent is also considered, and the effect of ethane accumulation and thermal control of the system on the higher heating value (HHV) of the discharged gas is shown.

2. Materials and Methods

2.1. Adsorbent and Adsorbed Gases

High purity methane (99.99%) and ethane (99.95%) produced by Linde Gas was used in adsorption data experiments. The properties of gases and their mixtures were determined using the CoolProp program [11] using NIST data.

High-moor peat of the wood group with a high degree of metamorphism was used as a raw material for the synthesis of the adsorbent.

The synthesis of nanoporous activated carbon ACPK from peat with a decomposition degree of more than 50% (H8 and higher on the Von Post scale) consisted of the stage of raw material preparation, carbonization without oxygen, thermochemical activation, washing, and drying. Thermochemical activation makes it possible to obtain a more uniform structure of narrow micropores.

The stage of preparation of raw materials included the processing and drying of peat, followed by grinding and sieving to separate the required fraction up to 2 mm in size. The carbonization stage was carried out in a muffle furnace without access to oxygen at a temperature of 800 °C with a heating rate of 10 °C/min. For activation, aqueous mixtures of carbonizate and KOH activator were prepared in ratios of 1:2, which were placed in a steel crucible and then in a furnace. Thermochemical activation was carried out at a temperature of 900 °C at a heating rate of 10 °C/min. After reaching the required temperature, the samples were kept for 1 h. After cooling the samples to ambient temperature, the samples were washed with distilled water to pH 8 and dried in an oven for 24 h at a temperature of 110 °C.

2.2. Characterization

The surface morphology and elemental composition of ACPK were examined by scanning electron microscopy (SEM) using a Quanta 650 FEG microscope (FEI Company, Hillsboro, OR, USA) equipped with an Oxford Energy Dispersive X-ray (EDX) detector operating at 15 kV accelerating voltage.

To study the phase composition of the studied adsorbent and the initial assessment of their physicochemical and adsorption properties, an X-ray powder diffraction study was carried out using an Empyrean Panalytical diffractometer in the range of scattering angles 2θ from 10 to 120°.

To evaluate the structural characteristics of the adsorbent, we used the small-angle X-ray scattering (SAXS) method on a SAXSess diffractometer (Anton Paar). Monochromatic radiation was obtained using a Cu-Kα filter (λ = 0.154 nm), and scattering was recorded on a two-dimensional Imaging Plate detector.

These methods provide a comprehensive understanding of the carbon framework and a primary understanding of the porous structure of the synthesized material ACPK. The detailed porous structure of ACPK was examined by N_2 adsorption–desorption isotherms at 77 K, which were performed on a Quantachrome Autosorb iQ multifunctional surface area analyzer. The specific volume of micropores W_0 (cm$^3 \cdot$g^{-1}), standard characteristic energy of adsorption E_0 (kJ·mol^{-1}), and effective half-width of micropores x_0 (nm) were calculated by the Dubinin–Radushkevich (D-R) equation [12,13]. The Brunauer–Emmett–Teller (BET) method [14] using the criteria for microporous adsorbents [15] was also applied to evaluate the specific surface area, S_{BET} (cm$^2 \cdot$g^{-1}). The Kelvin equation [16] was used to calculate the specific surface area of mesopores, S_{ME} (cm$^2 \cdot$g^{-1}), respectively. The specific mesopore volume was calculated as $W_{ME} = W_S - W_0$, where W_S (cm$^3 \cdot$g^{-1}) is the total pore volume obtained from the nitrogen adsorption at the relative pressure P/P_s = 0.99. The pore size distribution function in ACPK was calculated using the Quenched Solid Density Functional Theory (QSDFT) developed for micro-mesoporous adsorbents [17].

2.3. Single-Component Adsorption

Single-component adsorption isotherms of C_2H_6 and CH_4 were collected on purpose-designed adsorptions benches, the schemes of which are reported in the previous works [18–20].

Methane adsorption was studied at temperatures of 213, 243, 273, 293, 333, and 393 K at pressures up to 20 MPa. Ethane adsorption was studied at temperatures of 273, 293, 313, and 333 K at pressures up to 120 kPa. The selected measurement ranges cover the most common field of practical application of ANG systems. Before gas adsorption measurements, the degas process was carried out under vacuum at 250 °C for 6 h.

Adsorption at temperatures different from the experimental ones was determined by the property of the linearity of the isosteres in the ln P–$1/T$ coordinates, which in turn were plotted from the experimental isotherms (at least three points). To expand the pressure range of the calculated isotherms, an additional extrapolation was carried out according to the equations that best describe the experimental adsorption isotherms: the Sips and the theory of volume filling of micropores (TVFM) equations [21]. The TVFM equation was used in the subcritical region, and the Sips equation in the supercritical. The study of the thermodynamic functions of the adsorption process was carried out using Bakaev's approach [22–24], which is based on the method of changing variables to establish a, P, T (adsorption, pressure, and temperature, respectively) parameters of adsorption equilibrium in a wide range of pressures and temperatures, taking into account the nonideality of the gas phase in a thermodynamic system defined in accordance with the Guggenheim approach [25].

2.4. Mixture Adsorption Calculation

To determine the mixture adsorption of methane and ethane by the IAST method [26–28], the system of equations was solved:

$$\frac{\pi}{RT} = \int_0^{P_i^*} \frac{a_i^*}{P_i} dP_i \tag{1}$$

$$PY_i = P_i^* X_i \tag{2}$$

$$\sum_{i=1}^{N} X_i = 1 \tag{3}$$

$$\frac{1}{a_\Sigma} = \sum_{i=1}^{N} \frac{X_i}{a_i^*}, \tag{4}$$

where π is the spreading pressure, (Pa); T is the temperature, (K); R is the universal gas constant, (J/(mol K)); a_i^* is the adsorption of the pure component i of the mixture, (mmol/g), at T and the equilibrium pressure of the gas phase P_i^* (Pa), which corresponds to π; X_i is the mole fraction of component i in the adsorbed phase; Y_i is the mole fraction of component i in the gas phase; a_Σ is the adsorption of the mixture in mmol/g at its equilibrium pressure P, (Pa).

The integral reduced (to the mass of the adsorbent) enthalpy of the mixed adsorbate was calculated as the sum of the integral reduced enthalpies of the adsorbate components at spreading pressure π and, accordingly, a_i^*, corrected for the actual filling of the pore with this component a_i:

$$h_a = \sum_{i=1}^{N} \frac{h_i(a_i^*, T)}{a_i^*} a_i = \sum_{i=1}^{N} h_i(a_i^*, T) \frac{X_i \cdot a_\Sigma}{a_i^*}, \tag{5}$$

where h_i is the integral reduced (to the mass of the adsorbent) enthalpy of the pure component i of the adsorbate at adsorption a_i^* and temperature T, (J/kg).

2.5. Molecular Modeling

The adsorption of a methane–ethane mixture was modeled by molecular dynamics using the Dynamic software module from the Tinker molecular modeling software package [29]. To reduce the cost of computer time, the universal atom–atom force field OPLSAA was used [30], which simulates the total interaction potential. It was shown in [31] that among the existing universal potentials, this force field is best suited for describing model adsorption systems consisting of carbon structures and hydrocarbon molecules.

The simulation was carried out in the canonical (N, V, T are the number of molecules, volume, and temperature, respectively) ensemble. The simulation cell was a parallelepiped. Its height corresponded to the width of the pore plus 2 radii of carbon atoms. Micropores with a width in the range from 0.6 to 1.8 nm with a step of 0.2 nm were studied. The selected range of micropore widths corresponds to the most common widths for industrial carbon adsorbents. The walls of micropore are formed by single-layer graphene. The side faces of the simulation cell were 10 nm. The edges of the simulation cell are limited by periodic boundary conditions. The temperature of the numerical experiment was 293 K. Andersen's thermostat was used to thermostat the simulation system [32]. Studies of the molecular dynamics trajectory were carried out in the time interval 2×10^{-9} s. The elementary step of integrating the equation of motion was 10^{-15} s. Averaging of the system parameters for processing the results of a numerical experiment was carried out every 10^{-12} s. The time to reach the equilibrium states of the studied systems was estimated from the change in the total energy with time; it was no more than 10^{-9} s.

2.6. Adsorption Storage

The paper considers a model of an adsorption storage with a volume of 1 m^3, limited by an aluminum shell. The calculation uses 5005-H18 alloy, which has average characteristics for aluminum alloys and is comparable in strength to popular stainless steels. With a shell mass of 250 kg or 500 kg, the maximum safe pressure inside is 5.0 and 10.0 MPa, respectively, however, taking into account possible temperature changes during gas storage, charging should be limited to pressures of 3.5 and 7.0 MPa, respectively. The storage is completely filled with a monolith adsorbent formed from ACPK active carbon and a polymeric binder with mass fractions of 95% and 5%, respectively. The density of the monolith adsorbent in the dried state is 720 kg/m^3 (corresponds to the indicators achieved in practice). The storage model is lumped, zero-dimensional, with a uniform distribution of temperatures, pressure, and adsorption of each component of the mixture.

A simplified scheme of the adsorption storage is shown in Figure 1. The storage can consist of several tanks A1 ... AN, charging up to a pressure of 3.5 or 7.0 MPa (before the pressure regulator PR1) using a compressor C1 or directly from the gas pipeline, while the temperature of the supplied gas is constant (in all calculations 293 K). Gas is supplied to the consumer at a pressure of 0.1 MPa (after the PR2 regulator); if necessary, the pressure of the gas supplied to the consumer is increased using the C2 compressor. Constant pressure at the inlet (before PR1) and outlet (after PR2) allows us to assume that no technical work is done on the system, and the change in the enthalpy of the adsorption storage occurs due to the enthalpy of the inlet or outlet gas and heat supply or removal. The model is based on the assumption that there is no natural heat exchange with the environment: heat flow only occurs forcibly and is controlled using the TCU–thermal control unit. In practice, thermal control is provided using built-in heat exchangers [22,33,34] or gas circulation through the adsorbent bed with cooling/heating in an external heat exchanger [35–37].

The paper analyzes the cyclic charging and discharging processes in the adsorption storage in three different modes:

(a) in "isothermal" mode, which can be oriented as a base case;
(b) in a mode without heat exchange with the environment, which can be considered conditionally "adiabatic" (if we do not take into account the mass transfer due to the charging and discharging of gas);
(c) in "thermo-controlled" mode with cooling at gas charging and heating at gas discharging. In terms of temperature change, this mode is opposite to the "adiabatic" mode.

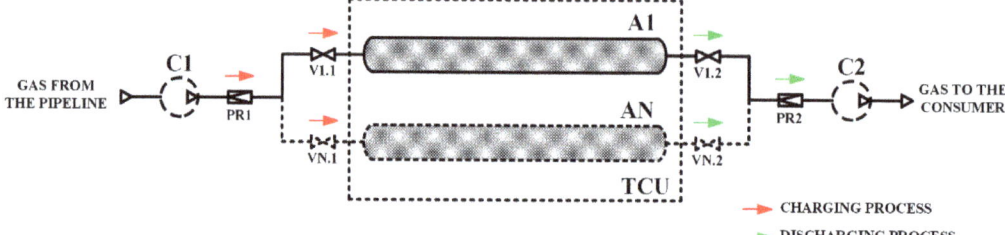

Figure 1. Scheme of the adsorption storage. C1, C2–natural gas compressors; A1 ... AN–ANG storage systems (adsorption tanks); TCU–thermal control unit; V1.1 ... VN.1 and V1.2 ... VN.2–stop valves; PR1-2–pressure regulators.

Cycling calculations are based on the heat balance of the system and the mass balance for each component.

The mass of each component of the mixture in the adsorption storage is determined by the expression, (kg):

$$M_i = \left(\rho_p(1 - x_b)a_i\mu_i + Y_{mi}\varepsilon\rho_{g\Sigma}\right)V_s , \qquad (6)$$

where ρ_p is the packing density of the adsorbent, (kg/m³); $x_b = 0.05$ is the mass fraction of the polymer binder; μ_i is the molar mass of component i, (kg/mol); $\rho_{g\Sigma}$ is the density of the gas phase as a mixture, (kg/m³) at pressure P and temperature T in storage; ε is the porosity, the fraction of space free for the gas phase; Y_{mi} is the mass fraction of component i in the gas phase; V_s is the volume of adsorption storage, (m³).

The enthalpy of the entire storage is the sum of the enthalpies of its individual components, (J):

$$H_\Sigma = \left(h_a\rho_p(1 - x_b) + h_{g\Sigma}\varepsilon\rho_{g\Sigma}\right)V_s + H_b + H_s , \qquad (7)$$

where $h_{g\Sigma}$ is the specific enthalpy of the gas phase as a mixture, (J/kg); H_b is the enthalpy of the binder, (J); H_s—is the enthalpy of the storage metal shell, (J).

3. Results and Discussion

3.1. Characterization

ACPK carbon adsorbent is made from highly decomposed peat, the content of carbon and mineral impurities in which can vary over a wide range. As a result, the adsorbent is heterogeneous in content. Part of the carbon remained in its original form of large nanocrystals of ordered graphite, for which narrow reflections (002), (10), (100), (004), and (11) are characteristic, Figure 2. An increase in the background to small angles indicates the content of high-carbon radicals of the amorphous phase in the structure of the adsorbent. In addition to reflexes related to graphite deposits, a number of extraneous peaks are also observed, indicating the presence of mineral inclusions.

Based on the experimental data of SAXS (Figure 3), for the region of inhomogeneities III (1–2 nm), which correspond to scattering in micropores, the radius of gyration (inertia) R_G was calculated using the method in [38]. This integral parameter conditionally characterizes the average sizes of micropores, and thus is most indicative in the analysis of the porous structure of carbon adsorbents [39]. Based on the obtained values of R_G, the sizes of model micropores of different shaped adsorbents were determined: spherical R_S and cylindrical R_T. The results of determining the parameters of the model adsorbent pores and gyration radius are shown in Table 1.

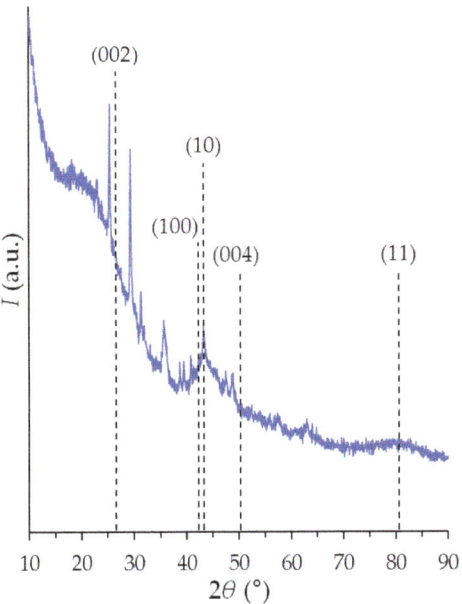

Figure 2. X-ray diffraction pattern of the carbon adsorbent ACPK. Graphite reflections are (002), (10), (100), (004), and (11).

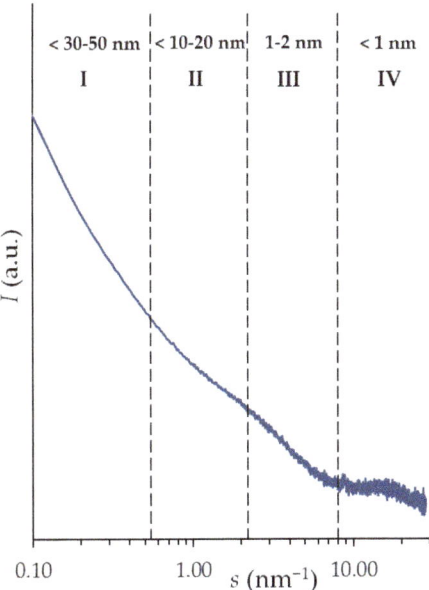

Figure 3. Dependence of the small-angle X-ray scattering intensity on the carbon adsorbent ACPK on the scattering vector at small and medium angles. I–IV are the characteristic X-ray scattering data areas used to obtain the structural parameters of the adsorbents.

Table 1. Parameters of the microporous structure of the ACPK carbon adsorbent calculated from SAXS data for various pore models.

Pore Model	Spherical		Cylindrical	
Model characteristics	R_G	R_S, nm	R_G	R_T, nm
ACPK	0.40	0.52	0.22	0.31

Pictures of the surface by scanning electron microscopy (SEM) are shown in Figure 4. The structure of the adsorbent has a granular type, and an inhomogeneous appearance in shape and size. In places, grains of a layered type are observed, similar to carbon graphite.

(a) (b)

Figure 4. Electron microscopy images of the surface of the adsorbent from peat raw material ACPK: 50 µm (a); 5 µm (b).

The presence of a large number of mineral impurities on the surface of the adsorbent is probably due to the fact that the raw in the form of fossil peat and coal dust contains a significant proportion of various elements. High content of impurities, up to 13.3% wt. (Table 2), may qualitatively indicate that the contribution to the energetics of adsorption of peat coals is made by surface chemistry, and not only by the width of micropores and their size distribution.

Table 2. Elemental chemical composition of the ACPK surface.

Content	C	O	Other Impurities
% at.	75.6	18.3	6.1
% wt.	65.6	21.1	13.3

Adsorption–desorption isotherms of standard nitrogen vapor at 77 K of the studied adsorbent are shown in Figure 5. As follows from Figure 5, the isotherm in the coordinates $a = f(P/P_S)$ has an L-shaped form of type I [40] in the initial region of the isotherm up to 0.3 P/P_S, which indicates the presence of a developed volume of micropores in the porous structure of the adsorbent. At pressures close to the saturated vapor pressure, the adsorption–desorption isotherms show a capillary-condensation hysteresis loop of the H4 type [40], which is characteristic of a mesoporous structure.

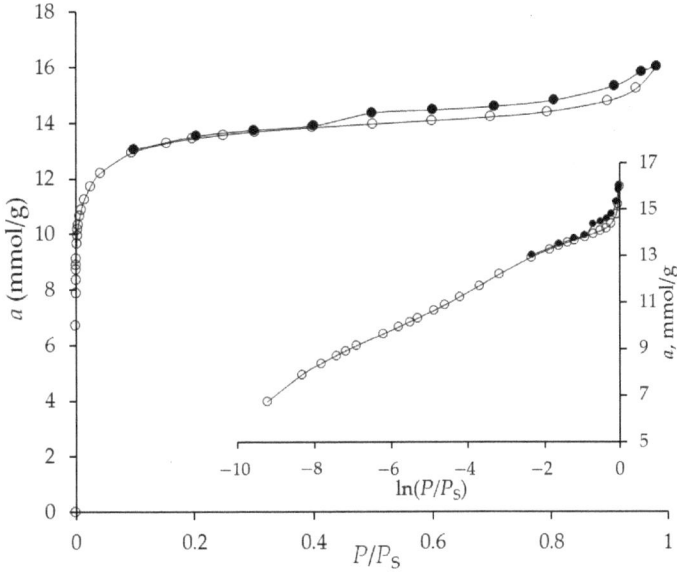

Figure 5. Adsorption (light symbols) and desorption (solid symbols) isotherms of nitrogen on the microporous carbon adsorbent ACPK at 77 K in linear and semi-logarithmic coordinates. Lines–spline approximation.

Table 3 presents the parameters of the porous structure of the ACPK adsorbent. As follows from Table 3, the mesopore volume of the adsorbent is about 25% of the total pore volume, and the mesopore surface is 40 m²/g. Thus, the adsorbent contains a significant proportion of mesopores, the adsorption of gases in which must be taken into account.

Table 3. Parameters of the porous structure of the carbon nanoporous adsorbent ACPK.

Method	BET	TVFM	TVFM	TVFM	TVFM	TVFM	BET	QSDFT
Adsorbent	S_{BET}, m²/g	W_0, cm³/g	x_0, nm	$E_{0(N2)}$, kJ/mol	W_S, cm³/g	W_{ME}, cm³/g	S_{ME}, m²/g	r_{max}, nm
ACPK	1105	0.44	0.52	7.63	0.56	0.12	40	0.59

To estimate the size distribution of micropores, we used the approach of the density functional theory QSDFT for a spherical pore model as the most probable pore model for a given adsorbent according to X-ray diffraction data. The distribution curves of micropores by size $dW_0/dd = f(d)$ are shown in Figure 6. The maximum on the micropore size distribution curve corresponds to r_{max} = 0.56 nm.

The pore parameters determined from the results of XRD (Table 1) coincide with the determination by the TVFM [41] (Figure 5, Table 3) and are close to the pore size distribution determined by the QSDFT method (Figure 6, Table 3). In this case, the adsorbent has a second, less pronounced maximum on the distribution curve, with a radius corresponding to the distribution maximum of 0.92 nm. The ratio of pore volumes for the two distribution maxima is about 75/25% for narrow and wide pores, respectively.

3.2. Equilibrium Adsorption Tests

Methane adsorption on ACPK was measured in the temperature range from 213 to 393 K, and ethane in the temperature range from 273 to 333 K. The results are shown in Figure 7. The isotherms of both gases are type I isotherms [40]. At a pressure of 3.5 MPa and 293 K, the adsorption of methane on the synthesized adsorbent ACPK is 6.55 mmol/g or 10.5% wt., and at a pressure of 7.0 MPa and 293 K, respectively, 7.80 mmol/g and 12.5% wt. These adsorption properties make it possible to use this adsorbent for the purpose of methane (natural gas) storage, but under the condition of a high packing density of the adsorbent, for example in the form of monoliths [34,42–45].

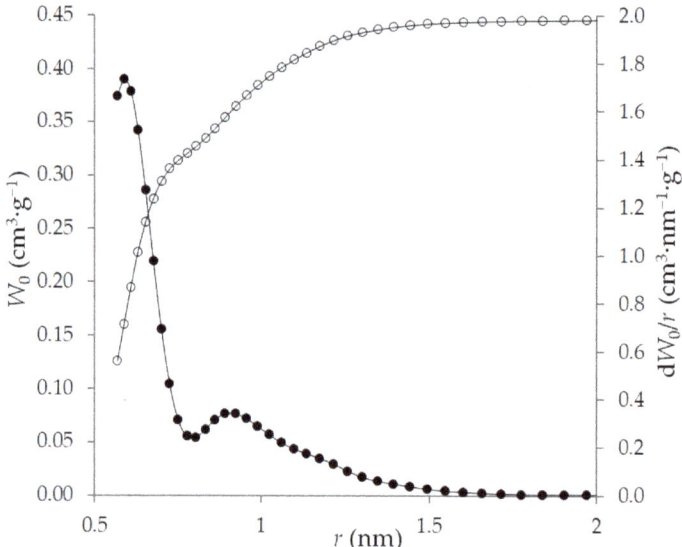

Figure 6. The QSDFT cumulative W_0 (light symbols) and differential dW_0/r (dark symbols) pore volume distributions calculated for ACPK from the nitrogen adsorption at 77 K for a sphere pore model.

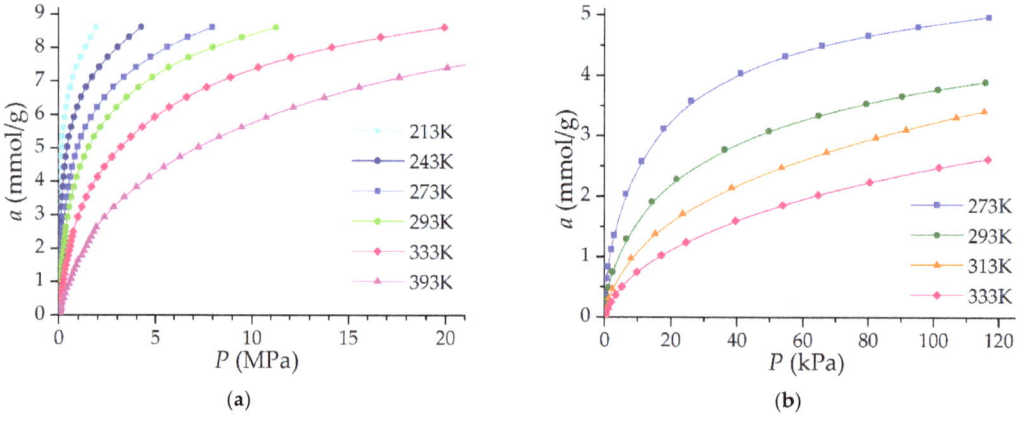

Figure 7. Adsorption isotherms of methane (**a**) and ethane (**b**) on the ACPK adsorbent. Symbols are an experiment. The lines are a spline approximation.

Figure 8 shows the dependence of the differential molar isosteric heat of adsorption of methane and ethane on the nanoporous carbon adsorbent ACPK at a temperature of 293 K. The plotting area for C_2H_6 is limited by the area of reliable experimental measurements of adsorption with the possibility of isostere linearization by at least three points. Heat is determined without taking into account adsorption-stimulated and thermal deformation. The value of carbon adsorbents deformation under the described conditions does not exceed 0.3–0.4%, while the contribution of deformation to the heat of adsorption at a temperature of 293 K does not exceed 2–5% [23,46,47]. The absence of an abrupt jump in the heats of adsorption in the initial region of filling indicates a weak inhomogeneity of the adsorbent sorption surface. A barely noticeable extremum at 0.2 mmol/g is observed on ethane, but this is probably due to experimental errors due to the proximity to the boundary of the measurement region. Otherwise, the dependences of the heats of adsorption of methane and ethane in the studied area repeat the shape of each other.

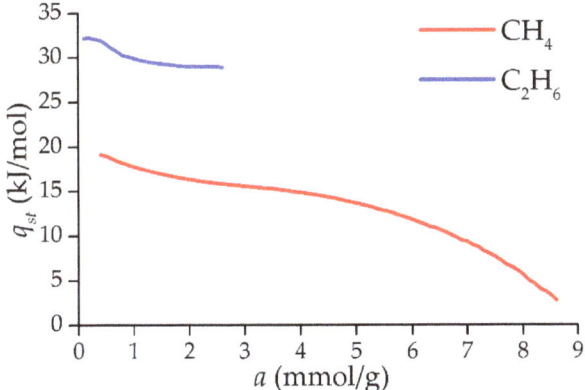

Figure 8. Differential molar isosteric heat of adsorption of methane and ethane on the microporous carbon adsorbent ACPK at a temperature of 293 K, determined from experimental data.

3.3. Mixture Adsorption

The IAST results of determining the adsorption of a mixture and individual components in a mixture of CH_4 and C_2H_6 are shown in Figure 9. With a content of 2% mol. ethane in the gas phase, its contribution to the total adsorption is much more significant: from 14.7% mol. at 3.5 MPa and 333 K up to 23.8% mol. when the temperature decreases to 273 K. At the same time, methane remains the main adsorbed gas in the entire considered pressure range from 0.05 to 7 MPa. At a content of 10% mol. ethane in the gas phase, the situation is reversed: ethane is adsorbed in even larger quantities than methane at a temperature of 273 K (59.5% mol. of ethane in the total adsorption at 3.5 MPa) to approximately equal adsorption with methane at a temperature of 333 K.

It is obvious that such active absorption of ethane can adversely affect the active capacity of the adsorption storage due to the fact that ethane is less efficiently desorbed upon depressurization compared to methane. On the other hand, as can be seen from Figure 9, the total molar adsorption of the mixture varies slightly with a change in the ethane content in the gas phase, however, the higher heating value of 1 mol of ethane is 75% higher than that of methane. Thus, the energy capacity of the adsorption storage increases with an increase in the ethane concentration: at a pressure of 3.5 MPa and 293 K, the higher heating value of the adsorbate at an ethane concentration of 10% mol. in the gas phase is 8.57 MJ per 1 kg of adsorbent, while in the case of pure methane this heat is 5.84 MJ/kg–1.47 times less–while the higher heating value of the gas phase differs by only 7.5%. In some cases, this effect can be useful; for example, by integrating the storage into a flowing gas pipeline network: the storage capacity will increase significantly due to the capture of ethane and heavy hydrocarbons from the flow. However, the extraction of this excess energy is complex due to the difficulty of extracting ethane (and other heavy hydrocarbons, if present in natural gas). The advantages and disadvantages of the presence of impurity hydrocarbons in natural gas depend on the form of the gas charging and discharging processes organization.

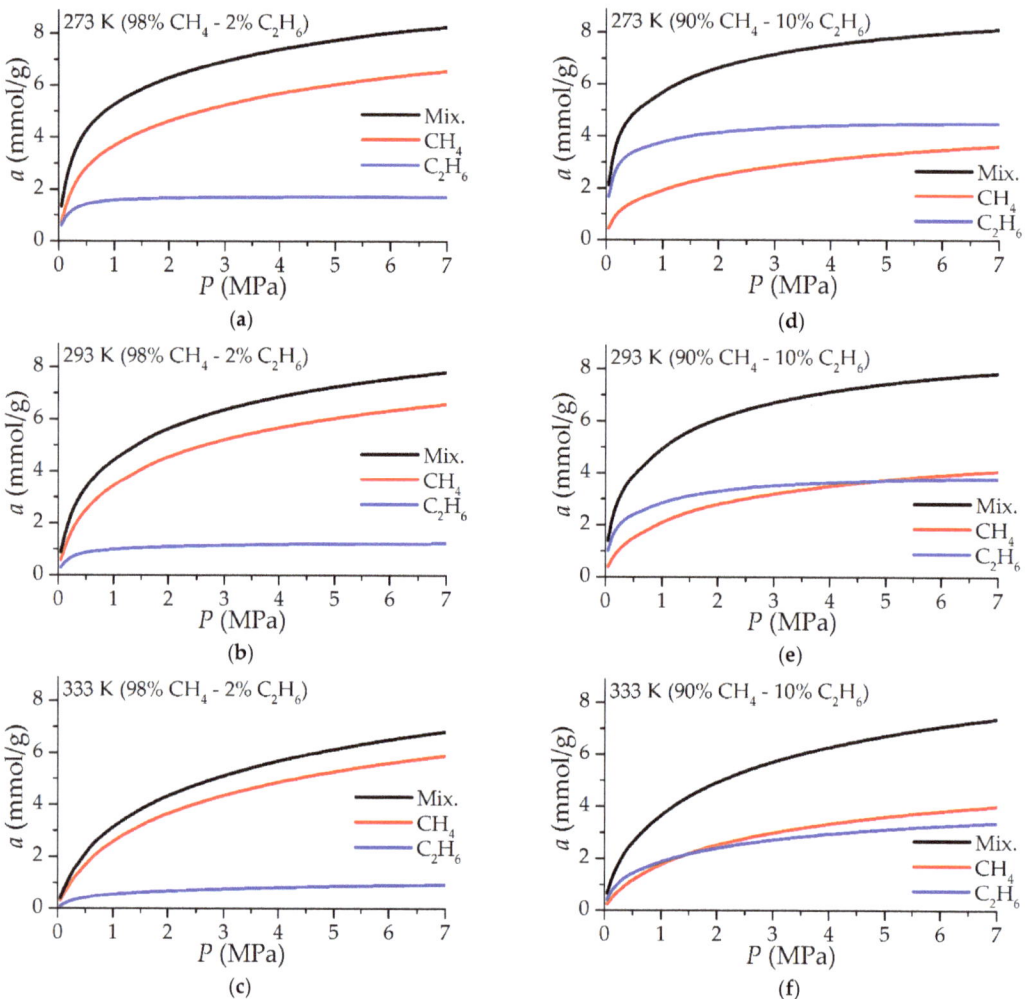

Figure 9. Equilibrium adsorption of a binary mixture of methane and ethane depending on the pressure of a gas mixture with a molar content of 2% (**a–c**) and 10% (**d–f**) ethane at temperatures of 273 K (**a,d**), 293 K (**b,e**), 333 K (**c,f**). The calculation method is IAST using experimental isotherms of pure components.

3.4. Molecular Modeling

For a more detailed understanding of the mechanism of mixture adsorption of methane and ethane, a numerical simulation in model pores of a carbon adsorbent of various widths H_S from 0.6 to 1.8 nm with a step of 0.2 nm was performed using the molecular dynamics method. For the study, a slit-like micropore model was used as it is one of the most common models for the formation of pores in carbon adsorbents. [48,49]. Carbon adsorbents, in general, consist of an ordered part of carbon crystallites and an amorphous part of high-carbon radicals [50]. The ratio of the amorphous and crystalline parts of carbon is determined by the raw and carbonization conditions. When using organic carbon-containing raw materials, the structure of crystallites resembles graphite, but they themselves are packed less regularly in the volume of carbon. At the same time, an increase in activation leads to a decrease in the fraction of crystallites and an increase in the amorphous fraction. The amorphous part may contain micropores of various shapes, including those close to spherical. According to the results of the adsorbent structure study (Tables 1 and 3 and Figure 6), the ACPK adsorbent contains a significant part of amorphous carbon with a pore shape close to spherical. On the other hand, X-ray diffraction analysis (Figure 2) revealed the presence of peak characteristics of graphite-like microporous structures. Thus, the use of the slit-like micropores of graphite-like crystallites model of a carbon adsorbent in a numerical simulation is legitimate. A numerical experiment was carried out for a mixture 95% CH_4 and 5% C_2H_6. The temperature of the experiments was 293 K. The concentration of the mixture for the study was selected based on the data on the average concentration of C_{2+} hydrocarbons.

Figure 10 shows the dependencies of the probability of the mass center location of a methane or ethane molecule in a pore for various pore widths and their filling. In the process of adsorption, for narrow pores 0.6, 0.8, and 1 nm wide, the concentration displacement of ethane is shown. For pores with a width of 1 nm, close to the size of ACPK micropores according to X-ray diffraction analysis (Table 1) and the first maximum on the QSDFT pore size distribution curve (Figure 6), the mixture is almost completely sorbed in the pore at low fillings (up to 100 molecules in the modeling system), forming molecular complexes near the graphene walls, Figure 10a. Figure 10a shows that the centers of mass of ethane molecules are shifted relative to the centers of mass of methane molecules, which can be explained by the fact that linear ethane molecules tend to occupy a perpendicular position to the micropore walls. With an increase in the number of molecules in the simulation system, methane molecules completely fill the molecular complexes located in the potential minima of the model micropore near the graphene surface, displacing ethane molecules outside the pore, Figure 10b.

For micropores with a width of more than 1.2 nm, the effect of concentration displacement is not observed. Figure 10c,d show the simulation results for a wide model micropore with a width of 1.8 nm, which is close in value to the wide ACPK pores corresponding to the second distribution maximum in Figure 6. At low fillings, the behavior of adsorbate molecules is similar to that observed in narrow pores, but at high fillings, ethane molecules are present in the micropore, forming a "layered" structure with methane molecules, while ethane molecules are displaced from molecular complexes closer to the center of the micropore.

Figure 11 shows instantaneous snapshots of the molecular dynamics trajectory of the simulation system for micropores of different widths. These images clearly demonstrate the dependences shown in Figure 10, while expanding our understanding of the behavior of methane and ethane molecules in a micropore. Thus, in a narrow micropore 1.0 nm wide, sorbed ethane molecules are located closer to the edges of graphenes, Figure 11a, probably due to the predominant interaction of methane molecules with each other in the field of dispersion forces of the adsorbent than the interactions of methane–ethane. As a result, ethane molecules are forced out of the narrow micropore.

In the case of a wide pore 1.8 nm wide, the situation is different: Figure 11b shows that methane and ethane molecules together form molecular complexes in accordance with Figure 10c. With further filling of a wide micropore, ethane is not completely displaced from the pore, Figure 11c: in the center of the micropore, the density of molecules is low, which allows two more layers of ethane molecules and even a third layer of methane between them to be accommodated, Figure 10d.

Figure 12 shows the dependence of the number of adsorbed methane and ethane molecules on the number of molecules in the ratio of 95% methane and 5% ethane in simulation cells with a width of 1.0 nm and 1.8 nm at a constant temperature of 293 K, the molecular adsorption isotherm of the mixture.

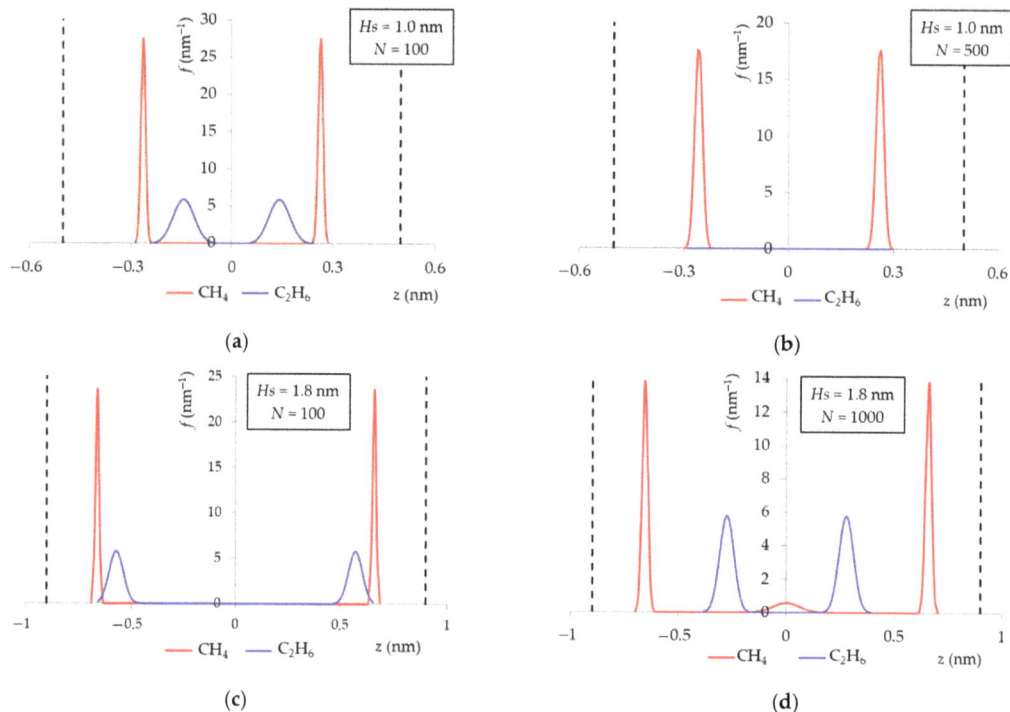

Figure 10. Probability density of the mass center location of a methane and ethane molecule in a model pore of a carbon adsorbent with a width of H_S = 1.0 nm (**a**,**b**) and 1.8 nm (**c**,**d**) for different numbers of molecules N in the simulation cell, corresponding to the region of averages (**a**,**c**) and maximum (**b**,**d**) fillings of micropores. The number of molecules of the mixture N (the ratio of the molecules number of methane and ethane is 95:5) in the modeling system, pcs: 100 (**a**); 500 (**b**); 100 (**c**); 1000 (**d**). The temperature is 293 K. The ordinate axis intersects the abscissa axis at point z = 0, corresponding to the symmetry plane of the pore. The dotted line is the boundary of the model micropore.

In a narrow micropore 1.0 nm wide, Figure 12a, in the region of up to 100 molecules (95 methane molecules and 5 ethane molecules), a classical increase in the amount of adsorbed methane and ethane is observed in the system. Moreover, almost all ethane molecules and most of the methane molecules in this region from the free phase enter the micropore. With an increase in the number of molecules in the simulation system, the number of ethane molecules in the pore gradually decreases due to its displacement by methane. At 160 molecules of the mixture (an instantaneous snapshot was presented in Figure 11a) in the simulation system, ethane is completely displaced by methane from micropores: an interesting behavior is observed that if there are ethane molecules in the free phase (analogue of the gas phase), they are not in the adsorption pore. This behavior is probably explained by the difference in size and shape of methane and ethane molecules. Small and spherically symmetric methane molecules fill the micropore into two adsorption layers, forming a more energetically favorable state of the adsorbate than ethane molecules, which, due to their linear shape, tend to occupy a perpendicular position in the narrow micropore and thereby prevent the formation of the second adsorption layer. Figure 12b shows the results of modeling the mixture adsorption in wider pores, using the example of a micropore 1.8 nm wide. In this case, ethane adsorption continuously increased as the number of molecules in the system increased. From the previously considered

Figures 10 and 11, it is already known that ethane is not completely replaced by methane in a wide pore.

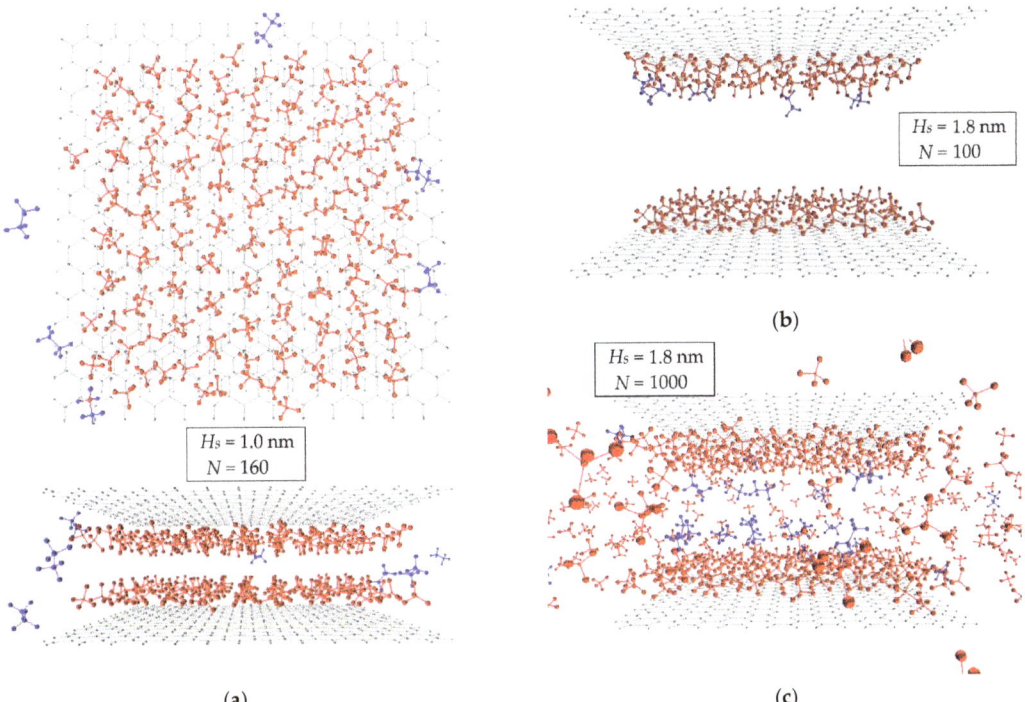

Figure 11. Instantaneous snapshot of the molecular dynamics trajectory of a simulation system consisting of two graphenes at a distance of H_S = 1.0 nm (**a**) and 1.8 nm (**b**,**c**) at different numbers of N molecules (the ratio of the number of molecules of methane and ethane is 95:5) in simulation cell: 160 (**a**); 100 (**b**); 1000 (**c**). The bonds of atoms forming graphene are indicated by solid lines. Atoms are shown as unscalable red (methane) and blue (ethane) spheres.

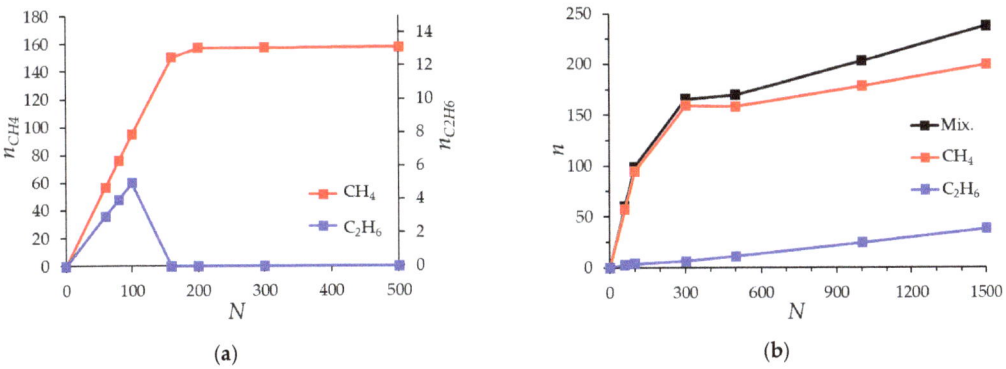

Figure 12. Dependence of the number of adsorbed molecules n of methane CH_4 and ethane C_2H_6 and their total number (Mix.) on the number of molecules in the simulation cell N (ratio of the number of molecules of methane and ethane 95:5) for a pore with a width of 1.0 nm (**a**) and 1.8 nm (**b**). The temperature is 293 K.

The obvious differences between the molecular isotherms shown in Figure 12 and the adsorption isotherms of the mixture calculated using the IAST model (Figure 9) are explained by the fact that the IAST method is based on the isotherms of the pure components, which are in fact integral isotherms over all pores of a real adsorbent with a wide pore size distribution. Figure 13 shows an attempt by the authors to evaluate the integral adsorption isotherm of a mixture obtained by summarizing molecular isotherms for pores of various widths, the weight contribution of which is determined by the share of pores of the corresponding width in the pore size distribution defined by the QSDFT method (Figure 6) for the ACPK adsorbent. As can be seen from Figure 13, the integral molecular adsorption isotherms of the mixture and their components behave similarly to the isotherms calculated using the IAST model: thus, the adsorption of ethane in a wide range of fillings (from 100 to 500 molecules) changes relatively weakly, similar to the saturation of ethane in Figure 9. This is due to the simultaneous decrease and increase in the amount of sorbed ethane in pores of various sizes. A direct comparison of the integral molecular isotherm and IAST isotherms is difficult due to the correct consideration of the free (gas) phase, as well as differences in the pore structure of the real and model adsorbent, due to the presence of a significant fraction of the amorphous carbon phase in the structure of the real adsorbent.

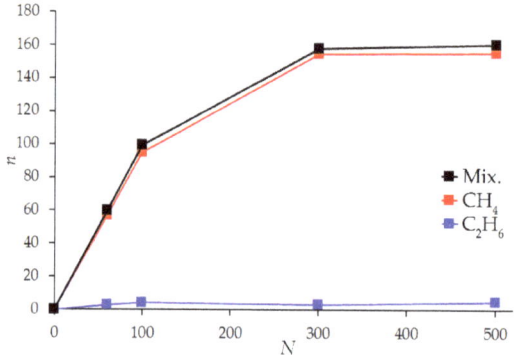

Figure 13. Integral dependence of the number of adsorbed molecules n of methane CH_4 and ethane C_2H_6 and their total number Mix. on the number of molecules N (ratio of the number of molecules of methane and ethane 95:5) in an array of simulation cells, quantitatively corresponding to the pore size distribution of the ACPK adsorbent. The temperature is 293 K.

Despite the fact that the IAST model and the molecular model can produce results similar in form, it is obvious that some molecular features of the filling of micropores of different sizes with certain substances cannot be taken into account in principle by the IAST model, even when applying this model to a specific model micropore. For example, the case of a slit-like micropore 1.0 nm wide: the adsorption of pure methane and ethane in it is not equal to zero, but the mixture adsorption, according to the results of molecular modeling, can lead to the complete displacement of one of the components, which is not observed in the IAST model.

3.5. Cycles in Adsorption Natural Gas Storage

To assess the influence of thermal effects and the presence of impurities (ethane in a binary mixture with methane), modeling of cyclic processes of gas charging and discharging was performed using the IAST method. Idle time and gas storing processes have been eliminated, charging starts immediately after discharging, and vice versa. In the initial state, the adsorption storage is filled with pure methane at a pressure of 0.1 MPa and a temperature of 293 K. The storage is charged up to 3.5 or 7.0 MPa. The supplied gas has a pressure of 3.5 or 7.0 MPa, respectively, and a temperature of 293 K. Gas leaves the storage under variable conditions corresponding to the conditions in the storage. Since in the considered zero-dimensional model (with lumped parameters) all processes are in equilibrium and there is no natural heat exchange with the environment, the time factor does not play a significant role. The time scale is replaced by the gas exchange scale: the total mass of gas entering and exiting the storage. If we assign certain flow rates for gas charging and discharging, then such a gas exchange scale is definitely translated into a time scale, but this does not play a role for the analysis of cycles.

Figure 14 shows the results of modeling the first 50 cycles of "adiabatic" (without heat exchange with the environment and thermal control) charging up to 3.5 MPa with gas containing 98% mol. methane and 2% mol. ethane and gas discharging up to 0.1 MPa from storage. The gradual cooling of the storage under "adiabatic" conditions is explained by mass exchange with the external environment: the incoming gas at a temperature of 293 K is colder than the outgoing gas in the first cycles, which ultimately leads to the stabilization of the storage temperature below the initially set 293 K. Despite the very low content of ethane in the inlet gas, it accumulates in the system until the outlet gas has, on average, the same concentration of ethane as the inlet gas. Therefore, in the first approximation, the effect of ethane can be estimated from the equilibrium adsorption isotherms of the mixture shown in Figure 9. However, it should be taken into account that the molar fraction of ethane changes during the process. The share of ethane in the adsorbate after 50 cycles reaches 41.5% mol. at the end of discharging (at the minimum pressure in the system) and about 21.5% mol. at the end of charging (at maximum pressure in the system). At the end of 50 cycles shown in Figure 14, the periodic mode is not yet steady. The periodic mode can be considered established approximately around 110 cycles of adiabatic charging and discharging, when the difference in the amount of charged and discharged ethane differs by less than 1%, and this error becomes comparable with the mathematical errors of the model itself.

To assess the effect of impurities in fuel storage systems, it should be taken into account that both considered gases are combustible and targeted. Therefore, the influence of impurities should be assessed from the position of loss in the energy capacity of the storage. The same approach makes it possible to take into account the thermal effects of adsorption and desorption, which are expressed in a lower active storage capacity. The usual approach to assessing the active capacity of an adsorption storage in terms of volume (m^3 of gas per m^3 of volume) is incorrect, since 1 m^3 of ethane under normal conditions (101,325 Pa and 293.15 K) has the higher heating value by 76% more than 1 m^3 of methane. The assessment by weight can be considered justified: the difference in the combustion energy of 1 kg of methane and ethane is only 7%, which with ethane content up to 10% mol. leads to a mass combustion energy range of 54.9–55.5 MJ/kg.

Figure 15 shows the dependencies of the higher heating value of the discharging gas by cycles for various types of charging up to a pressure of 3.5 MPa and discharging up to 0.1 MPa: "adiabatic", "isothermal", which is not ideal, but convenient as a reference point, and "thermocontrolled", the opposite of adiabatic, i.e., with cooling when charging up to 293 K and heating when discharging gas up to 333 K. The adiabatic mode significantly reduces the active energy capacity of the storage: by 27–33% compared to the isothermal process with the same composition of the supplied gas; this is the influence of the thermal effects of adsorption and desorption. Surprisingly, in the adiabatic mode, the accumulation of ethane has practically no effect on the active energy capacity of the storage (compared to operation on pure methane): losses in methane are compensated over time by a large amount of ethane released. The "harmful" accumulation of ethane is more noticeable in isothermal and thermocontrolled modes: over time, the active energy capacity of the storage decreases to a certain steady-state value. In isothermal mode 2% mol. the concentration of ethane in the gas supplied leads to a loss of efficiency of 5.7%, and at a concentration of 10% mol. losses already 16.6%. The accumulation of ethane has a greater effect on the isothermal mode.

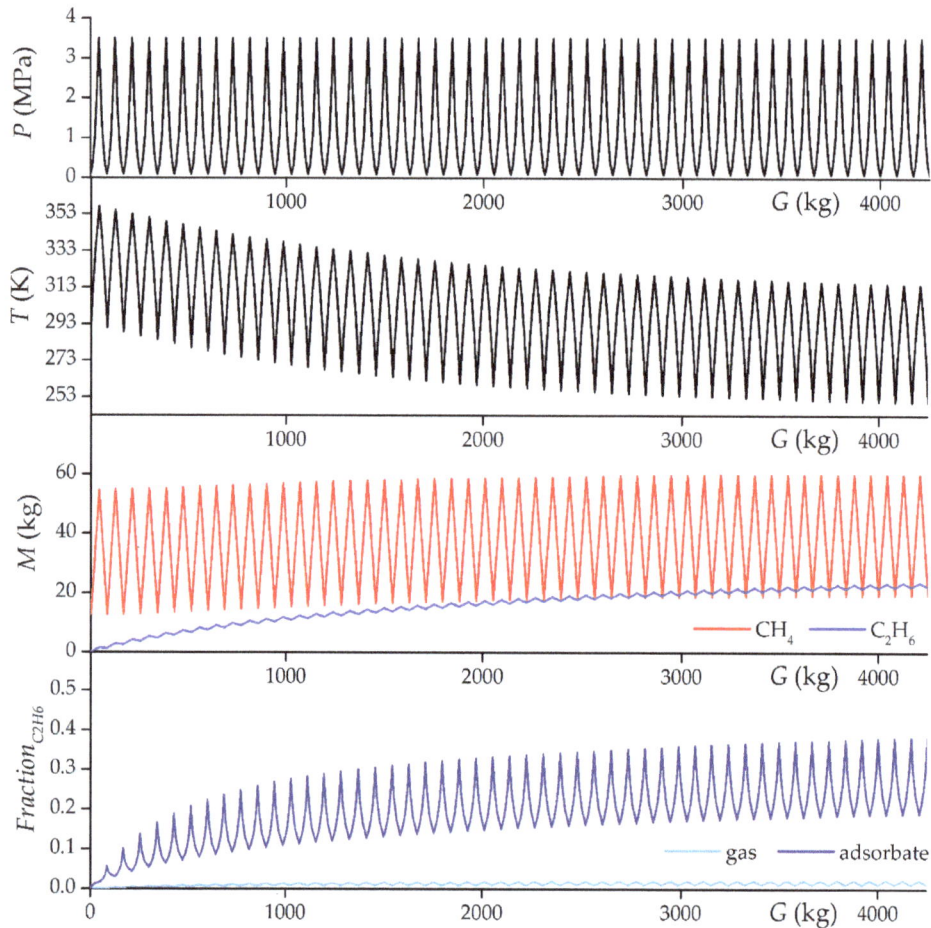

Figure 14. Dependencies of pressure, temperature, and mass of methane and ethane in the adsorption storage with a volume of 1 m^3 based on ACPK adsorbent and the mole fraction of ethane in the gas and adsorbed phases in the "adiabatic" cyclic charging/discharging process and supplied gas with 98% mol. methane and 2% mol. ethane using the IAST simulation results from the cumulative gas exchange since the beginning of the simulation. The first 50 cycles are shown.

The advantages of the thermocontrolled mode in the energy storage capacity compared to the isothermal mode depend on the composition of the supplied gas: with a large number of cycles, the advantage is 9.4% on pure methane (due to a higher temperature at the end of the discharging process), 11.7% at 2% mol. fraction of ethane in the gas supplied to the system, and 19.5% at 10% mol. Fraction of ethane, i.e., the temperature-controlled mode allows for more efficient removal of ethane and contributes to a reduction in its accumulation. It can also be seen from Figure 15 that thermocontrolling reduces the duration of reaching a stable periodic mode compared to the isothermal mode, while the opposite adiabatic mode can be unstable for quite a long time. The greater the proportion of impurities in the incoming gas, the faster a stable periodic process comes in isothermal and thermocontrolled modes: impurities accumulate faster to a steady value. In general, about 20 cycles in all considered modes are sufficient to achieve an approximately steady value of the combustion energy, however, the accumulation of ethane in the considered processes continues after 20 cycles.

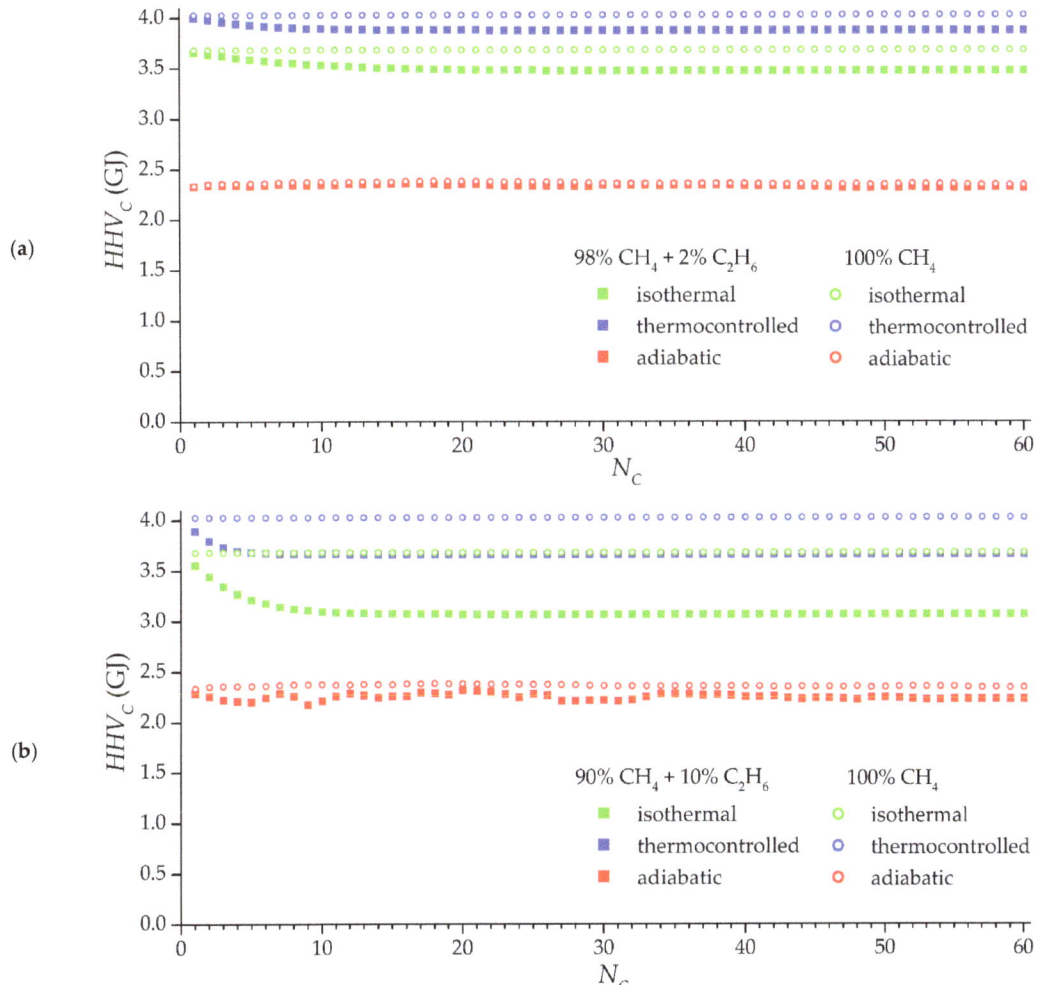

Figure 15. Dependence of the higher heating value of the mixed gas leaving for one cycle from a storage with a volume of 1 m³ based on the ACPK adsorbent, on the cycle number. Filled dots correspond to mixtures: 98% mol. methane and 2% mol. ethane (**a**); 90% mol. methane and 10% mol. ethane (**b**). Empty points correspond to pure methane.

These results are adequately consistent with the publications reviewed in the introduction. In most studies, the noticeable effect of ethane, which is the main impurity of natural gas, ended approximately in the region of 10–30 cycles [6,7,10]. In terms of the effect on the active capacitance, the results are also close to those published earlier: in [7], the effect of 10.3% mol. impurities (not only ethane) led to a loss of 20% capacity, and in [10] the effect of 10.4%mol. impurities to losses of 16%, which correlates with the simulation results. In general, such an approach can be considered justified, at least as a first approximation.

Figure 16 shows the dependencies of the steady-state values of the active energy capacity of the adsorption and conventional gas storage for one charging and discharging cycle on the ethane content in the supplied gas for different modes. As the ethane content increases, the energy capacity of the gas storage increases by increasing the density of the mixture–or by increasing the average molar energy of combustion. The efficiency of the adsorption storage, on the contrary, decreases. It is likely that there is a certain concentration of ethane at which adsorption storage will no longer provide an advantage over conventional gas storage. In Figure 16, this is clearly seen at a pressure of 7.0 MPa: with an ethane fraction of 10% mol. adsorption storage, which gives only a small advantage in active energy capacity compared to conventional gas storage: 29% in isothermal mode, although with pure methane the advantage is 71%. At a pressure of 3.5 MPa, the advantage of the adsorption storage is more obvious: even in the low-efficiency adiabatic mode, the energy capacity is higher than that of the conventional gas storage, and in the thermocontrolled mode, the active energy capacity is 2.5 times greater at an ethane concentration in the gas of 10% mol. and three times greater when operating on pure methane.

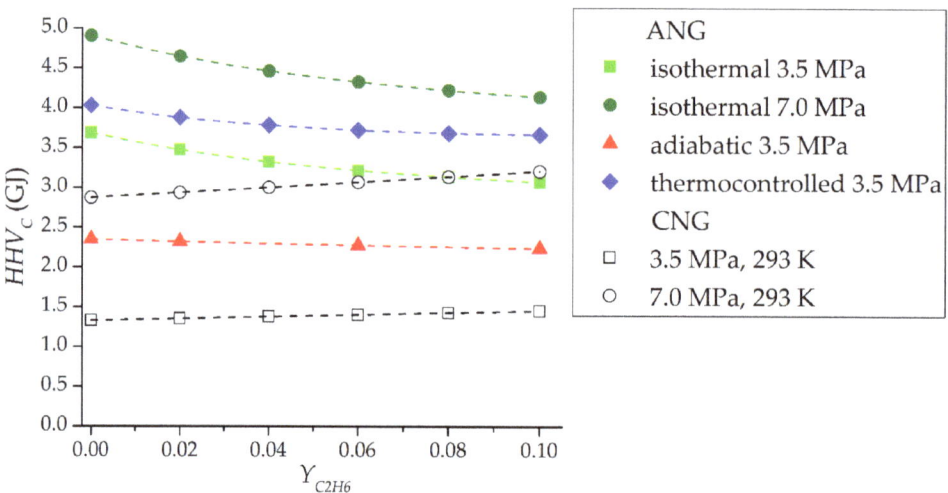

Figure 16. Dependence of the active energy capacity of the gas leaving the storage with a volume of 1 m^3 for 1 cycle based on ACPK adsorbent based on the results of modeling cyclic processes. ANG is the adsorption gas storage. CNG is the conventional gas storage in the "isothermal" mode, without taking into account non-equilibrium intake/exhaust processes.

The simulation results show that the area of effective application of adsorption storages on real natural gas with impurities is smaller than in the case of analyzing their operation on pure methane. Ignoring impurities during the design phase of the adsorption storage can lead to high and unrealistic expectations and will negatively affect real work. It should be noted that the dealing with impurities is possible not only due to the thermal management analyzed in the article, but also using various scheme solutions of the adsorption storage tank.

4. Conclusions

Synthesis of active carbon ACPK from peat raw materials was carried out. The resulting adsorbent has a wide pore size distribution with maxima in the region of a pore radius of 0.59 (main extremum) and 0.92 nm (secondary). The specific volume of micropores is 0.44 cm^3/g with a total pore volume of 0.56 cm^3/g. The adsorption of methane at 3.5 MPa and 293 K was 10.5% wt., which makes it possible to use this adsorbent for the purpose of methane storage under the condition of a high packing density of the adsorbent–for example, in the form of monoliths. Theoretical studies of a binary mixture of methane–ethane (with an ethane content of up to 10 mol.) adsorption were carried out by the IAST method based on experimental adsorption isotherms of pure components. It is shown that at a pressure of 3.5 MPa and 293 K, the higher heating value of the accumulated adsorbate at an ethane concentration of 10% mol. in the gas phase is 8.57 MJ per 1 kg of adsorbent, while in the case of pure methane this heat is 5.84 MJ/kg, 1.47 times less. Thus, capturing ethane (and other heavy hydrocarbons), for example, from a flowing natural gas stream can significantly increase the capacity of the adsorption storage, provided that the accumulated gases can later be recovered.

Numerical molecular modeling of the methane–ethane mixture adsorption in slit-like model micropores of various widths from 0.6 to 1.8 nm at a temperature of 293 K was carried out. It is shown that this mixture is characterized by the appearing characteristic molecular effects: the displacement of ethane by methane molecules from narrow 1.0 nm-wide micropores, as well as the formation of obvious layers of methane and ethane in wide micropores 1.8 nm wide. However, the integral molecular adsorption isotherm of the mixture, which includes adsorption in pores of various sizes, corresponding to the result of the QSDFT method, is similar in shape to the adsorption isotherm, determined by the IAST method. Thus, various molecular effects observed in individual model micropores together result in the classical adsorption isotherm.

Since the results of numerical molecular modeling showed no contradictions with the IAST model, this approach was used to simulate the cyclic processes of gas charging and discharging from a 1 m^3 adsorption storage filled with ACPK adsorbent. Three modes of operation were analyzed: adiabatic, isothermal, and thermocontrolled, which is the opposite of adiabatic. The results of the study showed that in the adiabatic mode, the accumulation of ethane can continue up to about 110 cycles, although the accumulation itself had very little effect on the higher heating value of the gas discharging in the cycle. To a much greater extent, the thermal effects of adsorption themselves were influential, reducing the energy capacity by 27–33%. The thermocontrolled process showed significant efficiency compared to the isothermal one: if on pure methane the advantage in energy capacity was 9.4%, then at 10% mol. ethane the share advantage increased by 19.5% due to a more efficient ethane recovery. Thus, thermal management (heating during discharging) of the adsorption storage is a viable way not only to directly increase the amount of outgoing gas, but also an effective means of dealing with the accumulation of ethane. The analysis of the active energy capacity of the adsorption storage in comparison with a conventional gas one showed that at a relatively high pressure of 7.0 MPa, the capacity of the adsorption storage is 71% higher in the isothermal mode on pure methane and only 29% higher when operating on natural gas with a content of 10% mol. ethane. Thus, it is not enough to evaluate the efficiency of the adsorption storage solely for pure methane; it is necessary to take into account the actual composition of the gas. At a pressure of 3.5 MPa, the efficiency of the adsorption storage is significantly higher than that of the conventional gas storage, especially in the thermocontrolled mode: the active energy capacity is 2.5 times higher at an ethane concentration in the gas of 10% mol. and three times higher when operating on pure methane.

Author Contributions: Conceptualization, A.V.S., E.M.S. and A.A.Z.; methodology, A.V.S., I.E.M. and E.M.S.; software, E.M.S.; validation, S.S.C., A.V.S. and V.V.G.; formal analysis, I.E.M., S.S.C. and A.A.Z.; investigation, I.E.M., A.V.S., E.M.S., S.S.C., V.V.G. and A.E.G.; resources, A.V.S., I.E.M. and A.A.Z.; data curation, A.V.S. and I.E.M.; writing—original draft preparation, E.M.S., S.S.C., A.V.S., I.E.M. and V.V.G.; writing—review and editing, E.M.S., S.S.C., A.V.S. and I.E.M.; visualization, E.M.S., S.S.C., V.V.G. and A.E.G.; supervision, A.A.Z. and A.V.S.; project administration, A.V.S. and A.A.Z.; funding acquisition, A.V.S. All authors have read and agreed to the published version of the manuscript.

Funding: The research was supported by a grant from the Russian Science Foundation (project No. 20-19-00421).

Data Availability Statement: Not applicable.

Conflicts of Interest: The authors declare no conflict of interest. The funders had no role in the design of the study; in the collection, analyses, or interpretation of data; in the writing of the manuscript, or in the decision to publish the results.

References

1. Golovoy, A. Sorbent-containing storage systems for natural gas powered vehicles. *SAE Technical Paper* **1983**, *831070*, 39–46.
2. Chugaev, S.S.; Strizhenov, E.M.; Zherdev, A.A.; Kuznetsov, R.A.; Podchufarov, A.A.; Zhidkov, D.A. Fire- and explosion-safe low-temperature filling of an adsorption natural gas storage system. *Chem. Pet. Eng.* **2017**, *52*, 846–854. [CrossRef]
3. Tsivadze, A.Y.; Aksyutin, O.E.; Ishkov, A.G.; Men'Shchikov, I.E.; Fomkin, A.A.; Shkolin, A.V.; Khozina, E.V.; Grachev, V.A. Porous carbon-based adsorption systems for natural gas (methane) storage. *Russ. Chem. Rev.* **2018**, *87*, 950–983. [CrossRef]
4. Biloe, S.; Goetz, V.; Mauran, S. Dynamic discharge and performance of a new adsorbent for natural gas storage. *AIChE J.* **2001**, *47*, 2819–2830. [CrossRef]
5. Mota, J.P.B. Impact of gas composition on natural gas storage by adsorption. *AIChE J.* **1999**, *45*, 986–996. [CrossRef]
6. Pupier, O.; Goetz, V.; Fiscal, R. Effect of cycling operations on an adsorbed natural gas storage. *Chem. Eng. Process.* **2005**, *44*, 71–79. [CrossRef]
7. Rios, R.B.; Bastos-Neto, M.; Amora, M.R., Jr.; Torres, A.E.B.; Azevedo, D.C.S.; Cavalcante, C.L., Jr. Experimental analysis of the efficiency on charge/discharge cycles in natural gas storage by adsorption. *Fuel* **2011**, *90*, 113–119. [CrossRef]
8. Romanos, J.; Rash, T.; Dargham, S.A.; Prosniewski, M.; Barakat, F.; Pfeifer, P. Cycling and regeneration of adsorbed natural gas in microporous materials. *Energy Fuels* **2017**, *31*, 14332–14337. [CrossRef]
9. Walton, K.S.; LeVan, M.D. Natural gas storage cycles: Influence of nonisothermal effects and heavy alkanes. *Adsorption* **2006**, *12*, 227–235. [CrossRef]
10. Prosniewski, M.; Rash, T.; Romanos, J.; Gillespie, A.; Stalla, D.; Knight, E.; Smith, A.; Pfeifer, P. Effect of cycling and thermal control on the storage and dynamics of a 40-L monolithic adsorbed natural gas tank. *Fuel* **2019**, *244*, 447–453. [CrossRef]
11. Bell, I.H.; Wronski, J.; Quoilin, S.; Lemort, V. Pure and pseudo-pure fluid thermophysical property evaluation and the open-source thermophysical property library CoolProp. *Ind. Eng. Chem. Res.* **2014**, *53*, 2498–2508. [CrossRef] [PubMed]
12. Dubinin, M.M.; Radushkevich, L.V. Equation of the characteristic curve of activated charcoal. *Proc. Acad. Sci. Phys. Chem. Sect. USSR* **1947**, *55*, 331–333.
13. Dubinin, M.M. Physical adsorption of gases and vapors in micropores. *Prog. Surf. Membr. Sci.* **1975**, *9*, 1–70.
14. Brunauer, S.; Emmett, P.H.; Teller, E. Adsorption of gases in multimolecular layers. *J. Am. Chem. Soc.* **1938**, *60*, 309–319. [CrossRef]
15. Rouquerol, J.; Llewellyn, P.; Rouquerol, F. Is the BET equation applicable to microporous adsorbents? *Stud. Surf. Sci. Catal.* **2007**, *160*, 49–56.
16. Gregg, S.J.; Sing, K.S.W. *Adsorption, Surface Area and Porosity*; Academic Press: London, UK; New York, NY, USA, 1982.
17. Neimark, A.V.; Lin, Y.; Ravikovitch, P.I.; Thommes, M. Quenched solid density functional theory and pore size analysis of micro-mesoporous carbons. *Carbon* **2009**, *47*, 1617–1628. [CrossRef]
18. Shkolin, A.V.; Fomkin, A.A. Measurement of carbon-nanotube adsorption of energy-carrier gases for alternative energy systems. *Meas. Tech.* **2018**, *61*, 395–401. [CrossRef]
19. Fomkin, A.A.; Shkolin, A.V.; Men'shchikov, I.E.; Pulin, A.L.; Pribylov, A.A.; Smirnov, I.A. Measurement of adsorption of methane at high pressures for alternative energy systems. *Meas. Tech.* **2016**, *58*, 1387–1391. [CrossRef]
20. Pribylov, A.A.; Serpinskii, V.V.; Kalashnikov, S.M. Adsorption of gases by microporous adsorbents under pressures up to hundreds of megapascals. *Zeolites* **1991**, *11*, 846–849. [CrossRef]
21. Salehi, E.; Taghikhani, V.; Ghotbi, C. Theoretical and experimental study on the adsorption and desorption of methane by granular activated carbon at 25 °C. *J. Nat. Gas Chem.* **2007**, *16*, 415–422. [CrossRef]
22. Men'shchikov, I.E.; Shkolin, A.V.; Strizhenov, E.M.; Khozina, E.V.; Chugaev, S.S.; Shiryaev, A.A.; Fomkin, A.A.; Zherdev, A.A. Thermodynamic behaviors of adsorbed methane storage systems based on nanoporous carbon adsorbents prepared from coconut shells. *Nanomaterials* **2020**, *10*, 2243. [CrossRef] [PubMed]
23. Shkolin, A.V.; Fomkin, A.A. Thermodynamics of methane adsorption on the microporous carbon adsorbent ACC. *Russ. Chem. Bull.* **2008**, *57*, 1799–1805. [CrossRef]
24. Strizhenov, E.M.; Zherdev, A.A.; Podchufarov, A.A.; Chugaev, S.S.; Kuznetsov, R.A.; Zhidkov, D.A. Capacity and thermodynamic nomograph for an adsorption methane storage system. *Chem. Pet. Eng.* **2016**, *51*, 812–818. [CrossRef]
25. Guggenheim, E.A. *Thermodynamics: An Advanced Treatise for Chemists and Physicists*; North-Holland: Amsterdam, The Netherlands, 1967.
26. Myers, A.; Prausnitz, J.M. Thermodynamics of mixed-gas adsorption. *AIChE J.* **1965**, *11*, 121–127. [CrossRef]
27. Walton, K.S.; Sholl, D.S. Predicting multicomponent adsorption: 50 years of the ideal adsorbed solution theory. *AIChE J.* **2015**, *61*, 2757–2762. [CrossRef]
28. Chen, J.; Loo, L.S.; Wang, K. An ideal absorbed solution theory (IAST) study of adsorption equilibria of binary mixtures of methane and ethane on a templated carbon. *J. Chem. Eng. Data* **2011**, *56*, 1209–1212. [CrossRef]
29. Rackers, J.A.; Laury, M.L.; Li, C.; Wang, Z.; Lagardère, L.; Piquemal, J.P.; Ren, P.; Jay, W. TINKER 8: A modular software package for molecular design and simulation ponder. *J. Chem. Theor. Comp.* **2018**, *14*, 5273–5289. [CrossRef]
30. Maxwell, D.S.; Rives, J.T. Development and testing of the OPLS All-Atom Force Field on Conformational Energetics and Properties of Organic Liquids. *J. Am. Chem. Soc.* **1996**, *118*, 11225–11236.
31. Tolmachev, A.M.; Anuchin, K.M.; Kryuchenkova, N.G.; Fomkin, A.A. Theoretical calculation of the isotherms of adsorption on active coals using the molecular dynamics method. *Prot. Met. Phys. Chem. Surf.* **2011**, *47*, 150–155. [CrossRef]
32. Andersen, H.C. Molecular dynamics simulations at constant pressure and/or temperature. *J. Chem. Phys.* **1980**, *72*, 2384. [CrossRef]

33. Strizhenov, E.M.; Chugaev, S.S.; Shelyakin, I.D.; Shkolin, A.V.; Men'shchikov, I.E.; Zherdev, A.A. Numerical modeling of heat and mass transfer in an adsorbed natural gas storage tank with monolithic active carbon during charging and discharging processes. *Heat Mass Transf.* **2022**, 1–14. [CrossRef]
34. Chugaev, S.S.; Strizhenov, E.M.; Men'shchikov, I.E.; Shkolin, A.V. Carbon nanoporous monoblocks for mobile natural gas adsorption storage systems operating under arctic conditions. *Nanobiotechnol. Rep.* **2022**, *17*, 541–549.
35. Strizhenov, E.M.; Chugaev, S.S.; Men'shchikov, I.E.; Shkolin, A.V.; Zherdev, A.A. Heat and mass transfer in an adsorbed natural gas storage system filled with monolithic carbon adsorbent during circulating gas charging. *Nanomaterials* **2021**, *11*, 3274. [CrossRef] [PubMed]
36. Chugaev, S.S.; Strizhenov, E.M.; Men'shchikov, I.E.; Shkolin, A.V. Experimental study of the thermal management process at low-temperature circulating charging of an adsorbed natural gas storage system. *J. Phys. Conf. Ser.* **2021**, *2116*, 012084. [CrossRef]
37. Strizhenov, E.M.; Chugaev, S.S.; Men'shchikov, I.E.; Shkolin, A.V.; Shelyakin, I.D. Experimental study of heat transfer in adsorbed natural gas storage system filled with microporous monolithic active carbon. *J. Phys. Conf. Ser.* **2021**, *2116*, 012085. [CrossRef]
38. Guinier, A. La diffraction des rayons X aux tres petits angles: Application a l'etude de phenomenes ultramicroscopiques. *Ann. Phys.* **1939**, *11*, 161–237. [CrossRef]
39. Dubinin, M.M.; Plavnik, G.M. Microporous structures of carbonaceous adsorbents. *Carbon* **1968**, *6*, 183–192. [CrossRef]
40. Thommes, M.; Kaneko, K.; Neimark, A.V.; Oliver, J.P.; Rodrigues-Reinoso, F.; Rouquerol, J.; Sing, K. Physisorption of gases, with special reference to the evaluation of surface area and pore size distribution (IUPAC Technical Report). *Pure Appl. Chem.* **2015**, *87*, 1051–1069. [CrossRef]
41. Shkolin, A.V.; Fomkin, A.A. Theory of volume filling of micropores applied to the description of methane adsorption on the microporous carbon adsorbent AUK. *Russ. Chem. Bull.* **2009**, *58*, 717–721. [CrossRef]
42. Strizhenov, E.M.; Zherdev, A.A.; Petrochenko, R.V.; Zhidkov, D.A.; Kuznetsov, R.A.; Chugaev, S.S.; Podchufarov, A.A.; Kurnasov, D.V. A study of methane storage characteristics of compacted adsorbent AU-1. *Chem. Pet. Eng.* **2017**, *52*, 838–845. [CrossRef]
43. Men'shchikov, I.E.; Fomkin, A.A.; Tsivadze, A.Y.; Shkolin, A.V.; Strizhenov, E.M.; Khozina, E.V. Adsorption accumulation of natural gas based on microporous carbon adsorbents of different origin. *Adsorption* **2017**, *23*, 327–339. [CrossRef]
44. Shkolin, A.V.; Fomkin, A.A.; Men'shchikov, I.E.; Strizhenov, E.M.; Pulin, A.L.; Khozina, E.V. Monolithic microporous carbon adsorbent for low-Temperature natural gas storage. *Adsorption* **2019**, *25*, 1559–1573. [CrossRef]
45. Solovtsova, O.V.; Chugaev, S.S.; Men'shchikov, I.E.; Pulin, A.L.; Shkolin, A.V.; Fomkin, A.A. High-density carbon adsorbents for natural gas storage. *Colloid J.* **2020**, *82*, 719–726. [CrossRef]
46. Shkolin, A.V.; Men'shchikov, I.E.; Khozina, E.V.; Yakovlev, V.Y.; Simonov, V.N.; Fomkin, A.A. Deformation of microporous carbon adsorbent sorbonorit-4 during methane adsorption. *J. Chem. Eng. Data* **2022**, *67*, 1699–1714. [CrossRef]
47. Men'shchikov, I.E.; Shkolin, A.V.; Fomkin, A.A.; Khozina, E.V. Thermodynamics of methane adsorption on carbon adsorbent prepared from mineral coal. *Adsorption* **2021**, *27*, 1095–1107. [CrossRef]
48. Inagaki, M. *New Carbons. Control of Structure and Functions*; Elsevier: Oxford, UK, 2000.
49. Dubinin, M.M. Microporous systems of carbon adsorbents. In *Carbon Adsorbents and Their Industrial Application*; Nauka: Moscow, Russia, 1983; pp. 100–115.
50. Men'shchikov, I.E.; Fomkin, A.A.; Shkolin, A.V.; Yakovlev, V.Y.; Khozina, E.V. Optimization of structural and energy characteristics of adsorbents for methane storage. *Russ. Chem. Bull.* **2018**, *67*, 1814–1822. [CrossRef]

Article

Extended Line Defect Graphene Modified by the Adsorption of Mn Atoms and Its Properties of Adsorbing CH$_4$

Chenxiaoyu Zhang, Shaobin Yang *, Xu Zhang, Yingkai Xia and Jiarui Li

College of Material Science and Engineering, Liaoning Technical University, Fuxin 123000, China; ZCXY202201@163.com (C.Z.); zhangxu220120@163.com (X.Z.); xiayingkai200719@126.com (Y.X.); LJR220128@163.com (J.L.)
* Correspondence: ysblgd@163.com; Tel.: +86-0418-5110098

Abstract: Extended line defect (ELD) graphene is a two-dimensional (2D) topologically defective graphene with alternate octagonal and quadrilateral carbon rings as basic defective units. This paper reports on the CH$_4$ adsorption properties of ELD graphene according to the first principles of density functional theory (DFT). The effects on the CH$_4$ adsorption of ELD graphene when modified by a single Mn atom or two Mn atoms were investigated, respectively. An ELD-42C graphene configuration consisting of 42 C atoms was first constructed. Then, the ELD-42C graphene configuration was used as a substrate, and a Mn-ELD-42C graphene configuration was obtained by modifying it with a single Mn atom. The results showed that the most stable adsorption site for Mn atoms was above the quadrilateral carbon ring. This Mn-ELD-42C graphene configuration could only stably adsorb up to 30 CH$_4$ molecules on each side, with an average adsorption energy of -0.867 eV/CH$_4$ and an adsorption capacity of 46.25 wt%. Three 2Mn-ELD-42C graphene configurations were then obtained by modifying the ELD-42C graphene substrate with two Mn atoms. When the two Mn atoms were located on either side of a 2Mn-ELD-42C graphene configuration and above the two octagonal carbon rings adjacent to the same quadrilateral carbon ring, it was able to adsorb up to 40 CH$_4$ molecules on each side, with an average adsorption energy of -0.862 eV/CH$_4$ and a CH$_4$ adsorption capacity of 51.09 wt%.

Keywords: first principles; graphene; extended line defect; CH$_4$ adsorption; Mn modification

1. Introduction

CH$_4$ is abundant in nature and has a higher energy density than fossil fuels such as petroleum and coal. It is also a relatively clean fuel, with the lowest rate of CO$_2$ emissions of all carbonaceous fuels [1,2]. It has therefore been widely recognized as a transitional resource until alternatives to oil and coal can be found and developed on a large enough scale [3,4]. As a result, studies on CH$_4$ adsorption and its storage have important practical significance for energy development and use as well as environmental protection.

Graphene is a new kind of two-dimensional (2D) honeycomb-shaped nanomaterial characterized by good mechanical properties, good hydrogen storage and a high sensitivity to as well as adsorption potential for certain types of gases [5]. Zhao et al. [6] found that pristine graphene has a weak adsorption capacity for CH$_4$, with an average adsorption energy of -0.227 eV/CH$_4$. Ghanbari et al. [7] found that the adsorption energy could be improved to -0.166 eV when the graphene was modified with Ag atoms (Ag-G). This implies that physical absorption occurs between Ag atoms and graphene. Xu et al. [8] used Ti atoms to modify graphene and found that the modified graphene was most stable when the Ti atoms were located above the top carbon rings. This had an average adsorption energy of -0.298 eV/CH$_4$. The United States Department of Energy (DOE) established a CH$_4$ storage objective for vehicles in 2012, where the goal was to have a CH$_4$ adsorption capacity of above 50 wt% under standard conditions [9]. Unfortunately, the majority of CH$_4$ storage materials still fail to meet this requirement.

Pristine graphene is composed of a single layer of carbon atoms in the form of sp2 hybrid orbits and has a perfect hexagonal carbon ring structure [10]. During its growth, graphene typically acquires defects such as monovacancy, divacancy, Stone–Wales (SW), and topological lines [11]. These defects have been proven to be conducive to the adsorption of CH_4. Xiong et al. [12] constructed an extended line defect (ELD) graphene periodically embedded with quadrilateral and octagonal carbon rings by means of surface synthesis. ELD graphene is a type of semiconductor with a unique carbon ring structure and special chemical, electronic and mechanical properties, with a reduced band gap [13]. A large number of studies have shown that the CH_4 adsorption properties of graphene can be further improved by doping atoms to introduce additional structural modifications. Existing studies on transition metal (TM)-modified ELD graphene have concentrated on the magnetic and electronic properties of the configurations after modification. Cheng et al. [14] found that a single TM atom is preferentially adsorbed at defect sites in ELD graphene where there is high chemical activity. When modified by TM atoms, an ELD graphene substrate displays magnetism and spin polarization. Yu et al. [15] investigated the 558-type ELD, which is composed of a periodic repetition of one octagonal and two pentagonal rings, embedded in the hexagonal lattice of a graphene sheet. They found that the magnetism of the TM atom depends on the adsorption sites and the type of the adatoms, which can be obtained by analyzing the underlying hybridization mechanism between 3d orbitals of the TM atom and the electronic states of the ELD. Guan et al. [16] calculated defective graphene nanoribbon and TM adsorption on a line-defective embedded graphene sheet. The results show that TM atom adsorption on graphene can introduce magnetism and spin polarization, which is at the ferromagnetic ground state and shows different electronic properties according to different metals.

Manganese (Mn) is a transition metal, with reserves of about 570 million tons worldwide, that can be made by aluminothermic reduction of soft manganese ore [17,18]. Manganese is easily oxidized to manganese dioxide [19]. Manganese dioxide is an excellent adsorbent material due to its large specific surface area and strong electrochemical properties [20]. Manganese is located in the fourth period of the chemical periodic table in Group VIIB. The valence electronic configuration of manganese atom is $3d^54s^2$, with more electrons and empty orbits at the 3d energy level, which helps to enhance chemical bonding between manganese atoms and other molecules, allowing manganese atoms to adsorb gaseous molecules more efficiently [21,22]. In the laboratory, potassium permanganate or manganese dioxide is used as a catalyst to prepare graphene, resulting in the produced graphene containing a certain number of Mn atoms in its structure [23,24].

This paper presents an analysis of the adsorption properties of ELD graphene for CH_4, working from first principles [25]. It looks at the CH_4 adsorption properties of ELD graphene modified with a single Mn atom or two Mn atoms and calculates the resulting CH_4 adsorption capacity. This study offers theoretical support for the preparation and industrial application of new CH_4 storage materials.

2. Calculation Methods and Models

This paper is based on a first-principles pseudopotential plane-wave (PSPW) method that comes from density functional theory (DFT) [26,27]. DFT calculations have been applied successfully to analyze the defective carbon-based graphene-like systems containing the same type of defects as in this paper, such as carbon-based fullerene-like sulfocarbide [28], fullerene-like phosphorus carbide [29] and graphene-like model systems based on coronene and corannulene molecules [30]. The goal of the study was to investigate the CH_4 adsorption properties of ELD graphene as well as ELD graphene when modified by a single Mn atom or two Mn atoms at an atomic level by using the Cambridge Sequential Total Energy Package (CASTEP) module in the Materials Studio software [31]. Perdew–Burke–Ernzerhof (PBE) and generalized gradient approximation (GGA) functionals were selected for the calculations [32], and the interaction between electrons and ions was approximately calculated using OTFG ultrasoft. If the adsorption energy calculated by the GGA functional is weak, the adsorption energy can be corrected by means of a DFT dispersion correction (van der Waals) functional (namely DFT-D) [33]. The convergence criteria to optimize the calculations for atoms within the objects of study were set as follows: a maximum stress of 0.05 eV/Å; a maximum displacement of 0.002 Å; a convergence energy of 2.0×10^{-5} eV/atom; and a self-consistent field (SCF tolerance) convergence threshold of 2.0×10^{-6} eV/atom. To ensure the calculation accuracy and reduce the calculation cost, the truncation energy was set at 450 eV, the K-point sampling was set at $5 \times 5 \times 1$, and the objects of study were integrated in a Brillouin zone with Monkhorst–Pack grids [34,35]. The periodic boundary conditions to be met for the calculation of ELD graphene unit cells and the vacuum layer were set at 30 Å to avoid mutual interference between layers.

An ELD-42C graphene configuration consisting of 42 C atoms was constructed using graphene unit cells, with a main body that was composed of continuous quadrilateral, hexagonal and octagonal carbon rings. Its geometrical configuration after structural optimization is shown in Figure 1. The length of all the C-C bonds that constitute pristine graphene is 1.42 Å. According to an analysis of the ELD-42C graphene configuration after optimization, the lengths of the three C-C bonds that constituted the octagonal carbon rings were 1.48 Å, 1.39 Å and 1.41 Å, respectively. For the two C-C bonds that constituted the quadrilateral carbon rings, the lengths were 1.39 Å and 1.47 Å, respectively. In comparison to pristine graphene, some of the C-C bonds (1.39 Å and 1.41 Å) of the ELD-42C graphene configuration were slightly compressed, and some (1.48 Å and 1.47 Å) were slightly stretched. This is consistent with the experimental results in Liu et al. [36] and Zhao et al. [37] and coincides closely with the simulation results in Ding et al. [38]. This confirms the validity of the ELD-42C graphene configuration design.

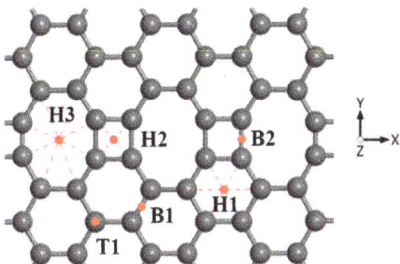

Figure 1. ELD-42C graphene configuration after structural optimization (the gray balls represent carbon atoms). H1, H2 and H3 indicate the hole sites; T1 stands for the top site; and B1 as well as B2 refer to the bridge sites.

For the iMn-ELD-42C graphene configurations obtained after modification by the Mn atoms, the binding energy, $E_{b_{iMn}}$, and average binding energy, $\overline{E_{b_{iMn}}}$, of the Mn atoms can be defined as follows:

$$E_{b_{iMn}} = E_{iMn-ELD-42C} - (E_{ELD-42C} + E_{iMn}) \tag{1}$$

$$\overline{E_{b_{iMn}}} = [E_{iMn-ELD-42C} - E_{ELD-42C} - E_{iMn}]/i \tag{2}$$

where i indicates the number of Mn atoms for modification; $E_{iMn-ELD-42C}$ refers to the total energy of the iMn-ELD-42C graphene configurations; $E_{ELD-42C}$ stands for the total energy of the ELD-42C graphene configurations; and E_{iMn} represents the total energy of the i-free Mn atom(s).

For the CH_4 molecules in the $jCH_4 \leftrightarrow iMn$-ELD-42C graphene adsorption configurations, the binding energy, E_{ad}, average binding energy, $\overline{E_{ad}}$, and PBW by percentage of weight can be defined as follows:

$$E_{ad} = E_{jCH_4 \leftrightarrow iMn-ELD-42C} - E_{(j-1)CH_4 \leftrightarrow iMn-ELD-42C} - E_{CH_4} \tag{3}$$

$$\overline{E_{ad}} = [E_{jCH_4 \leftrightarrow iMn-ELD-42C} - E_{iMn-ELD-42C} - jE_{CH_4}]/j \quad (4)$$

where j indicates the number of CH_4 molecule(s) adsorbed; $E_{jCH_4 \leftrightarrow iMn-ELD-42C}$ refers to the total energy of the $jCH_4 \leftrightarrow iMn$-ELD-42C graphene adsorption configurations; $E_{(j-1)CH_4 \leftrightarrow iMn-ELD-42C}$ stands for the total energy of the $(j-1)CH_4 \leftrightarrow iMn$-ELD-42C graphene adsorption configurations; jE_{CH_4} represents the total energy of the j-free CH_4 molecule(s); and $M_{r(CH_4)}$, $M_{r(Mn)}$ and $M_{r(ELD)}$ represent the weight of each CH_4 molecule, each Mn atom and the ELD system, respectively.

3. Results and Discussion

3.1. CH_4 Adsorption in the ELD-42C Graphene Configuration

Six typical CH_4 adsorption sites were selected to be studied on account of the symmetry of the geometric structure of the ELD-42C graphene configuration, as shown in Figure 1. H indicates a hole in a carbon ring, with H1, H2 and H3 representing the centroid sites of the hexagonal, quadrilateral and octagonal carbon rings, respectively. T1 stands for the top site of a C atom and B refers to the bridge site of a C-C bond, with B1 indicating the bridge sites in the hexagonal carbon rings and B2 representing the bridge sites in the defective rings (i.e., the quadrilateral and octagonal carbon rings). There were three adsorption forms for CH_4 molecules adsorbed on the ELD-42C graphene configuration according to the orientations of the four H atoms of the CH_4 molecules relative to the plane of the ELD-42C graphene configuration, as shown in Figure 2. Adsorption properties are affected by the adsorption sites for CH_4 molecules on the graphene, not the orientation of H atoms in the CH_4 molecules [39]. Therefore, the CH_4 adsorption form with three H atoms orientated to the plane of the ELD-42C graphene configuration (Figure 2c) was selected to study the adsorption properties at different sites. The CH_4 adsorption energy, E_{ad}; vertical distance, d, between the C atoms in the CH_4 molecules and the plane of the ELD-42C graphene configuration; and bond angle, $\angle$H-C-H, between the C atoms and the H atoms in the CH_4 molecules were calculated for a CH_4 molecule adsorbed at the six typical adsorption sites of the ELD-42C graphene configuration. The results are shown in Table 1.

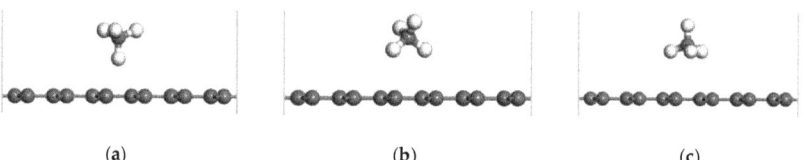

(a)   (b)   (c)

Figure 2. Three adsorption forms of CH_4 molecules in relation to the ELD-42C graphene configuration. (a–c) indicate that 1, 2 or 3 H atom(s) in the CH_4 molecule are oriented to the plane of the ELD-42C graphene configuration, respectively.

Table 1. Adsorption energy, E_{ad}, of the $CH_4 \leftrightarrow$ELD-42C graphene adsorption configuration; vertical distance, d, between the C atoms in the CH_4 molecules and the plane of the ELD-42C graphene configuration; and bond angle, $\angle$H-C-H, between the C atoms and H atoms in the CH_4 molecules.

The Absorption Point of CH_4	E_{ad} (eV)	d (Å)	$\angle$H-C-H (°)
H1	−0.824	3.291	109.507
H2	−0.833	3.222	109.878
H3	−0.847	3.081	109.551
T1	−0.835	3.278	109.554
B1	−0.832	3.251	109.811
B2	−0.830	3.251	109.560

Table 1 shows the adsorption energy released, E_{ad}, after one CH_4 molecule was adsorbed at the six adsorption sites in the ELD-42C graphene configuration. The larger its absolute value, the more energy released and the more stable the corresponding $CH_4 \leftrightarrow$

ELD-42C graphene adsorption configuration. It can be seen that the largest absolute value (-0.847 eV) was at H3, indicating that the ELD-42C graphene configuration was at its most stable at H3 when adsorbing CH_4 molecules, compared with the other five adsorption sites. Thus, the CH_4 molecules tended to stay above the octagonal carbon ring, as shown in Figure 3. For free CH_4 molecules, the length of bonds between the C and H atoms is 1.110 Å, and the bond angle is 109.381°. According to [40], the adsorption height of gas molecules is about three times their bond length, so the adsorption height of the CH_4 molecules was preset as 3.28 Å. After structural optimization, the vertical distance, d, between the CH_4 molecules at the six adsorption sites and the plane of the ELD-42C graphene configuration was analyzed. It was found that the difference between the stable adsorption height and the preset adsorption height was between 0.002 Å and 0.199 Å, indicating that the preset adsorption height was reasonable. The bond angle between the C and H atoms was close to that of CH_4 molecules adsorbed by the ELD-42C graphene configuration in their free state [41,42], so the adsorption for CH_4 on the ELD-42C graphene configuration was physical.

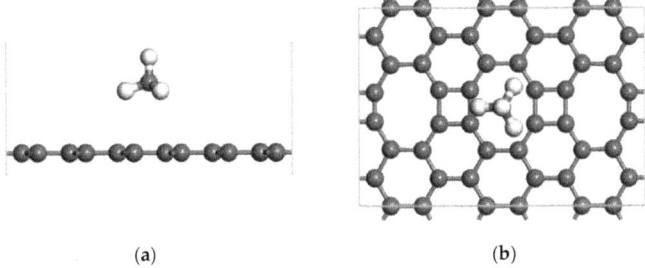

(a) (b)

Figure 3. $CH_4 \leftrightarrow$ELD-42C graphene adsorption configuration with 1 CH_4 molecule being adsorbed at H3 (gray balls represent carbon atoms, and white balls represent hydrogen atoms). (**a**) Front view; (**b**) top view.

CH_4 molecules were preferentially adsorbed by the ELD-42C graphene configuration at H3. They were then adsorbed at T1 after being fully adsorbed at H3. The ELD-42C graphene configuration could stably adsorb 26 CH_4 molecules at most on each side, with an average adsorption energy of -0.842 eV/CH_4 and an adsorption configuration similar to the one shown in Figure 4. According to the technical standards for natural gas adsorption systems issued by the DOE [43], the adsorption capacity of CH_4 storage materials should not be less than 50 wt%. For 26$CH_4 \leftrightarrow$ELD-42C graphene adsorption configurations, the CH_4 adsorption capacity was 45.26 wt%, which is lower than the technical standard; therefore, the ELD-42C graphene configuration was still not suitable for practical applications.

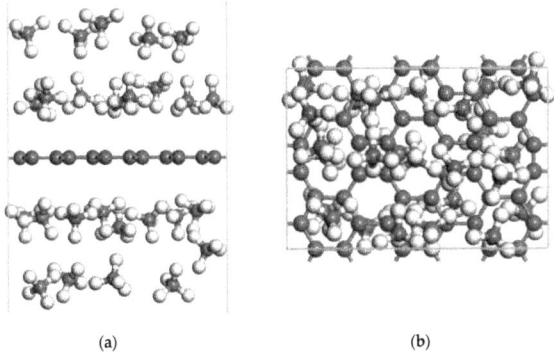

(a) (b)

Figure 4. 8$CH_4 \leftrightarrow$ELD-42C graphene adsorption configuration (gray balls represent carbon atoms, and white balls represent hydrogen carbons). (**a**) Front view; (**b**) top view.

3.2. CH_4 Adsorption in the Mn-ELD-42C Graphene Configuration

3.2.1. Modification of the ELD-42C Graphene Configuration by a Single Mn Atom

Jia et al. [2] found that pristine graphene doped with heteroatoms or pristine graphene with structural defects is significantly better able to adsorb gas molecules. Xu et al. [44] believed that the adsorption properties of graphene for gas molecules can be best improved by modifying graphene with alkali metals, alkaline earth metals and TM. Mn is an important TM element that is widely distributed throughout the Earth's crust. Its valence electron configuration is $3d^54s^2$, and chemical bonds can be easily formed between Mn and carbon atoms [45–47]. Due to its weak adsorption properties for CH_4 molecules, the ELD-42C graphene configuration was modified with TM Mn atoms to construct iMn-ELD-42C graphene configurations (where i indicates the number of Mn atoms, i = 1, 2). Their adsorption properties for CH_4 were then studied.

When the ELD-42C graphene configuration was modified by a single Mn atom, there were six optional adsorption sites for the Mn atoms: the hole sites, H1, H2 and H3; the bridge sites, B1 and B2; and the top site, T1 (see Figure 1). A single Mn atom, respectively placed at T1, B1 and B2 during the construction of the Mn-ELD-42C configurations, always moved to the top of the adjacent carbon ring under the action of the chemical bonds as the structure was optimized. This is in line with the optimal adsorption site of TM atoms determined by Zhao et al. [6] and Liu et al. [48]. The adsorption characteristics of the Mn atoms at H1, H2 and H3 were calculated, and the results are given in Table 2.

Table 2. Adsorption characteristics of a single Mn atom at H1, H2 and H3 on Mn-ELD-42C graphene configurations.

The Absorption Point of a Single Mn Atom	$E_{b_{Mn}}$ (eV)	Distance (Å)				$\Delta\rho$ (e)
		BL1	BL2	BL3	BL4	
H1	−2.922	2.037	2.037	2.104	2.105	0.29
H2	−3.453	2.967	2.974	2.974	2.981	0.39
H3	−3.218	2.113	2.128	2.430	2.431	0.33

In the table, $E_{b_{Mn}}$ (eV) indicates the binding energy of a single Mn atom; BL1, BL2, BL3 and BL4, respectively, refer to the length of the bonds between the Mn atoms and C atoms; and $\Delta\rho(e)$ represents the charge transfer between the Mn atoms and the ELD-42C graphene configuration. It can be seen from Table 2 that the binding energy of a single Mn atom was different at H1, H2 and H3. At H2 it was −3.453 eV. This was the largest absolute value out of the three hole sites. The binding energy of a single Mn atom at H1 was −2.922 eV, which was the smallest absolute value. These results indicated that Mn atoms adsorbed above H2 were the most stable, while Mn atoms adsorbed above H1 were the least stable. During modification, four Mn-C chemical bonds were formed between the Mn atoms and the four carbon atoms at H2, with lengths of 2.967 Å, 2.974 Å, 2.974 Å and 2.981 Å, respectively. This suggests that most of the Mn atoms were adsorbed on the central axis of H2. The charge transferred from the Mn atoms to the ELD-42C graphene configuration was 0.33 e at H3 and 0.29 e at H1. Therefore, the interaction between the ELD-42C graphene configuration and Mn atoms adsorbed at H1 and H3 was weaker than it was at H2. The most stable Mn-ELD-42C graphene system configuration (Figure 5) was therefore obtained via structural optimization after it had been modified by a single Mn atom at H2.

The Mulliken layout of the Mn-ELD-42C graphene configuration before and after adsorbing a single CH_4 molecule was analyzed. It was found that the charge transferred from the Mn atoms to the ELD-42C graphene configuration was 0.69 e, indicating a strong electrostatic effect between the two. The partial density of states (PDOS) for the Mn-ELD-42C graphene configuration is shown in Figure 6 (partial). There were resonance peaks between the d orbit of the Mn atoms and the p orbit of the C atoms within the range of −1.958 to −1.381 eV, confirming that there was interaction between the two orbits. As a result, the valence band of the Mn-ELD-42C graphene configuration largely derives from the interaction between the d orbit of the Mn atoms and the *p* orbit of the C atoms. This is similar to the results obtained by Wu et al. [49] and Zhao et al. [37], who modified graphene substrates by using TM atoms as a doping agent. As extra electrons were provided to the ELD-42C graphene configuration by the Mn atoms, the overall conduction band of the configuration moved to the Fermi level, where the conduction band intersected with the valence band and endowed the Mn-ELD-42C graphene configuration with typical metallic-phase characteristics.

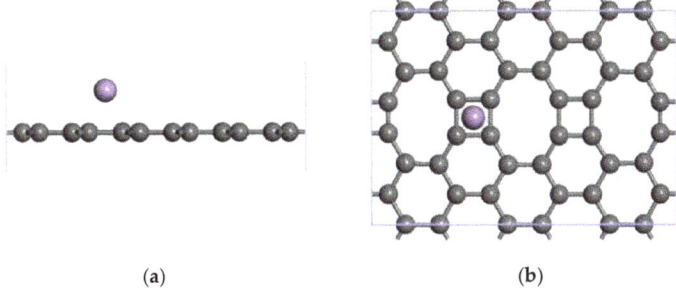

(a) (b)

Figure 5. The most stable Mn-ELD-42C graphene configuration was obtained by modifying the hole site H2 with a single Mn atom (gray balls represent carbon atoms, and purple balls represent Mn atoms). (**a**) Front view; (**b**) top view.

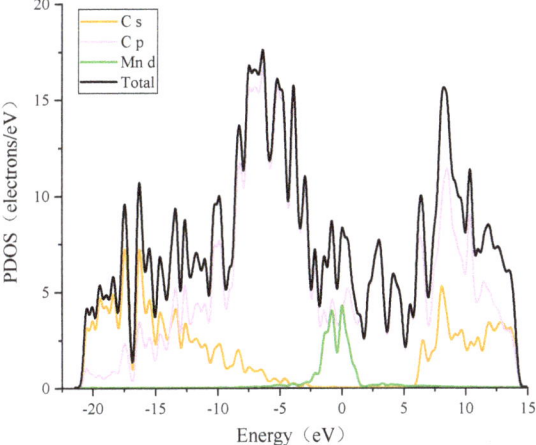

Figure 6. PDOS diagram of the Mn-ELD-42C graphene system (partial).

3.2.2. Adsorption of CH_4 by the Mn-ELD-42C Graphene Configuration

DFT was used to study the CH_4 adsorption capacity of the Mn-ELD-42C graphene configuration by adding CH_4 molecules to one side. A stable $CH_4 \leftrightarrow$ Mn-ELD-42C adsorption configuration (Figure 7a) was obtained after the Mn-ELD-42C graphene configuration with the first adsorbed CH_4 molecule had been optimized. The first CH_4 molecule was located above the Mn atom, proving that this was where the adsorption energy was the largest. The adsorption energy of this configuration was -1.717 eV, which is larger than that of the ELD-42C graphene configuration for CH_4 molecules (-0.847 eV), of Li-modified carbon nanotubes for CH_4 (-0.464 eV) [50] and of Pt-modified graphene for CH_4 (-0.488 eV) [51]. Apparently, modifying the ELD-42C graphene configuration with a single Mn atom improved its adsorption properties for CH_4 molecules. A $2CH_4 \leftrightarrow$ Mn-ELD-42C adsorption configuration was obtained after a second CH_4 molecule had been added (Figure 7b). Here, both the first and the second CH_4 molecules were located above the Mn atoms and close to the Mn-ELD-42C graphene configuration. The combined action of the mutual repulsion of the CH_4 molecules and their adsorption by the Mn-ELD-42C graphene configuration enabled a third CH_4 molecule to be adsorbed above the hexagonal carbon ring that was close to the Mn atoms (Figure 7c). A fourth CH_4 molecule was adsorbed at T1 above the C atoms (Figure 7d), and a fifth above the octagonal carbon ring (Figure 7e). Limited by the adsorption space, the repulsive force between the molecules gradually increased as more CH_4 molecules were adsorbed. The adsorption configuration began to arc when an eighth molecule was adsorbed (Figure 7f), and there was a stratification phenomenon when the ninth molecule was adsorbed (Figure 7g). Due to their layered adsorption, the distances between the 9th–16th CH_4 molecules and the Mn atom became larger, and the adsorption energy was reduced. The 16th CH_4 molecule was nowhere near the Mn atom and was the most distant from the Mn-ELD-42C graphene configuration. It also had the lowest adsorption energy (-0.755 eV). When a 17th CH_4 molecule was placed on one side of the configuration, the calculated adsorption energy became positive, indicating that the gas molecule had not been adsorbed. This proved that the Mn-ELD-42C graphene configuration could only stably adsorb up to 16 CH_4 molecules on each side. The geometrical configuration is shown in Figure 7h.

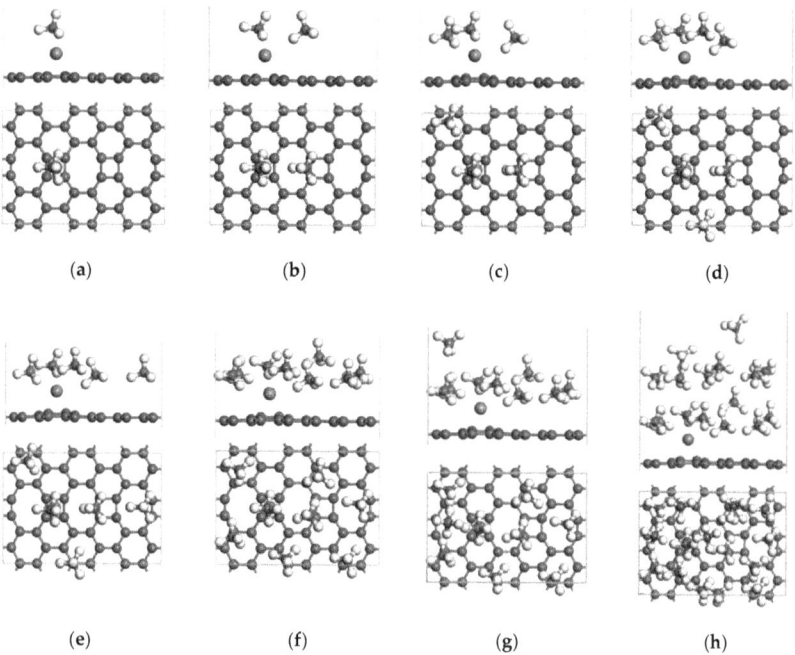

Figure 7. $jCH_4 \leftrightarrow$ Mn-ELD-42C adsorption configurations (j = 1, 2, ... , 16). Figures (**a**–**h**) respectively demonstrate the Mn-ELD-42C graphene configurations for 1–16 CH_4 molecule(s) adsorbed on one side.

Table 3 shows the average adsorption energy, $\overline{E_{ad}}$, and the adsorption energy, E_{ad}, of the $jCH_4 \leftrightarrow$ Mn-ELD-42C $\leftrightarrow jCH_4$ adsorption configuration for CH_4 adsorption on one side and both sides; the distance, d_{CH_4-S}, between the CH_4 molecules and the plane of the Mn-ELD-42C graphene configuration; the distance, d_{CH_4-Mn}, between the CH_4 molecules and the Mn atoms; and the adsorption capacity (PBW) of the Mn-ELD-42C graphene configuration for CH_4 in the $jCH_4 \leftrightarrow$ Mn-ELD-42C $\leftrightarrow jCH_4$ adsorption configuration. Analysis of these data reveals that the absolute value of the average adsorption energy, $\overline{E_{ad}}$, of the CH_4 gradually decreased as j, the number of CH_4 molecules adsorbed, increased. The adsorption energy, E_{ad}, of 16 CH_4 molecules adsorbed on one side was compared. When the gas molecules were not stratified (the first–eighth molecules), the first, third, and seventh CH_4 molecules presented a higher adsorption energy than the other CH_4 molecules because their adsorption sites were close to Mn atoms and their distance from the plane of the graphene configuration was relatively small. When the gas molecules were stratified (the ninth–sixteenth molecules), the interaction between the CH_4 and Mn steadily decreased as the distance between them increased, leading to reduced adsorption properties.

Table 3. Adsorption energy and related parameters for CH_4 molecules in the Mn-ELD-42C graphene configuration.

Number of CH_4 Molecules	$\overline{E_{ad}}$ (eV/CH_4)	E_{ad} (eV)	d_{CH_4-S} (Å)	d_{CH_4-Mn} (Å)	PBW (wt%)
1	−1.717	−1.717	3.824	1.967	2.79
2	−1.289	−0.862	3.789	3.614	5.43
3	−1.177	−0.953	3.590	4.483	7.92
4	−1.103	−0.879	4.100	4.956	10.29
5	−1.047	−0.824	3.551	7.299	12.54
6	−1.022	−0.897	3.448	7.293	14.68
7	−1.018	−0.978	3.397	3.928	16.72
8	−0.995	−0.832	4.840	5.319	18.66
9	−0.971	−0.775	7.176	5.724	20.52
10	−0.956	−0.825	7.156	5.777	22.29
11	−0.943	−0.810	7.329	5.612	23.98
12	−0.927	−0.755	7.460	8.228	25.60
13	−0.918	−0.806	7.831	9.377	27.16
14	−0.913	−0.856	8.035	7.837	28.65
15	−0.907	−0.816	9.132	7.743	30.08
16	−0.897	−0.755	11.550	10.780	31.46
30	−0.867				46.25

The above results indicate that the adsorption properties of a Mn-ELD-42C graphene configuration are affected by Mn atom modification and that this can play an important role in CH_4 adsorption. The adsorption energy was also affected by the distance between the CH_4 molecules and the plane of the graphene configuration. The adsorption distance, d_{CH_4-S}, between the 16th CH_4 molecule and the Mn-ELD-42C graphene configuration was 11.550 Å, which was the largest out of the 16 CH_4 molecules. At this point, both the average adsorption energy, $\overline{E_{ad}}$, and the adsorption energy, E_{ad}, of the configuration were at their lowest (−0.897 eV/CH_4 and −0.755 eV, respectively).

Up to 16 CH_4 molecules could be stably adsorbed on one side of the Mn-ELD-42C graphene configuration, with an average adsorption energy of −0.897 eV/CH_4. On this basis, it can be calculated that the Mn-ELD-42C graphene configuration is able to stably adsorb up to 14 CH_4 molecules on the other side, making a total of 30 CH_4 molecules overall (Figure 8), with an average adsorption energy of −0.867 eV/CH_4 and an adsorption capacity of 46.25 wt%. This is much closer to the proposed DOE standard (50 wt%) [9]. The adsorption capacity of the Mn-ELD-42C graphene configuration was 1.02 times that of the basic ELD-42C graphene configuration (45.26 wt%). This makes it clear that the adsorption capacity for CH_4 molecules can be effectively improved by the modification of Mn atoms.

Table 4 gives the Mulliken layout of the Mn-ELD-42C graphene configuration before and after adsorbing one CH_4 molecule, where H_1, H_2, H_3 and H_4 stand for the H atoms and C represents the C atom of the CH_4 molecule. For the CH_4 molecule adsorbed above the Mn atom, H_1, H_2 and H_3 faced the plane of the Mn-ELD-42C graphene configuration, while H_4 faced away from it (Figure 6a). The charge for H_4 was 0.27 e before the CH_4 molecule was adsorbed and 0.36 e after the CH_4 molecule was adsorbed, with 0.09 e of charge having been lost. For free CH_4 molecules, the C atom is negatively charged, and the four peripheral H atoms are positively charged. This results in a strong repulsive force between CH_4 molecules, making it difficult for multiple CH_4 molecules to gather at the same adsorption site. For the Mn-ELD-42C graphene configuration, the ELD-42C graphene substrate was negatively charged, allowing the positively charged CH_4 molecules on the outer surface to be adsorbed more easily via electrostatic interaction. In the CH_4 molecules adsorbed on the Mn-ELD-42C graphene configuration, both H_1 and H_2 lost their partial positive charge because they received equal numbers of electrons. This reduced the surface area of the positively charged CH_4 molecule, weakening the repulsive force between the CH_4 molecules. In addition, before and after a single CH_4 molecule had been adsorbed by the Mn-ELD-42C graphene configuration, a relatively large charge transfer occurred with the Mn atoms, with 0.29 e of charge being lost. When CH_4 molecules were adsorbed, the electrons of the Mn atoms were transferred to the CH_4 molecules; therefore, a strong Coulomb force was produced between the two, creating favorable conditions for CH_4 adsorption.

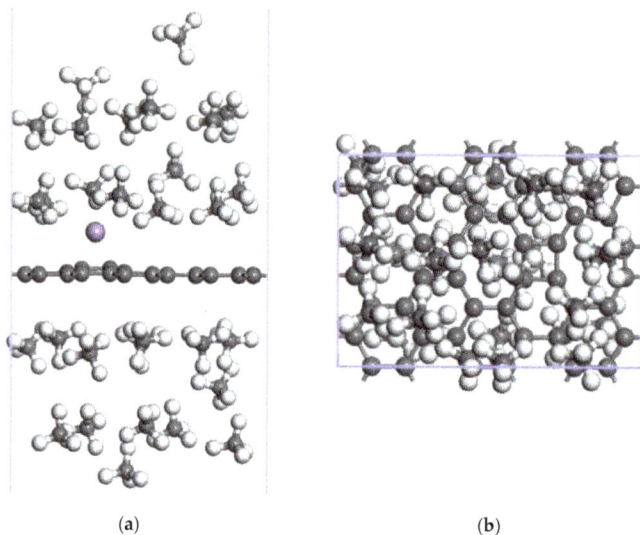

(a) (b)

Figure 8. $16CH_4 \leftrightarrow$ Mn-ELD-42C $\leftrightarrow 14CH_4$ adsorption configuration with 16 and 14 CH_4 molecules, respectively, adsorbed on each side. (a) Front view; (b) top view (gray balls represent carbon atoms, white balls stand for hydrogen atoms and purple balls represent Mn atoms).

Table 4. Mulliken layout of the Mn-ELD-42C graphene configuration before and after adsorbing a single CH_4 molecule.

Atom	Before Adsorption (e)				After Adsorption (e)			
	s	p	d	Charge	s	p	d	Charge
H_1	0.73			0.27	0.87			0.13
H_2	0.73			0.27	0.87			0.13
H_3	0.73			0.27	0.72			0.28
H_4	0.73			0.27	0.64			0.36
C	1.51	3.59		−1.10	1.49	3.61		−1.09
C1	1.03	3.00		−0.03	1.05	3.03		−0.08
C2	1.03	3.00		−0.04	1.05	3.03		−0.09
C3	1.05	3.03		−0.09	1.05	3.04		−0.10
C4	1.05	3.04		−0.09	1.05	3.04		−0.10
Mn	2.00	6.00	6.21	0.69	2.01	6.00	6.13	0.98

Figure 9 illustrates the charge density difference for the CH$_4$↔Mn-ELD-42C adsorption configuration. This directly reveals the charge transfer between the Mn atoms and CH$_4$ molecules. The blue elements are the electron gain zone, where the CH$_4$ molecules obtained electrons, and the yellow elements are the electron loss zone, where the Mn atoms lost electrons. As the large charge transfer between the Mn atoms and CH$_4$ molecules produced a Coulomb force between them, the Mn atoms had a significant effect on CH$_4$ adsorption. This is consistent with the analysis of the Mulliken layout in Table 4.

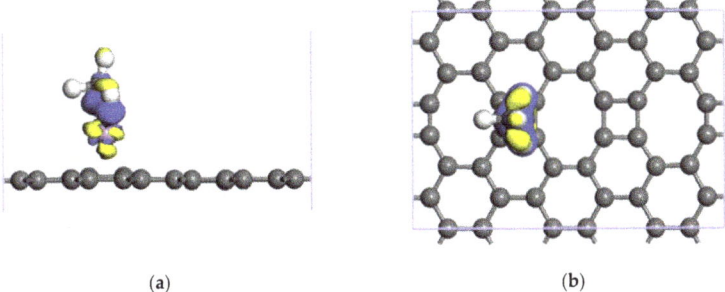

Figure 9. Charge density difference for the CH$_4$↔Mn-ELD-42C adsorption configuration. (**a**) Front view; (**b**) top view.

The interaction between the Mn-ELD-42C graphene configuration and CH$_4$ molecules was also analyzed in terms of the PDOS of the CH$_4$ molecules. Figure 10a shows the PDOS of the CH$_4$↔Mn-ELD-42C adsorption configuration after adsorbing a single CH$_4$ molecule. The density of states (DOS) peak of the Mn atoms increased from 4.031 eV (before adsorption, as shown in Figure 6) to 4.763 eV (after adsorption), and the energy range enlarged from (−4.107, 1.547 eV) before adsorption to (−4.341, 1.637 eV) after adsorption. As a consequence, the CH$_4$ adsorption enhanced the interaction between the Mn atoms and the Mn-ELD-42C graphene configuration, which is in accord with the analysis of the Mulliken layout in Table 4. After a single CH$_4$ molecule had been adsorbed, the DOS valence band peak of the CH$_4$↔Mn-ELD-42C adsorption configuration improved because of the hybridization between the 3d orbit of the Mn atoms and the 1s orbit of the H atoms. The DOS of the C atom in the Mn-ELD-42C graphene configuration also changed slightly.

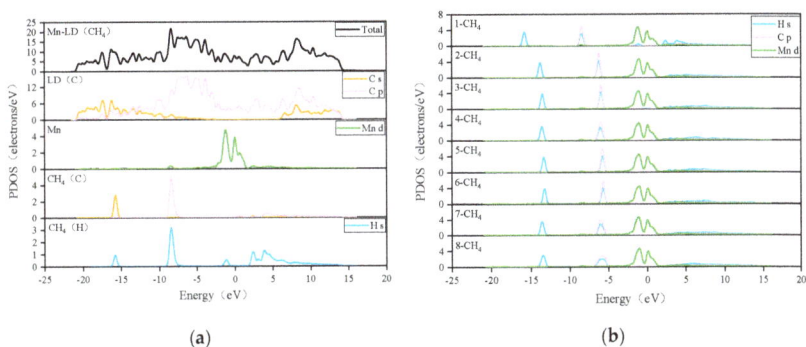

Figure 10. PDOS of the CH$_4$ molecules in the jCH$_4$↔Mn-ELD-42C adsorption configuration. (**a**) For the CH$_4$↔Mn-ELD-42C adsorption configuration; (**b**) for the (1–8)CH$_4$↔Mn-ELD-42C adsorption configurations.

Figure 10b was used to analyze the interaction between the d orbit of the Mn atoms, the s orbit of the H atoms and the p orbit of the C atom on the eight unstratified CH$_4$ molecules adsorbed on one side of the Mn-ELD-42C graphene configuration. It can be seen that the s orbit of the H atoms of the first CH$_4$ molecule overlapped with the 3d orbit of the Mn atoms near −16.0 eV and −8.0 eV. This suggests that there is an interaction between the first CH$_4$ molecule and the Mn atoms. Compared with the first CH$_4$ molecule, the 1s orbit of the second CH$_4$ molecule had shifted to the right, indicating that the interaction between the second CH$_4$ molecule and the Mn atoms had weakened, making the adsorption energy of the second CH$_4$ molecule smaller than that of the first CH$_4$ molecule. The displacement of the PDOS peak for the CH$_4$ molecules correlated with changes in the adsorption energy, with the PDOS peak moving to the left when the adsorption energy increased and to the right when the adsorption energy reduced. As the number of CH$_4$ molecules adsorbed increased, the PDOS peak reduced and moved to the right, showing that the interaction between the CH$_4$ molecules and Mn atoms was gradually weakening. In the interval [−6.989, −4.893 eV], the PDOS peak of the eighth CH$_4$ molecule was significantly lower than that of the other seven CH$_4$ molecules, indicating that the interaction between the 1s orbit of the H atoms of the eight CH$_4$ molecule and the 3d orbit of the Mn atoms was the weakest. This is consistent with the gradual decrease in the average adsorption energy when the CH$_4$ molecules were adsorbed by the Mn-ELD-42C graphene configuration.

3.3. CH$_4$ Adsorption in the 2Mn-ELD-42C Graphene Configuration

3.3.1. Modification of the ELD-42C Graphene Configuration by Two Mn Atoms

To further study the effect of Mn modification on the CH$_4$ storage properties of the ELD-42C graphene configuration, the ELD-42C graphene configuration was modified by two Mn atoms. This enabled three stable structures (Figure 11) to be obtained after optimization, with their structural symmetry being taken into account. In Figure 11a, the two Mn atoms were located on the same side of the ELD-42C graphene configuration. This made it difficult to increase the adsorption capacity for CH$_4$ molecules because of the limited adsorption space. As each modifying Mn atom forms an active adsorption site, the adsorption capacity for CH$_4$ can be increased more effectively by placing the two Mn atoms on either side of the ELD-42C graphene configuration. Figure 11b shows the two Mn atoms placed on either side of the same quadrilateral carbon ring of the ELD-42C graphene configuration, with an average binding energy of −2.816 eV. In Figure 11c, the two Mn atoms were placed on either side of two quadrilateral carbon rings separated by an octagonal carbon ring, with an average binding energy of −3.451 eV. This was slightly lower than the graphene configuration shown in Figure 11a (−3.556 eV) but produced a more stable structure. In the structure shown in Figure 11c, the binding energy of the first atom (−3.453 eV) was close to that of the second Mn atom (−3.449 eV), proving that the interaction between the two Mn atoms differed to a lesser degree. On the basis of this analysis, the 2Mn-ELD-42C graphene configuration shown in Figure 11c was selected for the study of its CH$_4$ adsorption properties.

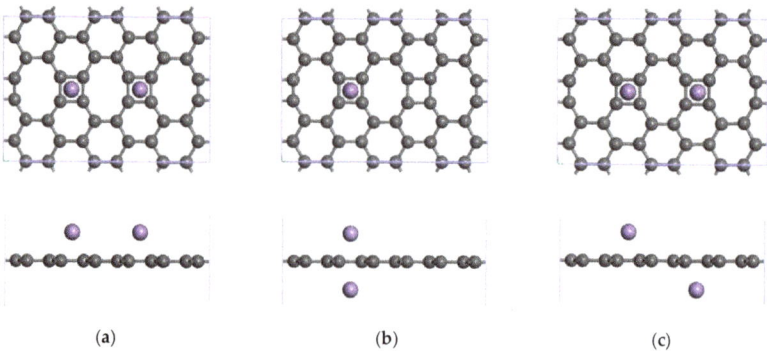

Figure 11. 2Mn-ELD-42C graphene configurations (gray balls represent carbon atoms, and purple balls represent Mn atoms). (**a**) The two Mn atoms were located on the same side of the ELD-42C

graphene configuration; (**b**) the two Mn atoms were placed on either side of the same quadrilateral carbon ring of the ELD-42C graphene configuration; (**c**) the two Mn atoms were placed on either side of two quadrilateral carbon rings separated by an octagonal carbon ring.

3.3.2. Adsorption of CH_4 by the 2Mn-ELD-42C Graphene Configuration

Twenty CH_4 molecules at most could be stably adsorbed on one side of the 2Mn-ELD-42C graphene configuration. The jCH$_4$↔2Mn-ELD-42C adsorption configuration (j = 1, 2, ..., 20) after optimization is shown in Figure 12. When the ninth CH_4 molecule was adsorbed by the 2Mn-ELD-42C graphene configuration, stratification occurred due to the repulsion between the positively charged surfaces of the CH_4 molecules and the limited adsorption space near the Mn atoms (Figure 7e). In other words, the preceding eight CH_4 molecules were adsorbed in the first layer (Figure 7a–d), near the Mn atoms. This was similar to the monolayer gas adsorption by the Mn-ELD-42C graphene configuration modified by a single Mn atom. When the second Mn atom was added, the increased CH_4 adsorption sites made the interaction between the Mn atoms and CH_4 molecules exceed the mutual repulsion between the CH_4 molecules, so five CH_4 molecules were able to be adsorbed in the third layer. As noted above, the 2Mn-ELD-42C graphene configuration could stably adsorb twenty CH_4 molecules at most on a single side, with the adsorption substrate of the optimized 20CH$_4$↔2Mn-ELD-42C adsorption configuration forming an arc (Figure 7h). Clearly, the 2Mn-ELD-42C graphene configuration obtained after the second Mn atom was added improved the CH_4 adsorption capacity of the substrate and increased the number of adsorption sites, effectively remedying the issue with the Mn-ELD-42C graphene configuration when adsorbing CH_4 molecules at a greater distance from the Mn atoms.

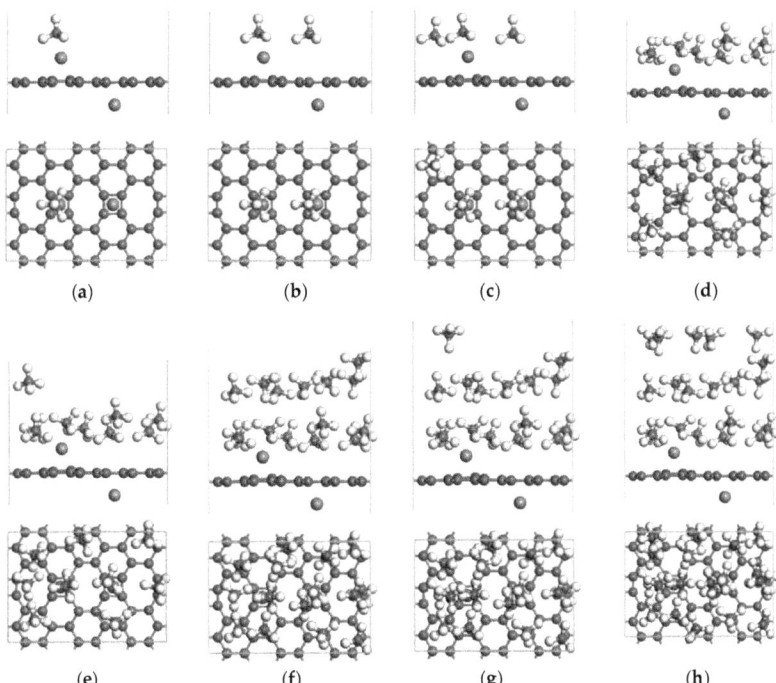

Figure 12. jCH$_4$↔2Mn-ELD-42C adsorption configurations (j = 1, 2, ..., 20). (**a**–**h**) respectively show the 2Mn-ELD-42C graphene configuration with 1–20 CH_4 molecule(s) adsorbed on one side.

Table 5 gives the average adsorption energy, $\overline{E_{ad}}$, and the adsorption energy, E_{ad}, of the jCH$_4$↔2Mn-ELD-42C adsorption configuration for CH$_4$ adsorption on one side and on both sides; the distance, d_{CH_4-S}, between the CH$_4$ molecules and the 2Mn-ELD-42C graphene configuration; the distance, d_{CH_4-Mn}, between the CH$_4$ molecules and Mn atoms; and the CH$_4$ adsorption capacity (PBW) of the 2Mn-ELD-42C graphene configuration. It can be seen that the absolute value of the average adsorption energy decreased with an increase in j, the number of adsorbed CH$_4$ molecules. Comparing the adsorption energy, E_{ad}, of the eight CH$_4$ molecules adsorbed on one side and in a single layer, the first and fourth CH$_4$ molecules were closer to the Mn atoms and had smaller adsorption distances from the plane of the graphene configuration, so there was a higher adsorption energy. The adsorption distance between the 20th CH$_4$ molecule and the 2Mn-ELD-42C graphene configuration was the largest out of the 20 CH$_4$ molecules (11.600 Å), with the average adsorption energy, $\overline{E_{ad}}$, and adsorption energy, E_{ad}, also being the smallest (−0.868 eV/CH$_4$ and −0.726 eV, respectively). Layered adsorption occurred when the CH$_4$ molecules were adsorbed by the 2Mn-ELD-42C graphene configuration, causing the 20th CH$_4$ molecule to be the farthest away from Mn atoms; there was repulsion between the positively charged surfaces of the CH$_4$ molecules, making the adsorption energy the lowest. When a 21st CH$_4$ molecule was placed on one side of the configuration, the calculated adsorption energy became positive, indicating that the gas molecule had not been adsorbed.

Table 5. Adsorption energy and related parameters for CH$_4$ molecules in the 2Mn-ELD-42C graphene configuration.

Number of CH$_4$ Molecules	$\overline{E_{ad}}$ (eV/CH$_4$)	E_{ad} (eV)	d_{CH_4-S} (Å)	d_{CH_4-Mn} (Å)	PBW (wt%)
1	−1.726	−1.726	3.789	1.943	2.55
2	−1.296	−0.866	3.828	4.045	4.96
3	−1.149	−0.855	3.590	4.406	7.27
4	−1.105	−0.974	3.483	3.996	9.46
5	−1.058	−0.869	4.196	7.647	11.55
6	−1.030	−0.891	3.937	4.558	13.55
7	−1.015	−0.926	3.379	7.596	15.46
8	−0.993	−0.835	4.498	5.836	17.28
9	−0.967	−0.760	7.187	5.871	19.03
10	−0.950	−0.796	7.167	5.827	20.71
11	−0.936	−0.797	7.385	6.282	22.32
12	−0.926	−0.814	7.574	6.176	23.86
13	−0.917	−0.806	7.885	9.321	25.35
14	−0.909	−0.813	7.931	8.754	26.77
15	−0.900	−0.777	9.242	11.033	28.15
16	−0.889	−0.727	11.369	9.523	29.47
17	−0.882	−0.756	11.336	9.868	30.75
18	−0.875	−0.769	11.578	10.493	31.98
19	−0.871	−0.787	11.450	10.427	33.16
20	−0.868	−0.726	11.600	11.992	34.31
40	−0.862				51.09

Up to 40 CH$_4$ molecules could be stably adsorbed on both sides of the 2Mn-ELD-42C graphene configuration (see Figure 13), with an average adsorption energy of −0.862 eV/CH$_4$ and an adsorption capacity of 51.09 wt%. Note that, although the average adsorption energy of the 2Mn-ELD-42C graphene configuration (−0.862 eV/CH$_4$) was close to that of the Mn-ELD-42C graphene configuration (−0.867 eV/CH$_4$), the CH$_4$ adsorption capacity of the 2Mn-ELD-42C graphene configuration (51.09 wt%) was 1.11 times that of the Mn-ELD-42C graphene configuration.

Although the CH$_4$ adsorption capacity of the 2Mn-ELD-42C graphene configuration is lower than some references, as tabulated in Table 6, it goes beyond the specified DOE standard (50 wt%), proving that 2Mn-ELD-42C graphene configurations may offer good opportunities for industrial development and are worth further research and development in the future.

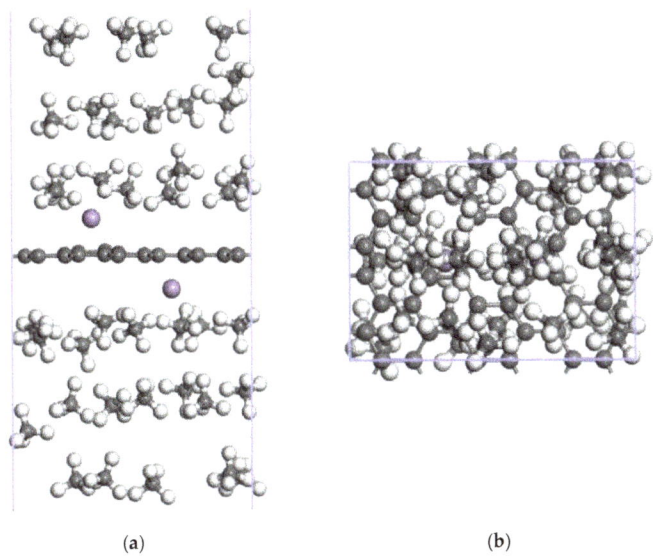

Figure 13. 20CH$_4$↔2Mn-ELD-42C↔20CH$_4$ adsorption configuration with 20 CH$_4$ molecules being adsorbed on each side. (**a**) Front view; (**b**) top view (the gray balls represent carbon atoms, the white balls are hydrogen atoms and the purple balls represent Mn atoms).

Table 6. Comparison of CH$_4$ adsorption capacity on various carbonaceous structure.

Adsorption Structure	Modified Elements	Temp; Pressure	PBW (wt%)	Interpretation	Ref.
2Mn-ELD-42C	Mn	-	51.09	ELD: Extended line defect	This work
2Ti-GDY	Ti	-	55.24	GDY: graphdiyne	[44]
CNT-PG	-	298 k; 40 bar	44.70	CNT-PG: carbon nanotube-porous graphene	[52]
SWNT	-	303 k; 3.55 MPa	19.80	SWNT: single-walled carbon nanotubes	[53]
2Ti-GY	Ti	-	48.40	GY: graphyne	[8]
2Mn-GR	Mn	-	32.93	GR: graphene	[6]
2Mn-N-GDY	Mn, N	-	58.50	GDY: graphdiyne	[54]
Ti-GDY	Ti	-	63.54	Co-mixing H$_2$ and CH$_4$	[55]
GRHA/ACNF	-	298 k; 12 bar	66.40	GRHA: graphene-derived rice husk ashes; ACNF: activated carbonNanofibers	[56]

4. Conclusions

This paper explored the CH$_4$ adsorption properties of ELD-42C, Mn-ELD-42C and 2Mn-ELD-42C graphene configurations using DFT to work from first principles. The study results showed that the CH$_4$ adsorption of a basic ELD-42C graphene configuration is weak. Here, the average adsorption energy was −0.842 eV/CH$_4$, and the CH$_4$ adsorption capacity was 45.26 wt%. Sixteen CH$_4$ molecules could be stably adsorbed on the Mn-doped side of the Mn-ELD-42C graphene configuration, and fourteen on the other side, with an average adsorption energy of −0.867 eV/CH$_4$ and an adsorption capacity of 46.25 wt%. Thus, the CH$_4$ adsorption capacity of Mn-ELD-42C graphene configurations can be effectively enhanced by modification with a single Mn atom. This configuration further improved the CH$_4$ adsorption and increased the number of adsorption sites, with it being able to stably adsorb up to 40 CH$_4$ molecules across the two sides, with an average adsorption energy of -0.862 eV/CH_4 and an adsorption capacity of 51.09 wt%. The adsorption capacity of the 2Mn-ELD-42C graphene configuration was 1.13 times that of the ELD-42C graphene configuration and exceeded the proposed DOE standards (50 wt%). Together, these results indicate that 2Mn-ELD-42C graphene configurations have great potential for the development of industrially viable CH_4 storage materials. The DFT calculation results in this paper only illustrate the CH_4 adsorption properties of ELD graphene configurations when modified by a single Mn atom or two Mn atoms. In subsequent studies, the effects of environmental factors such as temperature and pressure on the CH_4 adsorption properties of ELD-42C, Mn-ELD-42C and 2Mn-ELD-42C graphene configurations will be further investigated using molecular dynamics methods.

Author Contributions: Conceptualization, C.Z. and S.Y.; methodology, C.Z. and X.Z.; software, C.Z. and X.Z.; validation, S.Y.; formal analysis, S.Y.; investigation, C.Z.; resources, C.Z. and S.Y.; data curation, J.L.; writing—original draft preparation, C.Z.; writing—review and editing, S.Y.; visualization, Y.X.; supervision, S.Y.; project administration, X.Z.; funding acquisition, S.Y. All authors have read and agreed to the published version of the manuscript.

Funding: This research was funded by the National Natural Science Foundation of China, grant number 52174253.

Institutional Review Board Statement: Not applicable.

Informed Consent Statement: Not applicable.

Data Availability Statement: The data presented in this study are available on reasonable request from the corresponding author.

Acknowledgments: The authors are grateful for the support of the National Natural Science Foundation of China (grant No. 52174253).

Conflicts of Interest: The authors declare no conflict of interest.

References

1. Pantha, N.; Ulman, K.; Narasimhan, S. Adsorption of methane on single metal atoms supported on graphene: Role of electron back-donation in binding and activation. *J. Chem. Phys.* **2020**, *153*, 244701. [CrossRef] [PubMed]
2. Jia, X.; Zhang, H.; Zhang, Z.; An, L. First-principles investigation of vacancy-defected graphene and Mn-doped graphene towards adsorption of H2S. *Superlattices Microstruct.* **2019**, *134*, 106235. [CrossRef]
3. Vekeman, J.; Cuesta, I.G.; Faginas-Lago, N.; Wilson, J.; Sánchez-Marín, J.; de Merás, A.S. Potential models for the simulation of methane adsorption on graphene: Development and CCSD(T) benchmarks. *Phys. Chem. Chem. Phys.* **2018**, *20*, 25518–25530. [CrossRef] [PubMed]
4. Hasnan, N.S.N.; Timmiati, S.N.; Lim, K.L.; Yaakob, Z.; Kamaruddin, N.H.N.; Teh, L. Recent developments in methane decomposition over heterogeneous catalysts: An overview. *Mater. Renew. Sustain. Energy* **2020**, *9*, 1–18. [CrossRef]
5. Hassani, A.; Mosavian, M.T.H.; Ahmadpour, A.; Farhadian, N. Improvement of methane uptake inside graphene sheets using nitrogen, boron and lithium-doped structures: A hybrid molecular simulation. *Korean J. Chem. Eng.* **2017**, *34*, 876–884. [CrossRef]
6. Zhao, Y.; Chen, Y.; Song, M.; Liu, X.; Xu, W.; Zhang, M.; Zhang, C. Methane Adsorption Properties of Mn-Modified Graphene: A First-Principles Study. *Adv. Theory Simul.* **2020**, *3*, 2000035. [CrossRef]
7. Ghanbari, R.; Safaiee, R.; Golshan, M. A dispersion-corrected DFT investigation of CH4 adsorption by silver-decorated monolayer graphene in the presence of ambient oxygen molecules. *Appl. Surf. Sci.* **2018**, *457*, 303–314. [CrossRef]
8. Xu, W.; Chen, Y.; Zhao, Y.; Zhang, M.; Tian, R.; Zhang, C. First-principles study on the methane adsorption properties by Ti-modified graphyne. *Int. J. Quantum Chem.* **2021**, *121*, e26811. [CrossRef]
9. Mahmoud, E. Recent advances in the design of metal–organic frameworks for methane storage and delivery. *J. Porous Mater.* **2020**, *28*, 213–230. [CrossRef]
10. Zhang, X.; Wang, S. Interfacial Strengthening of Graphene/Aluminum Composites through Point Defects: A First-Principles Study. *Nanomaterials* **2021**, *11*, 738. [CrossRef]
11. Feicht, P.; Eigler, S. Defects in Graphene Oxide as Structural Motifs. *ChemNanoMat* **2018**, *4*, 244–252. [CrossRef]
12. Xiong, L.; Gong, B.; Peng, Z.; Yu, Z. Spin-Seebeck effect and thermoelectric properties of one-dimensional graphene-like nanoribbons periodically embedded with four- and eight-membered rings. *Phys. Chem. Chem. Phys.* **2021**, *23*, 23667–23672. [CrossRef] [PubMed]
13. Zhao, Q.; Zhao, Y.; Chen, Y.; Ju, J.; Xu, W.; Zhang, M.; Sang, C.; Zhang, C. First-principles study on methane storage properties of porous graphene modified with Mn. *Appl. Phys. A* **2021**, *127*, 1–12. [CrossRef]

14. Cao, C.; Wu, M.; Jiang, J.; Cheng, H.-P. Transition metal adatom and dimer adsorbed on graphene: Induced magnetization and electronic structures. *Phys. Rev. B* **2010**, *81*, 205424. [CrossRef]
15. Yu, G.; Zhu, M.; Zheng, Y. First-principles study of 3d transition metal atom adsorption onto graphene: The role of the extended line defect. *J. Mater. Chem. C* **2014**, *2*, 9767–9774. [CrossRef]
16. Guan, Z.; Ni, S.; Hu, S. First-Principles Study of 3d Transition-Metal-Atom Adsorption onto Graphene Embedded with the Extended Line Defect. *ACS Omega* **2020**, *5*, 5900–5910. [CrossRef] [PubMed]
17. Ghosh, S.K. Diversity in the Family of Manganese Oxides at the Nanoscale: From Fundamentals to Applications. *ACS Omega* **2020**, *5*, 25493–25504. [CrossRef] [PubMed]
18. Kudyba, A.; Akhtar, S.; Johansen, I.; Safarian, J. Aluminothermic reduction of manganese oxide from selected MnO-containing slags. *Materials* **2021**, *14*, 356. [CrossRef]
19. Luo, X.; Peng, C.; Shao, P.; Tang, A.; Huang, A.; Wu, Q.; Sun, L.; Yang, L.; Shi, H.; Luo, X. Enhancing nitrate removal from wastewater by integrating heterotrophic and autotrophic denitrification coupled manganese oxidation process (IHAD-MnO): Internal carbon utilization performance. *Environ. Res.* **2021**, *194*, 110744. [CrossRef]
20. Ma, J.; Wang, C.; Xi, W.; Zhao, Q.; Wang, S.; Qiu, M.; Wang, J.; Wang, X. Removal of Radionuclides from Aqueous Solution by Manganese Dioxide-Based Nanomaterials and Mechanism Research: A Review. *ACS ES&T Eng.* **2021**, *1*, 685–705. [CrossRef]
21. Mansouri, M. Effects of Vacancy-Defected, Dopant and the Adsorption of Water upon Mn_2O_3 and Mn_3O_4 (001) Surfaces: A First-Principles Study. *Acta Phys. Pol. A* **2018**, *133*, 1178–1185. [CrossRef]
22. Yan, H.; Hu, W.; Cheng, S.; Xia, H.; Chen, Q.; Zhang, L.; Zhang, Q. Microwave-assisted preparation of manganese dioxide modified activated carbon for adsorption of lead ions. *Water Sci. Technol.* **2020**, *82*, 170–184. [CrossRef]
23. Shen, L.; Zhang, L.; Wang, K.; Miao, L.; Lan, Q.; Jiang, K.; Lu, H.; Li, M.; Li, Y.; Shen, B.; et al. Analysis of oxidation degree of graphite oxide and chemical structure of corresponding reduced graphite oxide by selecting different-sized original graphite. *RSC Adv.* **2018**, *8*, 17209–17217. [CrossRef]
24. Jiao, S.; Li, T.; Xiong, C.; Tang, C.; Li, H.; Zhao, T.; Dang, A. A Facile Method to Prepare Silver Doped Graphene Combined with Polyaniline for High Performances of Filter Paper Based Flexible Electrode. *Nanomaterials* **2019**, *9*, 1434. [CrossRef]
25. Ghashghaee, M.; Ghambarian, M. Methane adsorption and hydrogen atom abstraction at diatomic radical cation metal oxo clusters: First-principles calculations. *Mol. Simul.* **2018**, *44*, 850–863. [CrossRef]
26. Chen, X.; Liang, Q.H.; Jiang, J.; Wong, C.K.Y.; Leung, S.Y.; Ye, H.; Yang, D.G.; Ren, T.L. Functionalization-induced changes in the structural and physical properties of amorphous polyaniline: A first-principles and molecular dynamics study. *Sci. Rep.* **2016**, *6*, 20621. [CrossRef]
27. Zhao, S.; Larsson, K. First Principle Study of the Attachment of Graphene onto Different Terminated Diamond (111) Surfaces. *Adv. Condens. Matter Phys.* **2019**, *2019*, 9098256. [CrossRef]
28. Goyenola, C.; Stafstrom, S.; Hultman, L.; Gueorguiev, K.G. Structural patterns arising during synthetic growth of fuller-ene-like sulfocarbide. *J. Phys. Chem. C* **2012**, *116*, 21124–21131. [CrossRef]
29. Gueorguiev, G.K.; Czigány, Z.; Furlan, A.; Stafström, S.; Hultman, L. Intercalation of P atoms in Fullerene-like CPx. *Chem. Phys. Lett.* **2011**, *501*, 400–403. [CrossRef]
30. Gueorguiev, G.; Goyenola, C.; Schmidt, S.; Hultman, L. CFx: A first-principles study of structural patterns arising during synthetic growth. *Chem. Phys. Lett.* **2011**, *516*, 62–67. [CrossRef]
31. Zhu, Y.-q.; Su, H.; Jing, Y.; Guo, J.; Tang, J. Methane adsorption on the surface of a model of shale: A density functional theory study. *Appl. Surf. Sci.* **2016**, *387*, 379–384. [CrossRef]
32. Perdew, J.; Burke, K.; Ernzerhof, M. Generalized gradient approximation made simple. *Phys. Rev. Lett.* **1996**, *77*, 3865, Erratum in *Phys. Rev. Lett.* **1997**, *78*, 1396. [CrossRef]
33. Björkman, T.; Gulāns, A.; Krasheninnikov, A.; Nieminen, R.M. van der Waals Bonding in Layered Compounds from Advanced Density-Functional First-Principles Calculations. *Phys. Rev. Lett.* **2012**, *108*, 235502. [CrossRef] [PubMed]
34. Wei, L.; Liu, G.-L.; Fan, D.-Z.; Zhang, G.-Y. Density functional theory study on the electronic structure and optical properties of S absorbed graphene. *Phys. B Condens. Matter* **2018**, *545*, 99–106. [CrossRef]
35. Akay, T.I.; Toffoli, D.; Ustunel, H. Combined effect of point defects and layer number on the adsorption of benzene and toluene on graphene. *Appl. Surf. Sci.* **2019**, *480*, 1063–1069. [CrossRef]
36. Liu, M.; Liu, M.; She, L.; Zha, Z.; Pan, J.; Li, S.; Li, T.; He, Y.; Cai, Z.; Wang, J.; et al. Graphene-like nanoribbons periodically embedded with four- and eight-membered rings. *Nat. Commun.* **2017**, *8*, 14924. [CrossRef]
37. Zhao, Y.; Chen, Y.; Xu, W.; Zhang, M.; Zhang, C. First-Principles Study on Methane Adsorption Performance of Ti-Modified Porous Graphene. *Phys. Status Solidi B* **2021**, *258*, 2100168. [CrossRef]
38. Ding, Y. Research on the Mechanism of CH_4, CO_2, H_2O and O_2 Adsorption on Coal Molecule. Master's Thesis, North China Electric Power University, Baoding, China, 2018.
39. Li, K.; Li, H.; Yan, N.; Wang, T.; Zhao, Z. Adsorption and dissociation of CH4 on graphene: A density functional theory study. *Appl. Surf. Sci.* **2018**, *459*, 693–699. [CrossRef]
40. Yuan, W.H. *Study on the Adsorption Properties of Formaldehyde-Methane on Graphene*; Beijing Institute of Technology: Beijing, China, 2016.
41. Zhou, J.; Xuan, H.Y.; Xiao, R.J.; Zhang, X.Y.; Man, M.O.; Fang, Z.J. First-principles study on adsorption property of carbon surface to methane. *J. Guangxi Univ. Nat. Sci. Ed.* **2018**.

42. Liu, J.; Chen, Y.Z.; Xie, Y. The calculation of adsorption properties of H_2S, CO_2 and CH_4 on Fe doped MoS_2 based on first-principles. *J. At. Mol. Phys.* **2020**, *37*, 501–507.
43. Kumar, K.V.; Preuss, K.; Titirici, M.; Rodriguez-Reinoso, F. Nanoporous Materials for the Onboard Storage of Natural Gas. *Chem. Rev.* **2017**, *117*, 1796–1825. [CrossRef]
44. Xu, W.; Chen, Y.-H.; Song, M.; Liu, X.; Zhao, Y.; Zhang, M.; Zhang, C.-R. First-Principles Study on Methane (CH_4) Storage Properties of Graphdiyne. *J. Phys. Chem. C* **2020**, *124*, 8110–8118. [CrossRef]
45. Zhou, X.; Gall, D.; Khare, S.V. Mechanical properties and electronic structure of anti-ReO3 structured cubic nitrides, M3N, of d block transition metals M: An ab initio study. *J. Alloy. Compd.* **2014**, *595*, 80–86. [CrossRef]
46. Tawfik, S.A.; Cui, X.Y.; Ringer, S.P.; Stampfl, C. Multiple CO_2 capture in stable metal-doped graphene: A theoretical trend study. *RSC Adv.* **2015**, *5*, 50975–50982. [CrossRef]
47. Hu, X.; Meng, F. Structure and gap opening of graphene with Fe doped bridged trivacancy. *Comput. Mater. Sci.* **2016**, *117*, 65–70. [CrossRef]
48. Liu, X. Theoretical Studies of Interaction between Metal and Graphene and Molecular Clusters. Master's Thesis, Jilin University, Changchun, China, 2011.
49. Wu, M.; Cao, C.; Jiang, J.Z. Electronic structure of substitutionally Mn-doped graphene. *New J. Phys.* **2010**, *12*, 63020. [CrossRef]
50. Chen, J.-J.; Li, W.-W.; Li, X.-L.; Yu, H.-Q. Improving Biogas Separation and Methane Storage with Multilayer Graphene Nanostructure via Layer Spacing Optimization and Lithium Doping: A Molecular Simulation Investigation. *Environ. Sci. Technol.* **2012**, *46*, 10341–10348. [CrossRef]
51. Rad, A.S.; Pazoki, H.; Mohseni, S.; Zareyee, D.; Peyravi, M. Surface study of platinum decorated graphene towards ad-sorption of NH3 and CH4. *Mater. Chem. Phys.* **2016**, *182*, 32–38. [CrossRef]
52. Jiang, H.; Cheng, X.-L. Simulations on methane uptake in tunable pillared porous graphene hybrid architectures. *J. Mol. Graph. Model.* **2018**, *85*, 223–231. [CrossRef] [PubMed]
53. Tanaka, H.; El-Merraoui, M.; Steele, W.; Kaneko, K. Methane adsorption on single-walled carbon nanotube: A density functional theory model. *Chem. Phys. Lett.* **2002**, *352*, 334–341. [CrossRef]
54. Xu, W.; Chen, Y.; Zhao, Y.; Zhang, M.; Tian, R.; Zhang, C. Methane adsorption properties of N-doped graphdiyne: A first-principles study. *Struct. Chem.* **2021**, *32*, 1517–1527. [CrossRef]
55. Zhao, Q.; Chen, Y.; Xu, W.; Ju, J.; Zhao, Y.; Zhang, M.; Sang, C.; Zhang, C. First-principles study of the impact of hydrogen on the adsorption properties of Ti-decorated graphdiyne storage methane. *Chem. Phys. Lett.* **2022**, *790*, 139329. [CrossRef]
56. Othman, F.E.C.; Yusof, N.; Harun, N.Y.; Bilad, M.R.; Jaafar, J.; Aziz, F.; Salleh, W.N.W.; Ismail, A.F. Novel Activated Carbon Nanofibers Composited with Cost-Effective Graphene-Based Materials for Enhanced Adsorption Performance toward Methane. *Polymers* **2020**, *12*, 2064. [CrossRef] [PubMed]

Review

Comprehensive Review on Zeolite-Based Nanocomposites for Treatment of Effluents from Wastewater

Veena Sodha [1], Syed Shahabuddin [1,*], Rama Gaur [1], Irfan Ahmad [2], Rajib Bandyopadhyay [1,*] and Nanthini Sridewi [3,*]

[1] Department of Chemistry, School of Technology, Pandit Deendayal Energy University, Knowledge Corridor, Raisan, Gandhinagar 382426, Gujarat, India
[2] Department of Clinical Laboratory Sciences, College of Applied Medical Sciences, King Khalid University, Abha 61421, Saudi Arabia
[3] Department of Maritime Science and Technology, Faculty of Defence Science and Technology, National Defence University of Malaysia, Kuala Lumpur 57000, Malaysia
* Correspondence: syedshahab.hyd@gmail.com or syed.shahabuddin@sot.pdpu.ac.in (S.S.); rajib.bandyopadhyay@sot.pdpu.ac.in (R.B.); nanthini@upnm.edu.my (N.S.); Tel.: +91-858-593-2338 (S.S.); +60-124-675-320 (N.S.)

Abstract: All humans and animals need access to clean water in their daily lives. Unfortunately, we are facing water scarcity in several places around the world, and, intentionally or unintentionally, we are contaminating the water in a number of ways. The rise in population, globalization, and industrialization has simultaneously given rise to the generation of wastewater. The pollutants in wastewater, such as organic contaminants, heavy metals, agrochemicals, radioactive pollutants, etc., can cause various ailments as well as environmental damage. In addition to the existing pollutants, a number of new pollutants are now being produced by developing industries. To address this issue, we require some emerging tools and materials to remove effluents from wastewater. Zeolites are the porous aluminosilicates that have been used for the effective pollutant removal for a long time owing to their extraordinary adsorption and ion-exchange properties, which make them available for the removal of a variety of contaminants. However, zeolite alone shows much less photocatalytic efficiency, therefore, different photoactive materials are being doped with zeolites to enhance their photocatalytic efficiency. The fabrication of zeolite-based composites is emerging due to their powerful results as adsorbents, ion-exchangers, and additional benefits as good photocatalysts. This review highlights the types, synthesis and removal mechanisms of zeolite-based materials for wastewater treatment with the basic knowledge about zeolites and wastewater along with the research gaps, which gives a quality background of worldwide research on this topic for future developments.

Keywords: zeolite; wastewater treatment; photocatalysis; nanocomposites

1. Introduction

Water is an integral part of all living organisms; it is crucial for humans and the environment. Water becomes contaminated after being used for numerous reasons such as bathing, washing, cooking, and manufacturing and is then dumped back into water sources after treatment. Since it is difficult for the wastewater treatment plants to treat the pollutants of emerging industries and the majority of industries levy fees for the same, it is less expensive for enterprises to treat or pre-treat the wastewater before discharging it into sewers. Figure 1 depicts how water is collected from resources, utilized, and then discharged into water bodies [1].

Wastewater can promote diseases such as polio, cholera, vomiting, diarrhoea, nausea, and even cancer in the human body [2,3]. If wastewater is discharged into water bodies, the pollutants in it can inhibit the establishment of marine plants [4]. There are numerous types of water pollutants for which multiple treatment methods have been described, which includes chemical, biological, and physical methods, such as adsorption [5], photocatalysis [6,7], ultrafiltration [8], and biofiltration [9]. In the area of wastewater treatment, zeolites, which are aluminosilicates with porous structures, are generally utilized as adsorbents [10,11], ion-exchangers [12], and photocatalysts [13,14]. The negatively charged structure of a zeolite attracts a variety of cationic pollutants to it. The use of zeolites in the removal of pollutants is not just restricted to the adsorption of cationic pollutants; by modifying it in various ways, its affinity for anionic pollutants can also be improved [15]. Photoactivity in zeolites has been enhanced by the addition of heteroatom to their framework, as in titanium silicates, exposing it to larger applications [16–18]. Composites are materials made by combining two or more materials known as parent materials. The term zeolite-based composite refers to the coupling of zeolites with other materials to form binary, ternary, and so on composites. The field of study on zeolite-based composites as a pollutant removal medium is broadening owing to their tuneable pore size [16,19], enhanced photoactivity [19], and easy operation [20]. For example, combining zeolites with materials that have a positively charged framework and an affinity towards anionic pollutants, may result in a composite that can be applied for the removal of both cationic and anionic contaminants. Generally, carbon-based materials [21,22], metal oxides [23,24], polymers [25,26], and clay compounds [27] are incorporated with zeolites for a variety of applications including fuel cells [28], catalysis [29], sorption [4], and others. Researchers have demonstrated the removal of contaminants from various model solutions, such as dyes [30,31], heavy metals [32], herbicides [33], etc., using zeolites and zeolite-based materials [34]. These models can also be applied to treat the real wastewater from industries. The synthesis, as well as adsorption and photocatalytic studies, on zeolite-based composites described in this review may help researchers in the treatment of real wastewater samples from industries. In addition, this review not only gives information about the zeolite-based composites, but also gives basic knowledge about wastewater and zeolites. The physicochemical properties of zeolites, synthesis of zeolite-based materials, and their mechanism in adsorption and photocatalysis are explained, which gives basic research background to early researchers and to scientists who aim to devise zeolite-based materials for pollutant remediation. The adsorption and photocatalytic research of zeolite-based composites are more thoroughly examined for the purpose of building a photoactive device for wastewater treatment.

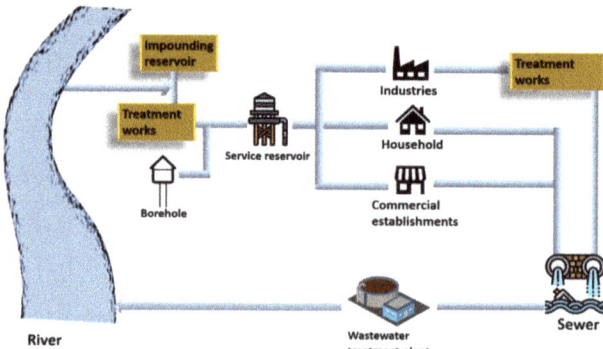

Figure 1. Cycle of water service.

1.1. Wastewater

The combination of the waterborne or liquid wastes removed from institutions, residences, and industrial and commercial establishments is wastewater [35]. Everything that is discharged into the sewers subsequently gets treated in a wastewater treatment plant [1]. It includes pollutants from various domestic activities, such as bathing, cleaning clothes and utensils, and flushing toilets; industrial activities, such as textiles, mining, and manufacturing; commercial activities, such as beauty salons and car washing; agricultural facilities; energy units. Apart from these, a variety of other events, such as surface run-offs, floods, and storms also produce wastewater. Additionally, if the sewer becomes damaged, groundwater will sweep in, increasing the volume of wastewater. Figure 2 represents major sources of wastewater.

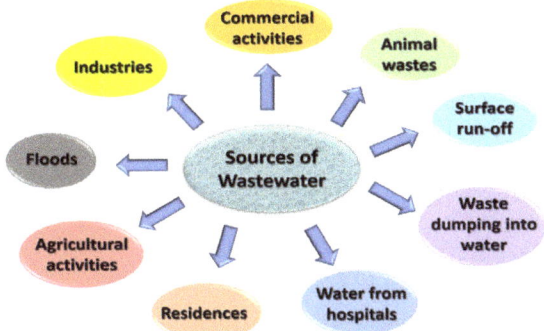

Figure 2. Sources of wastewater.

Wastewater contains a variety of pollutants including organic, inorganic, toxic, non-toxic, thermal, and suspended solids from industries, residences, commercial activities, etc. [36]. Refer to Figure 3.

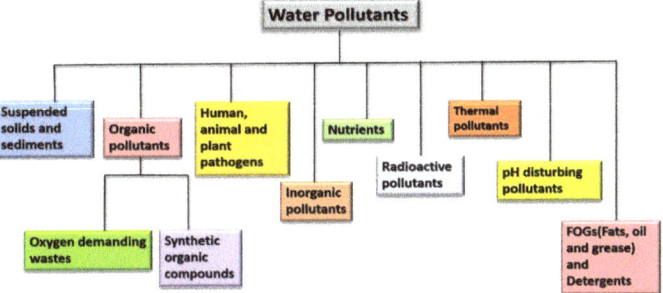

Figure 3. Flowchart representing various pollutants in wastewater.

1.1.1. Total Suspended Solids (TSS)

Suspended solids are particulate matter with a diameter of less than 62 μm. These are too small to settle down and too large to float, hence, they remain suspended in water. Generally, a water stream contains some SS, but an excessive amount might cause issues [37]. These exist in two forms: (i) inert and (ii) oxidizable solids. Sand particles and eroded minerals are examples of inert solids. These are sourced from mining, coal washing, construction sites, etc. Oxidizable solids settle out similar to inert solids but get decomposed on deposition releasing toxic compounds, such as methane, ammonia, and sulphides, causing higher oxygen demand in localized areas of water. These reduce the penetration of light into the water, inhibit the growth of filter feeders, cause temperature change, and other issues [1,37]. Their removal techniques include gravity settling [38], centrifugation [39], and filtration [40], followed by disinfection to remove floating bacteria and pathogens [41].

1.1.2. Organic Pollutants

Organic pollutants can be classified based on two aspects: (i) the nature of the pollutant, i.e., natural or synthetic, and (ii) its persistence, i.e., a persistent or non-persistent organic pollutant. Natural organic pollutants include oxygen demanding wastes, which decrease the oxygen levels of water. Synthetic organic compounds include chemicals from industries, agricultures, etc.

Some organic pollutants are persistent and some are non-persistent. Persistent organic pollutants (POPs) are chemicals that live in the environment for prolonged periods as these are resistant to biochemical and photolytic processes [42]. Moreover, these are lipophilic and hydrophobic pollutants that are receptive to long-range transport and bioaccumulation and are prone to enter the food chain as well [43]. Consider Figure 4 [43].

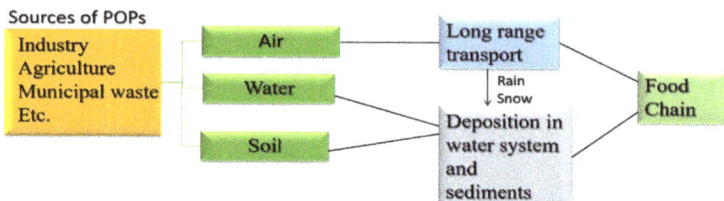

Figure 4. Flowchart representation of flow to POPs in the environment and food chain. Adapted with permission from Ref. [43]. Copyright 2020 Springer Nature.

Locally discharged POPs spread pollution far from its source. Sources of persistent organic pollutants (POPs) include volcanic activity and forest fires that produce dioxins and dibenzofurans. Other sources include agricultural pesticides and industries, such as dichlorodiphenyltrichloroethane (DDT), polychlorinated biphenyls (PCB), perfluoro octane sulfonic acid (PFOS), perfluorooctanoic acid (PFOA), and brominated flame retardants, etc. [44]. In adult studies, POPs have been implicated in a variety of adverse health impacts, such as thyroid and endocrine-related cancers, diabetes, obesity, and reproductive concerns in both males and females [45].

1.1.3. Inorganic Pollutants

Heavy metals, inorganic salts, mineral acids, trace elements, metals, and their complexes with organic compounds are examples of inorganic pollutants. Metals with a density higher than 5 g/cm^3 are classified as heavy metals [46]. Titanium, cobalt, manganese, iron, nickel, nickel, copper, zinc, arsenic, silver, gold, and mercury are commonly occurring heavy metals in everyday life. Few heavy metals are essential elements in our life but when present in a large amount they can be toxic. Natural deposits of heavy metals can be discovered in the Earth's crust, hence, one of their sources in wastewater includes surface run-off. Apart from that, metal-based industries, automobiles, roadworks, and metal leaching are the major sources of heavy metals in wastewater. Heavy metal exposure in humans can result in cellular function loss, cell damage, and potentially carcinogenic effects [46].

1.1.4. Radioactive Pollutants

Approximately 11% of the world's electricity is generated by nuclear power plants [47]. The nuclear fission process produces no carbon dioxide, which is a plus, but we still have to deal with nuclear waste. The release of radionuclides in the environment caused disaster during 1986 in the Chernobyl and Fukushima Daiichi plant in 2011 [48]. Radioactive pollutants can cause plant mutations and serious health damage to aquatic life. Their influence on humans may be mild or fatal depending on the magnitude and duration of exposure. When humans have a short time of exposure to a lower level of radioactive pollutants, it can cause mild skin irritations; on the other hand, prolonged exposure at low-intensity causes diarrhea, nausea, vomiting, hair loss, etc. Prolonged exposure to high levels of radiation will lead to some irreversible DNA damages. Apart from that, it can cause several other diseases, such as lung, thyroid, and skin cancers [49]. Several removal methods, including physical, chemical, and biological, are used. (i) Physical methods: evaporation, distillation, dumping; (ii) chemical methods: acid digestion wet oxidation, precipitation; (iii) biological: microbial remediation and plant remediation [3]. For further understanding of various types of pollutants refer to Table 1.

Table 1. Pollutants, their sources, adverse effects, and removal techniques.

Sr. No.	Type of Pollutant	Examples	Sources	Adverse Effects	Removal Techniques	Ref.
1	Suspended solids and sediments.	Sand particles, eroded minerals, etc.	Mining, coal washing, etc.	Reduced light penetration in water, temperature change, etc.	Gravity settling, centrifuge, etc.	[38,39]
2	Oxygen demanding wastes.	Dissolved organic matter.	Domestic wastes, pulp and paper mill, wastes from food processing plants, animal sewage, slaughter houses, agricultural runoffs, etc.	Reduce the oxygen levels in water.	Carbon oxidation process, co-agulation, fluctuation.	[50]
3	Organic dyes.	Methylene blue, methyl orange, crystal violet, rhodamine B, etc.	Textile, pharmaceutical, food, laser printing industries.	Diarrhea, breathing difficulty, nausea, vomiting, gastrointestinal issues, carcinogenic and oncogenic effects.	Adsorption, membrane separation, advanced oxidation processes (AOPs), biological decolorization.	[2,51,52]
4	Pesticides.	Organochlorines such as DDT, chlordane, aldrin, organofluorines such as PFOA, PFOS.	Agricultural activities.	Respiratory and skin conditions, cancer, reproduction issues, endocrine disruption, etc.	Advanced oxidation processes (AOPs), activated sludge treatment, adsorption, membrane technologies, etc.	[53,54]
5	Heavy metals.	Zinc (Zn), mercury (Hg), lead (Pb), arsenic (As), iron (Fe), cadmium (Cd), etc.	Surface run-off, roadworks, metal-based industries, automobiles, etc.	Haze, corrosion, eutrophication, and can lead to acid rain, also inhibits the biodegradation of organochlorines.	Adsorption, ion exchange.	[55,56]
6	Radioactive pollutants.	Caesium, strontium, uranium etc.	Nuclear power plants	DNA damage, thyroid, lung and skin cancers, hair loss etc.	Precipitation, distillation, microbial remediations, etc.	[3]
7	Nutrients.	Nitrogen, phosphorus, etc.	Agricultural fertilizers, run-offs from storm water, household detergents, human wastes, etc.	Eutrophication, reduction in oxygen level of water, reduced sunlight penetration, affect the growth of plants, etc.	To remove phosphorus biological nutrient removal (BNR) and nitrification and denitrification to remove to remove nitrogen.	[57]
8	Human, animal, and plant pathogens (pathogen microorganisms).	Different types of bacteria, viruses, protozoa, and helminths.	Animal and human fecal wastes, household and laundry wastewater.	Waterborne diseases such as polio, hepatitis cholera, anemia, typhoid, gastroenteritis, etc.	Natural elimination by temperature or prolonged life or adsorption to particles and sedimentation disinfection by chlorination and UV radiation.	[58–60]

1.2. Zeolites

Zeolites are crystalline three-dimensional, porous, aluminosilicates with organized structures and building blocks of tetrahedral units TO_4 with an O atom bridged between them, as given in Figure 5, where T denotes Si or Al atom [61]. In the zeolite framework, the gap between the huge cavities holds water and interchangeable cations [15]. The basic chemical formula for zeolites is represented as:

$$M_{a/n}\left[Al_aSi_bO_{2(a+b)}\right] \cdot qH_2O;$$

where M stands for [Sr, Ba, Ca, Mg] and/or [Li, K, Na], and cation charge is symbolized by n. The values of b/a range from 1 to 6 while q/a range from 1 to 4 [15].

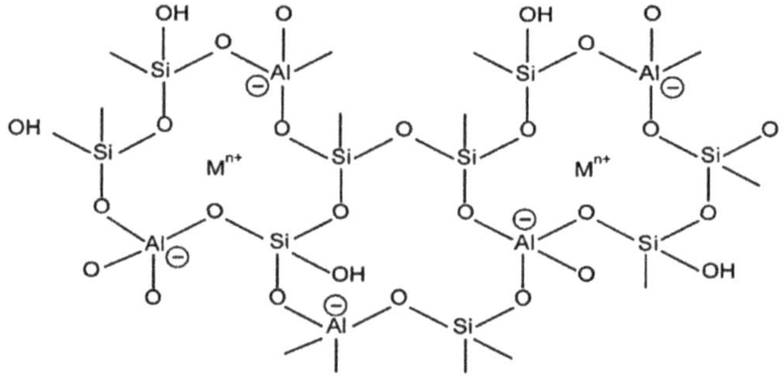

Figure 5. A two-dimensional illustration of the zeolite framework. Reprinted with permission from Ref. [62]. Copyright 2006 Elsevier.

Zeolites occur naturally as well as being synthesized chemically. The source of natural zeolites is volcanogenic sedimentary rocks. Clinoptilolite, phillipsite, mordenite, chabazite, stilbite, analcime, and laumontite are abundant among natural zeolites, while barrerite, offerite, and paulingite are rare.

Some of the natural zeolites with their chemical formula are given in Table 2.

Table 2. Name and chemical formula of some natural zeolites. Reprinted with permission from Ref. [15]. Copyright 2010 Elsevier.

Name of Zeolite	Chemical Formula
Clinoptilolite	$(Na_2, K_2\ Ca)_3Al_6Si_{30}O_{72} \cdot 21H_2O$
Phillipsite	$K_2(Na_2, Ca)_2Al_8Si_{10}O_{32} \cdot 12H_2O$
Mordenite	$(Na_2, Ca)_4Al_8Si_{40}O_{96} \cdot 28H_2O$
Stilbite	$Na_2Ca_4Al_{10}Si_{26}O_{72} \cdot 30H_2O$
Chabazite	$(Na_2, Ca, K_2)_2Al_4Si_8O_{24} \cdot 12H_2O$
Analcime	$Na_{16}Al_{16}Si_{32}O_{96} \cdot 16H_2O$
Ferrierite	$(Na_2, K_2, Ca, Mg)_3Al_6Si_{30}O_{72} \cdot 20H_2O$
Laumontite	$Ca_2Al_8S_{16}O_{48} \cdot 16H_2O$
Scolecite	$Ca_4Al_8Si_{12}O_{40} \cdot 12H_2O$
Heulandite	$Al_8Si_{28}Ca_4O_{68} \cdot 24H_2O$

Classification of Zeolites

Zeolites may be categorized as per their occurrence, Si-Al ratio, pore size, crystal structure, and other factors [63]. Figure 6 shows the broad classification of zeolites. Tables 3–5 show how zeolites are classified according to their pore size, silica to alumina ratio, and structure type [63,64]. Some of the many classes of framework types according to the website of international zeolitic association (IZA) are depicted in Table 5.

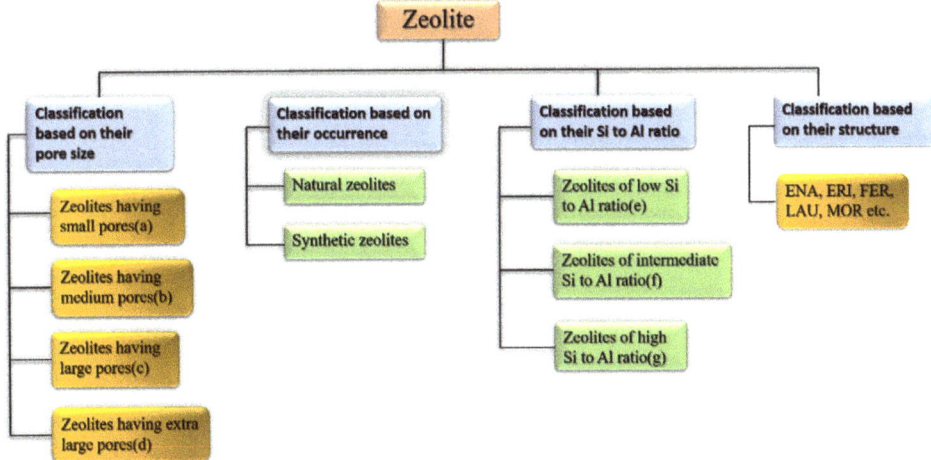

Figure 6. Broad classification of zeolites.

Table 3. Classification of zeolites according to their pore size.

Class of Zeolite	Number of Rings	Free Pore Diameter (nm)
(a) Zeolites with small pores	8	0.3 to 0.45
(b) Zeolites with medium pores	10	0.45 to 0.6
(c) Zeolites with large pores	12	0.6 to 0.8
(d) Zeolites with extra-large pores	14	0.8 to 1.0

Table 4. Classification of zeolites considering Si to Al ratio.

Class of Zeolite	Range of Si:Al Ratio
(e) Zeolites with small Si:Al ratio	1.0 to 1.5
(f) Zeolites with intermediate Si:Al ratio	2 to 5
(g) Zeolites with large Si:Al ratio	10 to several thousands

Table 5. Classification of zeolites as per their structure type [64].

Framework Type Code	Symmetry	Channel Dimensionality	Framework Density (Å³)	Total Volume (Å³)	Accessible Volume (%)	Order	Reference Material
ANA	Cubic	3D	19.2 T/1000	2497.2	0.00	Fully ordered	Analcime
BEA	Tetragonal	3D	15.3 T/1000	4178.4	20.52	Partially disordered	Beta polymorph A
CHA	Trigonal	3D	15.1 T/1000	4178.4	20.52	Fully ordered	Chabazite
DFT	Tetragonal	3D	17.7 T/1000	451.7	6.58	Fully ordered	DAF-2
ERI	Hexagonal	3D	16.1 T/1000	2239.5	15.10	Fully ordered	Erionite
FAU	Cubic	3D	13.3 T/1000	14,428.8	27.42	Fully ordered	Faujasite
FER	Orthorhombic	2D	17.6 T/1000	2051.3	10.01	Fully ordered	Ferrierite
HEU	Monoclinic	2D	17.5 T/1000	2054.8	9.42	Fully ordered	Heulandite
LAU	Monoclinic	1D	18.0 T/1000	1333.6	9.57	Fully ordered	Laumontite
MFI	Orthorhombic	3D	18.4 T/1000	5211.3	9.81	Fully ordered	ZSM-5
MOR	Orthorhombic	2D	17.0 T/1000	2827.3	12.27	Fully ordered	Mordenite
MRE	Orthorhombic	1D	19.7 T/1000	2442.5	6.55	Partially disordered	ZSM-48
NAT	Tetragonal	3D	16.2 T/1000	1231.5	9.06	Fully ordered	Natrolite
PHI	Orthorhombic	3D	16.4 T/1000	1953.7	9.89	Fully ordered	Phillipsite

Catalysis, cation exchange, sorption, and molecular sieving are all physiochemical characteristics of zeolites. The tetrahedrons of $[SiO_4]^{4-}$ and $[AlO_4]^{5-}$ are linked together in the zeolitic framework to build cages linked via precise and molecular-sized pores. Existence of $[AlO_4]^{5-}$ gives a negative charge to zeolite structure, which is stabilized by positively charged ions, such as K^+, Na^+, and Ca^{2+}. These ions are responsible for ion-exchange processes in zeolites [15]. The porous structure of a zeolite has been proven to have excellent adsorption efficiency for heavy metals, including mercury, fluoride, arsenic, and organic dyes.

1.3. Adsorption

One of the most common wastewater treatment methods is adsorption since it is simple to use and effective [65]. It is a surface phenomenon that occurs when the molecules from fluid bulk come into contact with a solid surface either by physical forces or chemical bonds. Usually, adsorption is a reversible process. Reversible of adsorption is when the adsorbent begins to release the adsorbed molecules; this is referred to as desorption [66]. There are two types of adsorptions: (i) physical adsorption, i.e., adsorption under the influence of physical forces such as weak van der Waals attractions, hydrogen bonding, etc., and (ii) chemical adsorption, i.e., adsorption by chemical bonds [67].

Activated carbon [68], industrial solid wastes [69], biomaterials [70], clay minerals [71], and zeolites [11,14] are among the most commonly used materials in wastewater treatment. In the year 1785, adsorption was first discovered by Lowitz, after that, it was used in sugar refining processes to remove color [72]. Subsequently, in American treatment plants, inactivated charcoal filters were employed for water purification [72]. For the first time, granular activated carbon (GAC) was used for adsorption in 1929 in Hamm, Germany, and Bay City, Michigan, 1930 [73]. When studied, modified and synthetic zeolites showed better adsorption and ion exchange capacities among synthetic, modified, and natural zeolites [72]. Zeolites are mostly used for the adsorption of heavy metals [32], dyes [74], ammonium ions [75], etc. Natural turkey clinoptilolite exhibited low absorptivity for three azo dyes (Everzol black, Everzol red, Everzol yellow) as examined by Armagan et al. [76]. Adsorption capacities of natural zeolite were improved greatly via modification with quaternary amines [76]. Figure 7 elaborates on the adsorption of methylene blue on zeolites which occurs via electrostatic attractions between negatively charged Al ions in the zeolite framework and positively charged nitrogen atom in methylene blue.

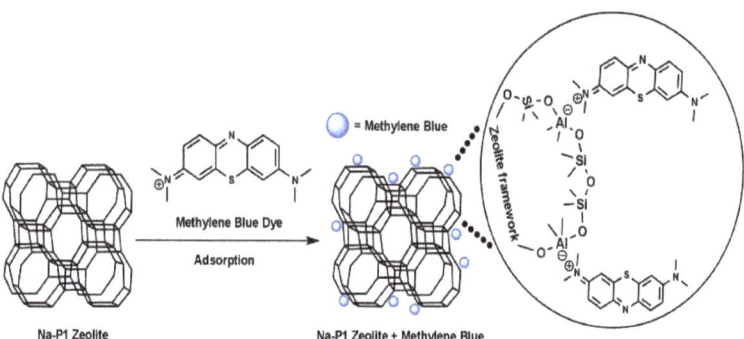

Figure 7. Adsorption of methylene blue on Na-P1 zeolite. Reprinted from Ref. [10].

1.4. Photocatalysis

Photocatalysis is the process in which a photon of light catalyzes a reaction. Materials with photocatalytic characteristics are known as semiconductors [77]. The conduction band and valance band are two different energy bands in a semiconductor. The bandgap is the energy difference between the two above-stated bands. When a photon of light strikes a semiconducting material, electrons (e^-) are excited from the valance band to the conduction band and positively charged holes (h^+) are left behind, as shown in Figure 8. The photogenerated electron-hole pairs develop the active oxidizing species, which then cause organic pollutants, such as dyes, in wastewater to degrade [43].

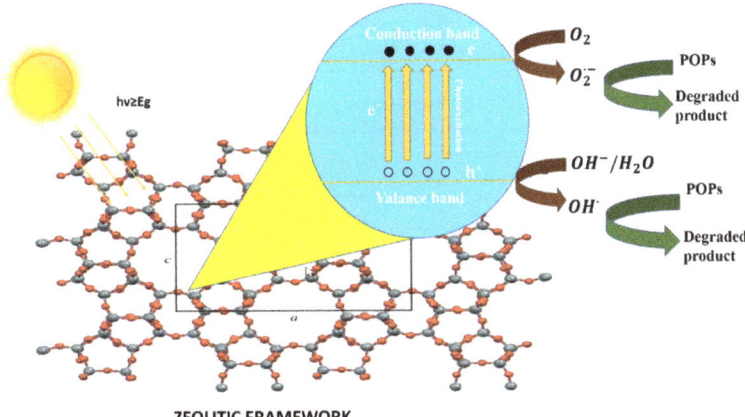

Figure 8. Schematic diagram of photocatalysis in zeolite (image of zeolite framework is reprinted from Ref. [78]).

Photocatalysis can be categorized into two types: (i) homogenous and (ii) heterogeneous photocatalysis. If in a photocatalytic reaction, both photocatalyst and reactant are in a different phase, then it is said to be heterogeneous photocatalysis. On the other hand, if photocatalyst and reactants and all other species, such as photosensitizers, are in the same phase, the system is called homogenous photocatalysis [79]. Due to its great stability, ease of separation, and photocatalyst regeneration, heterogeneous catalysis is superior to homogeneous catalysis [80].

The following steps are part of the heterogeneous photocatalysis mechanism [43].

$$\text{Semiconductor photocatalyst} \overset{h\nu \geq E_g}{\rightarrow} e^- + h^+$$
$$h^+ + H_2O \rightarrow OH^\bullet + H^+$$
$$h^+ + OH^- \rightarrow OH^\bullet \quad e^- + O_2 \rightarrow O_2^{\bullet -}$$
$$O_2^{\bullet -} + H^+ \rightarrow OOH^\bullet$$
$$2OOH^\bullet \rightarrow H_2O_2 + O_2$$
$$H_2O_2 + O_2^{\bullet -} \rightarrow OH^- + OH^\bullet + O_2$$
$$POPs + \left(h^+, OH^\bullet, O_2^{\bullet -}, OOH^\bullet \text{ or } H_2O_2\right) \rightarrow \text{Degraded products}$$

Photocatalytic activities of zeolite alone are rarely discussed. Research on the photocatalytic activity of zeolites is scarce. Rather, their cavities behave as hosts for semiconducting materials in photocatalysis. In 2003, Krisnandi et al. investigated zeolite ETS-10, a titanosilicate, microporous zeolite with liner Ti-O-Ti-O-chains in the framework as a photocatalyst for oxidation of ethane to CO_2 and water [79]. Before this work, some preliminary analyses have been conducted on the photocatalytic activity of ETS-10 [81]. The basic mechanism in ETS-10 is that the titanium sites present in Ti-O-Ti-O trap the electrons and undergo photoreduction in the existence of ethane. These trapped electrons quickly get shifted to oxygen and generate active oxidation species [79]. In 2020, Aguiñaga et al. reported clinoptilolite–mordenite, natural zeolite as an efficient self-photocatalyst [13]. The obtained results were compared with that of titanium dioxide particles, and it was found that under similar conditions, both zeolite and TiO_2 required the same time for the complete degradation of caffeine.

The NH abbreviation is used for the hydrogenated form of natural zeolite (NZ) and NFe denotes the ion-exchanged version of NZ. The symbol SH stands for synthetic zeolite clinoptilolite–mordenite. Aguiñaga et al. recorded the diffuse reflectance spectra of synthetic and natural zeolites as given in Figure 9. A bandgap analysis was conducted with the Tauc plot. Based on spectroscopic analysis of ZSM-5, it was estimated that bands developed in the range from 200–500 nm are attributed to different states of iron, and in synthetic zeolites, the presence of iron is usual. The aluminosilicate framework has a bandgap of approximately 7 eV. The bandgap of natural mordenite was estimated as 2.63 eV. In the case of synthetic clinoptilolite C, the analyzed band gap was 4.26 and 4.46 eV for direct and indirect transitions, respectively. In the same way, for synthetic mordenite, it was estimated at 3.26 and 3.45 eV. Hence, the wide bandgap of zeolite resembles the semiconductors and its application as a photocatalyst can be of interest in future research [13].

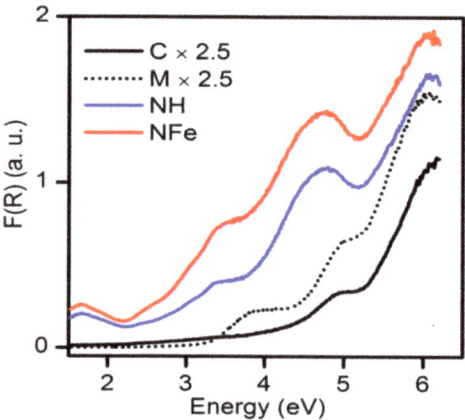

Figure 9. Diffuse reflectance spectra of synthetic and natural zeolites. Reprinted from Ref. [13].

1.5. Ion Exchange

Ion exchange can be defined as a reversible process in which exchangeable ions in an insoluble exchange material replaces similarly charged ions in the solution [82]. Electrostatic attractions are employed between ionic functional groups, which is the driving force of a typical ion exchange reaction. It can easily be employed for the extraction of heavy metals, such as cadmium, chromium, barium, arsenic, silver, lead, as well as nitrates from water [83]. Furthermore, ion exchange is the best process for the removal of radioactive nuclide in small systems [84]. Ion-exchangers can be categorized as cation and anion exchange resins. Cation exchangers are those that interchange their cations, and anion exchangers interchange their anions with the solution [84]. For further clarification refer Figure 10.

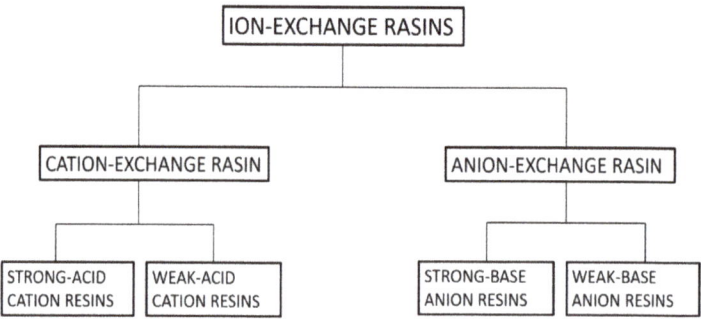

Figure 10. Classification of ion-exchange resins.

Natural inorganic zeolites, clays, and synthetic organic resins are extensively used ion-exchangers [85,86]. One of the major drawbacks that are associated with ion-exchangers is that their ion-exchange capacity is easily reduced with contamination via organic substances. Luca et al. used ETS-10 zeolite to remove metals originating from zinc ferrite [12]. A titanosilicate, ETS-10, is microporous zeolite with liner Ti-O-Ti-O-chains in its framework. ETS-10 is extensively employed as an ion-exchanger [87]. The major advantage of using ETS-10 as an ion-exchanger is that it is easy to regenerate and is thermally stable up to 550 °C [88]. Hence, if contaminated with organic pollutants, it can easily be regenerated by calcination. Based on ICP-MS elemental analysis, it was estimated that zinc ferrite releases high concentrations of Fe, Zn, Pb, Ca, and Mn. ETS-10 removed all metal ions present very efficiently. ETS-10 showed better cation exchange capacity (CEC) than commercial zeolite A for manganese, zinc, and lead. Nearly 100% removal was observed within 30 min [12]. Figure 11 gives a basic idea about how the ion-exchange mechanism works in zeolites. A natural zeolite, clinoptilolite, which has Ca^{+2} ions trapped in its cavity, is used as an ion-exchanger. Ca^{+2} ions are replaced with Cs^{+2} via ion exchange as depicted in Figure 11.

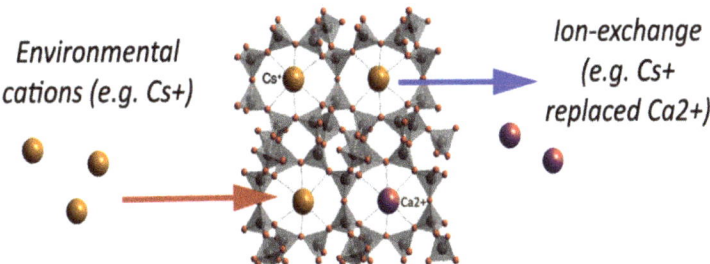

Figure 11. Ion exchange in naturally occurring zeolite, clinoptilolite. Reprinted from Ref. [89].

The adsorption properties of zeolites are influenced by their chemical and structural makeup. The cation exchange capacity (CEC) of zeolites varies with zeolitic framework structure, the density of anionic framework, size, and shape of foreign ions, etc. In raw natural zeolites, the pores are clogged with impurities and there is no uniform pore distribution, crystal structure, or chemical composition throughout the framework [90]. Additionally, raw natural zeolites have a negatively charged surface, hence, they only attract cationic pollutants, such as cationic dyes and heavy metal ions. They have a very low or little affinity toward anionic pollutants, such as anionic dyes and organic pollutants in aqueous media. Their efficiency can be increased by their modification. Table 6 shows some of the modification processes [34,91].

Table 6. Modification of zeolites, their advantages and disadvantages. Reprinted from Ref. [91].

Modification Method	Process	Advantages	Disadvantages	Ref.
Acid/base treatment.	Simple ion exchange using dilute acid solution.	Pore volume and electrostatic surface area is increased.	Decrease in CEC due to dealumination, decrease in thermal stability.	[92]
Surfactant modification.	By introducing cationic organic surfactants such as tetramethylammonium, hexadecyl trimethyl ammonium (HDTMA), n-cetyl pyridinium (CPD), etc.	Increases the hydrophobicity of zeolite, making it appropriate for the adsorption of a wide range of organic pollutants.	Complicated functional groups are formed for cationic exchange sites due to formation of admicelle.	[93–95]
Ultrasonic modification.	By sonicating with a solvent with help of an ultrasonicator bath.	Impurities are removed from the channel and the surface area is increased.	Always used in combination with other methods, inefficient.	[96]
Thermal modification.	By heating in the oven or muffle furnace.	Evaporation of water, removal of contaminants from the channel, and expansion of the pore diameter.	Uneven heating.	[97]

We observed that raw, natural zeolites have limited adsorption capacity and carry contaminants. Synthetic zeolites, on the other hand, have a consistent pore distribution across the framework, as well as improved adsorption and ion exchange behavior. Kozera-Sucharda et.al. conducted an experiment on the removal Cd^{+2} and Pb^{+2} by natural and synthetic zeolites [98]. They witnessed faster and efficient removal of Cd^{+2} and Pb^{+2} from multicomponent solutions with synthetic zeolites [98]. Zeolites can be synthesized by using raw natural materials, such as natural silica sources, as well as using synthetic precursors as well. Zeolite produced using natural precursors is inexpensive but lacks precise pore structures and contains contaminants. Zeolite produced from synthetic precursors has a precise structure and fewer imperfections but is expensive.

The challenge of reusability with nanosized synthesized zeolites is another reason that reduces their effectiveness. Zeolites demonstrate improved physiochemical stability, greater adsorption capacity, and simpler reusability when utilized in the form of their composites. Additionally, the characteristics of the materials used to create composites have significant advantages of their own. Most frequently, it has been found that the zeolite-based composites show better optical properties and pore distributions than zeolites themselves, increasing the application of zeolites in adsorption, photocatalysis, and other processes. Zeolites function in photocatalytic reactions in two ways: either as a host for semiconducting materials or by cooperating with those materials' electron transfer processes to significantly reduce the chance of electron–hole recombination.

2. Zeolite-Based Composites

In general, zeolite-based composites for pollutant removal from wastewaters are made by incorporating metal oxide nanoparticles, carbon-based materials, clay compounds, and polymers into zeolites.

2.1. Synthetic Approaches

General methods for the synthesis of zeolite/metal oxide composites are (i) sol-gel, (ii) hydrothermal, (iii) solvothermal, (iv) co-precipitation, (v) ultrasonic, and (vi) microwave. Figure 12 summarizes the techniques used for the preparation of zeolite/metal oxide composites [34].

Zeolite/carbon-based materials are generally synthesized by the conventional synthesis method of zeolites, such as hydrothermal, solvothermal, sol-gel, etc. During the initial stage in the synthesis of zeolites, carbon-based material is added along with the precursors of zeolites, leading to the formation of a zeolite/carbon-based material composite, as given in Figure 13a.

Zeolite/polymer composites are generally fabricated by in situ polymerization. In this process, monomer and zeolite are mixed together followed by polymerization, which leads to the formation of a zeolite/polymer composite. Figure 13b shows the schematic of the synthesis of a zeolite/polymer composite.

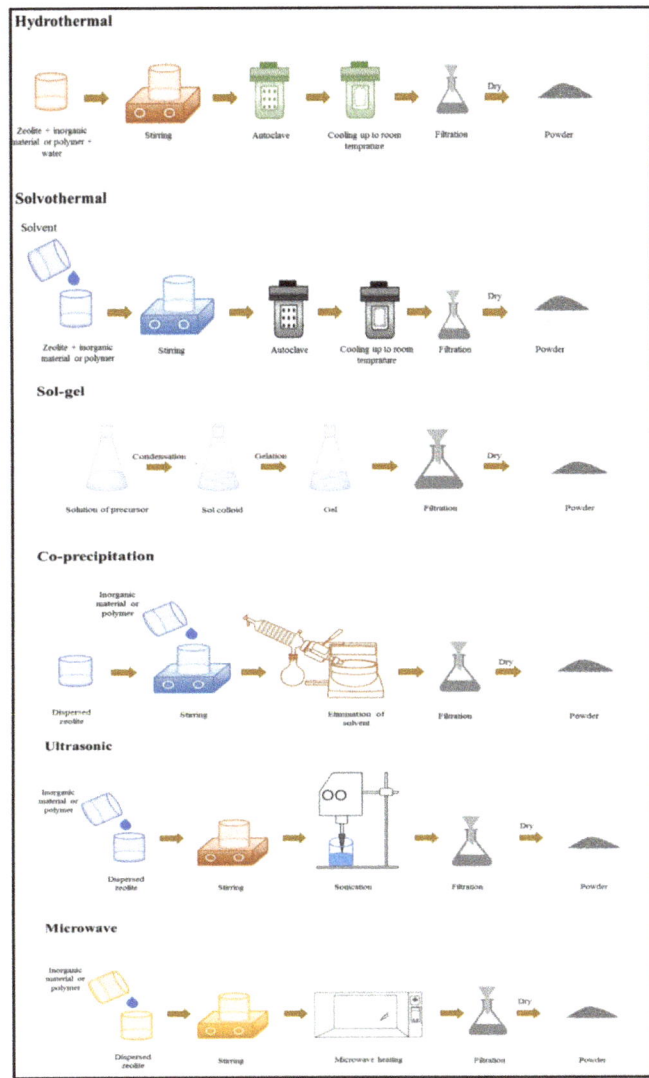

Figure 12. Synthesis processes of zeolite/metal oxide composites. Reprinted with permission from Ref. [34]. Copyright 2021 Elsevier.

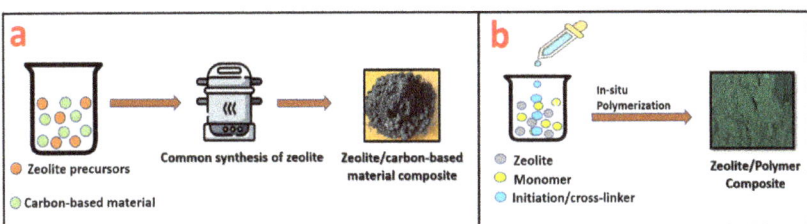

Figure 13. Schematic representation of (**a**) the synthesis of a zeolite/carbon-based material composite and (**b**) a zeolite/polymer composite.

2.2. Removal Process

When working with zeolites and zeolite-based composites, adsorption, ion-exchange, and photocatalysis are the general pollutant removal processes in wastewater treatment. In these processes, a known amount of catalyst/adsorbent is added into the model contaminant solution, i.e., dyes, heavy metals, agrochemicals, etc. Then, this suspension is kept under the treatment process and treated samples are centrifuged and analyzed using a UV–Vis spectroscope. Zeolite-based materials are also used in the fixed-bed reactors to remove pollutants from water as depicted in Figure 14.

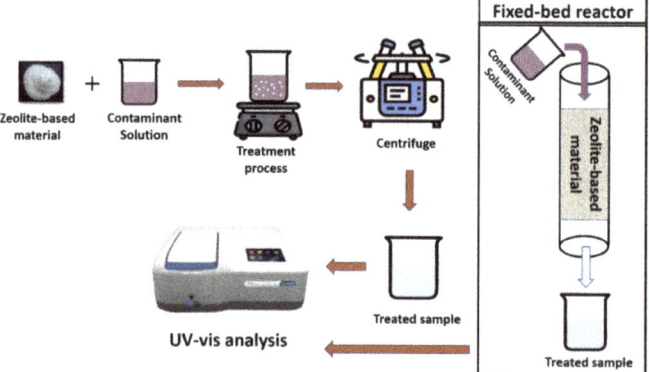

Figure 14. Schematic representation of pollutant removal process using zeolite-based materials.

2.3. Zeolite/Metal Oxide Composites

Metal oxides are employed in a variety of applications, such as adsorption, photocatalysis, energy storage, etc., due to their tunable size and morphology [99]. When employed in their nano form, metal oxides have a large surface area and exhibit excellent adsorption capabilities. Metal oxides are an ideal photocatalyst due to their distinct physicochemical properties, which include shape, size, morphology, composition dependence, and light sensitivity [100]. When doping metal oxides with zeolites, the cavities of zeolites behave as a support for metal oxides, increasing the surface area of metal oxides, thereby improving adsorption and photocatalytic properties.

Alswata et al. synthesized zeolite/ZnO nanocomposites using the co-precipitation method. Prepared samples were examined for the adsorption of lead Pb(II) and arsenic As(V) from its synthetic solution [101]. The FE-SEM images of bare and ZnO-doped zeolite are given in Figure 15. From the FE-SEM, it is evident that zeolite has a cubic shape with a smooth surface, while some granular doping can be observed on the surface of ZnO-doped zeolite. Zeolite's cubic shape remains as it is after doping of ZnO NPs.

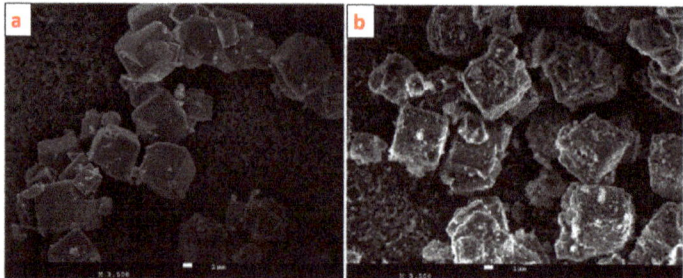

Figure 15. FE-SEM images of (**a**) zeolite and (**b**) ZnO nanoparticle-doped zeolite. Reprinted from Ref. [101].

Under similar experimental conditions, the zeolite/ZnO nanocomposite showed better adsorption capacity than zeolite alone for the removal of both arsenic (As) and lead (Pb). When combined with ZnO, zeolite eliminated 92% of the lead and 85.7% of the arsenic, respectively, compared to 43.6% and 32.3% for pure zeolite [101].

Sacco et al. integrated semiconducting ZnO into zeolite cavities by wet impregnation method and prepared ZnO/ZeO pellets [24]. These pellets were evaluated for the removal of caffeine by a simultaneous process of photocatalysis and adsorption. The studies were performed in two stages, i.e., by adsorption and by adsorption assisted photocatalysis. At the initial stage, the adsorption kinetics were found to be faster in the case of zeolite as compared to the composite. This might be due to decreased mesoporous surface area of the composite, though, the total adsorption was approximately 60% for both the ZnO and ZnO/ZeO composite. When the removal of caffeine by adsorption and adsorption/photocatalysis were examined, it was shown that UV irradiation resulted in a significantly higher total removal of caffeine. Almost 100% of caffeine was removed within 120 min of reaction time by using the ZnO/ZeO composite in adsorption/photocatalysis, while only adsorption gave 69% removal in 120 min. Under UV irradiation, ZnO can degrade the adsorbed caffeine and its chemical intermediates, creating active sites for the adsorption of any leftover caffeine molecules in the liquid medium [24].

Mahalakshmi et al. fabricated the zeolite-supported TiO_2 composite by using the H-form of zeolite Y, β, and ZSM-5 and labeled them as HY, Hβ, and H-ZSM-5, respectively [102]. The prepared materials were investigated for adsorption and photocatalytic degradation of propoxur, an N-methylcarbamate pesticide. According to the experimental data, the adsorption of propoxur was better over TiO_2/Hβ than HY/ TiO_2 and H-ZSM-5/ TiO_2. Propoxur degradation efficiency was found to be better in TiO_2/Hβ with optimal TiO_2 loading (7 wt%) than in pure TiO_2. The limited surface area of H-ZSM-5 and the hydrophilic character of HY were responsible for their poor adsorption capacity. The existence of acid sites in Hβ with high acid strength might be another factor in propoxur adsorption [102].

Liu et al. evaluated the TiO_2/zeolite composite for the removal of sulfadiazine (SDZ) via adsorption and photocatalysis under UV light [23]. The composite material was synthesized via the sol-gel method. FTIR analysis indicated the formation of the Ti-O-Si bond in the composite. In 60-min dark studies performed, a small amount of adsorption of SDZ was reported. UV light studies showed the 32.76% degradation of SDZ in 120 min without the aid of a catalyst. The prepared TiO_2/zeolite composite removed 93.31% of SDZ in the presence of UV light within 120 min. The general mechanism of degradation is as given in Figure 16 [23].

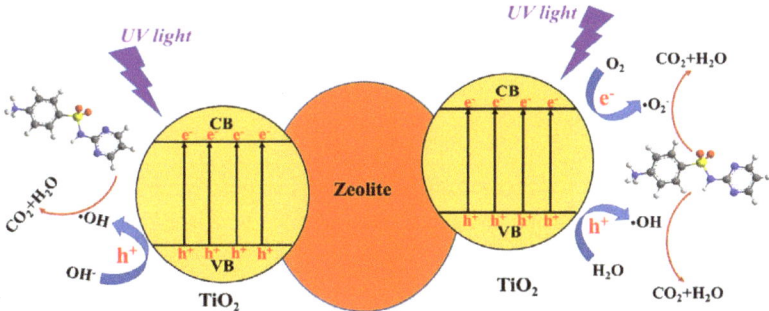

Figure 16. Mechanism of sulphadiazine removal by using zeolite/TiO2 composite. Reprinted with permission from Ref. [23]. Copyright 2018 Elsevier.

The participation of reactive oxygen species (ROS), i.e., e^-, h^+, $OH^\bullet$, $O_2^{-\bullet}$, $^1O_2$ (singlet oxygen), were examined by performing the scavenger studies. It was concluded that ROS contribution in the zeolite/TiO_2 composite follows the order of $OH^\bullet > h^+ > O_2^{-\bullet} > {}^1O_2$.

Additionally, the HPLC–MS/MS study of reaction intermediates was used to hypothesize the four potential degradation pathways [23].

D Mirzaei et al. synthesized the NaX/MgO–TiO$_2$ zeolite nanocomposite by using the ultrasound-assisted dispersion method [103]. Anionic dye methyl orange was used to investigate adsorption on the prepared composites. To obtain maximum MO adsorption yield from the aqueous solution, different parameters such as initial dye concentration, adsorbent dosage, pH, adsorbent type, and contact time have been assisted and optimized. To estimate the adsorption–desorption isotherm, Freundlich, Temkin, and Langmuir models were used. For subsequent processes, chemical parameters such as 0.3 g L^{-1} adsorbent dosage, 6.5 pH, contact time of 35 min, and temperature of 45 °C were evaluated as the optimized conditions. Under identical experimental conditions, it is determined that the NaX/MgO–TiO$_2$ nanocomposite led to the highest 95% adsorption efficiency from aqueous solution among NaX, MgO, TiO$_2$, MgO–TiO$_2$, and NaX/MgO–TiO$_2$ adsorbents, and MO adsorption efficiencies over MgO, TiO$_2$, NaX, MgO–TiO$_2$, and NaX/MgO–TiO$_2$ were greater than 30%, 46%, 40%, 68%, and 95%, respectively [103].

A.A. Alswat et al. used a co-precipitation approach to make zeolite/iron oxide (Fe$_3$O$_4$) and zeolite/copper oxide (CuO) nanocomposites (NCs) [104]. The adsorption efficiencies were 97.2% and 96.8% for Pb and As, respectively, by zeolite/iron oxide (Fe$_3$O$_4$) NCs, and 83.7% and 81.3% for Pb and As, respectively, by zeolite/ copper oxide (CuO) NCs at a pH of between 4 and 6 when these composites were kept for 40 min at room temperature and pressure. The Langmuir isotherm model was well followed by the adsorption data [105].

Kong et al. used the co-precipitation method to prepare nanosized Fe-Al bimetallic oxide-doped zeolite spheres and used it to remove Cr(VI) ions from constructed wetlands [106]. The pseudo-second order model was found to be perfectly fitted to the removal of Cr(VI). The composite zeolite spheres outperformed standard fillers in terms of removal, with excellent adsorption across a wide pH range. The Cr(VI) was absorbed and fixed by the composite zeolite spheres, and then it was reduced to Cr(III) using the Fe-Al oxide. Through co-precipitation and ion exchange, the Cr(III) made Cr(OH)$_3$ and Cr$_x$Fe$_{1-x}$(OH)$_3$ precipitates [106].

Zhang et al. used a one-step hydrothermal method to fabricate TiO$_2$/MoS$_2$ photocatalysts supported on zeolite utilizing micrometer-MoS$_2$ as the sensitizer [107]. Under simulated solar-light irradiation, the synthesized photocatalyst TiO$_2$/MoS$_2$/zeolite had significantly higher photocatalytic response than the Degussa P25 photocatalyst. The recombination of photogenerated electrons and holes is one of the major factors that limit the efficiency of a photocatalyst. According to Zang et al., during the fabrication procedure, the Z-scheme photocatalyst of TiO$_2$/MoS$_2$ was developed where MoS$_2$ acted as an electron donor in interfacial charge conduction, thereby improving the charge separation. In addition, the generation of superoxide anion radicals ($O_2^{-\bullet}$), major oxidation species in photocatalytic reactions, can be aided by the micro/nano-MoS$_2$ generated via the hydrothermal process [107].

D. Wang et al. successfully prepared Cr-doped TiO$_2$ photocatalysts supported on natural zeolite [104]. Because Ti^{4+} and Cr^{3+} have similar ionic radii, Cr ions can be integrated into the TiO$_2$ lattice by taking the place of Ti^{4+} sites. As the calcination temperature rises, the bond strength of Cr-O-Ti increases. Cr dopant is found as Cr^{6+} (81.2%) and Cr^{3+} (19.8%) species. In comparison to undoped TiO$_2$/zeolite, the band gap energy (eV) of 10 mol% Cr/TiO$_2$/zeolite decreases dramatically from 2.84 eV to 1.70 eV. The percentage degradation of methyl orange by the calcined 10% Cr/TiO$_2$/zeolite reaches 41.73%, after 5 h of illumination, which is 17.9% higher than the degradation efficiency of undoped TiO$_2$/zeolite [104].

Italia et al. prepared two composites of zeolite and bentonite separately bonded with titanium and was evaluated for the adsorption of phosphate. Titanium/zeolite and titanium/bentonite composites removed up to 83% and 84% of phosphate at 3 pH [108].

However, it is not always observed that the composites exhibit superior removal efficiencies than the zeolites alone. A. Alcantara-Cobos et al. compared ZnO nanoparticles and the ZnO-zeolite composite for tetrazine removal [20]. The zeolite-ZnO composite was prepared by the chemical precipitation method. Both adsorption and photocatalysis were working mechanisms behind the removal of tetrazine. The adsorption with ZnO nanoparticles was faster than the ZnO-zeolite composite. The degradation reported followed by adsorption under UV light radiations was 81% and 87% for the zeolite-ZnO composite and ZnO nanoparticles, respectively; although, the latter was difficult to remove from the aqueous solution. Additionally, the ZnO nanoparticles show low toxicity towards Lactuca sativa when kept with the dye solution and diluted aqueous solutions [20].

Jaramillo-Fierro et al. synthesized extruded semiconducting $ZnTiO_3/TiO_2$ supported on zeolite and its precursor clay [109]. Zeolites were synthesized by using two types of Ecuadorian clays via hydrothermal treatment and the method of alkali fusion, i.e., R-clay and G-clay. Zeolite prepared using R-clay was labeled as R-zeolite and was mostly of the Na-LTA type with a trace quantity of Na-FAU type. Zeolite prepared using G-clay was labeled as G-zeolite and was made up primarily of Na-FAU type zeolite with residues of Na-P1 type zeolite. The semiconducting support $ZnTiO_3/TiO_2$ was prepared separately via the sol-gel method. The composites were prepared by mixing zeolites, precursor clays, and $ZnTiO_3/TiO_2$ in different ratios. The reported order of adsorption of MB on parent material was G-zeolite > R-zeolite > G-clay > R-clay > $ZnTiO_3/TiO_2$. Despite having a higher dye removal capacity than mixed oxide $ZnTiO_3/TiO_2$ and precursor clays, the capacity of the extruded composites to remove MB was not increased by zeolites. Additionally, extruded zeolites are less capable of removing the color than powdered zeolites because they have a lower specific surface area [109].

2.4. Zeolite/Carbon-Based Material Composites

Currently, carbon nanomaterials are considered to be the most adaptable materials that can be employed to improve wastewater treatment techniques. Innovative carbon materials have been discovered as a result of extensive research conducted globally and effectively used in wastewater remediation and environmental safety technologies [110,111]. SWNT (single-walled carbon nanotubes), MWNT (multi-walled carbon nanotubes), G (graphene), and GO (graphene oxide) are among the most frequently studied carbon-based nanomaterials. These materials may be employed in their natural forms or as complex hybrid substances [111,112].

Zeolites have been modified by the addition of a heteroatom to their structural framework, giving them new and fascinating qualities, including photoactivity [17,18,113]. Ren et al. produced new generation photocatalysts by combining functional inorganic nanomaterials (such as zeolitic TS-1) with graphene and carbon-nanotube (CNTs) [16]. The performance of these photocatalysts were outstanding owing to (i) the synergistic effect on the basis of interfacial charge and heat transfer reactions and (ii) graphene's ability to modify the shape and size of TS-1. The few layers of graphene were first synthesized via applying direct current discharge to graphene. Zeolite TS-1 was synthesized via the sol-gel process. Zeolite/graphene and zeolite/CNT hybrids were prepared by combining in situ to the common sol-gel synthesis of TS-1. The photocatalytic behavior of these materials was examined through dye degradation in the presence of low-intensity UV radiations. It was observed that the photocatalytic activity of TS-1 zeolite increased 27–28 times with graphene loading and only 4–5 times by CNT loading. The reason is the interfacial charge transfer between the conduction band of zeolite and nanocarbon. Moreover, the high reactivity of edge atoms in graphene might be responsible for high photocatalytic activity [16].

W.A. Khanday et al. synthesized a zeolite/activated carbon composite from oil palm ash using a two-step method, i.e., fusion followed by the hydrothermal process and labeled as Z–AC composite [114]. The adsorption results of MB adsorption on the Z–AC composite were compared with those of activated and non-activated oil palm ash and oil palm ash zeolite. The highest adsorption capacity observed for the Z–AC composite was approximately 90% [114].

H. Li et al. prepared zeolite by granulation, calcination of coal gangue, followed by the hydrothermal process and marked as ZMC [19]. Since the carbon content was removed at the calcination step, prepared ZMC is pure zeolite. By modifying the above-mentioned process, they synthesized a carbon retaining and extra carbon-containing zeolite-activated carbon composite and marked them as ZTC and ZAC, respectively. The prepared sample were examined for the adsorption of Cu^{+2} and rhodamine-B(Rh-B). The adsorption capacities for Cu^{+2} was observed to be 98.2%, 97.1%, 92.8% for ZMC, ZTC, and ZAC, respectively. Similarly, the adsorption capacities for RB were 17.0%, 41.3%, and 94.2% for ZMC, ZTC, and ZAC, respectively. Adsorption capacity for Cu^{+2} decreases upon increasing carbon content. Zeolite has a uniform pore distribution with a pore size of 0.41 nm, which can easily hold small Cu^{+2} ions. On the other hand, activated carbon has variable pore sizes across the network, including micro, meso, and macro pores; high pore sizes are incompatible with holding tiny Cu^{+2} ions and are suitable for adsorption of large organic molecules; hence, adsorption capacity for Rh-B increasing upon increasing carbon content [19].

M.A. Farghali et al. synthesized a mesoporous zeolite A/reduced graphene oxide nanocomposite [31]. For this, zeolite was first surface-modified with the help of 3-aminopropyl-trimethoxy silane (APTMS), which is used as a binding and mesopore generating agent. Then, using a hydrothermal technique, reduced graphene oxide was added to zeolite-A to create modified mesoporous zeolite-A/reduced graphene oxide NCs (MZ-A/RGO). Following that, the synthesized material was employed to remove lead ions (Pb^{+2}) and methylene blue at the same time. The synthesized composite removed 98% of methylene blue and 93.9% Pb^{+2} [31].

Mahmoodi et al. immobilized the laccase enzyme onto a zeolite and graphene oxide composite via the covalent bond and prepared a biocatalyst for the removal of direct red 23 dye [22]. They used the hydrothermal method for the preparation of zeolite and the Hummers method for the preparation of graphene oxide. The composites were prepared by taking different weight ratios of graphene oxide followed by laccase immobilization. It was observed that the dye degradation increased upon increasing the loading ratio of graphene oxide. Graphene oxides increase the electron transfer between dye and enzyme, thereby enhancing the oxidation ability of the enzyme [22].

Huang T et al. evaluated magnetic graphene oxide-modified zeolite for uptake of methylene blue from an aqueous solution [21]. The magnetic $MnFe_2O_3$ nanoparticles were synthesized by the co-precipitation method. The Cu-zeolite was made separately as Cu-Z. The Cu-Z-GO-M composite was prepared using the solid-state dispersion method. The composite offered the highest adsorption capacity, i.e., 97.346 mg/g at 318 °C [21].

Using two-step alkali fusion and hydrothermal treatment, Zhao et al. generated a honeycomb-activated carbon-zeolite composite (CZC) using coal fly ash (FA) and utilized it to adsorb Pb^{+2} from an aqueous solution [4]. The pre-treated coal fly ash was heated in a muffle furnace to 750°C as part of the activation process to produce activated carbon. Activated carbon zeolite CZC owns a specific surface area that is approximately six times greater than FA's, and its average pore size is enlarged from 3.4 to 12.7 µm. At pH 7, CZC demonstrated 185.68 mg/g of Pb(II) absorption after 40 min of contact time. According to kinetics studies, Pb(II) ion adsorption onto the surface of CZC is more consistent with pseudo-second order kinetics [4].

2.5. Zeolite/Polymer Composites

Polymers are the organic compounds with a variety of exceptional properties, including high mechanical strength, extraordinary flexibility, large surface area, and chemical stability. Owing to these characteristics, polymers can serve as a host for many inorganic and organic compounds [115]. The application of composite materials, which incorporate both organic and inorganic components, has recently attracted more attention. Combining the materials results in several beneficial properties that the separate materials could not produce. For instance, the elasticity and easy processing of polymers and mechanical properties of inorganic constituents are integrated [116].

Being an amino polysaccharide, chitosan contains several OH and NH_2 groups that might serve as coordination and reaction sites. The fact that chitosan is made from chitin, the second-most abundant natural polymer after cellulose, makes it an extremely abundant and inexpensive material. Chitosan's use as a material in real world applications is, nevertheless, constrained by its weak mechanical characteristics, which can be overcome via crosslinking with high mechanical strength materials, such as zeolites [116]. Khanday WA et al. synthesized composite beads by cross-linking chitosan and zeolite derived from oil palm ash and labeled it as Z-AC/C [117]. The activated oil palm ash was first hydrothermally treated before being beaded with chitosan. The effect of weight percentage of chitosan and Z-AC on the adsorption of MB (methylene blue) and AB-29 (acid blue-29) dyes was studied. Increase in the adsorption of AB29 and decrease in MB adsorption was observed by increasing the percentage of chitosan and vice versa. The Z-AC/C composite with 50:50 weight ratios of chitosan and zeolite well-adsorbed both AB-29 and MB dyes and was used for further studies. For the Z-AC/C composite, the adsorption capacities at 30 °C, 40 °C, and 50 °C were 212.76 mg/g, 238.09 mg/g, and 270.27 mg/g for AB29, 151.51 mg/g, 169.49 mg/g, and 199.20 mg/g for methylene blue, and 212.76 mg/g, 238.09 mg/g, and 270.27 mg/g for acid blue-19, respectively [117].

pH plays a crucial role in adsorption processes as the surface charge, speciation, and degree of ionization of adsorbate are influenced by the pH of the solution. pH influence on the adsorption of AB29 and MB on the Z-AC/C composite was studied in the pH range from 3–13 with 100 ppm initial dye concentration at 30 °C. It was observed that at a pH from 3–5, adsorption of AB29 was great and decreased linearly at pH 13. In the case of MB, the reverse phenomena were reported. At a pH from 3–5, adsorption of MB was less and linearly increased up to a certain pH and stayed constant afterward. At a low pH, electrostatic attractions between the adsorbent's negatively charged surface and the positively charged H^+ ions of AB29 dye are responsible for faster adsorption. As the pH of the solution increases, the hydrogen ions get diminished, lowering the attraction between the dye molecule and composite resulting in decreased adsorption of AB29. In case of MB, at a low pH, the adsorption is low due to the competition between MB and protons at binding sites. At a higher pH, the surface attains a negative charge, hence, the adsorption increases. Similarly, the initial dye concentration effect on the adsorption was also examined and it was observed that low concentrations reached equilibrium faster than high concentrations in the case of both dyes; the number of unoccupied active sites per dye molecule is low and the motion of dye molecules toward binding sites gets hindered [117].

Pizarro et al. prepared a composite using natural zeolite (NZ) and commercial cationic polymer, Polyammonium cation (SC—581), and labeled it as MZ (Modified zeolite) [118]. The material was investigated for adsorption of sulphate, a pollutant originating from processes of sulphate mining. Modification of zeolite with cationic polymer induces a positive charge on the adsorbent surface, which helps to bind sulphate ions. Also, the positive charge of MZ remains intact above pH 4, while the NZ holds the negative charge. The adsorption capacities of NZ were almost doubled with polymer impregnation. The effect of ion strength was also investigated, indicating that the adsorbent functions well below a KCl concentration of 0.050 mol/liter [118].

Senguttuvan et al. synthesized the zeolite/polypyrrole composite for the removal of reactive blue and reactive red dyes [25]. By performing oxidative polymerization of pyrrole in the presence of zeolite, the composite nano-adsorbent was fabricated. The FE-SEM and TEM images of PPy/Ze composites are shown in Figure 17. The particles are mostly agglomerated with a spherical shape and an average particle size between 40 and 80 nm. Even after adsorption of RB and RR, the PPy/Ze nanocomposites' morphology did not change, which indicates the uniform dye adsorption on the surface of PPy/Ze nanocomposites.

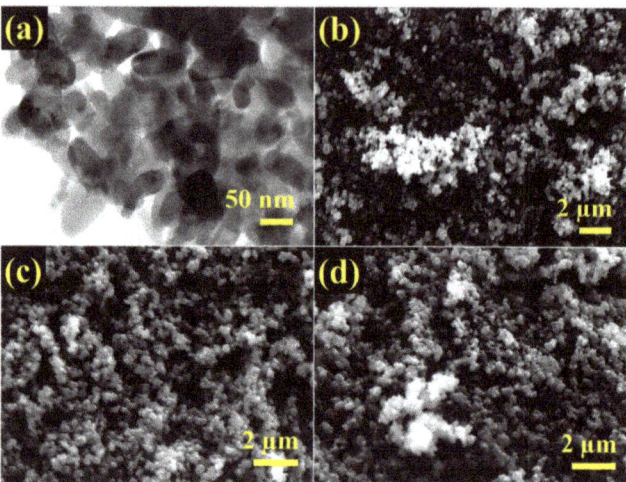

Figure 17. (**a**) TEM images of PPy/Ze NCs; SEM images of (**b**) PPy/Ze NCs, (**c**) PPy/Ze NCs after RB adsorption, (**d**) PPy/Ze NCs after RR adsorption. Reprinted with permission from Ref. [25]. Copyright 2022 Elsevier.

With a 1.8 mg/mL catalyst dosage, and 75 ppm initial dye concentration, the nanocomposite adsorbed 88.3% of RR and 86.2% of RB within 75 min. Interaction between the dyes and PPy/Ze NCs was mostly mediated by aromatic, hydroxyl, and amide functional groups. The adsorption mechanism involved hydrogen bonding and electrostatic interactions. Senguttuvan et al. also reported the same PPy/Ze NCs for 83.5% removal of Cr(VI) in 50 min [119].

Yigit et al. fabricated beads of natural composite material using clinoptilolite, a natural zeolite, and alginate, a naturally occurring polymer and labeled it A-C (alginate-clinoptilolite) [120]. These A-C beads were employed in the mixture of heavy metals carrying copper (Cu^{+2}), lead (Pb^{+2}), and cadmium (Cd^{+2}) ions. For 10 ppm of initial metal concentration, a constant operation with 2 mL/min flow rate revealed 98% of lead uptake. The repeatability test revealed that the efficiency of the adsorbent was up to the mark until 3 cycles with regeneration via HNO_3 washing [120].

Conducting polymers has attracted a lot of interest because of their appealing features, including electrical conductivity and optical qualities. Due to its fascinating qualities, such as excellent chemical stability, cost-effectiveness, and facile synthesis, Polyaniline (PANI) is a unique and amazing polymer of the conducting polymer family. Additionally, upon visible light excitation, PANI can donate the electrons and act as a good hole transporter [121,122]. Hence, in addition to being adsorbent, PANI can also act as a photocatalyst and degrade organic pollutants.

To make zeolite/conducting polymer-based (nano-)composites, four alternative routes can be used [123]. (i) The organic solvent is enclosed in the zeolite cavities first; subsequently, oxidative polymerization is performed to produce polymeric chains in the zeolitic cavities [124]. (ii) Zeolites containing oxidant ions, such as Fe(III) and Cu(II) are reacted with monomer and acid vapors [125]. (iii) In the presence of zeolite, by performing in situ polymerization of the monomer, polymers can be developed inside or outside the channels of zeolite [126,127]. (iv) Powdered zeolite and conducting polymer are mixed mechanically [128]. Methods (ii) and (ii) of the previously outlined techniques hold the most interest since they produce nanoscale polymeric chains that are embedded in zeolite cavities. Therefore, as the polymeric chains are arranged at the nanoscale, the mechanical, electrical, chemical, and optical characteristics may be enhanced [123].

Abukhadra et al. synthesized the heulandite/polyaniline composite for the efficient removal of light green SF (LGSF), methylene blue (MB), and Congo red dyes from water [129]. Abukhadra et al. prepared the heulandite/polyaniline composite by the mechanical mixing of heulandite, natural zeolite, and synthesized conducting polymer, polyaniline. According to Tauc plot calculations, the optical band gaps of HU/PANI and PANI were 1.69 eV and 2.98, respectively. To examine the photocatalytic activities, studies were carried out both in dark conditions and with artificial visible light. The pseudo-second order and the Elovich model both provided good fits for the kinetic results. The Langmuir isotherm model provided a good description of the adsorption process in the dark, and the estimated q_{max} was 44.6 mg/g. The experimental data were well-fitted to the Freundlich and Temkin models under visible light illumination compared to the Langmuir model. This shows the function of photodegradation by the HU/PANI composite in improving the multilayer adsorption. The quadratic programming projected that the ideal conditions for the highest elimination percentage in the dark, i.e., 70.9%, 5.5 ppm dye concentration, pH 3, 24 mg dosage of Hu/PANI, and contact duration of 430 min. Whereas, in the presence of visible light, at 15 mg catalyst dosage, 15 ppm dye concentration, contact time of 589 min, pH 3, and 97% dye removal is possible [129].

Milojević-Rakić investigated the polyaniline/FeZSM-5 composite for the removal of glyphosate, a herbicide via oxidative degradation [33]. Different weight ratios of aniline/FeZSM-5 was used. The method employed for synthesis was the oxidation polymerization of FeZSM-5 with ammonium peroxy disulphate as an oxidant, with and without using acid (H_2SO_4). 1/1 and 1/5 weight ratios of aniline/FeZSM-5 were used to make the composites. The composite with the 1/1 and 1/5 ratios of aniline/FeZSM-5 and synthesized without using acid was labeled as PFeZ1/1 and PFeZ1/5, respectively. Similarly, the composite with the 1/5 ratio of aniline/FeZSM-5 and synthesized using acid was labeled as PFeZ1/1S and PFeZ1/5S, respectively. The NH_4OH treated (deprotonated) forms of PFeZ1/1, PFeZ1/5, PFeZ1/1S, and PFeZ1/5S were labeled as PFeZ1/1d, PFeZ1/5d, PFeZ1/1Sd, and PFeZ1/5Sd. Polyaniline was also synthesized and treated similarly and labeled as PANI, PANI/S, PANId, and PANIS/d, the same way as composites. The morphological analysis can be seen in Figure 18.

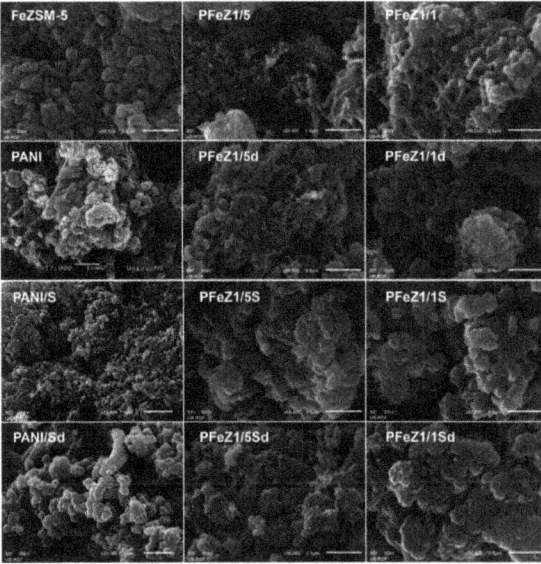

Figure 18. SEM images of PANI, FeZSM-5, and their composites. Reprinted with permission from Ref. [33]. Copyright 2018 Elsevier.

By performing the conductivity analysis, it was observed that the composites synthesized with the use of acid have higher conductivity in comparison to composites synthesized without acid. Additionally, the decrease in conductivity was observed by increasing the loading of zeolite; this elucidates the lower conducting nature of zeolites. Moreover, with the help of thermogravimetric analysis, the weight content of polyaniline and FeZSM-5 in the composites were evaluated. Table 7 gives the weight content of PANI and FeZSM-5 as well as the percentage degradation and removal of glyphosate from the solution [33].

Table 7. Weight contents, percentage removal, degradation, and adsorption in composites. Reprinted with permission from Ref. [33]. Copyright 2018 Elsevier.

Sample	Weight Content		FeZSM/PANI Weight Ratio	% Removal	% Degradation	% Adsorption
	FeZSM-5	PANI				
PANI	-	-	-	12.4	11.6	0.8
ZSM-5	-	-	-	12.8	7.1	5.7
PFeZ1/1	43.1	52.2	0.83	40.8	31.9	8.9
PFeZ1/5	77.9	18.1	4.30	80.4	66.5	13.9
PANI/S	-	-	-	10.2	9.7	0.5
PFeZ1/1S	42.2	51.6	0.82	26.8	22.6	4.2
PFeZ1/5S	76.9	16.8	4.58	18.1	13.6	4.5
PANId	-	-	-	56.6	54.4	2.2
PFeZ1/1d	47.4	48.4	0.98	13.6	8.4	5.2
PFeZ1/5d	78.9	15.0	5.26	20.4	14.6	5.8
PANI/Sd	-	-	-	6.6	2.8	3.8
PFeZ1/1Sd	47.2	48.4	0.98	7.4	1.1	6.3
PFeZ1/5Sd	79.4	15.5	5.12	3.8	0	3.8

The composites prepared without added acids showed better removal and degradation than those synthesized via adding acids. The deprotonation significantly reduced the degradation and removal efficiencies of all catalysts. From the data given in the above Table 7, it can be concluded that composite PFeZ1/5 showed the higher removal efficiency among all other catalysts. The PFeZ1/5 has an increased number of iron sites; hence, an increased amount of degradation is observed [33].

Milojevic'-Rakic et al. also prepared the ZSM-5/polyaniline composite by glyphosate adsorption [26]. The ZSM-5/polyaniline composite was prepared by performing the oxidative polymerization of aniline with and without added acid. Similar to the above-reported studies, the deprotonation of samples was also conducted. Among all the adsorbents, the deprotonated form of PANI prepared via acid synthesized polyaniline showed the maximum adsorption of glyphosate, i.e., (98.5 mg/g). The decrease in adsorption capacities was observed by increasing the zeolite loadings. The poor adsorption properties of ZSM-5 and composites containing the high weight ratios of ZSM-5 were most likely caused by its high microporosity, which is unfavorable for adsorption of comparatively huge glyphosate molecules. The linear, regular, and defect-free emeraldine base structure of PANI/Sd chains results in effective interaction with the glyphosate molecules. This was attributed to superior glyphosate adsorption characteristics of PANI/ Sd [26].

3. Comparison of Removal Efficiencies

The brief literature survey on zeolite and its composites for wastewater treatment is compiled in Table 8.

Table 8. Comparison of effluent removal efficiencies of zeolites and zeolite-based composites.

Material	Source/Synthesis Approach	Band Gap (eV)	Contaminant	Mechanism	Dosage (mg/mL)	Concentration (ppm)	Contact Time	Removal	Light Source	Published Year	Ref
Zeolite	Indonesia commercial		NH_4^+	Adsorption, ion exchange	0.00152	12.9	134.89 min	98%		2020	[76]
Natural zeolite	Chinese commercial		NH_4^+	Adsorption	0.048	80	180 min	96%		2010	[11]
Natural zeolite	Australia commercial		Methylene blue	Adsorption	0.25	3.55	200 h	6.8×10^{-5} mol/g		2005	[75]
Natural zeolite	Australia commercial		Rhodamine B	Adsorption	0.25	3.55	50 h	2.1×10^{-5} mol/g		2005	[75]
Hydrogenated form of natural zeolite	Carranco Blanco	2.63	Caffeine	Photocatalysis	10	50	4 h	99%	UV	2020	[13]
Synthetic zeolite	Hydrothermal treatment using aluminum iso propoxide	3.29	Methylene blue	Photocatalysis	2	10	180 min	85%	UV	2020	[14]
Titania-Supported zeolite	In situ using $TiCl_4$ impregnation	3.31	Methylene blue	Photocatalysis	0.33	30	60 min	40%	UV	2010	[130]
Titania-supported zeolite	In situ using $TiCl_4$ impregnation	3.31	Direct blue 71	Photocatalysis	0.33	30	60 min	55%	UV	2010	[130]
Titania-supported zeolite	In situ using $TiCl_4$ impregnation	3.31	Direct yellow 8	Photocatalysis	0.33	30	60 min	62.5%	UV	2010	[130]
Heulandite/Polyaniline/nickel oxide	In situ polymerization followed by Ni_2O_3 impregnation	1.42	Safranin T	Photocatalysis	0.35	5	1 min	100%	Solar irradiation	2018	[30]
bentonite/PANI/Ni_2O_3	In situ polymerization followed by Ni_2O_3 impregnation	1.61	Safranin-O	Photocatalysis	0.5	5	90 min	100%	Sunlight	2018	[131]
Heulandite/polyaniline	Mechanical mixing	1.69	LCSF	Photocatalysis	0.3	15	589 min	97%	VIS	2018	[129]
Heulandite/polyaniline	Mechanical mixing	1.69	MB	Photocatalysis	0.2	20	589 min	68.77%	VIS	2018	[129]
Fe-Al bimetallic oxide loaded zeolite	Co-precipitation		Cr(IV)	Adsorption	40	20	300 min	84.9%		2020	[106]
Zeolite/ZnO	Wet impregnation method		Caffeine	Adsorption/photocatalysis	500	25	120 min	100%	UV	2018	[24]
Zeolite/ZnO	Co-precipitation		Pb(II)	Adsorption	3	100	30 min	92%		2017	[101]
Zeolite/ZnO	Co-precipitation		As(V)	Adsorption	3	10	30 min	85.7%		2017	[101]
Zeolite/TiO_2	Sol-gel		Sulfadiazine	Photocatalysis	1	1	120 min	93.31%	UV	2018	[23]
Zeolite/Activated carbon	Hydrothermal		Cu^{+2}	Adsorption	2	240	60 min	92.8%		2020	[19]
Zeolite/Activated carbon	Hydrothermal		Rhodamine B	Adsorption	2	240	60 min	94.2%		2020	[19]
Zeolite/Activated carbon	Fusion/Hydrothermal		Methylene blue	Adsorption	1	100	30 h	83%		2016	[114]
Zeolite/Poly pyrrole	In situ polymerization		Reactive Red	Adsorption	1.8	75	75 min	88.3%		2022	[25]
Zeolite/Poly pyrrole	In situ polymerization		Reactive Blue	Adsorption	1.8	75	75 min	86.2%		2022	[25]

4. Conclusions, Challenges, and Future Perspectives

Zeolites are the materials widely used as adsorbents and ion-exchangers for the remediation of pollutants. This study describes the synthesis, removal process, mechanism, and application of zeolite-based materials in wastewater treatment, which gives first-hand information for researchers who want to explore zeolite-based materials.

Researchers have doped various materials, such as metal oxides, polymers, and carbon-based materials, with zeolites to make zeolite-based composites. In most cases, these composites exhibit higher removal efficiency than bare zeolites. Additionally, the composites also enable zeolites to function as an effective photocatalyst for the total breakdown of contaminants, which is noticeably less apparent in bare zeolites. Composites of zeolite with semi-conducting materials, such as semiconducting metal oxides, graphene, CNTs, and conducting polymers, are of great interest because of the photodegradation of the contaminants.

The low efficiency, lack of homogeneous properties, lack of accessibility, and high levels of impurities in natural zeolite led to the synthetic preparation of zeolites. Major challenges are faced in the synthesis of zeolites, such as high-cost, the tedious and time-consuming synthesis and filtration process, generation of alkaline wastewater, etc. Research needs to be conducted for the synthesis of zeolites with homogenous properties, cost minimization, and easy processibility in order to establish the material for its easy application on the industrial level.

The raw natural zeolites, i.e., pristine zeolites without any modifications, have a negative charge across its framework and are only capable of attracting cationic contaminants. Some modifications expand its application for the elimination of anionic pollutants as well. Zeolites can also be modified through doping with foreign materials.

This review motivates researchers to further investigate the photocatalytic behavior of zeolites and zeolite-based materials, since very few studies have been undertaken for the estimation of degradation pathways, contribution of ROS species (scavenger studies), and calculation of band edge positions. In future research, these aspects should also be considered for the better understanding of photodegradation in zeolite-based materials. The photocatalytic devices of zeolite-based materials for wastewater treatment are still not available in the market. Future interest should be on the preparation of zeolite-based wastewater purifiers that work on both the concept of photocatalysis and adsorption, which are easy to synthesize and can be realized in a much cheaper way so they can be easily commercialized and become feasible on an industrial scale.

Author Contributions: Conceptualization, V.S., S.S. and R.G.; methodology, V.S. and S.S.; software, V.S. and S.S.; validation, V.S. and S.S.; formal analysis, V.S.; resources, S.S., N.S. and I.A.; writing—original draft preparation, V.S. and S.S.; writing—review and editing, V.S., S.S., R.G., R.B., I.A. and N.S.; supervision, S.S. and R.B.; funding acquisition, N.S., I.A. and S.S. All authors have read and agreed to the published version of the manuscript.

Funding: The authors would like to thank Pandit Deendayal Energy University, Scientific Research Deanship at King Khalid University, Abha, Saudi Arabia through the Large Research Group Project under grant number (RGP.02/219/43) and the Marine Pollution Special Interest Group, the National Defence University of Malaysia via SF0076-UPNM/2019/SF/ICT/6, for providing research facilities and funding.

Data Availability Statement: Not applicable.

Conflicts of Interest: The authors declare no conflict of interest.

References

1. Gray, N. *Water Science and Technology: An Introduction*; CRC Press: Boca Raton, FL, USA, 2017.
2. Moradihamedani, P. Recent advances in dye removal from wastewater by membrane technology: A review. *Polym. Bull.* **2021**, *79*, 2603–2631. [CrossRef]
3. Natarajan, V.; Karunanidhi, M.; Raja, B. A critical review on radioactive waste management through biological techniques. *Environ. Sci. Pollut. Res.* **2020**, *27*, 29812–29823. [CrossRef] [PubMed]

4. Zhao, M.; Ma, X.; Chen, D.; Liao, Y. Preparation of Honeycomb-Structured Activated Carbon–Zeolite Composites from Modified Fly Ash and the Adsorptive Removal of Pb(II). *ACS Omega* **2022**, *7*, 9684–9689. [CrossRef] [PubMed]
5. Li, Y.; Yu, H.; Liu, L.; Yu, H. Application of co-pyrolysis biochar for the adsorption and immobilization of heavy metals in contaminated environmental substrates. *J. Hazard. Mater.* **2021**, *420*, 126655. [CrossRef]
6. Soni, V.; Singh, P.; Quang, H.H.P.; Khan, A.A.P.; Bajpai, A.; Van Le, Q.; Thakur, V.K.; Thakur, S.; Nguyen, V.H.; Raizada, P. Emerging architecture titanium carbide (Ti_3C_2Tx) MXene based photocatalyst toward degradation of hazardous pollutants: Recent progress and perspectives. *Chemosphere* **2022**, *293*, 133541. [CrossRef]
7. Shahabuddin, S.; Sarih, N.M.; Ismail, F.H.; Shahid, M.M.; Huang, N.M. Synthesis of chitosan grafted-polyaniline/Co_3O_4 nanocube nanocomposites and their photocatalytic activity to-ward methylene blue dye degradation. *RSC Adv.* **2015**, *5*, 83857–83867. [CrossRef]
8. Zhang, L.; Wang, L.; Zhang, Y.; Wang, D.; Guo, J.; Zhang, M.; Li, Y. The performance of electrode ultrafiltration membrane bioreactor in treating cosmetics wastewater and its anti-fouling properties. *Environ. Res.* **2021**, *206*, 112629. [CrossRef]
9. Lu, W.; Zhang, Y.; Wang, Q.; Wei, Y.; Bu, Y.; Ma, B. Achieving advanced nitrogen removal in a novel partial denitrification/anammox-nitrifying (PDA-N) biofilter process treating low C/N ratio municipal wastewater. *Bioresour. Technol.* **2021**, *340*, 125661. [CrossRef]
10. Prajaputra, V.; Abidin, Z.; Widiatmaka; Suryaningtyas, D.T.; Rizal, H. Characterization of Na-P1 zeolite synthesized from pumice as low-cost materials and its ability for methylene blue adsorption. In *IOP Conference Series: Earth and Environmental Science*; IOP Publishing: Bristol, UK, 2019.
11. Huang, H.; Xiao, X.; Yan, B.; Yang, L. Ammonium removal from aqueous solutions by using natural Chinese (Chende) zeolite as adsorbent. *J. Hazard. Mater.* **2010**, *175*, 247–252. [CrossRef]
12. De Luca, P.; Bernaudo, I.; Elliani, R.; Tagarelli, A.; Nagy, J.B.; Macario, A. Industrial Waste Treatment by ETS-10 Ion Exchanger Material. *Materials* **2018**, *11*, 2316. [CrossRef]
13. Alvarez-Aguinaga, E.A.; Elizalde-González, M.P.; Sabinas-Hernández, S.A. Unpredicted photocatalytic activity of clinoptilo-lite–mordenite natural zeolite. *RSC Adv.* **2020**, *10*, 39251–39260. [CrossRef] [PubMed]
14. Nassar, M.Y.; Abdelrahman, E.A. Hydrothermal tuning of the morphology and crystallite size of zeolite nanostructures for simultaneous adsorption and photocatalytic degradation of methylene blue dye. *J. Mol. Liq.* **2017**, *242*, 364–374. [CrossRef]
15. Wang, S.; Peng, Y. Natural zeolites as effective adsorbents in water and wastewater treatment. *Chem. Eng. J.* **2010**, *156*, 11–24. [CrossRef]
16. Ren, Z.; Kim, E.; Pattinson, S.W.; Subrahmanyam, K.S.; Rao, C.N.R.; Cheetham, A.K.; Eder, D. Hybridizing photoactive zeolites with graphene: A powerful strategy towards superior photocatalytic properties. *Chem. Sci.* **2011**, *3*, 209–216. [CrossRef]
17. Notari, B. Titanium silicalite: A new selective oxidation catalyst. In *Studies in Surface Science and Catalysis*; Elsevier: Amsterdam, The Netherlands, 1991; pp. 343–352.
18. Hashimoto, S. Zeolite photochemistry: Impact of zeolites on photochemistry and feedback from photochemistry to zeolite science. *J. Photochem. Photobiol. C Photochem. Rev.* **2003**, *4*, 19–49. [CrossRef]
19. Li, H.; Zheng, F.; Wang, J.; Zhou, J.; Huang, X.; Chen, L.; Hu, P.; Gao, J.-M.; Zhen, Q.; Bashir, S.; et al. Facile preparation of zeolite-activated carbon composite from coal gangue with enhanced adsorption performance. *Chem. Eng. J.* **2020**, *390*, 124513. [CrossRef]
20. Alcantara-Cobos, A.; Gutiérrez-Segura, E.; Solache-Ríos, M.; Amaya-Chávez, A.; Solís-Casados, D. Tartrazine removal by ZnO nanoparticles and a zeolite-ZnO nanoparticles composite and the phytotoxicity of ZnO nanoparticles. *Microporous Mesoporous Mater.* **2020**, *302*, 110212. [CrossRef]
21. Huang, T.; Yan, M.; He, K.; Huang, Z.; Zeng, G.; Chen, A.; Peng, M.; Li, H.; Yuan, L.; Chen, G. Efficient removal of methylene blue from aqueous solutions using magnetic graphene oxide modified zeolite. *J. Colloid Interface Sci.* **2019**, *543*, 43–51. [CrossRef]
22. Mahmoodi, N.M.; Saffar-Dastgerdi, M.H. Clean Laccase immobilized nanobiocatalysts (graphene oxide-zeolite nanocomposites): From production to detailed biocatalytic degradation of organic pollutant. *Appl. Catal. B Environ.* **2019**, *268*, 118443. [CrossRef]
23. Liu, X.; Liu, Y.; Lu, S.; Guo, W.; Xi, B. Performance and mechanism into TiO_2/Zeolite composites for sulfadiazine adsorption and photodegradation. *Chem. Eng. J.* **2018**, *350*, 131–147. [CrossRef]
24. Sacco, O.; Vaiano, V.; Matarangolo, M. ZnO supported on zeolite pellets as efficient catalytic system for the removal of caffeine by adsorption and photocatalysis. *Sep. Purif. Technol.* **2018**, *193*, 303–310. [CrossRef]
25. Senguttuvan, S.; Janaki, V.; Senthilkumar, P.; Kamala-Kannan, S. Polypyrrole/zeolite composite—A nanoadsorbent for reactive dyes removal from synthetic solution. *Chemosphere* **2021**, *287*, 132164. [CrossRef] [PubMed]
26. Milojević-Rakić, M.; Janošević, A.; Krstić, J.; Vasiljević, B.N.; Dondur, V.; Ćirić-Marjanović, G. Polyaniline and its composites with zeolite ZSM-5 for efficient removal of glyphosate from aqueous solution. *Microporous Mesoporous Mater.* **2013**, *180*, 141–155. [CrossRef]
27. Kamali, M.; Esmaeili, H.; Tamjidi, S. Synthesis of Zeolite Clay/Fe-Al Hydrotalcite Composite as a Reusable Adsorbent for Adsorption/Desorption of Cationic Dyes. *Arab. J. Sci. Eng.* **2022**, *47*, 6651–6665. [CrossRef]
28. Baglio, V.; Arico, A.S.; Di Blasi, A.; Antonucci, P.L.; Nannetti, F.; Tricoli, V.; Antonucci, V. Zeolite-based composite membranes for high temperature direct methanol fuel cells. *J. Appl. Electrochem.* **2005**, *35*, 207–212. [CrossRef]
29. Pan, X.; Jiao, F.; Miao, D.; Bao, X. Oxide–Zeolite-Based Composite Catalyst Concept That Enables Syngas Chemistry beyond Fischer–Tropsch Synthesis. *Chem. Rev.* **2021**, *121*, 6588–6609. [CrossRef]

30. Abukhadra, M.R.; Shaban, M.; El Samad, M.A.A. Enhanced photocatalytic removal of Safranin-T dye under sunlight within minute time intervals using heulandite/polyaniline@ nickel oxide composite as a novel photocatalyst. *Ecotoxicol. Environ. Saf.* **2018**, *162*, 261–271. [CrossRef]
31. Farghali, M.A.; Abo-Aly, M.M.; Salaheldin, T.A. Modified mesoporous zeolite-A/reduced graphene oxide nanocomposite for dual removal of methylene blue and Pb^{2+} ions from wastewater. *Inorg. Chem. Commun.* **2021**, *126*, 108487. [CrossRef]
32. Baker, H.M.; Massadeh, A.M.; Younes, H. Natural Jordanian zeolite: Removal of heavy metal ions from water samples using column and batch methods. *Environ. Monit. Assess.* **2008**, *157*, 319–330. [CrossRef]
33. Milojević-Rakić, M.; Bajuk-Bogdanović, D.; Vasiljević, B.N.; Rakić, A.; Škrivanj, S.; Ignjatović, L.; Dondur, V.; Mentus, S.; Ćirić-Marjanović, G. Polyaniline/FeZSM-5 composites–Synthesis, characterization and their high catalytic activity for the oxida-tive degradation of herbicide glyphosate. *Microporous Mesoporous Mater.* **2018**, *267*, 68–79. [CrossRef]
34. Rad, L.R.; Anbia, M. Zeolite-based composites for the adsorption of toxic matters from water: A review. *J. Environ. Chem. Eng.* **2021**, *9*, 106088.
35. Tchobanoglus, G.; Burton, F.; Stensel, H.D. Wastewater engineering. *Management* **1991**, *7*, 201.
36. Ilyas, S.; Bhatti, H.N. Microbial Diversity as a Tool for Wastewater Treatment. In *Advanced Materials for Wastewater Treatment*; John Wiley & Sons: Hoboken, NJ, USA, 2017; p. 171.
37. Bilotta, G.; Brazier, R. Understanding the influence of suspended solids on water quality and aquatic biota. *Water Res.* **2008**, *42*, 2849–2861. [CrossRef] [PubMed]
38. Chenglin, Z.; Jing, Y.; Yulei, Z.; Fan, W.; Hao, X.; Shi, C.; Qi, N.; Wei, L. Design and performance of Multiway Gravity Device on removing suspended solids in aquaculture water. *Trans. Chin. Soc. Agric. Eng.* **2015**, *31*, 53–60.
39. Wang, D.; Sun, Y.; Tsang, D.C.; Hou, D.; Khan, E.; Alessi, D.S.; Zhao, Y.; Gong, J.; Wang, L. The roles of suspended solids in persulfate/Fe^{2+} treatment of hydraulic fracturing wastewater: Synergistic interplay of inherent wastewater components. *Chem. Eng. J.* **2020**, *388*, 124243. [CrossRef]
40. Lohani, S.P.; Khanal, S.N.; Bakke, R. A simple anaerobic and filtration combined system for domestic wastewater treatment. *Water-Energy Nexus* **2020**, *3*, 41–45. [CrossRef]
41. Huang, J.; Chen, S.; Ma, X.; Yu, P.; Zuo, P.; Shi, B.; Wang, H.; Alvarez, P.J. Opportunistic pathogens and their health risk in four full-scale drinking water treatment and distribution systems. *Ecol. Eng.* **2021**, *160*, 106134. [CrossRef]
42. Srinivas, N.; Malla, R.R.; Kumar, K.S.; Sailesh, A.R. Environmental carcinogens and their impact on female-specific cancers. In *A Theranostic and Precision Medicine Approach for Female-Specific Cancers*; Elsevier: Amsterdam, The Netherlands, 2021; pp. 249–262.
43. Paumo, H.K.; Das, R.; Bhaumik, M.; Maity, A. Visible-Light-Responsive Nanostructured Materials for Photocatalytic Degradation of Persistent Organic Pollutants in Water. In *Green Methods for Wastewater Treatment*; Springer: Berlin/Heidelberg, Germany, 2020; pp. 1–29.
44. Guo, Y.; Kannan, K. Analytical methods for the measurement of legacy and emerging persistent organic pollutants in complex sample matrices. In *Comprehensive Analytical Chemistry*; Elsevier: Amsterdam, The Netherlands, 2015; pp. 1–56.
45. Silver, M.K.; Meeker, J.D. Endocrine Disruption of Developmental Pathways and Children's Health. In *Endocrine Disruption and Human Health*; Elsevier: Amsterdam, The Netherlands, 2022; pp. 291–320.
46. Briffa, J.; Sinagra, E.; Blundell, R. Heavy metal pollution in the environment and their toxicological effects on humans. *Heliyon* **2020**, *6*, e04691. [CrossRef]
47. Prăvălie, R.; Bandoc, G. Nuclear energy: Between global electricity demand, worldwide decarbonisation imperativeness, and planetary environmental implications. *J. Environ. Manag.* **2018**, *209*, 81–92. [CrossRef] [PubMed]
48. Halevi, O.; Chen, T.-Y.; Lee, P.S.; Magdassi, S.; Hriljac, J.A. Nuclear wastewater decontamination by 3D-Printed hierarchical zeolite monoliths. *RSC Adv.* **2020**, *10*, 5766–5776. [CrossRef]
49. Kautsky, U.; Lindborg, T.; Valentin, J. Humans and Ecosystems Over the Coming Millennia: Overview of a Biosphere Assessment of Radioactive Waste Disposal in Sweden. *Ambio* **2013**, *42*, 383–392. [CrossRef] [PubMed]
50. Ødegaard, H. Advanced compact wastewater treatment based on coagulation and moving bed biofilm processes. *Water Sci. Technol.* **2000**, *42*, 33–48. [CrossRef]
51. Gouamid, M.; Ouahrani, M.; Bensaci, M. Adsorption Equilibrium, Kinetics and Thermodynamics of Methylene Blue from Aqueous Solutions using Date Palm Leaves. *Energy Procedia* **2013**, *36*, 898–907. [CrossRef]
52. Ginimuge, P.R.; Jyothi, S. Methylene blue: Revisited. *J. Anaesthesiol. Clin. Pharmacol.* **2010**, *26*, 517. [CrossRef]
53. Zacharia, J.T. Ecological Effects of Pesticides. In *Pesticides in the Modern World-Risks and Benefits*; InTech Publisher: London, UK, 2011; pp. 129–142.
54. Jatoi, A.S.; Hashmi, Z.; Adriyani, R.; Yuniarto, A.; Mazari, S.A.; Akhter, F.; Mubarak, N.M. Recent trends and future challenges of pesticide removal techniques—A comprehensive review. *J. Environ. Chem. Eng.* **2021**, *9*, 105571. [CrossRef]
55. Pratush, A.; Kumar, A.; Hu, Z. Adverse effect of heavy metals (As, Pb, Hg, and Cr) on health and their bioremediation strategies: A review. *Int. Microbiol.* **2018**, *21*, 97–106. [CrossRef]
56. Darban, Z.; Shahabuddin, S.; Gaur, R.; Ahmad, I.; Sridewi, N. Hydrogel-Based Adsorbent Material for the Effective Removal of Heavy Metals from Wastewater: A Comprehensive Review. *Gels* **2022**, *8*, 263. [CrossRef]
57. Gerardi, M.H. *Troubleshooting the Sequencing Batch Reactor*; John Wiley & Sons: Hoboken, NJ, USA, 2011.
58. Cai, L.; Zhang, T. Detecting Human Bacterial Pathogens in Wastewater Treatment Plants by a High-Throughput Shotgun Sequencing Technique. *Environ. Sci. Technol.* **2013**, *47*, 5433–5441. [CrossRef]

59. Chahal, C.; Van Den Akker, B.; Young, F.; Franco, C.; Blackbeard, J.; Monis, P. Pathogen and particle associations in wastewater: Significance and implications for treatment and disinfection processes. *Adv. Appl. Microbiol.* **2016**, *97*, 63–119.
60. Karim, M.R.; Manshadi, F.D.; Karpiscak, M.M.; Gerba, C.P. The persistence and removal of enteric pathogens in constructed wetlands. *Water Res.* **2004**, *38*, 1831–1837. [CrossRef]
61. Auerbach, S.M.; Carrado, K.A.; Dutta, P.K. (Eds.) *Handbook of Zeolite Science and Technology, 1st ed*; CRC Press: Boca Raton, FL, USA, 2003.
62. Baile, P.; Fernández, E.; Vidal, L.; Canals, A. Zeolites and zeolite-based materials in extraction and microextraction techniques. *Analyst* **2018**, *144*, 366–387. [CrossRef] [PubMed]
63. Ramesh, K.; Reddy, D.D. Zeolites and Their Potential Uses in Agriculture. *Adv. Agron.* **2011**, *113*, 219–241.
64. Available online: https://asia.iza-structure.org/IZA-SC/ftc_table.php (accessed on 1 August 2022).
65. Ma, Y.; Qi, P.; Ju, J.; Wang, Q.; Hao, L.; Wang, R.; Sui, K.; Tan, Y. Gelatin/alginate composite nanofiber membranes for effective and even adsorption of cationic dyes. *Compos. Part B Eng.* **2019**, *162*, 671–677. [CrossRef]
66. Artioli, Y. Adsorption. In *Encyclopedia of Ecology Fath, SEJD*; Academic Press: Oxford, UK, 2008.
67. Hu, H.; Xu, K. Physicochemical technologies for HRPs and risk control. In *High-Risk Pollutants in Wastewater*; Elsevier: Amsterdam, The Netherlands, 2020; pp. 169–207.
68. Wang, X.; Zhu, N.; Yin, B. Preparation of sludge-based activated carbon and its application in dye wastewater treatment. *J. Hazard. Mater.* **2008**, *153*, 22–27. [CrossRef] [PubMed]
69. Soliman, N.; Moustafa, A. Industrial solid waste for heavy metals adsorption features and challenges; a review. *J. Mater. Res. Technol.* **2020**, *9*, 10235–10253. [CrossRef]
70. Rangabhashiyam, S.; Suganya, E.; Selvaraju, N.; Varghese, L.A. Significance of exploiting non-living biomaterials for the biosorption of wastewater pollutants. *World J. Microbiol. Biotechnol.* **2014**, *30*, 1669–1689. [CrossRef]
71. Bhattacharyya, K.G.; Sen, G.S. Adsorption of chromium (VI) from water by clays. *Ind. Eng. Chem. Res.* **2006**, *45*, 7232–7240.
72. Yuna, Z. Review of the Natural, Modified, and Synthetic Zeolites for Heavy Metals Removal from Wastewater. *Environ. Eng. Sci.* **2016**, *33*, 443–454. [CrossRef]
73. Hawari, A.H.; Mulligan, C.N. Biosorption of lead (II), cadmium (II), copper (II) and nickel (II) by anaerobic granular biomass. *Bio-Resour. Technol.* **2006**, *97*, 692–700. [CrossRef]
74. Wang, S.; Zhu, Z. Characterisation and environmental application of an Australian natural zeolite for basic dye removal from aqueous solution. *J. Hazard. Mater.* **2006**, *136*, 946–952. [CrossRef]
75. Khamidun, M.; Fulazzaky, M.A.; Al-Gheethi, A.; Ali, U.; Muda, K.; Hadibarata, T.; Razi, M.M. Adsorption of ammonium from wastewater treatment plant effluents onto the zeolite; A plug-flow column, optimisation, dynamic and isotherms studies. *Int. J. Environ. Anal. Chem.* **2020**, 1–22. [CrossRef]
76. Armağan, B.; Turan, M.; Elik, M.S. Equilibrium studies on the adsorption of reactive azo dyes into zeolite. *Desalination* **2004**, *170*, 33–39. [CrossRef]
77. Ameta, R.; Solanki, M.S.; Benjamin, S.; Ameta, S.C. Photocatalysis. In *Advanced Oxidation Processes for Waste Water Treatment*; Elsevier: Amsterdam, The Netherlands, 2018; pp. 135–175.
78. Pirc, G.; Stare, J. Titanium Site Preference Problem in the TS-1 Zeolite Catalyst: A Periodic Hartree-Fock Study. *Acta Chim. Slov.* **2008**, *55*, 951–959.
79. Krisnandi, Y.K.; Southon, P.D.; Adesina, A.A.; Howe, R.F. ETS-10 as a photocatalyst. *Int. J. Photoenergy* **2003**, *5*, 131–140. [CrossRef]
80. Tahir, M.B.; Rafique, M.; Rafique, M.S. Nanomaterials for photocatalysis. In *Nanotechnology and Photocatalysis for Environmental Applications*; Elsevier: Amsterdam, The Netherlands, 2020; pp. 65–76.
81. Fox, M.A.; Doan, K.E.; Dulay, M.T. The effect of the "inert" support on relative photocatalytic activity in the oxidative decomposition of alcohols on irradiated titanium dioxide composites. *Res. Chem. Intermed.* **1994**, *20*, 711–721. [CrossRef]
82. Arthur, J.D.; Langhus, B.G.; Patel, C. *Technical Summary of Oil & Gas Produced Water Treatment Technologies*; All Consulting LLC: Tulsa, OK, USA, 2005.
83. Shahryar, J. 6-Treatment of Oily Wastewater. In *Petroleum Waste Treatment and Pollution Control*; Shahryar, J., Ed.; Butterworth-Heinemann: Oxford, UK, 2017; pp. 185–267.
84. Artegiani, A. Ion Exchange and Demineralization. In *Tech Brief A National Drinking Water Clearing House Fact Sheet*; National Drinking Water Clearinghouse: Morgantown, WV, USA, 1997.
85. Guida, S.; Potter, C.; Jefferson, B.; Soares, A. Preparation and evaluation of zeolites for ammonium removal from municipal wastewater through ion exchange process. *Sci. Rep.* **2020**, *10*, 12426. [CrossRef]
86. Moran, S. Chapter 7—Clean water unit operation design: Physical processes. In *An Applied Guide to Water and Effluent Treatment Plant Design*; Moran, S., Ed.; Butterworth-Heinemann: Oxford, UK, 2018; pp. 69–100.
87. De Raffele, G.; Aloise, A.; De Luca, P.; Vuono, D.; Tagarelli, A.; Nagy, J.B. Kinetic and thermodynamic effects during the adsorption of heavy metals on ETS-4 and ETS-10 microporous materials. *J. Porous Mater.* **2016**, *23*, 389–400. [CrossRef]
88. Anderson, M.; Terasaki, O.; Ohsuna, T.; Philippou, A.; MacKay, S.P.; Ferreira, A.; Rocha, J.; Lidin, S. Structure of the microporous titanosilicate ETS-10. *Nature* **1994**, *367*, 347–351. [CrossRef]
89. Pavelić, S.K.; Medica, J.S.; Gumbarević, D.; Filošević, A.; Przulj, N.; Pavelić, K. Critical Review on Zeolite Clinoptilolite Safety and Medical Applications in vivo. *Front. Pharmacol.* **2018**, *9*, 1350. [CrossRef]

90. De Magalhães, L.F.; da Silva, G.R.; Peres, A.E.C. Zeolite application in wastewater treatment. *Adsorpt. Sci. Technol.* **2022**, *2022*, 4544104. [CrossRef]
91. Shi, J.; Yang, Z.; Dai, H.; Lu, X.; Peng, L.; Tan, X.; Shi, L.; Fahim, R. Preparation and application of modified zeolites as adsorbents in wastewater treatment. *Water Sci. Technol.* **2018**, *2017*, 621–635. [CrossRef] [PubMed]
92. Cakicioglu-Ozkan, F.; Ulku, S. The effect of HCl treatment on water vapor adsorption characteristics of clinoptilolite rich natural zeolite. *Microporous Mesoporous Mater.* **2005**, *77*, 47–53. [CrossRef]
93. Bouffard, S.C.; Duff, S.J. Uptake of dehydroabietic acid using organically-tailored zeolites. *Water Res.* **2000**, *34*, 2469–2476. [CrossRef]
94. Cortés-Martínez, R.; Solache-Ríos, M.; Martínez-Miranda, V.; Alfaro-Cuevas, V.R. Sorption Behavior of 4-Chlorophenol from Aqueous Solutions by a Surfactant-modified Mexican Zeolitic Rock in Batch and Fixed Bed Systems. *Water Air Soil Pollut.* **2007**, *183*, 85–94. [CrossRef]
95. Ghiaci, M.; Abbaspur, A.; Kia, R.; Seyedeyn-Azad, F. Equilibrium isotherm studies for the sorption of benzene, toluene, and phenol onto organo-zeolites and as-synthesized MCM-41. *Sep. Purif. Technol.* **2004**, *40*, 217–229. [CrossRef]
96. Zieliński, M.; Zielińska, M.; Dębowski, M. Ammonium removal on zeolite modified by ultrasound. *Desalination Water Treat.* **2015**, *57*, 8748–8753. [CrossRef]
97. Bagheri, S.; Hir, Z.A.M.; Yousefi, A.T.; Hamid, S.B.A. Progress on mesoporous titanium dioxide: Synthesis, modification and applications. *Microporous Mesoporous Mater.* **2015**, *218*, 206–222. [CrossRef]
98. Kozera-Sucharda, B.; Gworek, B.; Kondzielski, I.; Chojnicki, J. The Comparison of the Efficacy of Natural and Synthetic Aluminosilicates, Including Zeolites, in Concurrent Elimination of Lead and Copper from Multi-Component Aqueous Solutions. *Processes* **2021**, *9*, 812. [CrossRef]
99. Wang, L.; Shi, C.; Pan, L.; Zhang, X.; Zou, J.-J. Rational design, synthesis, adsorption principles and applications of metal oxide adsorbents: A review. *Nanoscale* **2020**, *12*, 4790–4815. [CrossRef]
100. Khan, M.M.; Adil, S.F.; Al-Mayouf, A. Metal oxides as photocatalysts. *J. Saudi Chem. Soc.* **2015**, *19*, 462–464. [CrossRef]
101. Alswata, A.A.; Bin Ahmad, M.; Al-Hada, N.M.; Kamari, H.M.; Bin Hussein, M.Z.; Ibrahim, N.A. Preparation of Zeolite/Zinc Oxide Nanocomposites for toxic metals removal from water. *Results Phys.* **2017**, *7*, 723–731. [CrossRef]
102. Mahalakshmi, M.; Priya, S.V.; Arabindoo, B.; Palanichamy, M.; Murugesan, V. Photocatalytic degradation of aqueous propoxur solution using TiO_2 and Hβ zeolite-supported TiO_2. *J. Hazard. Mater.* **2009**, *161*, 336–343. [CrossRef] [PubMed]
103. Mirzaei, D.; Zabardasti, A.; Mansourpanah, Y.; Sadeghi, M.; Farhadi, S. Efficacy of Novel NaX/MgO–TiO_2 Zeolite Nanocomposite for the Adsorption of Methyl Orange (MO) Dye: Isotherm, Kinetic and Thermodynamic Studies. *J. Inorg. Organomet. Polym. Mater.* **2020**, *30*, 2067–2080. [CrossRef]
104. Wang, C.; Shi, H.; Li, Y. Synthesis and characterization of natural zeolite supported Cr-doped TiO_2 photocatalysts. *Appl. Surf. Sci.* **2012**, *258*, 4328–4333. [CrossRef]
105. Alswat, A.A.; Bin Ahmad, M.; Saleh, T.A. Zeolite modified with copper oxide and iron oxide for lead and arsenic adsorption from aqueous solutions. *J. Water Supply Res. Technol.* **2016**, *65*, 465–479. [CrossRef]
106. Kong, F.; Zhang, Y.; Wang, H.; Tang, J.; Li, Y.; Wang, S. Removal of Cr(VI) from wastewater by artificial zeolite spheres loaded with nano Fe–Al bimetallic oxide in constructed wetland. *Chemosphere* **2020**, *257*, 127224. [CrossRef]
107. Zhang, W.; Xiao, X.; Zheng, L.; Wan, C. Fabrication of TiO_2/MoS_2@ zeolite photocatalyst and its photocatalytic activity for degradation of methyl orange under visible light. *Appl. Surf. Sci.* **2015**, *358*, 468–478. [CrossRef]
108. Italiya, G.; Ahmed, H.; Subramanian, S. Titanium oxide bonded Zeolite and Bentonite composites for adsorptive removal of phosphate. *Environ. Nanotechnol. Monit. Manag.* **2022**, *17*, 100649. [CrossRef]
109. Jaramillo-Fierro, X.; González, S.; Montesdeoca-Mendoza, F.; Medina, F. Structuring of $ZnTio_3$/Tio_2 adsorbents for the removal of methylene blue, using zeolite precursor clays as natural additives. *Nanomaterials* **2021**, *11*, 898. [CrossRef]
110. Madima, N.; Mishra, S.B.; Inamuddin, I.; Mishra, A.K. Carbon-based nanomaterials for remediation of organic and inorganic pollutants from wastewater. A review. *Environ. Chem. Lett.* **2020**, *18*, 1169–1191. [CrossRef]
111. Piaskowski, K.; Zarzycki, P.K. Carbon-Based Nanomaterials as Promising Material for Wastewater Treatment Processes. *Int. J. Environ. Res. Public Health* **2020**, *17*, 5862. [CrossRef] [PubMed]
112. Zarzycki, P.K. *Pure and Functionalized Carbon Based Nanomaterials: Analytical, Biomedical, Civil and Environmental Engineering Applications*; CRC Press: Boca Raton, FL, USA, 2020.
113. Krissanasaeranee, M.; Wongkasemjit, S.; Cheetham, A.K.; Eder, D. Complex carbon nanotube-inorganic hybrid materials as next-generation photocatalysts. *Chem. Phys. Lett.* **2010**, *496*, 133–138. [CrossRef]
114. Khanday, W.; Marrakchi, F.; Asif, M.; Hameed, B. Mesoporous zeolite–activated carbon composite from oil palm ash as an effective adsorbent for methylene blue. *J. Taiwan Inst. Chem. Eng.* **2017**, *70*, 32–41. [CrossRef]
115. Berber, M.R. Current Advances of Polymer Composites for Water Treatment and Desalination. *J. Chem.* **2020**, *2020*, 7608423. [CrossRef]
116. Alver, E.; Metin, A.Ü.; Çiftçi, H. Synthesis and characterization of chitosan/polyvinylpyrrolidone/zeolite composite by solution blending method. *J. Inorg. Organomet. Polym. Mater.* **2014**, *24*, 1048–1054. [CrossRef]
117. Khanday, W.; Asif, M.; Hameed, B. Cross-linked beads of activated oil palm ash zeolite/chitosan composite as a bio-adsorbent for the removal of methylene blue and acid blue 29 dyes. *Int. J. Biol. Macromol.* **2017**, *95*, 895–902. [CrossRef] [PubMed]

118. Pizarro, C.; Escudey, M.; Bravo, C.; Gacitua, M.; Pavez, L. Sulfate Kinetics and Adsorption Studies on a Zeolite/Polyammonium Cation Composite for Environmental Remediation. *Minerals* **2021**, *11*, 180. [CrossRef]
119. Senguttuvan, S.; Janaki, V.; Senthilkumar, P.; Kamala-Kannan, S. Biocompatible polypyrrole/zeolite composite for chromate removal and detoxification. *Mater. Lett.* **2021**, *308*, 131290. [CrossRef]
120. Yiğit, M.Y.; Baran, E.S.; Moral, Ç.K. A polymer—Zeolite composite for mixed metal removal from aqueous solution. *Water Sci. Technol.* **2021**, *83*, 1152–1166. [CrossRef]
121. Shahabuddin, S.; Sarih, N.M.; Mohamad, S.; Ching, J.J. SrTiO3 Nanocube-Doped Polyaniline Nanocomposites with Enhanced Photocatalytic Degradation of Methylene Blue under Visible Light. *Polymers* **2016**, *8*, 27. [CrossRef]
122. Parekh, K.; Shahabuddin, S.; Gaur, R.; Dave, N. Prospects of conducting polymer as an adsorbent for used lubricant oil reclamation. *Mater. Today Proc.* **2022**, *62*, 7053–7056. [CrossRef]
123. Jaymand, M. Conductive polymers/zeolite (nano-)composites: Under-exploited materials. *RSC Adv.* **2014**, *4*, 33935–33954. [CrossRef]
124. Flores-Loyola, E.; Cruz-Silva, R.; Romero, J.; Angulo-Sánchez, J.; Castillon, F.; Farías, M. Enzymatic polymerization of aniline in the presence of different inorganic substrates. *Mater. Chem. Phys.* **2007**, *105*, 136–141. [CrossRef]
125. Nascimento, G.M.D.; Temperini, M.L.A. Structure of polyaniline formed in different inorganic porous materials: A spectroscopic study. *Eur. Polym. J.* **2008**, *44*, 3501–3511. [CrossRef]
126. Sakellis, I.; Papathanassiou, A.; Grammatikakis, J. Effect of composition on the dielectric relaxation of zeolite-conducting polyaniline blends. *J. Appl. Phys.* **2009**, *105*, 64109. [CrossRef]
127. Malkaj, P.; Dalas, E.; Vitoratos, E.; Sakkopoulos, S. pH electrodes constructed from polyaniline/zeolite and polypyrrole/zeolite conductive blends. *J. Appl. Polym. Sci.* **2006**, *101*, 1853–1856. [CrossRef]
128. Vitoratos, E.; Sakkopoulos, S.; Dalas, E.; Malkaj, P.; Anestis, C.D.C. conductivity and thermal aging of conducting zeolite/polyaniline and zeolite/polypyrrole blends. *Curr. Appl. Phys.* **2007**, *7*, 578–581. [CrossRef]
129. Abukhadra, M.R.; Rabia, M.; Shaban, M.; Verpoort, F. Heulandite/polyaniline hybrid composite for efficient removal of acidic dye from water; kinetic, equilibrium studies and statistical optimization. *Adv. Powder Technol.* **2018**, *29*, 2501–2511. [CrossRef]
130. Petkowicz, D.I.; Pergher, S.B.; da Silva, C.D.S.; da Rocha, Z.N.; dos Santos, J.H. Catalytic photodegradation of dyes by in situ zeolite-supported titania. *Chem. Eng. J.* **2010**, *158*, 505–512. [CrossRef]
131. Abukhadra, M.R.; Shaban, M.; Sayed, F.; Saad, I. Efficient photocatalytic removal of safarnin-O dye pollutants from water under sunlight using synthetic ben-tonite/polyaniline@ Ni_2O_3 photocatalyst of enhanced properties. *Environ. Sci. Pollut. Res.* **2018**, *25*, 33264–33276. [CrossRef]

Article

Tuning Particle Sizes and Active Sites of Ni/CeO$_2$ Catalysts and Their Influence on Maleic Anhydride Hydrogenation

Qiuming Zhang, Xin Liao, Shaobo Liu, Hao Wang *, Yin Zhang * and Yongxiang Zhao *

Engineering Research Center of Ministry of Education for Fine Chemicals,
School of Chemistry and Chemical Engineering, Shanxi University, Taiyuan 030006, China;
matthwwe@163.com (Q.Z.); lx2006294034@126.com (X.L.); lovog@163.com (S.L.)
* Correspondence: haowang@sxu.edu.cn (H.W.); sxuzhy@sxu.edu.cn (Y.Z.); yxzhao@sxu.edu.cn (Y.Z.)

Abstract: Supported metal catalysts are widely used in industrial processes, and the particle size of the active metal plays a key role in determining the catalytic activity. Herein, CeO$_2$-supported Ni catalysts with different Ni loading and particle size were prepared by the impregnation method, and the hydrogenation performance of maleic anhydride (MA) over the Ni/CeO$_2$ catalysts was investigated deeply. It was found that changes in Ni loading causes changes in metal particle size and active sites, which significantly affected the conversion and selectivity of MAH reaction. The conversion of MA reached the maximum at about 17.5 Ni loading compared with other contents of Ni loading because of its proper particle size and active sites. In addition, the effects of Ni grain size, surface oxygen vacancy, and Ni–CeO$_2$ interaction on MAH were investigated in detail, and the possible mechanism for MAH over Ni/CeO$_2$ catalysts was deduced. This work greatly deepens the fundamental understanding of Ni loading and size regimes over Ni/CeO$_2$ catalysts for the hydrogenation of MA and provides a theoretical and experimental basis for the preparation of high-activity catalysts for MAH.

Keywords: hydrogenation; particle size; maleic anhydride; Ni loading; CeO$_2$

Citation: Zhang, Q.; Liao, X.; Liu, S.; Wang, H.; Zhang, Y.; Zhao, Y. Tuning Particle Sizes and Active Sites of Ni/CeO$_2$ Catalysts and Their Influence on Maleic Anhydride Hydrogenation. *Nanomaterials* **2022**, *12*, 2156. https://doi.org/10.3390/nano12132156

Academic Editor: Maria Filipa Ribeiro

Received: 27 May 2022
Accepted: 17 June 2022
Published: 23 June 2022

Publisher's Note: MDPI stays neutral with regard to jurisdictional claims in published maps and institutional affiliations.

Copyright: © 2022 by the authors. Licensee MDPI, Basel, Switzerland. This article is an open access article distributed under the terms and conditions of the Creative Commons Attribution (CC BY) license (https://creativecommons.org/licenses/by/4.0/).

1. Introduction

Maleic anhydride (MA) is an important C4 fundamental material in the chemical industry that can be obtained by oxidation of coking benzene, butane, or biomass platform compounds. MA is a multifunctional and five-membered ring compound composed of one C=C, double C=O bonds, and one C–O–C functional group. A series of high-value-added fine chemicals such as succinic anhydride (SA), γ-butyrolactone (GBL), and tetrahydrofuran (THF) can be synthesized by maleic anhydride hydrogenation (MAH). These solvents and intermediates are widely used in the military, textile, pharmaceutical, and food industries [1–3]. The hydrogenation of MA involves C=C and C=O hydrogenation, and the investigation of hydrogenation mechanism for C=C and C=O bonds has been a hot topic. Until now, the catalysts used in the MAH have mainly been supported Ni-based catalysts, and the supports have mainly been metal or nonmetal oxides such as Al$_2$O$_3$, SiO$_2$, TiO$_2$, and CeO$_2$ [4–7].

For supported catalysts, the type of active metals, the acid and base properties of the surface, defect sites, and metal–support interactions have important effects on the adsorption and activation forms, hydrogenation path, and product selectivity of MAH. Among these factors, the particle size of the active metal plays a crucial role in the catalytic performance of catalysts [8–10]. The geometrical structure, electronic structure, and dispersion of metal particles change dynamically with changes in particle size, and these changes lead to variation in the active sites on the catalyst surface, which significantly affects the catalytic activity of the catalyst [11–14]. Zhao et al. [15] discovered that for the Ni/SiO$_2$ catalytic system, when Ni species were fine clusters, the product GBL was obtained from the hydrogenation of MA because of the strong interaction between Ni and the support. However, succinic anhydride (SA) was obtained when the Ni species was in a crystalline state and had weak interaction with the support. Li et al. [16] found that the selectivity of MAH was closely related to the grain size of the active metal, Ni. For a Ni/HY–Al$_2$O$_3$ catalyst, smaller sizes of Ni nanoparticles were favorable for the formation of SA, while as the Ni loading amount increased, the particle size of Ni and the selectivity of GBL increased. Meyer et al. [17] observed that NiO had a stronger interaction with the support when the Ni loading was lower (less than 8 wt%) and that the Ni nanoparticles were conducive to the generation of SA. However, when the Ni loading was gradually increased, NiO particles tended to aggregate on the surface of the support, which reduced the interaction between NiO and the support until more GBL products were finally obtained. Bertone et al. [3] found that compared with a Ni/SiO$_2$ catalyst, a Ni/SiO$_2$–Al$_2$O$_3$ catalyst had smaller grain size of Ni on the surface and showed higher GBL selectivity. They speculated that the Lewis acid on the surface of the SiO$_2$–Al$_2$O$_3$ support promoted the formation of GBL. Ma et al. [18] prepared Pd/CeO$_2$ catalysts with different Pd particle sizes on a CeO$_2$ carrier and found that the CeO$_2$-supported Pd single atomic catalyst showed the best activity for CO oxidation reaction. In addition, in recent works [4,19,20], we synthesized a series of Ni/CeO$_2$ catalysts under different conditions and investigated deeply the important role of CeO$_2$ in MA hydrogenation. These works will be very helpful for investigating the effect of the particle size and active sites of metal on MAH. On the basis of regulating the particle size and active sites of metal on CeO$_2$ support, they provided a new opportunity to comprehensively understand the interaction between the active metal and support and systematically study the change in the active sites of catalysts in heterogeneous catalysis.

Based on the above discussion, in this paper, Ni-supported catalysts with different Ni loading were prepared by the impregnation method using CeO$_2$ as support, and the hydrogenation performance of the catalysts was investigated carefully. It was found that changes in Ni loading caused changes in the metal particle size and active sites, which significantly affected the conversion and selectivity of MAH reaction. In this work, the effects of Ni grain size, dispersion, surface oxygen vacancy, and Ni–CeO$_2$ interaction on the hydrogenation of MA were investigated in detail, and the synthesis process of metal-supported catalysts was optimized. This paper provides a theoretical and experimental basis for the preparation of MAH catalysts with higher activity and selectivity.

2. Experimental Section

2.1. Catalysts Preparation

The chemicals, including Ce(NO$_3$)$_3$·6H$_2$O, Ni(NO$_3$)$_2$·6H$_2$O, and NaOH, were purchased from the Sinopharm Chemical Reagent Co., Ltd. (Shanghai, China) and used without any purification. A CeO$_2$ support was prepared by the sol–gel method. First, 5.00 g Ce(NO$_3$)$_3$·6H$_2$O was dissolved into 20 mL distilled water, and then, 6.56 g citric acid (CA) was added and stirred. After the cerium salt and citric acid were completely dissolved, the solution was heated in a water bath at 80 °C until the dry sol was formed. After drying at 120 °C for 8 h in the oven, the dry sol formed a spongy material. It was then moved to a muffle oven and calcined at 500 °C for 3 h to finally obtain the CeO$_2$ support. xNi/CeO$_2$ (x: mass content of Ni) catalysts with different loading contents were prepared by citric acid assisted over-volume impregnation method. For the 5Ni/CeO$_2$ catalyst, 0.505 g Ni(NO$_3$)$_2$·6H$_2$O was dissolved in a mixture of 10 mL ethanol and deionized water (volume ratio 1:1), followed by 2.00 g solid CeO$_2$ and 0.139 g citric acid (mole ratio 1:1). After stirring at room temperature for 30 min, the mixture was placed in a water bath at 80 °C to volatilize the solvent. After the solvent was completely volatilized, the sample was transferred to a drying oven at 120 °C for 8 h. The obtained samples were calcinated at 450 °C for 3 h (heating rate of 3 °C/min) and then reduced for 3 h at 350 °C with H$_2$ at a flow rate of 50 mL/min to prepare the catalyst, which was used for subsequent characterization and evaluation. According to different content of Ni, the catalysts were labeled as 5Ni/CeO$_2$, 10Ni/CeO$_2$, 15Ni/CeO$_2$, 17.5Ni/CeO$_2$, 20Ni/CeO$_2$, and 30Ni/CeO$_2$.

2.2. Catalyst Characterizations and Tests

X-ray diffraction (XRD) was performed using a Bruker D8 Advanced X-ray diffractometer (Billerica, MA, USA). The instrument used Cu Kα1 radiation ($\lambda = 0.15418$ nm) as an X-ray source and was supplied with a Ni filter and Vantec detector. The scanning range was 10~80°, and the scanning rate was 30°/min. The average crystallite size was calculated by the Scherrer formula, $D = K\lambda/\beta\cos\theta$, where K is Scherrer's constant (0.89). The characterizations of H_2-TPR (hydrogen temperature programed reduction) and H_2-TPD (hydrogen temperature programmed desorption) were determined using a Micromeritics AutoChem II 2950 chemisorption apparatus. Raman spectroscopy (Raman) was performed on a Raman spectrometer with a laser wavelength of 532 nm (HORIBA, Tokyo, Japan). X-ray photoelectron spectroscopy (XPS) was recorded using a SCIENTIFIC ESCALAB 250 X-ray photoelectron spectrometer (Thermo Company, Waltham, MA, USA) with a standard Al-Kα (h = 1486.6 eV). The spectra were calibrated according to standard C 1s (284.6 eV).

The catalytic performance of the catalyst was evaluated in a 100 mL stainless steel autoclave. First, 0.1 g catalyst was added together with 4.9 g MA and 40 mL THF into the reactor. The N_2 was passed through to replace the air in the reactor 5 times, and then H_2 was passed through 5 times to replace N_2. Then, the reaction system was heated to 210 °C with stirring at 500 rpm, and the pressure was kept at 5.0 MPa. The product was analyzed using an Agilent 7890A gas chromatograph. To verify precise separation of each component in the products, the programmed temperature was selected. The primary temperature of the oven was increased to 120 °C from 100 °C at a ramp of 5 °C min^{-1}, and the temperatures of the detector and injector were 190 °C and 260 °C, respectively. The conversion and selectivity of MA to the product were calculated according to the following equations [20]:

$$X_{MA} (\%) = \frac{C_{GBL} + C_{SA}}{C_{GBL} + C_{SA} + C_{MA}} \times 100\%$$

$$S_{SA} (\%) = \frac{C_{SA}}{C_{SA} + C_{GBL}} \times 100\%$$

where C_{MA}, C_{SA}, and C_{GBL} represent the percent content of the reactant and the two products in the reaction, respectively, and X_{MA} and S_{SA} represent the conversion of MA and selectivity of SA.

3. Results and Discussion

3.1. Catalyst Characterization

Figure 1 shows the XRD patterns of xNiO/CeO$_2$ samples with different Ni contents. As shown in Figure 1A, after metal Ni loading, the CeO$_2$ support still maintained the crystal structure of fluorite cubic phase (JCPDS File 34-0394), similarly to pure CeO$_2$ [21]. The enlarged pattern (shown in Figure 1B) revealed that the diffraction peaks of the CeO$_2$ (111) crystal plane in xNiO/CeO$_2$ samples moved to a higher angle, indicating that the crystal cell parameters of CeO$_2$ shrank after Ni loading. This may have been due to the Ni^{2+}, with its smaller ionic radius (R = 0.72 nm), replacing Ce^{4+} (R = 0.81 nm) in the CeO$_2$ lattice, which resulted in reductions in the cell parameters of CeO$_2$ [22]. The average crystal sizes of NiO and CeO$_2$ in the xNiO/CeO$_2$ samples were calculated by the Scherrer formula, and the results are listed in Table 1. Compared with pure CeO$_2$ support, the grain size of CeO$_2$ increased after Ni loading, which may have been caused by sintering during the thermal calcination or the lattice distortion of CeO$_2$ caused by Ni species [23].

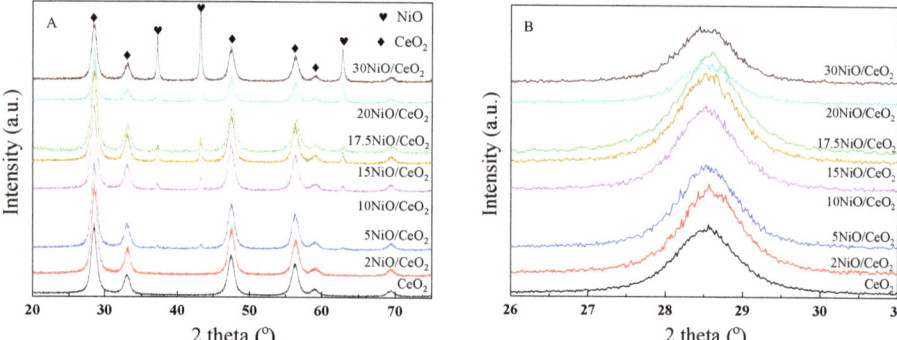

Figure 1. (**A**) XRD patterns of calcined xNiO/CeO$_2$ samples; (**B**) the enlarged pattern at the range of 26–31°.

Table 1. The surface areas (S$_{BET}$), Ni loading, and average crystallite sizes of CeO$_2$, NiO, and metallic Ni in the reduced catalysts.

Sample	D(CeO$_2$) (nm)	Surface Area (m^2/g)	Ni Loading (wt%)	D(NiO) (nm)	D(Ni) (nm)
CeO$_2$	11.6	35.6	-	-	-
5Ni/CeO$_2$	13.4	35.8	4.8	10.1	-
10Ni/CeO$_2$	13.8	37.5	10.2	21.6	10.7
15Ni/CeO$_2$	13.3	34.2	15.3	24.7	15.2
17.5Ni/CeO$_2$	13.8	32.1	17.2	28.2	17.5
20Ni/CeO$_2$	13.7	28.4	20.5	30.5	18.9
30Ni/CeO$_2$	13.1	22.2	29.1	35.1	36.6

As shown in Figure 1A, the XRD diffraction peaks at 37.0°, 43.0°, and 62.9° corresponded to the characteristic diffraction peaks of NiO's (111), (200), and (220) crystal planes (JCPDS 47-1049), respectively. As Ni loading increased, the intensity of the NiO diffraction peak gradually increased, indicating that NiO particles aggregated on the surface of the catalyst and the grain size gradually grew. The particle sizes of NiO are also listed in Table 1, revealing that as Ni loading increased, the particle size of NiO increased from about 10.1 nm to 35.1 nm. The change in NiO grain size led to a change in the interaction between NiO and CeO$_2$, which may have affected the reduction behavior of NiO and the structural difference of the catalyst surface.

Figure 2 shows the XRD patterns of xNi/CeO$_2$ catalysts after reduction at 350 °C. As shown in Figure 2, the CeO$_2$ support maintained a fluorite cubic structure after reduction, and the characteristic peak of NiO disappeared, while the characteristic diffraction peak of the Ni (111) plane appeared at 44.6° (JCPDS 01-1258), indicating that NiO was reduced to metallic Ni. However, for the 5Ni/CeO$_2$ catalyst, the diffraction peak of the metal Ni was not observed, which may have been due to the high dispersion of amorphous Ni species on the catalyst surface or the smaller particle size of Ni (<4 nm). The crystal sizes of Ni in xNi/CeO$_2$ catalysts with different loading content were calculated by the Scherrer formula and are listed in Table 1. As the Ni loading content increased, the metal Ni aggregates on the surface of the catalyst increased, and the average grain size increased gradually from about 10.7 nm to 36.6 nm.

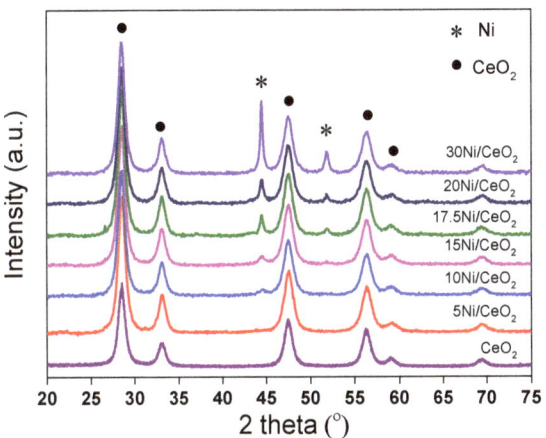

Figure 2. The XRD patterns of reduced xNi/CeO$_2$ catalysts.

Figure 3 shows the H$_2$-TPR spectrum of xNiO/CeO$_2$ samples. The peaks of H$_2$ consumption, named α, β, γ, and δ, were well fitted through a Gauss-type function for these samples. The α peak showed lower intensity and a broader shape at about 150 °C, which was attributed to the reduction of oxygen species adsorbed on the surface of CeO$_2$ [24,25]. It has been reported that parts of Ni^{2+} species could enter the CeO$_2$ lattice to replace Ce^{4+}, which resulted in the distortion of CeO$_2$ lattice and produced oxygen vacancies to balance charges [24]. Raman results also confirmed that the loaded NiO species promoted the formation of oxygen vacancies on the CeO$_2$ surface (Figure 4A). These oxygen vacancies could adsorb some small oxygen-containing molecules and generate reactive oxygen species, which can easily react with hydrogen [25]. The sharp β peak of H$_2$ consumption at about 200 °C could be attributed to the H$_2$ depletion caused by the dissociation and adsorption of H$_2$ onto the oxygen vacancies or the Ni–Ce interface and the formation of OH groups on the surface. A similar result was found in Ni–Ce solid solution [26]. As shown in Figure 3, as Ni loading increased, the β peak gradually moves towards higher temperatures, and the peak intensity decreased, indicating that the increase in Ni content inhibited the dissociation and adsorption of H$_2$ on the oxygen vacancies or the Ni–Ce interface, which may have been caused by the excessive Ni species masking the oxygen vacancies on the surface.

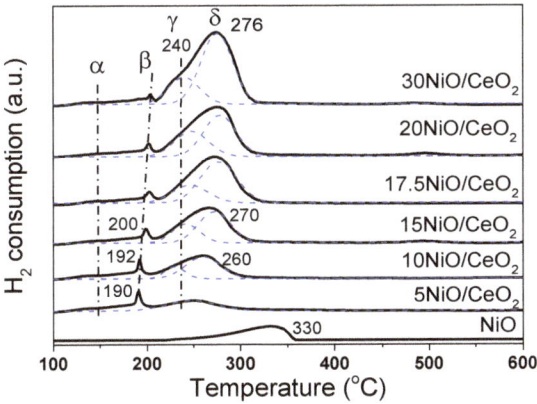

Figure 3. H$_2$-TPR profiles of xNiO/CeO$_2$ precursors.

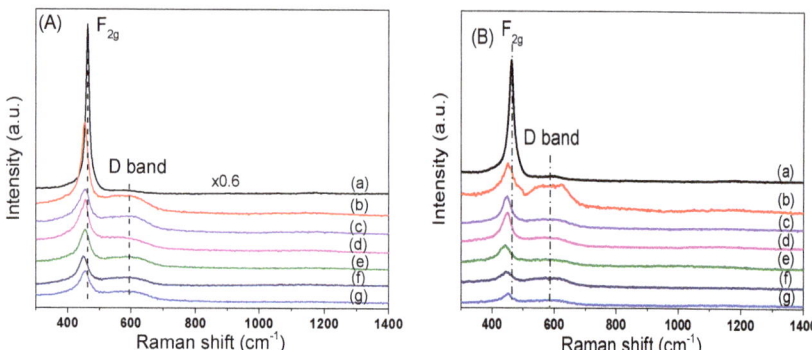

Figure 4. Raman spectra (**A**) xNiO/CeO$_2$ precursors and (**B**) reduced xNi/CeO$_2$ catalysts (a) CeO$_2$, (b) 5Ni/CeO$_2$, (c) 10Ni/CeO$_2$, (d) 15Ni/CeO$_2$, (e) 17.5Ni/CeO$_2$, (f) 20Ni/CeO$_2$, (g) 30Ni/CeO$_2$.

In general, the reduction of NiO species occurs in the temperature range of 200–300 °C. The asymmetric reduction peaks of NiO were deconvolved into two peaks for H$_2$ consumption, which are labeled γ and δ, respectively. The γ peak at 240 °C was attributed to the reduction of highly dispersed NiO species closely linked to the CeO$_2$ support. The stronger metal–support interaction promoted the reduction of NiO at lower temperatures [21]. The δ peak at high temperature (about 275 °C) was ascribed to the reduction of bulk NiO species aggregated on the CeO$_2$ surface. From Figure 3, the reduction temperature of NiO species on the CeO$_2$ surface was lower than that of bulk NiO. This was mainly because loaded NiO, with smaller size and larger surface area, could more easily contact with H$_2$, which resulted in the lower reduction temperature. Moreover, oxygen vacancies and preferential reduced Ni species on the surface of CeO$_2$ support (at 240 °C) promoted the dissociation and activation of H$_2$, and the overflow of H atoms to NiO with large particle size was favorable to the reduction of NiO at low temperature. It should be noted that as Ni loading increased, the δ peak moved towards high temperatures. A possible reason for this is that the activation and migration of H$_2$ may have been inhibited because of the increase in NiO particle size and the decrease in oxygen vacancy, thus retarding the reduction of NiO at low temperatures.

In order to study the effect of Ni loading on the surface structure of CeO$_2$, Raman characterizations for xNiO/CeO$_2$ samples were conducted, and the results are shown in Figure 4. The Raman peak intensity of CeO$_2$ in the figure was 0.6 times that of the original peak intensity in order to facilitate comparison of results. For CeO$_2$ support, a strong Raman vibration peak was observed at 466 cm^{-1}, corresponding to the F2g vibration mode for the Ce–O bond in the cubic fluorite structure of CeO$_2$ [27]. After the loading of NiO on the surface of CeO$_2$, the F2g peak intensity of CeO$_2$ decreased, the peak shape widened, and the peak position moved towards low wavelengths. This was because the strong interaction between NiO and CeO$_2$ led to lattice distortion of CeO$_2$, which reduced the symmetry of the Ce–O bond [25]. Besides the F2g vibration peak, the Raman vibration peak at 600 cm^{-1} was attributed to the vibration (D band) caused by defect sites on the CeO$_2$ surface [25]. Compared with that of the pure CeO2 support, the peak intensity of the D band of the xNiO/CeO$_2$ sample increased significantly, indicating that the existence of NiO promoted the formation of oxygen vacancies on the CeO$_2$ surface. However, the vibration peak of NiO at 520 cm^{-1} could not be observed and may be covered by the F2g vibration peak of CeO$_2$ [26]. Raman spectrum results for the xNiO/CeO$_2$ catalyst after reduction are shown in Figure 4B. Similarly to the xNiO/CeO$_2$ precursor, two Raman characteristic peaks were observed at 466 cm^{-1} and 600 cm^{-1}, corresponding to the F2g vibration of Ce–O bond for cubic fluorite CeO$_2$ and the D-band vibration induced by surface defects, respectively [25].

Figure 5 shows the variation trend of the I_D/I_{F2g} ratio with Ni content before and after reduction, which reflects the influence of Ni loading on the oxygen vacancy concentration on the catalyst surface [20]. As shown in Figure 5, the oxygen vacancy concentrations of all xNiO/CeO$_2$ samples loaded with Ni were higher than that of the CeO$_2$ support without Ni, which indicates that the addition of Ni was beneficial to the formation of oxygen vacancies on the surface of CeO$_2$. Among these NiO/CeO$_2$ samples, the I_D/I_{F2g} ratio of the 5NiO/CeO$_2$ sample is the highest, and then the I_D/I_{F2g} ratio decreased gradually as the Ni content increased, which means that the oxygen vacancy decreased as the Ni content increased. A possible reason for this is that the aggregation of NiO and the growth in particle size on the surface of CeO$_2$ weakened the interaction of NiO and CeO$_2$ and covered part of the oxygen vacancies on the surface, which resulted in a decrease in oxygen vacancies.

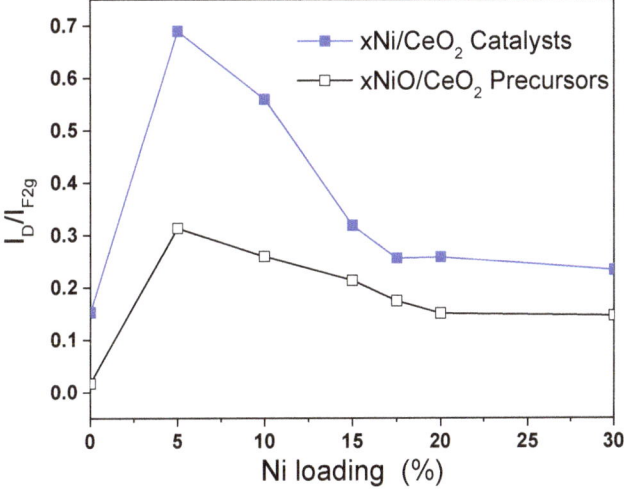

Figure 5. I_D/I_{F2g} ratios for the xNiO/CeO$_2$ precursors and reduced xNi/CeO$_2$ catalysts.

As shown in Figure 5, compared with the xNiO/CeO$_2$ samples, the I_D/I_{F2g} ratios for the xNi/CeO$_2$ catalysts increased significantly after reduction, indicating that the oxygen vacancy concentrations on the surface of the xNi/CeO$_2$ catalysts increased obviously after H$_2$ reduction. The oxygen vacancy increments of 5Ni/CeO$_2$ and 10Ni/CeO$_2$ were significantly larger than those of other catalysts with higher Ni loading, which suggests that lower Ni content was beneficial to the formation of oxygen vacancies on the surface of the catalyst. When the loaded content of Ni was low, Ni species and CeO$_2$ were in close contact and interacted strongly each other, which could have promoted the reduction of the CeO$_2$ surface and facilitated the formation of oxygen vacancies on the surface. However, as Ni loading increased, the active Ni species began to aggregate and cover the surface of CeO$_2$, which weakened the Ni–CeO$_2$ interaction and inhibited the reduction of CeO$_2$ surface.

In order to further study the effect of Ni content on the surface species of Ni/CeO$_2$, five samples of CeO$_2$, 5Ni/CeO$_2$, 10Ni/CeO$_2$, 17.5Ni/CeO$_2$, and 30Ni/CeO$_2$ were characterized by the XPS technique. Figure 6A shows the Ce 3d XPS spectra of the catalyst. The peak of Ce is deconvolved into five groups of characteristic peaks according to the literature [28,29]. The three characteristic peaks labeled u and v, u″ and v″, and u‴ and v‴ belong to the XPS peaks of $3d_{1/2}$ and $3d_{5/2}$ of Ce^{4+} 3d, while the two characteristic peaks of u′ and v′ and u$_0$ and v$_0$ belong to the $3d_{1/2}$ and $3d_{5/2}$ of Ce^{3+} 3d. Compared with pure CeO$_2$, the XPS peak of Ce^{4+} in the 5Ni/CeO$_2$ catalyst moved slightly towards the high-energy direction, indicating that the strong interaction between Ni and CeO$_2$ changed the electronic configuration of Ce on the surface. Similar phenomena were observed in

Pt/CeO$_2$ and Cu/CeO$_2$ catalysts, and the peak shift of Ce^{4+} should be caused by electron transfer from metal to CeO$_2$ [30,31]. According to the XPS peaks of Ce^{3+} and Ce^{4+}, the concentration of Ce^{3+} on the catalyst surface was estimated, and the results are listed in Table 2. Per Table 2, the amount of Ce^{3+} on the surface of the 5Ni/CeO$_2$ catalyst was the highest among these samples. As the Ni loading amount increased, the amount of Ce^{3+} on the surface gradually decreased and was even lower than that of pure CeO$_2$ after reduction for 17.5Ni/CeO$_2$ and 30Ni/CeO$_2$. This may have been caused by excessive Ni covering the Ce^{3+} on the surface of the catalyst.

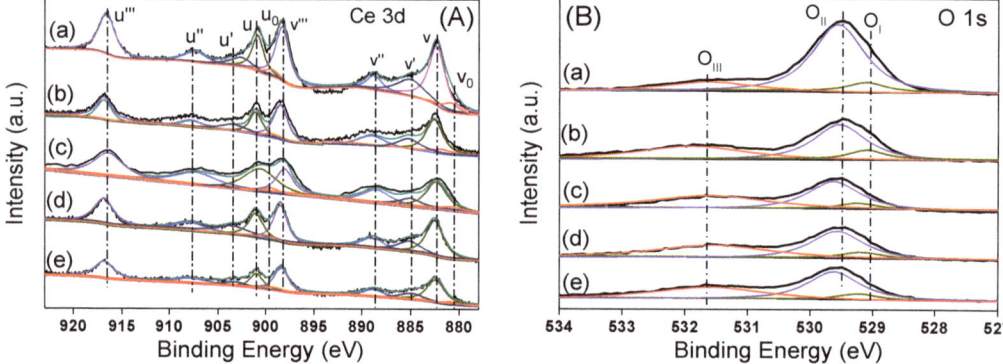

Figure 6. The (**A**) Ce 3d and (**B**) O 1s core-level XPS spectra of the xNi/CeO$_2$ catalysts (a) CeO$_2$, (b) 5Ni/CeO$_2$, (c) 10Ni/CeO$_2$, (d) 17.5Ni/CeO$_2$, and (e) 30Ni/CeO$_2$.

Table 2. The quantitative analysis of XPS for reduced xNi/CeO$_2$ catalysts.

Sample	Ce^{3+}/(Ce^{3+} + Ce^{4+})	O$_{III}$/(O$_I$ + O$_{II}$ + O$_{III}$)	Ni^0:Ni^{2+}:Ni^{3+}
CeO$_2$	0.162	0.2066	-
5Ni/CeO$_2$	0.178	0.414	0.18:0.57:0.25
10Ni/CeO$_2$	0.164	0.374	0.25:0.56:0.19
17.5Ni/CeO$_2$	0.146	0.364	0.34:0.45:0.20
30Ni/CeO$_2$	0.139	0.334	0.43:0.39:0.18

Figure 6B shows the O 1s XPS spectra of the reduced xNi/CeO$_2$ catalysts. After deconvolution, three groups of XPS peaks of O were observed, representing three types of O species. O$_I$ and O$_{II}$ represented O species with different coordination in the CeO$_2$ lattice, and O$_{III}$ represented oxygen species adsorbed at defect sites on the catalyst surface. The O$_I$ peak at 528.8 eV was the oxygen species coordinated with Ce^{3+} in the CeO$_2$ lattice, while the O$_{II}$ peak with slightly higher binding energy (529.4 eV) represented the oxygen species coordinated with Ce^{4+} [32]. The concentration of oxygen vacancies on the surface of the catalyst can be estimated by the ratio O$_{III}$/(O$_I$ + O$_{II}$ + O$_{III}$), and the results are listed in Table 2. Per Table 2, as the Ni content increased, the concentration of oxygen vacancies gradually decreased but was higher than that of the pure CeO$_2$, indicating that the introduction of Ni promotes the formation of oxygen vacancies on CeO$_2$ surface, which was consistent with the Raman results.

Figure 7 shows the XPS peaks of Ni 2$_{p3/2}$ for all catalysts. In addition to the satellite shake-up peak of Ni at about 861.0 eV, three fitting peaks represented three kinds of Ni species with different chemical states, namely α, β, and γ, which were assigned to Ni^0 (~852.4 eV), Ni^{2+} (~854.7 eV), and Ni^{3+} (~856.8 eV), respectively. Three kinds of Ni species coexisted on the surface of the Ni/CeO$_2$ catalysts. According to previous research [30,33,34], highly dispersed Ni clusters can interact with CeO$_2$ support to generate the Ni–O–Ce structure, in which case the outer electrons of Ni would transfer to the 4f orbital of Ce through the Ni–O–Ce bond, which would result in the formation of Ni^{2+} or $Ni^{\delta+}$. Ni^{3+} ions should come from Ni species that enter into the CeO_2 lattice and form a $Ni_xCe_{1-x}O_{2-y}$ solid solution with CeO_2 [35]. According to the peak area of different Ni species, the proportionate relationship among different Ni species was estimated, and the results are listed in Table 2. From Table 2, as Ni loading increased, the content of Ni^0 gradually increased, while the content of Ni^{2+} gradually decreased. This was due to the fact that when the content of loaded Ni was low, the Ni particles with smaller size were highly dispersed on the surface of the catalyst and had stronger interaction with CeO_2 support, which made the outer electrons of Ni easily transfer to CeO_2, thus forming more Ni^{2+}. However, the increase in Ni loading led to the growth of the Ni particle size, which weakened the electron induction effect of CeO_2 on Ni and led to the decrease in Ni^{2+} content. In addition, the relative content of Ni^{3+} was relatively low for all xNi/CeO_2 catalysts, which means that only a small amount of Ni formed a $Ni_xCe_{1-x}O_{2-y}$ solid solution with the CeO_2 support because of the limitation of the loading method (the impregnation method).

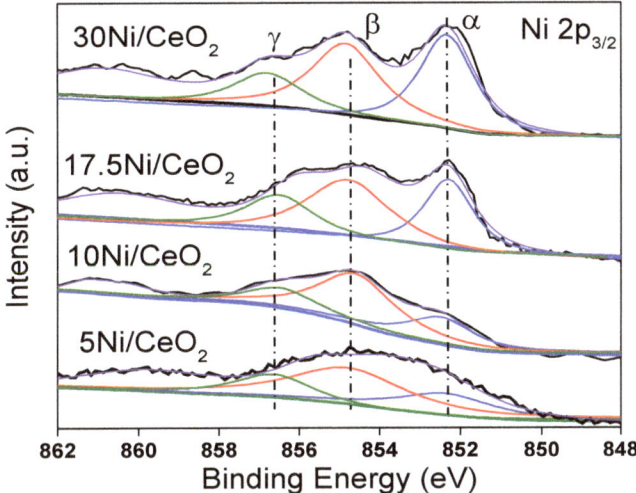

Figure 7. The Ni $2p_{3/2}$ core-level XPS spectra of the xNi/CeO_2 catalysts.

The H_2-TPD characterization results of the CeO_2 support and each catalyst are listed in Figure 8. As shown in Figure 8A, the CeO_2 support had the ability to activate and adsorb H_2 before and after reduction at 350 °C, and the oxygen vacancies had a great influence on the form of existence for the adsorption of H_2 [36]. In order to further study the hydrogen species on CeO_2 surface, H_2-TPD combined with a mass spectrometer (MS) was used to detect the desorbed H_2 species. Figure 9A shows that H_2 was desorbed in the form of H_2O in the range of 150–400 °C on the surface of unreduced CeO_2 (labelled CeO_2), indicating that the adsorption of H_2 on the surface was irreversible, and OH groups were generated on the surface of the support. In contrast, as shown in Figure 9B, on the surface of reduced CeO_2 (labelled CeO_2-350), the adsorbed atomic H was desorbed from the CeO_2 surface in the form of H_2 at about 80 °C, meaning that the oxygen vacancies on the surface of reduced CeO_2 were favorable for the reversible adsorption of H_2, which was consistent with literature reports [37].

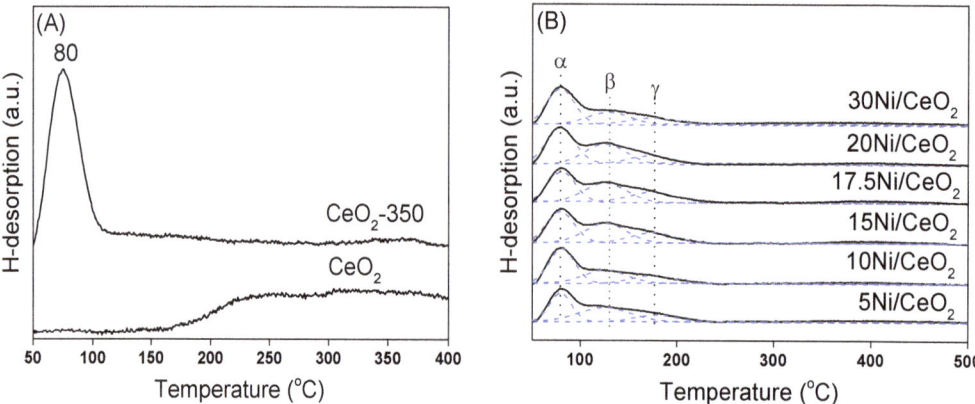

Figure 8. H_2-TPD profiles of (**A**) the unreduced CeO_2 and reduced CeO_2 support (CeO_2-350) and (**B**) xNi/CeO_2 catalysts.

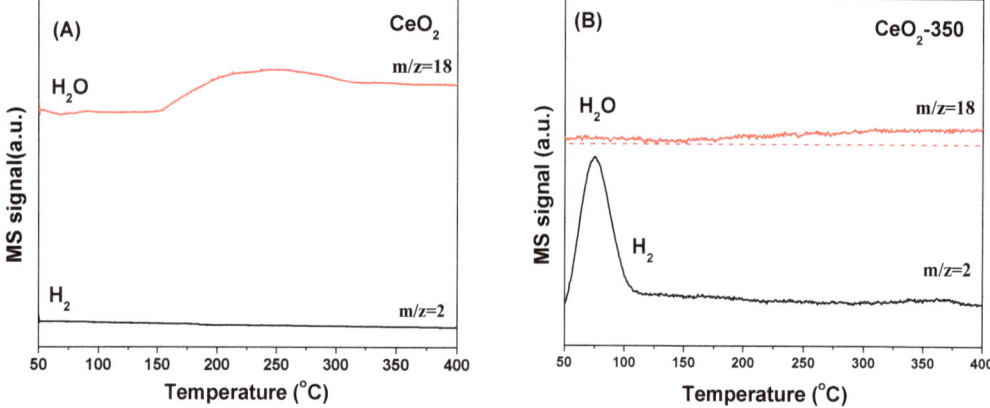

Figure 9. H_2-TPD-MS profiles of (**A**) unreduced CeO_2 (CeO_2) and (**B**) reduced CeO_2 (CeO_2-350) samples.

As shown in Figure 8B, xNi/CeO_2 catalysts with different Ni loadings had similar H_2-TPD spectra. The desorption peak was fitted into three desorption peaks by the Gaussian method. The desorption peak (α peak) at about 80 °C was similar to the H_2 desorption peak of CeO_2 after reduction and was attributed to the desorption of H_2 from the support surface. The desorption peaks β and γ were attributed to the desorption of H_2 adsorbed on different Ni species. The β peak could be assigned to the H_2 desorption of Ni species at the Ni–CeO_2 interface. The strong interaction between Ni and CeO_2 support weakened the binding ability of Ni species to H_2 and then lowered the energy barrier of H_2 desorption. The desorption peak γ (at 178 °C) was assigned to the desorption of hydrogen species adsorbed on the surface of Ni in bulk phase, which was similar to the H_2 desorption on the Ni surface in Ni/Al_2O_3 and Ni/SiO_2 systems and indicated that the support had little influence on the H_2 adsorption capacity on Ni species here [38]. It can be concluded that different Ni species on the xNi/CeO_2 surface had different adsorption and activation abilities for H_2.

Based on H_2-TPD results, H_2 adsorption volumes at different active sites were estimated and correlated with Ni loading. Figure 10A shows that compared with reduced CeO_2, the H_2 adsorption capacity of xNi/CeO_2 catalysts greatly increased, confirming that

Ni was the center of adsorption and activation of H_2. For all xNi/CeO$_2$ catalysts, as the Ni loading increased, the amount of adsorbed H_2 on the catalyst first increased and then decreased gradually. A possible reason for this is that as the Ni loading increased, the Ni particles aggregated on the surface of CeO$_2$, and the grain size became larger, which may have reduced the number of active sites for H_2 adsorption. Figure 10B shows that the peak area of α desorption for 5Ni/CeO$_2$ was the largest among the samples. The peak area of the other samples decreased as the Ni loading increased, which indicates that the 5Ni/CeO$_2$ catalyst possessed the highest concentration of oxygen vacancies for H_2 adsorption. According to Raman and XPS results, excessive Ni was not conducive to the formation of oxygen vacancies on the surface and inhibited the ability of oxygen vacancies to activate hydrogen [39]. In addition, as the Ni loading increased, the peak areas of β and γ increased gradually in the beginning and then decreased obviously after 17.5Ni/CeO$_2$. The results showed that a proper amount of Ni loading was helpful to increase the number of active sites on the support surface, while an excessive amount of Ni loading may have led to the aggregation and growth of Ni species, which could reduce the surface area of Ni particles and the number of active sites on the surface.

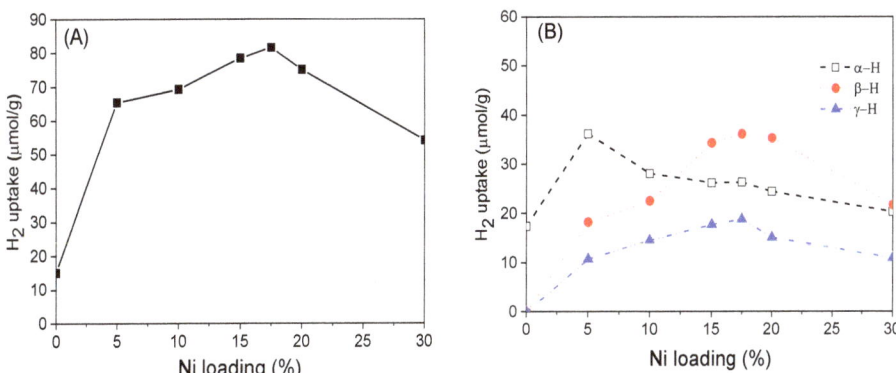

Figure 10. The total (**A**) and site-defined (**B**) hydrogen uptake on the xNi/CeO$_2$ catalyst.

3.2. Catalytic Performance

Figure 11 shows the conversion curves of maleic anhydride (MA) over xNi/CeO$_2$ catalysts and reduced CeO$_2$ support at 210 °C and 5 MPa. After a reaction time of 1 h, the conversion of MA for all xNi/CeO$_2$ catalysts was close to 100%, and the main product was succinic anhydride (SA), indicating that all xNi/CeO$_2$ catalysts showed high hydrogenation activity for the C=C bond. It is noteworthy that the reduced CeO$_2$ carrier also had a certain ability of MAH and that the conversion of MA was about 30% after 1 h under the same conditions. When Ni species were loaded on the surface of CeO$_2$, the activity of MA hydrogenation increased sharply, indicating that Ni was the main active site for the MAH reaction. For all xNi/CeO$_2$ catalysts, in the initial time, the catalytic activity for MAH increased gradually as Ni content increased until 17.5 wt%, and then the conversion of MA decreased slightly until 1 h.

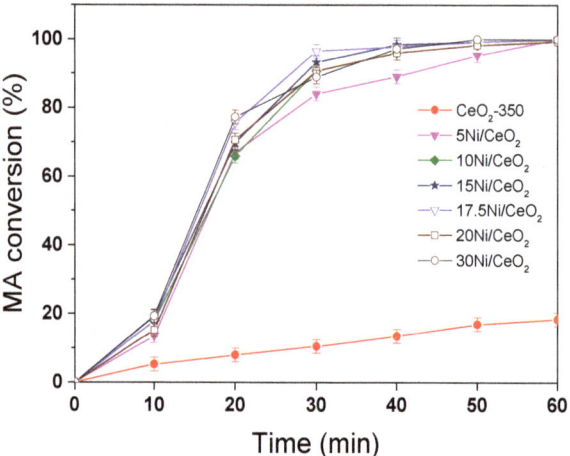

Figure 11. The conversion of MA on reduced the xNi/CeO$_2$ catalysts.

In order to further investigate the C=C hydrogenation performance of xNi/CeO$_2$ catalysts, the turnover frequency values for MA to SA (TOF$_{MA \to SA}$) over the active Ni were calculated and correlated with the oxygen vacancies, Ni species, and Ni loading content. As shown in Figure 12, the TOF$_{MA \to SA}$ of the xNi/CeO$_2$ catalysts decreased as the Ni content increased, which was consistent with the change trend of oxygen vacancies on the surface, indicating that the oxygen vacancies of the catalyst also played an important role in the C=C hydrogenation of MA. According to H$_2$-TPR and H$_2$-TPD results, oxygen vacancies not only improved the dissociation and adsorption capacity of H$_2$ on the catalyst but promoted the diffusion of active H on the catalyst surface, providing more active H species for the hydrogenation reaction [37]. Moreover, according to theoretical calculations, oxygen vacancies with rich electron structure can provide electrons to the active metal and enhance the electron-giving ability of the active metal, thus improving the C=C hydrogenation performance of the metal [40]. For the xNi/CeO$_2$ catalytic system, it can be speculated that the synergistic effect between active metal Ni and oxygen vacancies (Ovac) could have improved the C=C hydrogenation performance of Ni.

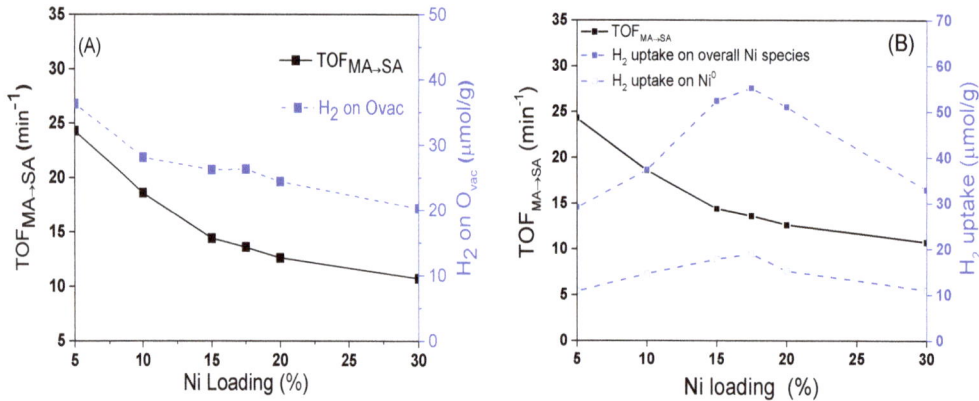

Figure 12. Effect of Ovac (**A**) and surface Ni species (**B**) on the TOF$_{MA \to SA}$ over the xNi/CeO$_2$ catalysts.

Figure 13A shows the trend of SA selectivity with reaction time on different catalysts. From Figure 13A and B, the selectivity of SA for all xNi/CeO$_2$ catalysts was around 100% at the initial reaction time of 40 min and then decreased gradually while the selectivity of γ-butyrolactone (GBL) increased gradually. The selectivity of GBL on the 17.5Ni/CeO$_2$ catalyst was the highest (about 35.7%) after 8 h compared with the other xNi/CeO$_2$ catalysts. In addition, the selectivity of SA on the CeO$_2$ support remained at 100% within 8 h of the reaction, indicating that the CeO$_2$ support had almost no hydrogenation activity for the C=O bond. The above results identify that the metal Ni was the active center for the hydrogenation of SA to GBL and that the content of Ni loading significantly affected the C=O hydrogenation over the catalyst. As for the stability of the xNi/CeO$_2$ catalysts, it should be noted that all samples showed good stability in the hydrogenation process. After a reaction time of 1 h, the conversion of MA for all xNi/CeO$_2$ catalysts was close to 100%, and the catalysts kept their high catalytic performance. Furthermore, after five cycles of use, all the catalysts kept their high activity and selectivity, and there was no obvious decrease in either. In addition, the stability of the 17.5Ni/CeO$_2$ catalyst had no obvious change compared with other catalysts in the MAH process.

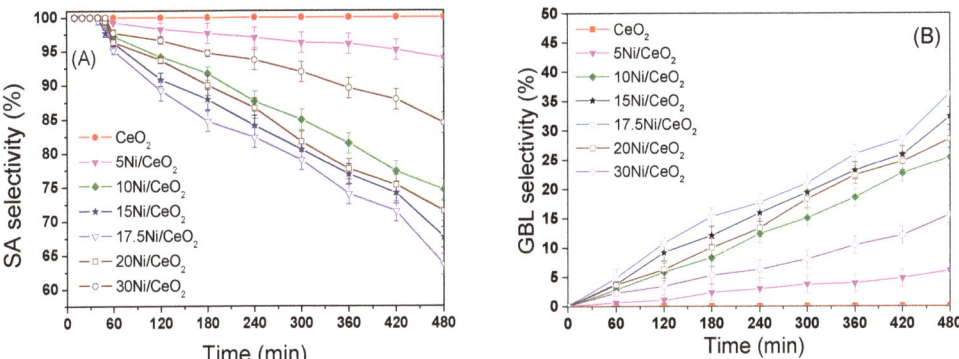

Figure 13. The selectivity of SA (**A**) and GBL (**B**) on reduced the xNi/CeO$_2$ catalysts.

It has been reported that the metal Ni with certain grain size is the active center of hydrogenation of SA to GBL. Meyer et al. [17] studied the effect of Ni loading on the hydrogenation of MA and found that Ni/SiO$_2$–Al$_2$O$_3$ catalyst had hydrogenation activity for C=O only when Ni loading was more than 8 wt%. They concluded that a certain size of Ni grain was the active center for the hydrogenation of SA to GBL. In this work, the selectivity of GBL also showed a strong dependence on the particle size of Ni. However, when the particle size of Ni exceeded a certain amount (17.5 nm), the hydrogenation activity of C=O started to decrease. For example, though the average sizes of Ni particles on the 20Ni/CeO$_2$ (18.9 nm) and 30Ni/CeO$_2$ catalysts (36.6 nm) were larger than that of the 17.5Ni/CeO$_2$ catalyst (17.5 nm), the selectivities of GBL were lower than that of 17.5Ni/CeO$_2$ (as shown in Figure 13B).

In order to understand deeply the influence of catalysts on the hydrogenation activity of C=O, the values of TOF$_{SA \to GBL}$ over different catalysts were calculated. As shown in Figure 14, as the Ni loading increased, the value of TOF$_{SA \to GBL}$ gradually increased. When the content of Ni was 17.5 wt%, the value of TOF$_{SA \to GBL}$ reached the maximum. It then rapidly decreased as the Ni loading increased further. At the same time, the H$_2$ concentration adsorbed on Ovac decreased monotonously as the Ni loading increased. The volcanic curve for TOF$_{SA \to GBL}$ showed that the hydrogenation of SA to GBL was structure sensitive, which is quite different from the trend of TOF$_{MA \to SA}$ in Figure 12A.

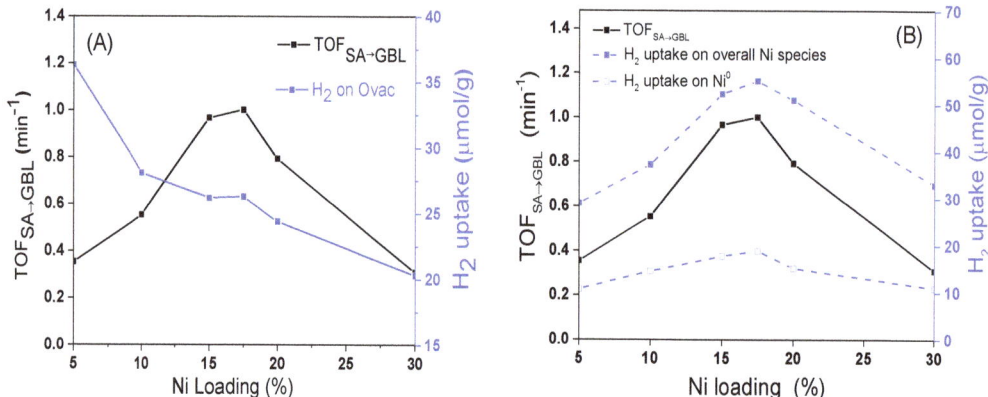

Figure 14. Effect of Ovac (**A**) and surface Ni species (**B**) on the TOF$_{SA \rightarrow GBL}$ over the xNi/CeO$_2$ catalysts.

According to previous research [41–43], for reducible supports, such as TiO$_2$ and CeO$_2$, the metal–support interface is considered to be the active site of C=O adsorption activation. It was found that the C=O functional groups could be adsorbed and polarized at the interfaces of Pt–TiOx and Ni–TiOx and that the catalytic activity of hydrogenation for crotonaldehyde to crotonyl alcohol was significantly improved [41]. Because of strong interaction with the carrier, the electronic configuration for most of the metal particles at the interface was in an ionic state (such as Ni^{2+} at the interface of Ni–TiO$_2$). These ionic metal particles could play the role of a Lewis acid and participate in the adsorption and activation of C=O functional groups [42]. In addition, in the hydrogenation reaction of citral, a small amount of Ni^{2+} at the Ni–TiO$_2$ interface promoted the adsorption and activation of C=O in the citral molecule and finally improved the selectivity of the hydrogenation of citral to citric alcohol [41,43].

As shown in Figure 14B, the change trend of TOF$_{SA \rightarrow GBL}$ was consistent with the change in the total amount of H$_2$ adsorbed on the active sites of interface Ni and bulk Ni, which indicates that both the interfacial Ni and bulk Ni^0 could catalyze the hydrogenation reaction of the C=O bond. According to the characterization results of H$_2$-TPR, H$_2$-TPD, and XPS, the Ni species at the interface showed a valence state of Ni$^{\delta+}$ because of the strong interaction with the CeO$_2$ support [34]. Therefore, it can be inferred that Ni$^{\delta+}$ at the interface could also promote the adsorption of C=O on the catalyst surface as the Lewis acid site. Based on the catalytic effect of metal Ni on the adsorption and activation of C atoms in C=O and subsequent C–O bond breaking [33], we propose the possible mechanism of the Ni$^{\delta+}$–Ni^0 synergistic effect on the hydrogenation reaction of C=O. As shown in Figure 15, first, the metal Ni^0 adsorbs and activates C atoms in the C=O functional group, and Ni$^{\delta+}$ at the interface acts as a Lewis acid to synergistically activate O atoms. Second, the synergistic effect of Ni$^{\delta+}$ and Ni^0 promotes the adsorption and activation of C=O, and the activated C=O group reacts with highly active hydrogen atoms on the surface of metal Ni, which results in the C=O bond hydrogenation and subsequent C–O fracture. According to this mechanism, if the particle size of Ni becomes larger, the distance between the top Ni^0 and the bottom Ni$^{\delta+}$ increases, which weakens the synergistic activation for C=O by Ni$^{\delta+}$–Ni^0. This constitutes a good explanation for the phenomenon in which the selectivity of GBL decreased as the average particle size of Ni increased beyond 17.5 nm.

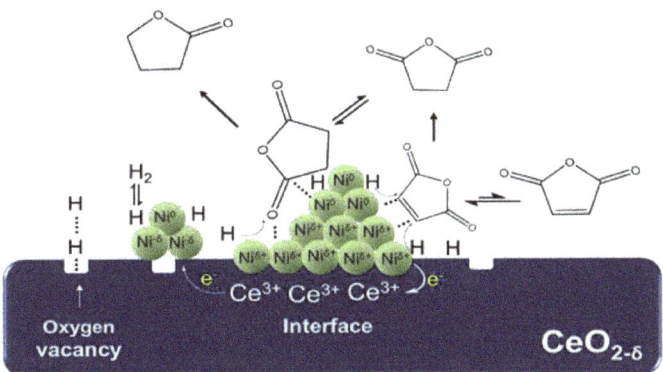

Figure 15. The synergy of $Ni^{\delta+}$–Ni^0 in C=O hydrogenolysis over Ni/CeO_2 catalyst.

4. Conclusions

In this work, Ni/CeO_2 catalysts were synthesized by the impregnation method, and a series of xNi/CeO_2 catalysts with different particle sizes and active sites were successfully prepared by changing the Ni loading. The effects of particle size and active sites of Ni/CeO_2 on the hydrogenation of MA were systematically studied. It was found that the catalytic activity of the xNi/CeO_2 catalysts was size dependent for MAH and that the metal Ni was the active center for the catalytic hydrogenation of C=C from MA to SA and of C=O from SA to GBL. In the beginning of the reaction, the hydrogenation activity of the catalyst increased as the Ni loading increased until $17.5Ni/CeO_2$ and then decreased gradually as the Ni loading increased further. The oxygen vacancies on the surface of Ni/CeO_2 could promote the adsorption and activation of H_2, and the synergistic effect of active metal Ni and oxygen vacancies could improve the hydrogenation of the C=C bond. The synergistic effect of $Ni^{\delta+}$ species obtained from the strong electronic attraction of the CeO_2 support and Ni^0 promoted the adsorption and activation of C=O in MAH. The current results confirmed that the particle size and catalytic ability of Ni/CeO_2 catalysts could be modulated through changing the Ni loading on the CeO_2 support. This work not only provides a deep understanding of MA hydrogenation over Ni/CeO_2 catalysts but highlights the potential of size-dependent catalysts in heterogeneous catalysis.

Author Contributions: Conceptualization, H.W., Y.Z. (Yin Zhang) and Y.Z. (Yongxiang Zhao); methodology, Q.Z.; formal analysis, S.L.; writing—original draft preparation, Q.Z. and X.L.; writing—review and editing, H.W. and Y.Z. (Yin Zhang); supervision, Y.Z. (Yongxiang Zhao); funding acquisition, Y.Z. (Yongxiang Zhao). All authors have read and agreed to the published version of the manuscript.

Funding: This research was funded by the National Natural Science Foundation (U1710221, 21303097, 22075167) and the Shanxi Province International Science and Technology Cooperation Program Project (201803D421074).

Institutional Review Board Statement: Not applicable.

Informed Consent Statement: Not applicable.

Data Availability Statement: Data presented in this article are available at request from the corresponding author.

Conflicts of Interest: The authors declare no conflict of interest.

References

1. Meyer, C.I.; Regenhardt, S.A.; Marchi, A.J.; Garetto, T.F. Gas phase hydrogenation of maleic anhydride at low pressure over silica-supported cobalt and nickel catalysts. *Appl. Catal. A* **2012**, *417–418*, 59–65. [CrossRef]
2. Papageorgiou, G.Z.; Grigoriadou, I.; Andriotis, E.; Bikiaris, D.N.; Panayiotou, C. Miscibility and Properties of New Poly(propylene succinate)/Poly(4-vinylphenol) Blends. *Ind. Eng. Chem. Res.* **2013**, *52*, 11948–11955. [CrossRef]
3. Bertone, M.E.; Meyer, C.I.; Regenhardt, S.A.; Sebastian, V.; Garetto, T.F.; Marchi, A.J. Highly selective conversion of maleic anhydride to γ-butyrolactone over Ni-supported catalysts prepared by precipitation–deposition method. *Appl. Catal. A Gen.* **2015**, *503*, 135–146. [CrossRef]
4. Liao, X.; Zhang, Y.; Hill, M.; Xia, X.; Zhao, Y.; Jiang, Z. Highly efficient Ni/CeO_2 catalyst for the liquid phase hydrogenation of maleic anhydride. *Appl. Catal. A* **2014**, *488*, 256–264. [CrossRef]
5. Gao, C.G.; Zhao, Y.X.; Zhang, Y.; Liu, D.S. Synthesis characterization and catalytic evaluation of Ni/ZrO_2/SiO_2 aerogels catalysts. *J. Sol.-Gel. Sci. Technol.* **2007**, *44*, 145–151. [CrossRef]
6. Yang, Y.P.; Zhang, Y.; Gao, C.G.; Zhao, Y.X. Selective Hydrogenation of Maleic Anhydride to γ-Butyrolactone over TiOx (x < 2) Surface-Modified Ni/TiO_2-SiO_2 in Liquid Phase. *Chin. J. Catal.* **2011**, *32*, 1768–1774.
7. Englisch, M.; Jentys, A.; Lercher, J.A. Structure Sensitivity of the Hydrogenation of Crotonaldehyde over Pt/SiO_2 and Pt/TiO_2. *J. Catal.* **1997**, *166*, 25–35. [CrossRef]
8. Che, M.; Bennett, C.O. The influence of particle size on the catalytic properties of supported metals. *Adv. Catal.* **1989**, *36*, 55–172.
9. Nørskov, J.K.; Bligaard, T.; Hvolbæk, B.; Abild-Pedersen, F.; Chorkendorff, I.; Christensen, C.H. The nature of the active site in heterogeneous metal catalysis. *Chem. Soc. Rev.* **2008**, *37*, 2163. [CrossRef]
10. Wang, H.; Lu, J. A review on particle size effect in metal-catalyzed heterogeneous reactions. *Chin. J. Chem.* **2019**, *38*, 1422–1444. [CrossRef]
11. Breejen, J.P.; Radstake, P.B.; Bezemer, G.L.; Bitter, J.H.; Frøseth, V.; Holmen, A.; de Jong, K.P. On the origin of the cobalt particle size effects in Fischer-Tropsch catalysis. *J. Am. Chem. Soc.* **2009**, *131*, 7197–7203. [CrossRef] [PubMed]
12. Chen, S.; Luo, L.; Jiang, Z.; Huang, W. Size-dependent reaction pathways of lowtemperature CO oxidation on Au/CeO_2 catalysts. *ACS Catal.* **2015**, *5*, 1653–1662. [CrossRef]
13. Li, J.; Chen, W.; Zhao, H.; Zheng, X.; Wu, L.; Pan, H.; Zhu, J.; Chen, Y.; Lu, J. Sizedependent catalytic activity over carbon-supported palladium nanoparticles in dehydrogenation of formic acid. *J. Catal.* **2017**, *352*, 371–381. [CrossRef]
14. Zhang, X.; Gu, Q.; Ma, Y.; Guan, Q.; Jin, R.; Wang, H.; Yang, B.; Lu, J. Support-induced unusual size dependence of Pd catalysts in chemoselective hydrogenation of para-chloronitrobenzene. *J. Catal.* **2021**, *400*, 173–183. [CrossRef]
15. Zhao, Y.X.; Qin, X.Q.; Wu, Z.G.; Xu, L.P.; Liu, D.S. Comparison of selective hydrogenation of maleic anhydride over NiO-SiO_2, NiO-Al_2O_3 and NiO-Al_2O_3-SiO_2 catalysts. *J. Fuel Chem. Technol.* **2003**, *31*, 263–266.
16. Li, J.; Tian, W.; Shi, L. Hydrogenation of Maleic Anhydride to Succinic Anhydride over Ni/HY-Al_2O_3. *Ind. Eng. Chem. Res.* **2010**, *49*, 11837–11840. [CrossRef]
17. Meyer, C.I.; Regenhardt, S.A.; Bertone, M.E. Gas-Phase Maleic Anhydride Hydrogenation Over Ni/SiO_2–Al_2O_3 Catalysts: Effect of Metal Loading. *Catal. Lett.* **2013**, *143*, 1067–1073. [CrossRef]
18. Ma, K.X.; Liao, W.Q.; Shi, W.; Xu, F.K.; Zhou, Y.; Tang, C.; Lu, J.Q.; Shen, W.J.; Zhang, Z.H. Ceria-supported Pd catalysts with different size regimes ranging from single atoms to nanoparticles for the oxidation of CO. *J. Catal.* **2022**, *407*, 104–114. [CrossRef]
19. Liao, X.; Zhang, Y.; Guo, J.; Zhao, L.; Hill, M.; Jiang, Z.; Zhao, Y. The Catalytic Hydrogenation of Maleic Anhydride on CeO_2-delta-Supported Transition Metal Catalysts. *Catalysts* **2017**, *7*, 272. [CrossRef]
20. Liu, S.B.; Liao, X.; Zhang, Q.M.; Zhang, Y.; Wang, H.; Zhao, Y.X. Crystal-Plane and Shape Influences of Nanoscale $CeO_2$2 on the Activity of Ni/CeO_2 Catalysts for Maleic Anhydride Hydrogenation. *Nanomaterials* **2022**, *12*, 762. [CrossRef]
21. Matte, L.P.; Kilian, A.S.; Luza, L.; Alves, M.C.M.; Morais, J.; Baptista, D.L.; Dupont, J.; Bernardi, F. Influence of the CeO2 Support on the Reduction Properties of Cu/CeO_2 and Ni/CeO_2 Nanoparticles. *J. Phys. Chem. C* **2015**, *119*, 26459–26470. [CrossRef]
22. Deraz, N.M. Effect of NiO content on structural, surface and catalytic characteristics of nano-crystalline NiO/CeO_2 system. *Ceram. Int.* **2012**, *38*, 747–753. [CrossRef]
23. Wang, N.; Qian, W.; Chu, W.; Wei, F. Crystal-plane effect of nanoscale CeO_2 on the catalytic performance of Ni/CeO_2 catalysts for methane dry reforming. *Catal. Sci. Technol.* **2016**, *6*, 3594–3605. [CrossRef]
24. Shan, W. Reduction property and catalytic activity of Ce1-XNiXO_2 mixed oxide catalysts for CH_4 oxidation. *Appl. Catal. A* **2003**, *246*, 1–9.
25. Wu, Z.L.; Li, M.J.; Howe, J.; Meyer, H.M.; Overbury, S.H. Probing defect sites on CeO_2 nanocrystals with well-defined surface planes by Raman spectroscopy and O_2 adsorption. *Langmuir* **2010**, *26*, 16595–16606. [PubMed]
26. Barrio, L.; Kubacka, A.; Zhou, G.; Estrella, M.; Arias, A.M.; Hanson, J.C.; García, M.F.; Rodriguez, J.A. Unusual Physical and Chemical Properties of Ni in $Ce_{1−x}Ni_xO_{2−y}$ Oxides: Structural Characterization and Catalytic Activity for the Water Gas Shift Reaction. *J. Phys. Chem. C* **2010**, *114*, 12689–12697. [CrossRef]
27. Spanier, J.E.; Robinson, R.D.; Zhang, F.; Chan, S.W.; Herman, I.P. Size-dependent properties of $CeO_{2−y}$ nanoparticles as studied by Raman scattering. *Phys. Rev. B* **2001**, *64*, 245407. [CrossRef]
28. Burroughs, P.; Hamnett, A.; Orchard, A.F.; Thornton, G. Satellite structure in the X-ray photoelectron spectra of some binary and mixed oxides of lanthanum and cerium. *J. Chem. Soc. Dalton Trans.* **1976**, *17*, 1686–1698. [CrossRef]

29. Nelson, A.E.; Schulz, K.H. Surface chemistry and microstructural analysis of $Ce_xZr_{1-x}O_{2-y}$ model catalyst surfaces. *Appl. Surf. Sci.* **2003**, *210*, 206–221. [CrossRef]
30. Campbell, C.T. Catalyst-support interactions: Electronic perturbations. *Nat. Chem.* **2012**, *4*, 597–598. [CrossRef]
31. Branda, M.M.; Hernández, N.C.; Sanz, J.F.; Illas, F. Density Functional Theory Study of the Interaction of Cu, Ag, and Au Atoms with the Regular CeO_2(111) Surface. *J. Phys. Chem. C* **2010**, *114*, 1934–1941. [CrossRef]
32. Soler, L.; Casanovas, A.; Urrich, A.; Angurell, I.; Llorca, J. CO oxidation and COPrOx over preformed Au nanoparticles supported over nanoshaped CeO_2. *Appl. Catal. B* **2016**, *197*, 47–55. [CrossRef]
33. Carrasco, J.; Rodriguez, J.A.; Lopez-Duran, D.; Liu, Z.Y.; Duchon, T.; Evans, J.; Senanayake, S.D.; Crumlin, E.J.; Matolin, V.; Ganduglia-Pirovano, M.V. In situ and theoretical studies for the dissociation of water on an active Ni/CeO_2 catalyst: Importance of strong metal-support interactions for the cleavage of O-H bonds. *Angew. Chem. Int. Ed.* **2015**, *54*, 3917–3921. [CrossRef]
34. He, L.; Liang, B.L.; Li, L.; Yang, X.F.; Huang, Y.Q.; Wang, A.Q.; Wang, X.D.; Zhang, T. Cerium-Oxide-Modified Nickel as a Non-Noble Metal Catalyst for Selective Decomposition of Hydrous Hydrazine to Hydrogen. *ACS Catal.* **2015**, *5*, 1623–1628. [CrossRef]
35. Tang, K.; Liu, W.; Li, J.; Guo, J.X.; Zhang, J.C.; Wang, S.P.; Niu, S.L.; Yang, Y.Z. The Effect of Exposed Facets of Ceria to the Nickel Species in Nickel-Ceria Catalysts and Their Performance in a NO+CO Reaction. *ACS Appl. Mater. Interfaces* **2015**, *7*, 26839–26849. [CrossRef]
36. Chen, B.H.; Ma, Y.S.; Ding, L.B.; Xu, L.S.; Wu, Z.F.; Yuan, Q.; Huang, W.X. Reactivity of Hydroxyls and Water on a CeO_2(111) Thin Film Surface: The Role of Oxygen Vacancy. *J. Phys. Chem. C* **2013**, *117*, 5800–5810. [CrossRef]
37. Wu, X.P.; Gong, X.Q.; Lu, G. Role of oxygen vacancies in the surface evolution of H at CeO_2(111): A charge modification effect. *Phys. Chem. Chem. Phys.* **2015**, *17*, 3544–3549. [CrossRef]
38. Znak, L.; Zieliński, J. Effects of support on hydrogen adsorption/desorption on nickel. *Appl. Catal. A Gen.* **2008**, *334*, 268–276. [CrossRef]
39. Hahn, K.R.; Seitsonen, A.P.; Iannuzzi, M.; Hutter, J. Functionalization of CeO_2(111) by Deposition of Small Ni Clusters: Effects on CO_2 Adsorption and O Vacancy Formation. *ChemCatChem* **2015**, *7*, 625–634. [CrossRef]
40. Zhang, S.; Li, J.; Xia, Z.; Wu, C.; Zhang, Z.; Ma, Y.; Qu, Y. Towards highly active Pd/CeO_2 for alkene hydrogenation by tuning Pd dispersion and surface properties of the catalysts. *Nanoscale* **2017**, *9*, 3140–3149. [CrossRef]
41. Dandekar, A.; Vannice, M.A. Crotonaldehyde Hydrogenation on Pt/TiO_2 and Ni/TiO_2 SMSI Catalysts. *J. Catal.* **1999**, *183*, 344–354. [CrossRef]
42. Rodriguez, J.A.; Grinter, D.C.; Liu, Z.Y.; Palomino, R.M.; Senanayake, S.D. Ceria-based model catalysts: Fundamental studies on the importance of the metal-ceria interface in CO oxidation, the water-gas shift, CO_2 hydrogenation, and methane and alcohol reforming. *Chem. Soc. Rev.* **2017**, *46*, 1824–1841. [PubMed]
43. Prakash, M.G.; Mahalakshmy, R.; Krishnamurthy, K.R.; Viswanathan, B. Selective hydrogenation of cinnamaldehyde on nickel nanoparticles supported on titania: Role of catalyst preparation methods. *Catal. Sci. Technol.* **2015**, *5*, 3313–3321. [CrossRef]

Article

Crystal-Plane and Shape Influences of Nanoscale CeO_2 on the Activity of Ni/CeO_2 Catalysts for Maleic Anhydride Hydrogenation

Shaobo Liu, Xin Liao, Qiuming Zhang, Yin Zhang *, Hao Wang * and Yongxiang Zhao *

Engineering Research Center of Ministry of Education for Fine Chemicals, School of Chemistry and Chemical Engineering, Shanxi University, Taiyuan 030006, China; lovog@163.com (S.L.); lx2006294034@126.com (X.L.); matthwwe@163.com (Q.Z.)
* Correspondence: sxuzhy@sxu.edu.cn (Y.Z.); haowang@sxu.edu.cn (H.W.); yxzhao@sxu.edu.cn (Y.Z.)

Abstract: Through use of the hydrothermal technique, various shaped CeO_2 supports, such as nanocubes (CeO_2-C), nanorods (CeO_2-R), and nanoparticles (CeO_2-P), were synthesized and employed for supporting Ni species as catalysts for a maleic anhydride hydrogenation (MAH) reaction. The achievements of this characterization illustrate that Ni atoms are capable of being incorporated into crystal lattices and can occupy the vacant sites on the CeO_2 surface, which leads to an enhancement of oxygen vacancies. The results of the MAH reaction show that the morphology and shape of CeO_2 play an important role in the catalytic performance of the MAH reaction. The catalyst for the rod-like CeO_2-R obtains a higher catalytic activity than the other two catalysts. It can be concluded that the higher catalytic performances of rod-like CeO_2-R sample should be attributed to the higher dispersion of Ni particles, stronger support-metal interaction, more oxygen vacancies, and the lattice oxygen mobility. The research on the performances of morphology-dependent Ni/CeO_2 catalysts as well as the relative reaction strategy of MAH will be remarkably advantageous for developing novel catalysts for MA hydrogenation.

Keywords: maleic anhydride; crystal-plane; oxygen vacancies; hydrogenation; Ni/CeO_2

Citation: Liu, S.; Liao, X.; Zhang, Q.; Zhang, Y.; Wang, H.; Zhao, Y. Crystal-Plane and Shape Influences of Nanoscale CeO_2 on the Activity of Ni/CeO_2 Catalysts for Maleic Anhydride Hydrogenation. *Nanomaterials* 2022, 12, 762. https://doi.org/10.3390/nano12050762

Academic Editors: Simon Freakley and Alberto Villa

Received: 22 January 2022
Accepted: 21 February 2022
Published: 24 February 2022

Publisher's Note: MDPI stays neutral with regard to jurisdictional claims in published maps and institutional affiliations.

Copyright: © 2022 by the authors. Licensee MDPI, Basel, Switzerland. This article is an open access article distributed under the terms and conditions of the Creative Commons Attribution (CC BY) license (https://creativecommons.org/licenses/by/4.0/).

1. Introduction

As high value-added solvents and intermediates, γ-butyrolactone (GBL), tetrahydrofuran (THF), and succinic anhydride (SA) are broadly employed in diverse industries including pesticides, machinery, military, plastics, and batteries. These fine chemicals can be prepared by means of catalytic hydrogenation of maleic anhydride (MA), which is a fundamental material that can be produced using butane in the petrochemical industry, using benzene from coal chemical's primary product, and using 5-hydroxymethyl-furfaldehyde or furfaldehyde oxidation from biomass platform compounds [1–4]. In general, the procedure of maleic anhydride hydrogenation (MAH) comprises two sorts of hydrogenation procedures, namely the catalytic hydrogenation of C=C and C=O. These two hydrogenation procedures include similar reaction conditions (pressure and temperature) and can generate a mixture of various fine chemicals [5,6]. Therefore, the novel catalysts with remarkable catalytic activity and selectivity are highly desirable in order to reduce the contamination and further purification costs for synthesizing high-quality fine chemicals.

As a fundamental support material, ceria is capable of promoting the catalytic performances of catalysts by increasing the dispersion of the active metal particles and enhancing the support-metal interactions [7,8]. Furthermore, ceria is known as an oxygen buffer because of a fast and reversible transition between Ce^{4+} and Ce^{3+}, which can provide various peculiar chemical and physical features. Hence, in a diverse range of catalytic reactions, including methane reforming, CO oxidation, NO catalytic reduction, and water-gas shift, ceria is extensively employed as an active catalyst or support material [9–13].

In our recent works [4,14], series Ni/CeO$_2$ catalysts were synthesized and implemented for in-depth understanding of the important role of CeO$_2$ in MA hydrogenation. By means of comparative research on support selection, the association of oxygen vacancies, Ni-CeO$_2$ interaction, interface of M-CeO$_2$ with the catalytic behaviors were thoroughly studied.

Very recently, a broad range of investigations has discovered that the feature of exposed lattice face has a notable impact on the catalytic behavior of CeO$_2$. Moreover, morphology-dependent CeO$_2$ has been deeply explored as a catalyst [15–23] and support [24–33] in various reactions, for instance in acetylene semi-hydrogenation, WGS reaction, and CO oxidation. Thus, the catalytic behaviors of CeO$_2$-supported catalysts could be optimized and improved through modification of CeO$_2$ morphology. Riley et al. [34] reported DFT calculation results on CeO$_2$ (111) surface with oxygen vacancies and proposed the doping of ceria by Ni as a means of creating oxygen vacancies and enhancing the catalytic performance of the hydrogenation of acetylene. Vilé et al. [35] discovered that the cubic CeO$_2$ was mostly exposed to (100) planes, while the particles of polyhedral CeO$_2$ were mostly exposed to (111) planes, and the particles of polyhedral CeO$_2$ demonstrated more considerable catalytic activity compared to the particles of cubic CeO$_2$ in the C$_2$H$_2$ semi-hydrogenation reaction. Moreover, the exposed (110) plane of rod-like CeO$_2$ illustrated more desirable catalytic performances in selective hydrogenation of nitroaromatic hydrocarbons [36]. Si and coworkers [24] prepared cubic, rod, and polyhedral shaped crystals of CeO$_2$ with the aid of precipitation and deposition approaches. Subsequently, the preparation of Au/CeO$_2$ catalysts was conducted by employing the synthesized CeO$_2$ as support and implemented to WGS reaction. The following trend exhibits the catalytic activities of the three considered catalysts in the WGS reactions: rod-shaped Au/CeO$_2$ > polyhedral Au/CeO$_2$ ≫ cubic Au/CeO$_2$. These investigations revealed that the exposed surface of CeO$_2$ represented a substantial impact on the catalytic behaviors of the active metals since the dispersion of metals on the surfaces of the supports and the strong interaction of metal–support were regarded to be the key factors for the alterations in the physicochemical characteristics of the catalyst. As a result, the Ni/CeO$_2$ catalyst possesses a remarkable catalytic activity in the hydrogenating process of maleic anhydride, and the morphology of CeO$_2$ support also has an important effect on the chemical features and catalytic behavior of the catalyst, although researchers' understanding of this aspect is limited.

In this study, based on the above discussions, three kinds of ceria, namely cube-like, rod-like, and particle-like, were prepared. By employing the impregnation technique, Ni species were dispersed on the surface of the particles of CeO$_2$. The fundamental purpose of the current research is to study the potential effects of the CeO$_2$ nano-crystal shape on the MAH reaction. It is revealed that the crystal plane and morphology of the CeO$_2$-support show an apparent effect on the selectivity and conversion of MAH reaction. The catalyst supported over rod-like CeO$_2$ obtains a substantially higher catalytic activity than the other two catalysts. The present research provides a deep understanding of nature and process of metal/oxide–carrier interactions and elucidates the optimization and synthesis technique of metal-supported catalysts; therefore, contributing to the preparation of MAH catalysts with higher product selectivity and catalytic activity.

2. Experimental Section

2.1. Catalysts Preparation

The utilized chemicals, namely, Ni(NO$_3$)$_2$·6H$_2$O, Ce(NO$_3$)$_3$·6H$_2$O, and NaOH were provided by the Sinopharm Chemical Reagent Co., Ltd. (Shanghai, China), and they were analytical grade and used as-purchased without pre-purification. The nanocrystals of CeO$_2$ with various morphologies were prepared in accordance with Mai et al.'s procedure [17]. In general, the dissolution of 16.88 g NaOH was fulfilled in 30 mL water (6.0 mol/L), and the dissolution of 1.96 g Ce(NO$_3$)$_3$·6H$_2$O was performed in 40 mL water (0.05 mol/mL). Subsequently, the solution of NaOH was increased dropwise into the solution of Ce(NO$_3$)$_3$ under stirring at ambient temperature. The prepared solution was sufficiently stirred for a further 30 min at ambient temperature and then transferred into a 100 mL Teflon bottle, which was tightly sealed and hydrothermally processed in a stainless-steel autoclave at 180 and 100 °C for 1 day accordingly for synthesizing CeO_2 cubes (demonstrated as CeO_2-C) and rods (demonstrated as CeO_2-R). Following the cooling procedure, the resulting precipitate was collected, washed out with water, and dehydrated in a vacuum for 16 h at 80 °C. Afterwards, in a muffle oven, the obtained yellowish powder was calcined for 3 h at 500 °C for synthesizing the nanocrystals of CeO_2-R and CeO_2-C. In the case of CeO_2 particles (denoted as CeO_2-P), the synthesized process was practically the same as CeO_2-R, except the added concentration of NaOH was about 0.01 mol/L. To prepare the Ni/CeO_2 catalyst, the acquired CeO_2 supports with three morphologies were moistened with corresponding volumes of the $Ni(NO_3)_2 \cdot 6H_2O$ solution for achieving the theoretical Ni-loading of 5 wt%. The hydrated precursor was dehydrated during the night hours at 120 °C and then through calcination in air for 3 h at 450 °C for the production of NiO/CeO_2 powder. Finally, the Ni (5 wt%)/CeO_2 catalysts were achieved by means of reducing the NiO/CeO_2 for 3 h at 350 °C, denoted as 5Ni/CeO_2-R, 5Ni/CeO_2-C, and 5Ni/CeO_2-P, respectively.

2.2. Catalyst Characterizations

The outcomes of X-ray diffraction were provided through the powder diffraction of X-ray implementing a D8 Advance (Bruker, Billerica, MA, USA) (λ = 0.15418 nm, Cu Kα1 radiation) supplied with a Ni filter and Vantec detector. The findings were obtained within the 2θ zone of 10–80° through a scan speed at 3 °/min. The average crystallite size of the selected samples was evaluated by implementing the formula of Scherrer $D = K\lambda/B\cos\theta$, where K is 0.9. The materials morphologies were visualized by employing a Transmission Electron Microscope (TEM) (JEOL, JEM-2010, Tokyo, Japan). By taking advantage of the ultrasonic method, the specimens were dispersed in ethanol. The loading of Ni species of the catalysts was executed with the help of an inductively coupled plasma (ICP) spectroscopy instrument (iCAP 7400 ICP-OES, Thermo Fisher Scientific, Waltham, MA, USA). To characterize the property of a surface, the spectrum of Raman was obtained by using a Raman microscope with a 532 nm laser wavelength (HORIBA, Tokyo, Japan). The pore distribution and specific surface area of the catalysts, supports, and oxide precursors were determined through physisorption of N_2 at −196 °C employing an ASAP-2020 device (Micromeritics, Norcross, GA, USA). The specimens were pre-degassed under vacuum at 250 °C prior to assessment. The specific surface area of the specimen was evaluated according to the Brunauer–Emmett–Teller (BET) method. The distribution of pore-size was assessed by considering the adsorption isotherms employing the Barett–Joyner–Halenda (BJH) approach.

The assessments of H_2-TPR (hydrogen temperature programed reduction) and H_2-TPD (Hydrogen temperature programmed desorption) were conducted on similar device employing the Micromeritics AutoChem II 2950 system supplemented with a thermal conductivity detector. X-ray photoelectron spectra (XPS) were recorded at ambient temperature employing a SCIENTIFIC ESCALAB 250 spectrometer supplemented with a standard Al-Kα (h = 1486.6 eV). By employing the carbon contamination (C1s, 284.6 eV), the calibration of the binding energies was accomplished. To fit XPS peaks, Lorentzian/Gaussian and Shirley functions were implemented together.

2.3. Catalytic Tests

The hydrogenation of MA at liquid was executed on the catalysts of Ni/CeO_2 in a batch reactor at 5 MPa and 180 °C. Prior to MA hydrogenating, through the flow of pure H_2 in a reactor at correspondent temperature, the catalysts were pre-reduced. The 4.9 g MA and 0.1 g reduced catalysts were mixed in a 100 mL autoclave comprising THF solvent; the mixed system was subsequently purged with N_2 for the elimination of air. The reaction system was next heated to 180 °C with stirring at 500 rpm to diminish any possible mass transfer restriction. The pressure of hydrogen was kept at 5.0 MPa during the MAH reaction at the same time. The products of the reaction were scrutinized by employing a gas chromatograph (7890A, Agilent, Palo Alto, CA, USA). To verify precise separation of each component in the products, the programmed temperature was selected. The primary temperature of the oven was increased to 120 °C from 100 °C at a ramp of 5 °C min^{-1}, and the temperatures of the detector and injector were 190 °C and 260 °C, respectively. The selectivity and conversion of MA to the product were evaluated considering the equations given below:

$$X_{MA}\ (\%) = \frac{C_{GBL} + C_{SA}}{C_{GBL} + C_{SA} + C_{MA}} \times 100\%$$

$$S_{SA}\ (\%) = \frac{C_{SA}}{C_{SA} + C_{GBL}} \times 100\%$$

where C_{MA}, C_{SA}, and C_{GBL} demonstrate the percent content of the reactant and products in the reaction sewage, respectively. S_{SA} and X_{MA} imply the SA selectivity and MA conversion.

3. Results and Discussion

3.1. Catalyst Characterization

Figure 1 demonstrates the images of HRTEM and TEM for the three nanomaterials studied. Figure 1A shows the images of TEM for the irregular nanoparticles of CeO_2 with a less-uniform length within 10–20 nm. The image of HRTEM in Figure 1D shows the clear (111), (220), (200), and (311) lattice fringes with the interplanar spacing of 0.32, 0.28, 0.20, and 0.16 nm, respectively, meaning that the particles of CeO_2-P are predominantly a hexahedral shape and surrounded by the (111) facet. Figure 1B exhibits the image of TEM for the CeO_2 nanorods, with a less-uniform length within 20–150 nm and a uniform diameter in 10 ± 1.0 nm. Figure 1E illustrates the image of HRTEM for a CeO_2 nanorod incorporated with a fast Fourier transform (FFT) assessment (inset). Considering the FTT assessment, two kinds of lattice fringe directions assigned to (200) and (220) are detected for the nanorods, which possess an interplanar spacing of 0.28 and 0.19 nm on the HRTEM image, respectively.

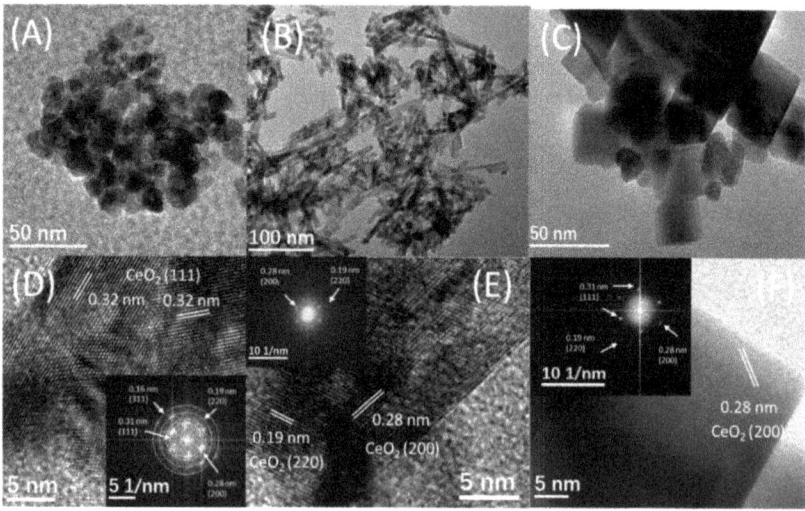

Figure 1. TEM, HRTEM, and FFT images of CeO_2-P (**A,D**), CeO_2-R (**B,E**), and CeO_2-C (**C,F**) supports; inset is a fast Fourier transform (FFT) analysis.

The nanorods demonstrate a 1D growth structure with a preferred growth direction along (220), and are surrounded by (200) planes, which are similar to the CeO_2 nanorods synthesized under similar hydrothermal conditions by Mai and colleagues [17]. It is worth mentioning that the surface of CeO_2 nanorods is rough, which implies the crystals have lower crystallinity and more defect sites on their surface. The image of TEM for the uniform nanocubes of CeO$_2$ with the size of 10–50 nm is illustrated in Figure 1C. The image of HRTEM in Figure 1F incorporated with FFT assessment (inset) shows the apparent (220), (200), and (111) lattice fringes with the interplanar spacing of 0.19, 028, and 0.31 nm, respectively, indicating that the nanocubes of CeO$_2$ are surrounded by the (200) planes.

The images of TEM in Figure 2 show that three various shaped nanomaterials of ceria maintain their intrinsic crystal shapes after the impregnation of Ni and further heat processing. According to the images of HRTEM, it could be observed that the species of Ni with exposed (111) planes are uniformly dispersed on the surface of CeO$_2$ supports for all samples. The average sizes of Ni nanoparticles in 5Ni/CeO$_2$-P, 5Ni/CeO$_2$-R, and 5Ni/CeO$_2$-C samples are 3.0, 2.0–3.0, and 5.0 nm, respectively. Therefore, most of the counted Ni in the samples are supported on the faces of the nanocrystals of CeO$_2$ in comparison to their truncated corners and edges.

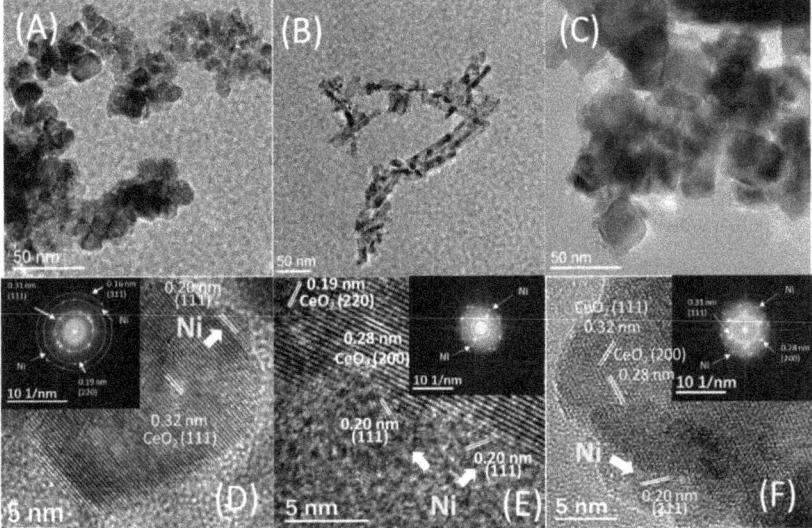

Figure 2. TEM, HRTEM, and FT images of 5Ni/CeO$_2$-P (**A,D**), 5Ni/CeO$_2$-R (**B,E**), and 5Ni/CeO$_2$-C (**C,F**) supports; inset is a fast Fourier transform (FFT) analysis.

The XRD patterns for the as-prepared supports of CeO$_2$ are shown in Figure 3A. The diffraction peaks related to Bragg for the CeO$_2$ specimen presented at 28, 33, 47, and 56◦ can be described as (111), (200), (220), and (311) planes, which should belong to the fluorite-type structure of ceria with cubic crystalline (Fd3m, JCPDS file 34-0394) [37]. The weaker and wider peaks of diffraction for CeO$_2$-Rs indicate a lower crystallinity and smaller crystallite size in comparison with the other samples. Owing to the impurity, no other diffraction peaks can be detected.

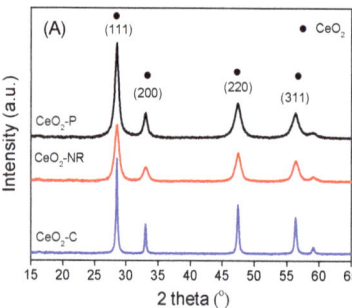
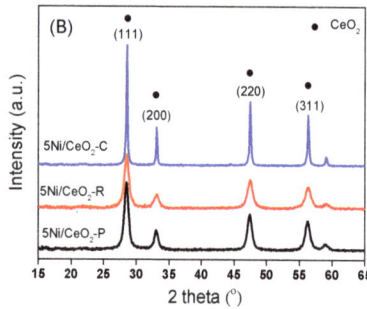

Figure 3. Patterns of XRD for different CeO_2 supports (**A**) and $5Ni/CeO_2$ samples (**B**).

Following the introduction of nickel, the ceria remained in the primary face-centered cubic structure and no Ni diffraction peaks appeared (Figure 2B), which reveals that the Ni species anchored on CeO_2 are smaller and highly dispersed, suggesting there is stronger interaction between CeO_2 and Ni. Moreover, in comparison with pure ceria structures, the crystallite size of ceria (D) over Ni/CeO_2 increases (Table 1), which is possibly correlated with the partial sintering during the procedure of thermal calcination. Based on the previous investigation [38], the microstrain (ε) of crystal is an assessment of lattice stress available in the materials due to lattice elongation, distortion, or contraction, which can be determined according to the broadening degree of XRD diffraction peak with pseudo-Voigt method. Thus, the number of inherent defects on the surface of three CeO_2 supports can be qualitatively analyzed by comparing the value of microstrain. As shown in Table 1, the CeO_2-R shows the highest value of microstrain both prior to and following the Ni loading, and the CeO_2-C has the least value of microstrain. This order should be consistent with the reducibility (the oxygen vacancies concentration) for diverse CeO_2 supports. Since there is a close association between the concentration of oxygen vacancies and the lattice strain, it is suggested that the asymmetrical five-coordinate structure of Ni/CeO_2-R with the greatest strain could be unstable, which is desirable for the surface oxygen mobility. In contrast, the stability of the symmetrical eight-coordination structure of Ni/CeO_2-C with the least strain could be considerable [38]. Due to the largest lattice strain and relatively highest concentration of oxygen defects, the $5Ni/CeO_2$-R should exhibit superior stability and catalytic activity compared with the other two samples for MA hydrogenation. Moreover, the specific surface areas achieved from the isotherms of N_2 adsorption–desorption are also illustrated in Table 1. It can be detected that the $5Ni/CeO_2$-R and CeO_2-R express the greatest S_{BET} values in comparison with the other samples. It should be noted that greater surface area of CeO_2-R is in favor of the Ni dispersion on the support.

Table 1. Physical parameters and structure of CeO_2 supports and $5Ni/CeO_2$ catalysts.

Sample	Ni Loading (%)	S_{BET} (m²/g)	D (CeO_2) (nm)	Microstrains (ε) (%)		
				(111)	(200)	(220)
CeO_2-P	-	60.3	12.8	0.98	0.67	0.58
CeO_2-R	-	80.2	10.6	1.04	0.79	0.62
CeO_2-C	-	20.4	24.5	0.43	0.29	0.26
$5Ni/CeO_2$-P	5.2	56.6	11.8	1.19	0.77	0.68
$5Ni/CeO_2$-R	4.7	78.7	11.2	1.24	0.89	0.72
$5Ni/CeO_2$-C	4.8	18.6	26.8	0.53	0.39	0.28

Note: The amount of metal loading was ascertained through ICP-OES. The specific surface area achieved from the isotherms of N_2 adsorption–desorption. D(CeO_2) represents the crystallite size of CeO_2 phase, evaluated employing the equation of Scherrer to the (111) plane of ceria.

Raman spectroscopy is implemented for exploring the surface structure of the 5Ni/CeO$_2$ catalysts and CeO$_2$ supports. As shown in Figure 4A, the strong vibration mode (F2g, ~460 cm^{-1}), because of the symmetrical stretching vibration of Ce-O bonds, predominates the Raman spectrum of CeO$_2$. In addition to the F2g band, two wide bands at the regions of 1162 cm^{-1} and 590 cm^{-1} are also detected for the 5Ni/CeO$_2$ and CeO$_2$ specimens, which could be ascribed to the second order longitudinal optical mode (2LO), and the Frenkel defect-induced modes (D band), respectively [16]. The 2LO and D bonds are relevant to the existence of oxygen vacancies (Ovac) because of the existence of Ce^{3+} ions in the lattice of ceria, and the comparative ratio of intensity for I_D/I_{F2g} reflects the oxygen vacancies concentration in ceria [39]. As shown in Figure 4, the values of I_D/I_{F2g} and $I_D + I_{2LO}/I_{F2g}$ for Ni/CeO$_2$ and CeO$_2$ with different morphologies reduces in the following trend: CeO$_2$-R > CeO$_2$-P > CeO$_2$-C. This represents that the amount of oxygen vacancies for the various ceria nanostructures differ as: nanorod > nanoparticle > nanocube.

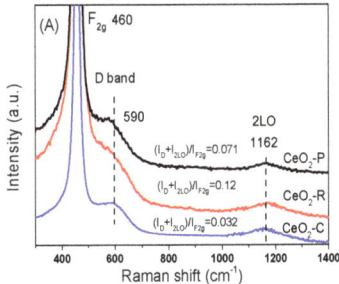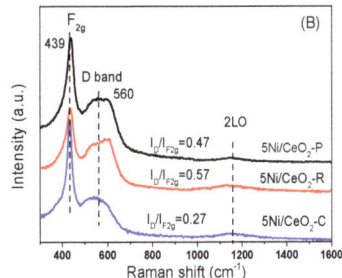

Figure 4. Raman spectra of different CeO$_2$ support (**A**) and 5Ni/CeO$_2$ catalysts (**B**).

Following loading the nickel (demonstrated in Figure 4B), the F2g modes red shifted from 460 cm^{-1} to 439 cm^{-1} with peak widening because of the strong interactions between CeO$_2$ and Ni that resulted in the distortion of the lattice of CeO$_2$ and generates electron-rich oxygen vacancies to maintain the system charge neutral [16]. In comparison with CeO$_2$, all the samples of 5Ni/CeO$_2$ show wider and stronger vibrations in the 2LO modes and D bands, and all of the ratios of intensity enhance sharply. Note that the incorporation of lower-valent cations such as Ni^{2+} in the CeO$_2$ lattice yields defects as oxygen vacancies and dopant cations: the defect-induced (D) mode includes both the contributions due to oxygen vacancies and to cation substitution in the lattice (D1 and D2 bands, respectively). As shown in Figure 4B, D1 and D2 components are clearly seen in the D band profile of Ni/CeO$_2$-R and Ni/CeO$_2$-P samples. The increase in (I_D/I_{F2g}) as well as the (I_{D1}/I_{D2}) intensity ratios, can be taken as an indication of a solid solution formation. This finding implies that the loading of Ni species facilitates the creation of oxygen vacancies (Ovac) thanks to metal substitution in the lattice of ceria, which is in a desirable consistency with the XRD achievements. Further, Ni/CeO$_2$-R exhibits the largest value of I_D/I_{F2g} among three samples, which subsequently reveals that the interactions between ceria rods, and nickel is stronger than other ceria structures. Thus, the morphology of CeO$_2$ demonstrates an important influence on the synergistic interactions between ceria and nickel.

The H$_2$-TPR files revealing the reducibility of catalysts are demonstrated in Figure 5. Four peaks of H$_2$ consumption, namely α, β, γ, and δ are well-fitted through a Gauss-type function for three specimens of Ni/CeO$_2$. The α and β peaks of H$_2$ consumption at low temperature of 160 and 200 °C could be attributed to the reduction of the surface adsorbed oxygen species attached to the surface oxygen vacancies and considerably dispersed nanocrystallites of NiO for the catalysts, respectively. As explained before, the oxygen vacancy could be produced through the incorporation of Ni cations with the lattice of CeO$_2$ and partial substitution of Ce^{4+} or/and Ce^{3+} cations to create solid solution. Moreover, Shan et al. and Li et al. successfully verified that the solid solution of Ni-Ce-O can be produced by means of the incorporation of the ions of Ni^{2+} into the lattice of CeO_2, leading to oxygen vacancies and easily reduced oxygen species [40,41]. The oxygen vacancies can lead to charge imbalance and lattice distortion that is capable of adsorbing oxygen molecules on the surface of oxide. The adsorbed oxygen molecules are considerably reactive oxygen species and can be reduced easily by H_2 at relatively lower temperatures in the range of 100 to 200 °C as shown in Figure 5.

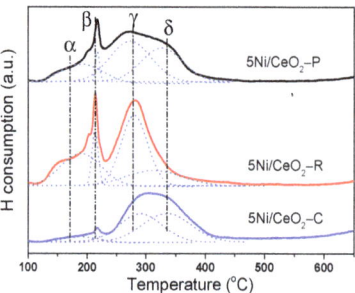

Figure 5. H_2-TPR profiles of three $5Ni/CeO_2$ catalysts.

The reduction peak γ could be attributed to the reduction of strongly interactive species of NiO with ceria support. It could be seen that the consumption of H_2 for the γ peak on the Ni/CeO_2-R is superior to that on the Ni/CeO_2-C and Ni/CeO_2-P catalysts, suggesting that a larger amount of strong interactive NiO is available on the surface of CeO_2-R. The interaction of NiO-CeO_2 obeys the following trend: Ni/CeO_2-R > Ni/CeO_2-P > Ni/CeO_2-C. The reduction peak δ exhibited in the profiles of Ni/CeO_2-C and Ni/CeO_2-P is earmarked to the single step reduction of free NiO species, which accumulates on the surface and possesses weakly interaction with ceria support [42]. However, for the Ni/CeO_2-R sample, the reduction behavior of the free NiO almost did not appear in the profile, which should be ascribed to the abundant defect sites on the CeO_2 surface and high dispersion of NiO on the support in accordance with the XRD and Roman analysis. Hence, it can be concluded that the interaction between ceria rods and nickel is stronger in comparison with the other ceria structures. Consequently, the morphology of ceria support has a significant influence on the synergistic interactions between ceria and nickel.

XPS is a powerful tool for characterizing the surface composition of the catalysts and the valance state of the constituent elements, and also exploring the availability of oxygen vacancies. Figure 6A,B show the Ce3d and O1s core level XPS spectra of the reduced catalyst of Ni/CeO_2. As shown in Figure 6A, the XPS spectrum for the Ce3d core level is deconvoluted into 10 Gaussian peaks and tagged according to the deconvolution conducted by Burroughs and colleagues [43]. The detected peaks labelled as U and V, U″ and V″ and U‴ and V‴ relate to $3d_{3/2}$ ($3d_{5/2}$) and are characteristic of the final state of Ce^{4+}3d, whilst U′ and V′ and U_0 and V_0 relate to $3d_{3/2}$ ($3d_{5/2}$) for the final states of Ce^{3+} 3d [4,44]. Thus, the chemical valance of Ce on the surface of the reduced specimens is mostly an oxidation state of Ce^{4+}, and a limited amount of Ce^{3+} co-existed.

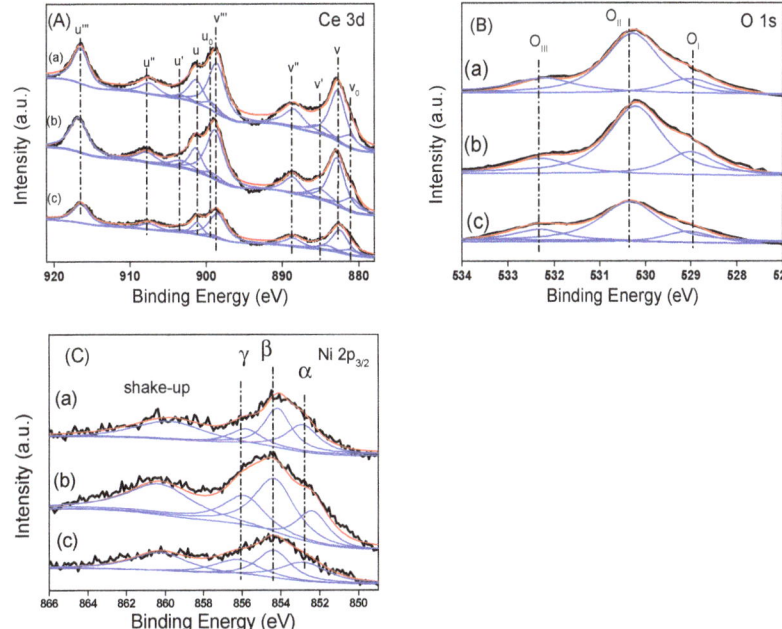

Figure 6. Spectra of XPS for (**A**) Ce 3d, (**B**) O 1s, and (**C**) Ni 2p$_{3/2}$ for reduced 5Ni/CeO$_2$. Catalysts, (a) 5Ni/CeO$_2$-P, (b) 5Ni/CeO$_2$-R, and (c) 5Ni/CeO$_2$-C. The black line is primary curve, and the red and blue lines are fitted curves.

Furthermore, it is obvious that there is an enhancement in the value of binding energy for the Ce3d$_{5/2}$ component (883.9 eV) in comparison with the pure CeO$_2$ (882.9 eV). This little shift could be described by the interactions between cerium oxide and nickel, meaning the incorporation of nickel into the cerium surface lattice [25]. Table 2 represents the comparative contributions of Ce^{4+} and Ce^{3+} evaluated by fitting the peaks and the areas under the fitted entities. The ratio of Ce^{3+}/(Ce^{4+} + Ce^{3+}) is observed to be dependent on the morphologies and the contents of Ce^{3+} in Ni/CeO$_2$ decrease in the following trend: Ni/CeO$_2$-R > Ni/CeO$_2$-P > Ni/CeO$_2$-C. In the previous reports [18,28], it has been reported that the existence of oxygen vacancies is capable of promoting the conversion of Ce^{4+} to Ce^{3+}. Hence, considering both the XRD and Raman results, it can be defined through the creation of more surface oxygen vacancies over Ni/CeO$_2$-R.

Table 2. The quantitative XPS assessment of the 5Ni/CeO$_2$ catalysts.

Sample	Ce^{3+}/(Ce^{4+} + Ce^{3+}) (%)	O$_I$/(O$_I$ + O$_{II}$ + O$_{III}$) (%)	O$_{III}$/(O$_I$ + O$_{II}$ + O$_{III}$) (%)
5Ni/CeO$_2$-P	15.5	18.6	13.5
5Ni/CeO$_2$-R	18.7	20.5	25.4
5Ni/CeO$_2$-C	14.6	12.3	9.4

As shown in Figure 6B, the O1s XPS is fitted into three peaks and summarized in Table 2. The two lesser peaks of energy placed at 528.9 eV and 530.4 eV are assigned to lattice oxygen entities (O^{2-}) binding to Ce^{4+} (O$_{II}$) and Ce^{3+} (O$_I$) [45], whilst the peak placed at 532.4 eV (O$_{III}$) is attributed to the adsorbed oxygen entities (C–O species and water) on the surface of CeO$_2$ [25]. The adsorption of CO$_2$ and CO on the reduced state Ce^{3+} sites demonstrate a greater thermal stability compared to that on the sites of Ce^{4+} [46]. Consequently, the adsorbed oxygen is originated from carbonate entities trapped with the aid of oxygen vacancies. The oxygen vacancies content could be achieved from XPS comparative percentage of adsorbed oxygen. Table 2 represents the XPS findings of the lattice oxygen (O_{II} and O_I) and the adsorbed oxygen (O_{III}). According to Table 2, the ratio of $O_I/(O_I + O_{II} + O_{III})$ of these specimens illustrate the following trend: $5Ni/CeO_2$-R > $5Ni/CeO_2$-P > $5Ni/CeO_2$-C, which is consistent with the trend of the Ce^{3+} content. The great concentration of oxygen species on $5Ni/CeO_2$-R is because of its (110) plane with considerable chemical activity, which are active sites for the chemisorption of oxygen from H_2O and CO_2. Moreover, the ratio of $O_{III}/(O_I + O_{II} + O_{III})$ is able to determine this point. Figure 6C shows the $Ni2p_{3/2}$ XPS spectra of reduced $5Ni/CeO_2$ catalysts. The $Ni2p_{3/2}$ region is fitted into three peaks firstly by curve fitting using peaks and associated satellites in Figure 6C, and then three peaks are named with α, β, and γ, respectively. According to previous research, the peak α is assigned to Ni^0, and peaks of β and γ refer to Ni^{2+}. It can be seen that both Ni^0 (α~852.7 eV) and Ni^{2+} (β~854.7 eV and γ~856.8 eV) coexist on the surface of Ni/CeO_2 catalysts. Note that the peak area (α) of Ni^0 for Ni/CeO_2-R is larger than other two samples, which means more Ni^{2+} is reduced to Ni^0 over the Ni/CeO_2-R compared with other two samples, indicating that the Ni/CeO_2-R has a higher reducibility of Ni species in this condition.

Hydrogen temperature programmed desorption (H_2-TPD) is an advantageous approach for gaining deep insight in metal-support interaction and hydrogen activation on the catalysts. Figure 7 exhibits the profiles of H_2-TPD for the $5Ni/CeO_2$ specimens with three configurations are deconvoluted into three peaks. The α peaks for the catalysts of Ni/CeO_2 emerge in a remarkably similar temperature region (at 80 °C), which are attributed to the desorption of H_2 up-taking surface oxygen vacancies in CeO_2. The γ and β peaks placed in 100–200 °C are because of the desorption of H_2 from the correspondent Ni entities. The β peaks could be assigned to ineffectively adsorbed dissociative H-entities taken up the interface of Ni-CeO_2, which is close to oxygen vacancy on the support [4]. The γ peaks are correlated with the dissociated hydrogen entities attaching to the free particles of Ni. In the higher temperature region, the catalyst of Ni/CeO_2 shows a single wide peak of H_2-TPD emerging at 300 °C and tailing to further than 500 °C. This H_2-TPD peak is because of the H_2 adsorption in the subsurface layers of Ni or/and to the spillover of H_2 [47].

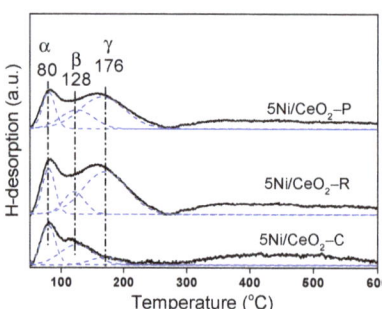

Figure 7. H_2-TPD profiles of $5Ni/CeO_2$ catalysts.

In order to meticulously understand the difference in the capability of hydrogen activation between three various catalysts, quantitative assessment of the desorption of H was conducted and the results are given in Table 3. From Table 3, the amount of H_2 desorption for the Ni/CeO_2-R specimen is superior to other samples (such as interface-adsorbed $H_β$, vacancy-adsorbed $H_α$, and chemically adsorbed $H_β$ on free Ni). The higher amount of H_2 uptake for the Ni/CeO_2-R explains it possesses higher dispersion and tinier size of Ni particle, which is in compliance with the results of XRD. Regarding the attribution of the peaks of H_2-TPD, the dispersion of Ni could be quantified by the metal-related hydrogen activation (sum of γ- and β-H_2), which is distinct from the total amount of H_2 activation for the catalysts. As given in Table 3, the values of Ni dispersion for the $5Ni/CeO_2$

catalysts are shown to be diminishing according to the following order: 5Ni/CeO$_2$-R > 5Ni/CeO$_2$-P > 5Ni/CeO$_2$-C. The results are in a favorable agreement with the TPR and Raman assessments. Furthermore, by means of comparative evaluations of H$_2$-TPD and H$_2$-TPR for three catalysts, we can deduce that the catalyst of Ni/CeO$_2$-R represents more active sites for reversible adsorption/desorption of H-species, leading to fast conversion between H$_2$ and the dissociated H species on the catalyst in MAH process and thus facilitating the MA conversion rate. Consequently, it can be estimated that the catalyst Ni/CeO$_2$-R with rod shape has more activity in the reaction of hydrogenation.

Table 3. H$_2$ uptake and Ni dispersion on the 5Ni/CeO$_2$ catalysts.

Sample	H$_\alpha$ (µmol/g)	H$_\beta$ (µmol/g)	H$_\gamma$ (µmol/g)	Ni Dispersion (%)
5Ni/CeO$_2$-P	29.8	29.7	84.5	25.7
5Ni/CeO$_2$-R	48.8	56.3	208.6	66.1
5Ni/CeO$_2$-C	21.7	30.6	13.5	10.9

H$_\alpha$, H$_\beta$, H$_\gamma$ represent the amount of H$_2$ desorbed at different temperatures. The Ni dispersion = (Ni$_{surf}$/Ni$_{total}$), Ni$_{total}$ implies the total amount of Ni in the catalysts, and Ni$_{surf}$ demonstrates the amount of surface-exposed Ni on the catalysts, considering H/Ni = 1 and Ni = 2 × (amount of H$_2$ desorption).

3.2. Catalytic Performance

Figure 8A shows the conversion of MA (X$_{MA}$) along the courses of hydrogenation on Ni/CeO$_2$ catalysts in a batch reactor at 180 °C and the pressure of hydrogen was considered of about 5.0 MPa. As shown in Figure 8A, the 5Ni/CeO$_2$-R catalyst is shown to present the highest activity for the conversion of MA, obtaining ~98% conversion of MA in 1 h. In the meantime, the conversion of MA on 5Ni/CeO$_2$-C and 5Ni/CeO$_2$-P in the same duration only obtains 90% and 92%, respectively. It is worth mentioning that the MA is completely converted into succinic anhydride (SA) on all three Ni/CeO$_2$ catalysts (not shown) and exhibits 100% selectivity to SA within the 1.0 h continuous MAH without other products observed in the reaction, indicating it is inert for SA hydrogenolysis to other products in the current condition. The considerable difference of the MA conversion on three kinds of 5Ni/CeO$_2$ catalysts can be attributed to the morphology and particle size, which can cause the difference of Ni dispersion and Ovac amount.

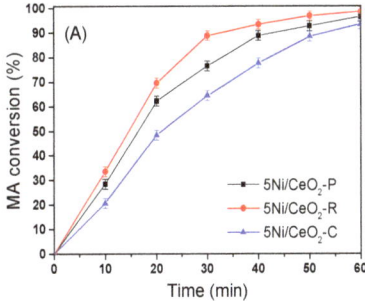
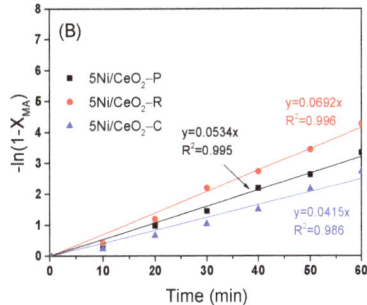

Figure 8. The conversion of maleic anhydride (MA) over 5Ni/CeO$_2$ catalysts (**A**) and their −ln(1 − X$_{MA}$) plots versus reaction time (**B**) at 180 °C and under 5 MPa of H$_2$.

Figure 8B represents the −ln(1 − X$_{MA}$) curves vs. time during the first 1 h, which are well-fitted conforming to the first-order kinetic law in respect of the conversion of MA on the metal-based catalysts [48]. The linear kinetic diagrams over three catalysts of Ni/CeO$_2$ define that the hydrogenation of C=C obeys the quasi-first order reaction in respect of the conversion of MA. The MA hydrogenation rate coefficients (k) on these catalysts are determined according to the gradients of their linear plots and summarized in Figure 8B. The k values for 5Ni/CeO$_2$-R, 5Ni/CeO$_2$-P, and 5Ni/CeO$_2$-C are around

0.0692, 0.0534, and 0.0415, respectively. The greater k value for 5Ni/CeO$_2$-R specimen in comparison with others indicates that the 5Ni/CeO$_2$-R is more reactive than other two catalysts.

Furthermore, the influence of the reaction temperature (393–453 K) on the activity of three catalysts in the hydrogenation of MA is carefully investigated. An Arrhenius diagram illustrating lnk vs. reaction temperature (1/T) is depicted in Figure 9. According to Figure 9, the constants of pseudo-first reaction rate lay entirely to the straight line, and an increase in activity with increasing temperature from 393 K to 453 K is observed. The apparent activation energies (Ea), evaluated from the slope of the Arrhenius plot straight line (shown in Figure 9), are revealed to be 47.93 ± 4.95 kJ/mol for 5Ni/CeO$_2$-R, 50.69 ± 3.33 kJ/mol for 5Ni/CeO$_2$-P and 58.22 ± 7.92 kJ/mol for 5Ni/CeO$_2$-C, respectively. The values of Ea follow the order of: 5Ni/CeO$_2$-R < 5Ni/CeO$_2$-P < 5Ni/CeO$_2$-C, which implies that the 5Ni/CeO$_2$-R can represent the higher ability for MA hydrogenation.

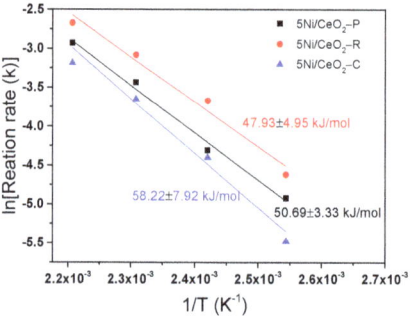

Figure 9. Arrhenius plot for the hydrogenation of maleic anhydride on three 5Ni/CeO$_2$ catalysts and the relationship between ln(k$_{MA}$) versus 1/T.

From the above results, it can be seen that the surface structure, namely the morphology (crystal plane) of CeO$_2$, is the key parameter influencing the interaction between Ni and CeO$_2$ support. As shown in TEM images (Figures 1 and 2), the CeO$_2$-C, CeO$_2$-R, and CeO$_2$-P samples expose predominantly the (100), (110 or 220), and (111) planes, respectively. Because the typical CeO$_2$ could be regarded as an array of cations creating a face-centered cubic lattice with oxygen ions situating at the tetrahedral interstitial sites, the Ni^{2+} is capable of penetrating easily into the lattices of (220) and (111) planes by placing in these sites, along with the capping oxygen atom to compensate the charge. This structural impact may lead to the differences in the synergistic interactions between CeO$_2$ and Ni, thus affecting the catalytic behavior of Ni/CeO$_2$ catalysts. The above analysis reasonably explained why Ni/CeO$_2$-R shows a higher catalytic property compared to that of Ni/CeO$_2$-P and Ni/CeO$_2$-C.

Besides the interaction between CeO$_2$ support and Ni, the oxygen vacancy (Ovac) is considered as another factor for enhancing the hydrogenation of C=C since the oxygen vacancy is capable of enriching the electron density of active metals, which promotes the electron donating capability and the dissociation of H2 [4,49]. In this study, the catalyst of Ni/CeO$_2$-R possesses richer oxygen vacancies and hence can donate more electrons to the metallic nickel compared with the Ni/CeO$_2$-C and Ni/CeO$_2$-P, which shows higher activity of Ni/CeO$_2$-R in the hydrogenation of MA. Thus, it can be concluded that the rod-shape of CeO$_2$ can enhance the dispersion of metallic nickel, presenting stronger interaction between support and Ni and more oxygen vacancies, as a result the catalyst of 5Ni/CeO$_2$-R shows higher reactivity in MAH compared with other two samples.

4. Conclusions

In this study, we have prepared three kinds of the catalysts of Ni/CeO$_2$ with various morphologies of CeO$_2$ support, namely cube-like, rod-like, and particle-like, and investigated the crystal-plane and morphology influences on catalytic behaviors in the reaction of maleic anhydride hydrogenation (MAH). The results of characterization demonstrate that nickel species can incorporate into the lattice of CeO_2, and cause an increase in oxygen vacancies by occupying the empty sites. The catalytic features relate to the exposed plane and shape of the CeO_2 supports. For all considered catalysts, the MA is able to be completely converted into succinic anhydride (SA), and represent 100% selectivity to SA in the current conditions. Among three catalysts, the Ni/CeO_2-R exhibits excellent catalytic performance in stability and catalytic behavior, which is because of the stronger anchoring influence of CeO_2 to nickel species. The oxygen vacancies concentration as well as the mobility of lattice oxygen within the Ni/CeO_2 indicate the morphology dependencies. With the aid of reactivity assessments, the redox features and crystal structures of the catalysts achieved from various characterization methods, it can be deduced that the desirable behavior of $5Ni/CeO_2$-R is strongly associated with the higher dispersion of metallic nickel species, the higher availability of oxygen vacancies, and the stronger interaction between CeO_2 and Ni. The obtained findings confirm that the catalytic performances and structures of Ni/CeO_2 catalysts could be modulated through modifying the CeO_2 support morphology. This investigation provides a deep understanding of the reaction MA hydrogenation by means of Ni/CeO_2 catalysts.

Author Contributions: Conceptualization, H.W., Y.Z. (Yin Zhang), and Y.Z. (Yongxiang Zhao); methodology, S.L.; formal analysis, Q.Z.; writing—original draft preparation, S.L. and X.L.; writing—review and editing, H.W. and Y.Z. (Yin Zhang); supervision, Y.Z. (Yongxiang Zhao); funding acquisition, Y.Z. (Yongxiang Zhao). All authors have read and agreed to the published version of the manuscript.

Funding: This research was funded by the National Natural Science Foundation (U1710221, 21303097, 22075167); Shanxi Province International Science and Technology Cooperation Program Project (201803D421074).

Data Availability Statement: Not Applicable.

Conflicts of Interest: The authors declare no conflict of interest.

References

1. Papageorgiou, G.Z.; Grigoriadou, I.; Andriotis, E.; Bikiaris, D.N.; Panayiotou, C. Miscibility and Properties of New Poly(propylene succinate)/Poly(4-vinylphenol) Blends. *Ind. Eng. Chem. Res.* **2013**, *52*, 11948–11955. [CrossRef]
2. Bertone, M.E.; Meyer, C.I.; Regenhardt, S.A.; Sebastian, V.; Garetto, T.F.; Marchi, A.J. Highly selective conversion of maleic anhydride to γ-butyrolactone over Ni-supported catalysts prepared by precipitation–deposition method. *Appl. Catal. A Gen.* **2015**, *503*, 135–146. [CrossRef]
3. Meyer, C.I.; Regenhardt, S.A.; Marchi, A.J.; Garetto, T.F. Gas phase hydrogenation of maleic anhydride at low pressure over silica-supported cobalt and nickel catalysts. *Appl. Catal. A* **2012**, *417–418*, 59–65. [CrossRef]
4. Liao, X.; Zhang, Y.; Hill, M.; Xia, X.; Zhao, Y.; Jiang, Z. Highly efficient Ni/CeO_2 catalyst for the liquid phase hydrogenation of maleic anhydride. *Appl. Catal. A* **2014**, *488*, 256–264. [CrossRef]
5. Regenhardt, S.A.; Meyer, C.I.; Garetto, T.F.; Marchi, A.J. Selective gas phase hydrogenation of maleic anhydride over Ni-supported catalysts: Effect of support on the catalytic performance. *Appl. Catal. A Gen.* **2012**, *449*, 81–87. [CrossRef]
6. Jung, S.M.; Godard, E.; Jung, S.Y.; Park, K.C.; Choi, J.U. Liquid-phase hydrogenation of maleic anhydride over $Pd-Sn/SiO_2$. *Catal. Today* **2003**, *87*, 171–177. [CrossRef]
7. Wang, N.; Shen, K.; Huang, L.; Yu, X.; Qian, W.; Chu, W. Facile Route for Synthesizing Ordered Mesoporous Ni-Ce-Al Oxide Materials and Their Catalytic Performance for Methane Dry Reforming to Hydrogen and Syngas. *ACS Catal.* **2013**, *3*, 1638–1651. [CrossRef]
8. Zhang, S.; Muratsugu, S.; Ishiguro, N.; Tada, M. Ceria-Doped Ni/SBA-16 Catalysts for Dry Reforming of Methane. *ACS Catal.* **2013**, *3*, 1855–1864. [CrossRef]
9. Guo, H.; He, Y.; Wang, Y.; Liu, L.; Yang, X.; Wang, S.; Huang, Z.; Wei, Q. Morphology-controlled synthesis of cage-bell $Pd@CeO_2$ structured nanoparticle aggregates as catalysts for the low-temperature oxidation of CO. *J. Mater. Chem. A* **2013**, *1*, 7494–7499. [CrossRef]
10. Zhu, Y.; Zhang, S.; Shan, J.; Nguyen, L.; Zhan, S.; Gu, X.; Tao, F. In Situ Surface Chemistries and Catalytic Performances of Ceria Doped with Palladium, Platinum, and Rhodium in Methane Partial Oxidation for the Production of Syngas. *ACS Catal.* **2013**, *3*, 2627–2639. [CrossRef]
11. Senanayake, S.D.; Stacchiola, D.; Evans, J.; Estrella, M.; Barrio, L.; Pérez, M.; Hrbek, J.; Rodriguez, J.A. Probing the reaction intermediates for the water-gas shift over inverse $CeO_x/Au(111)$ catalysts. *J. Catal.* **2010**, *271*, 392–400. [CrossRef]

12. Nga, N.L.T.; Potvin, C.; Djéga-Mariadassou, G.; Delannoy, L.; Louis, C. Catalytic reduction of nitrogen monoxide by propene in the presence of excess oxygen over gold based ceria catalyst. *Top. Catal.* **2007**, *42–43*, 91–94. [CrossRef]
13. Wang, N.; Qian, W.; Chu, W.; Wei, F. Crystal-plane effect of nanoscale CeO_2 on the catalytic performance of Ni/CeO_2 catalysts for methane dry reforming. *Catal. Sci. Technol.* **2016**, *6*, 3594–3605. [CrossRef]
14. Liao, X.; Zhang, Y.; Guo, J.; Zhao, L.; Hill, M.; Jiang, Z.; Zhao, Y. The Catalytic Hydrogenation of Maleic Anhydride on CeO_2-delta-Supported Transition Metal Catalysts. *Catalysts* **2017**, *7*, 272. [CrossRef]
15. Zhou, K.; Wang, X.; Sun, X.; Peng, Q.; Li, Y. Enhanced catalytic activity of ceria nanorods from well-defined reactive crystal planes. *J. Catal.* **2005**, *229*, 206–212. [CrossRef]
16. Wu, Z.; Li, M.; Howe, J.; Meyer, H.M.; Overbury, S.H. Probing defect sites on CeO_2 nanocrystals with well-defined surface planes by raman spectroscopy and O_2 adsorption. *Langmuir* **2010**, *26*, 16595–16606. [CrossRef]
17. Mai, H.-X.; Sun, L.-D.; Zhang, Y.-W.; Si, R.; Feng, W.; Zhang, H.-P.; Liu, H.-C.; Yan, C.-H. Shape-selective synthesis and oxygen storage behavior of Ceria Nanopolyhedra, Nanorods, and Nanocubes. *J. Phys. Chem. B* **2005**, *109*, 24380–24385. [CrossRef]
18. Liu, X.; Zhou, K.; Wang, L.; Wang, B.; Li, Y. Oxygen vacancy clusters promoting reducibility and activity of ceria nanorods. *J. Am. Chem. Soc.* **2009**, *131*, 3140–3141. [CrossRef]
19. Agarwal, S.; Lefferts, L.; Mojet, B.L. Ceria nanocatalysts: Shape dependent reactivity and formation of OH. *ChemCatChem* **2013**, *5*, 479–489. [CrossRef]
20. Wu, Z.; Li, M.; Overbury, S.H. On the structure dependence of CO oxidation over CeO_2 nanocrystals with well-defined surface planes. *J. Catal.* **2012**, *285*, 61–73. [CrossRef]
21. Agarwal, S.; Lefferts, L.; Mojet, B.L.; Ligthart, D.A.J.M.; Hensen, E.J.M.; Mitchell, D.R.G.; Erasmus, W.J.; Anderson, B.G.; Olivier, E.J.; Neethling, J.H.; et al. Exposed Surfaces on shape-controlled ceria nanoparticles revealed through AC-TEM and water-gas shift reactivity. *ChemSusChem* **2013**, *6*, 1898–1906. [CrossRef] [PubMed]
22. Han, W.-Q.; Wen, W.; Hanson, J.C.; Teng, X.; Marinkovic, N.; Rodriguez, J.A. Onedimensional ceria as catalyst for the low-temperature water-gas shift reaction. *J. Phys. Chem. C* **2009**, *113*, 21949–21955. [CrossRef]
23. Cao, T.; You, R.; Li, Z.; Zhang, X.; Li, D.; Chen, S.; Zhang, Z.; Huang, W. Morphology-dependent CeO_2 catalysis in acetylene semihydrogenation reaction. *Appl. Surf. Sci.* **2020**, *501*, 144120. [CrossRef]
24. Si, R.; Flytzani-Stephanopoulos, M. Shape and crystal-plane effects of nanoscale ceria on the activity of $Au-CeO_2$ catalysts for the water-gas shift reaction. *Angew. Chem. Int. Ed.* **2008**, *47*, 2884–2887. [CrossRef] [PubMed]
25. Liu, L.; Yao, Z.; Deng, Y.; Gao, F.; Liu, B.; Dong, L. Morphology and crystal-plane effects of nanoscale ceria on the activity of CuO/CeO_2 for NO reduction by CO. *ChemCatChem* **2011**, *3*, 978–989. [CrossRef]
26. Chang, S.; Li, M.; Hua, Q.; Zhang, L.; Ma, Y.; Ye, B.; Huang, W. Shape-dependent interplay between oxygen vacancies and $Ag-CeO_2$ interaction in Ag/CeO_2 catalysts and their influence on the catalytic activity. *J. Catal.* **2012**, *293*, 195–204. [CrossRef]
27. Wu, Z.; Schwartz, V.; Li, M.; Rondinone, A.J.; Overbury, S.H. Support shape effect in metal oxide catalysis: Ceria-nanoshape-supported Vanadia catalysts for oxidative dehydrogenation of isobutane. *J. Phys. Chem. Lett.* **2012**, *3*, 1517–1522. [CrossRef]
28. Du, X.; Zhang, D.; Shi, L.; Gao, R.; Zhang, J. Morphology dependence of catalytic properties of Ni/CeO_2 nanostructures for carbon dioxide reforming of methane. *J. Phys. Chem. C* **2012**, *116*, 10009–10016. [CrossRef]
29. Zhang, X.; You, R.; Li, D.; Cao, T.; Huang, W. Reaction sensitivity of ceria morphology effect on Ni/CeO_2 catalysis in propane oxidation reactions. *ACS Appl. Mater. Interfaces* **2017**, *9*, 35897–35907. [CrossRef]
30. Liu, Y.; Luo, L.; Gao, Y.; Huang, W. CeO_2 morphology-dependent $NbOx-CeO_2$ interaction, structure and catalytic performance of $NbOx/CeO_2$ catalysts in oxidative dehydrogenation of propane. *Appl. Catal. B* **2016**, *197*, 214–221. [CrossRef]
31. Wu, Z.; Li, M.; Overbury, S.H. A Raman spectroscopic study of the speciation of Vanadia supported on ceria nanocrystals with defined surface planes. *ChemCatChem* **2012**, *4*, 1653–1661. [CrossRef]
32. Gao, Y.; Wang, W.; Chang, S.; Huang, W. Morphology effect of CeO_2 support in the preparation, metal-support interaction, and catalytic performance of Pt/CeO_2 catalysts. *ChemCatChem* **2013**, *5*, 3610–3620. [CrossRef]
33. You, R.; Zhang, X.; Luo, L.; Pan, Y.; Pan, H.; Yang, J.; Wu, L.; Zheng, X.; Jin, Y.; Huang, W. $NbOx/CeO_2$-rods catalysts for oxidative dehydrogenation of propane: Nb-CeO_2 interaction and reaction mechanism. *J. Catal.* **2017**, *348*, 189–199. [CrossRef]
34. Riley, C.; Zhou, S.; Kunwar, D.; De La Riva, A.; Peterson, E.; Payne, R.; Gao, L.; Lin, S.; Guo, H.; Datye, A. Design of effective catalysts for selective alkyne hydrogenation by doping of ceria with a single-atom promotor. *J. Am. Chem. Soc.* **2018**, *140*, 12964–12973. [CrossRef]
35. Vilé, G.; Colussi, S.; Krumeich, F.; Trovarelli, A.; Pérez-Ramírez, J. Opposite face sensitivity of CeO_2 in hydrogenation and oxidation catalysis. *Angew. Chem. Int. Ed.* **2014**, *53*, 12069–12072. [CrossRef] [PubMed]
36. Zhu, H.-Z.; Lu, Y.-M.; Fan, F.-J.; Yu, S.-H. Selective hydrogenation of nitroaromatics by ceria nanorods. *Nanoscale* **2013**, *5*, 7219–7223. [CrossRef]
37. Matte, L.P.; Kilian, A.S.; Luza, L.; Alves, M.C.; Morais, J.; Baptista, D.L.; Dupont, J.; Bernardi, F. Influence of the CeO_2 Support on the Reduction Properties of Cu/CeO_2 and Ni/CeO_2 Nanoparticles. *J. Phys. Chem. C* **2015**, *119*, 26459–26470. [CrossRef]
38. Rodriguez, J.A.; Wang, X.; Liu, G.; Hanson, J.C.; Hrbek, J.; Peden, C.H.F.; Iglesias-Juez, A.; Fernández-Garc, M. Physical and chemical properties of $Ce_{1-x}Zr_xO_2$ nanoparticles and $Ce_{1-x}Zr_xO_2(111)$ surfaces: Synchrotron-based studies. *J. Mol. Catal. A Chem.* **2005**, *228*, 11–19. [CrossRef]
39. Francisco, M.S.P.; Mastelaro, V.R.; Nascente, P.A.P.; Florentino, A.O. Activity and Characterization by XPS, HR-TEM, Raman Spectroscopy, and BET Surface Area of CuO/CeO_2-TiO_2 Catalysts. *J. Phys. Chem. B* **2001**, *105*, 10515–10522. [CrossRef]

40. Li, Y.; Zhang, B.C.; Tang, X.L.; Xu, Y.D.; Shen, W.J. Hydrogen production from methane decomposition over Ni/CeO$_2$ catalysts. *Catal. Commun.* **2006**, *7*, 380–386. [CrossRef]
41. Shan, W.J.; Luo, M.F.; Ying, P.L.; Shen, W.J.; Li, C. Reduction property and catalytic activity of Ce1-XNiXO$_2$ mixed oxide catalysts for CH$_4$ oxidation. *Appl. Catal. A Gen.* **2003**, *246*, 1–9. [CrossRef]
42. Sa, J.; Kayser, Y.; Milne, C.J.; Abreu Fernandes, D.L.; Szlachetko, J. Temperature-programmed reduction of NiO nanoparticles followed by time-resolved RIXS. *Phys. Chem. Chem. Phys.* **2014**, *16*, 7692–7696. [CrossRef] [PubMed]
43. Burroughs, P.; Hamnett, A.; Orchard, A.F.; Thornton, G. Satellite structure in the X-ray photoelectron spectra of some binary and mixed oxides of lanthanum and cerium. *J. Chem. Soc. Dalton Trans.* **1976**, *17*, 1686–1698. [CrossRef]
44. Nelson, A.E.; Schulz, K.H. Surface chemistry and microstructural analysis of Ce$_x$Zr$_{1-x}$O$_{2-y}$ model catalyst surfaces. *Appl. Surf. Sci.* **2003**, *210*, 206–221. [CrossRef]
45. Santos, V.P.; Carabineiro, S.A.C.; Bakker, J.J.W.; Soares, O.S.G.P.; Chen, X.; Pereira, M.F.R.; Órfão, J.J.M.; Figueiredo, J.L.; Gascon, J.; Kapteijn, F. Stabilized gold on cerium-modified cryptomelane: Highly active in low-temperature CO oxidation. *J. Catal.* **2014**, *309*, 58–65. [CrossRef]
46. Boaro, M.; Giordano, F.; Recchia, S.; Dal Santo, V.; Giona, M.; Trovarelli, A. On the mechanism of fast oxygen storage and release in ceria–zirconia model catalysts. *Appl. Catal. B* **2004**, *52*, 225–237. [CrossRef]
47. Liu, Q.; Gao, J.; Zhang, M.; Li, H.; Gu, F.; Xu, G.; Zhong, Z.; Su, F. Highly active and stable Ni/γ-Al$_2$O$_3$ catalysts selectively deposited with CeO$_2$ for CO methanation. *RSC Adv.* **2014**, *4*, 16094–16103. [CrossRef]
48. Torres, C.C.; Alderete, J.B.; Mella, C.; Pawelec, B. Maleic anhydride hydrogenation to succinic anhydride over mesoporous Ni/TiO$_2$ catalysts: Effects of Ni loading and temperature. *J. Mol. Catal. A Chem.* **2016**, *423*, 441–448. [CrossRef]
49. Zhang, S.; Li, J.; Xia, Z.; Wu, C.; Zhang, Z.; Ma, Y.; Qu, Y. Towards highly active Pd/CeO$_2$ for alkene hydrogenation by tuning Pd dispersion and surface properties of the catalysts. *Nanoscale* **2017**, *9*, 3140–3149. [CrossRef]

Article

Transesterification of Glycerol to Glycerol Carbonate over Mg-Zr Composite Oxide Prepared by Hydrothermal Process

Yihao Li, Hepan Zhao, Wei Xue , Fang Li *, and Zhimiao Wang *

Key Laboratory of Green Chemical Technology and High Efficient Energy Saving of Hebei Province, Tianjin Key Laboratory of Chemical Process Safety, School of Chemical Engineering, Hebei University of Technology, Tianjin 300401, China; 201921503002@stu.hebut.edu.cn (Y.L.); hgdzhao16029@hebut.edu.cn (H.Z.); weixue@hebut.edu.cn (W.X.)
* Correspondence: lifang@hebut.edu.cn (F.L.); wangzhimiao@hebut.edu.cn (Z.W.)

Abstract: A series of Mg-Zr composite oxide catalysts prepared by the hydrothermal process were used for the transesterification of glycerol (GL) with dimethyl carbonate (DMC) to produce glycerol carbonate (GC). The effects of the preparation method (co-precipitation, hydrothermal process) and Mg/Zr ratio on the catalytic performance were systematically investigated, and the deactivation of the catalyst was also explored. The Mg-Zr composite oxide catalysts were characterized by XRD, TEM, TPD, N_2 adsorption-desorption, and XPS. The characterization results showed that compared with the co-precipitation process, the catalyst prepared by the hydrothermal process has a larger specific surface area, smaller grain size, and higher dispersion. Mg1Zr2-HT catalyst calcined at 600 °C in a nitrogen atmosphere exhibited the best catalytic performance. Under the conditions of reaction time of 90 min, reaction temperature of 90 °C, catalyst dosage of 3 wt% of GL, and GL/DMC molar ratio of 1/5, the GL conversion was 99% with 96.1% GC selectivity, and the yield of GC was 74.5% when it was reused for the fourth time.

Keywords: glycerol carbonate; transesterification; hydrothermal process; Mg-Zr oxides

1. Introduction

With the increasing consumption of fossil fuels and the consequent environmental problems, especially the threat of global warming, China has put forward the strategic goal of "carbon peaking and carbon neutralization". The realization of this goal needs to accelerate clean energy substitution and energy transformation. Biodiesel is a promising clean and renewable energy, and has become a hot spot for the sustainable development of global energy and the environment [1]. Biodiesel is obtained by transesterification of vegetable oil and waste oil with methanol or ethanol and will produce by-product glycerol (GL). By 2024, the global biofuel market is expected to reach US $153.8 billion, but for every 1000 kg biodiesel produced, there will be 100 kg GL [2].

In order to solve the problem of crude glycerol utilization, scientists have explored and developed different synthetic procedures to convert GL into high value-added derivatives, such as the steam reforming of glycerol [3–7], catalytic esterification of glycerol, catalytic hydrogenolysis of glycerol and among others. Among various GL derivatives, glycerol carbonate (4-hydroxymethyl-1,3-dioxolane-2-one, GC) has the advantages of low flammability, low toxicity, high boiling point, and biodegradability [8]. It is widely used as a solvent in the cosmetics industry, and can also be used in the manufacture of paint, fiber, plastic, coating, cement curing agent, biological lubricant, and so on [9].

At present, the routes for the synthesis of GC using GL as raw material mainly include phosgenaion [10], oxidative carbonylation [11], urea alcoholysis [12], and transesterification [13]. Among them, transesterification (Scheme 1) has the advantages of mild reaction conditions and simple operation, which is considered to be one of the most direct and feasible ways in the industry [14].

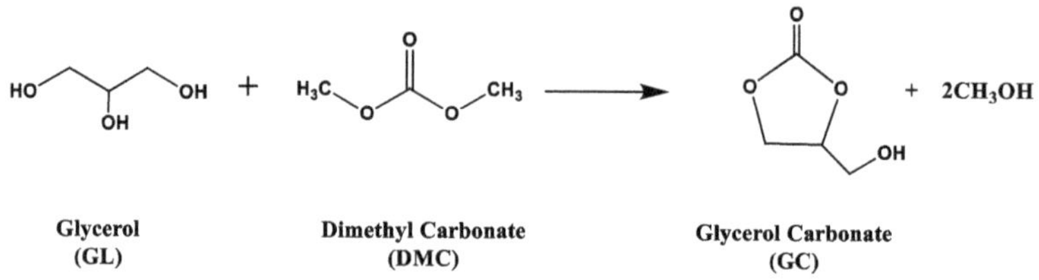

Scheme 1. Synthesis of GC by transesterification from GL and DMC.

In recent years, alkali catalysts such as MgO [15,16] and CaO [17] have been widely used in GC synthesis by transesterification. However, CaO can be dissolved into the reactant GL and form a calcium-glycerin bond [18]. Moreover, CaO may produce $CaCO_3$ with DMC in the presence of water [19], which reduces the catalytic activity, limiting the reuse of catalysts. In addition, single metal oxides such as MgO and CaO will react with water and CO_2 in the air during preparation and storage, and then deactivate [20]. In general, composite oxides have stronger acidity and basicity and larger specific surface areas than single metal oxides; the lattice structure can also be changed by doping metal cations with different electronegativity, thereby changing the acidity and basicity of the catalyst surface [21]. So it shows good application prospects in heterogeneous alkali catalytic reactions [22]. Zhang [23] prepared a large specific surface area $CaO-ZrO_2$ catalyst with the mesoporous structure for continuous transesterification synthesis of GC in a fixed bed reactor. Under the optimized conditions, the yield of GC can reach 90%. However, the catalysts prepared by the co-precipitation process have some disadvantages, such as easy loss of active components and deactivation due to carbon species deposited on the surface [24]. The hydrothermal process has been widely used in the synthesis of oxide nanoparticles in recent years. Compared with other preparation processes, hydrothermally synthesized nanoparticles have high purity, good dispersibility, and controllable grain size [25]. Cui prepared MgO nanosheets with a two-dimensional flaky porous structure by simple hydrothermal process, which has a larger specific surface area than commercial MgO nanoparticles [26]. The ZrO_2 nanocrystals prepared by hydrothermal synthesis of Akune [27] show high catalytic activity due to their high specific surface area and high crystallinity. Wang compared the Mg/Sn/W composite oxide catalysts prepared by co-precipitation process and hydrothermal process, and pointed out that the catalysts prepared by the hydrothermal process had smaller particles, higher thermal stability, and catalytic activity [28].

In this article, Mg-Zr composite oxide catalysts with different Mg/Zr molar ratios were prepared by hydrothermal process for the transesterification of GL and DMC to synthesize GC. The effects of the preparation method and Mg/Zr molar ratio were systematically investigated. The catalysts were characterized by XRD, N_2 adsorption-desorption, TEM, TPD, and XPS, and the structure-activity relationship of the Mg-Zr oxide catalysts was discussed. In addition, the transesterification reaction conditions were optimized, the reusability of the catalyst was investigated, and the deactivation reasons of the catalysts were explored.

2. Materials and Methods

2.1. Materials

Glycerol (99%, AR); methanol (99.5%, AR); magnesium nitrate hexahydrate (99%, AR) and sodium hydroxide were purchased from Kemio Reagent Co, Ltd. (Tianjin, China). Zirconium oxychloride octahydrate (99%, AR); dimethyl carbonate (99%, AR), and n-butanol (99%, AR) were purchased from Damao Chemical Reagent Factory (Tianjin, China). Glycidyl (98%) and glycerol carbonate (90%) were purchased from Aladdin Biochemical Technology Co, Ltd. (Shanghai, China).

2.2. Catalyst Preparation

A series of Mg-Zr composite oxide catalysts with different Mg/Zr ratio were prepared by hydrothermal process. Briefly, the typical preparation route could be described as follows: Mg $(NO_3)_2 \cdot 6H_2O$ (0.64 g, 2.5 mmol) and $ZrOCl_2 \cdot 8H_2O$ (1.61 g, 5 mmol) were dissolved in the deionized water at room temperature, and the solution was added drop by drop to the 500 mL flask with the suitable amount of 2 mol/L NaOH solution at the same time by the co-current-precipitation process under vigorous stirring. Until the pH of the solution reached 11, and then stirred continuously for 30 min. Subsequently, the suspension was hydrothermally treated in a Teflon-lined stainless-steel autoclave at 150 °C for 6 h, and then calcined at 600 °C in flowing nitrogen atmosphere for 3 h. Depending on the molar ratio of Mg/Zr used in the preparation step, the catalysts were marked as Mg1Zr3, Mg1Zr2, Mg1Zr1, Mg2Zr1 and Mg3Zr1. Single metal oxide catalysts ZrO_2 and MgO were synthesized in the same way as above composite oxide catalysts. The difference between co-precipitation process and hydrothermal process is that the mixture after stirring is allowed to stand at room temperature for 6 h without high-temperature treatment, and other steps remain unchanged. The catalyst samples prepared by hydrothermal method and coprecipitation method were named Mg_xZr_y-HT and Mg_xZr_y-CP respectively, where x/y was n(Mg)/n(Zr) molar ratio.

2.3. Characterization

The crystal phases of all samples were identified by powder X-ray diffractometer using D8 FOCUS (German Brook AXS Company, Karlsruhe, Germany) with Cu Kα radiation (40 kV) and a secondary beam graphite monochromator (SS/DS = 1°, RS 0.15 mm, counter SC). Talos F200s field emission transmission electron microscope (FEI company, Hillsboro, OR, USA) was used to observe the morphology and grain size of the catalysts. The strength and distribution of the basic/acid sites of the catalyst were determined by temperature programmed desorption of preadsorbed CO_2 or NH_3, which was performed using Auto Chem 2920 instrument. (Micromeritics, Norcross, GA, USA). The texture properties including the specific surface area, pore volume, and pore size of the catalysts were derived from N_2 adsorption-desorption technique using 3H-2000PS2 (Beishied, Beijing, China) at -196 °C. The catalysts were pretreated by outgassing in vacuum at 200 °C for 3 h before measurement. X-ray photoelectron spectroscopy (XPS) data were collected on a Thermo Scientific K-Alpha electron spectrometer (Thermo Fisher, Waltham, MA, USA) equipped with Al Kα radiation (hv = 1486.6 eV).

2.4. Catalytic Activity Test

Transesterification of GL to GC was carried out in a round bottom flask with reflux condenser at atmospheric pressure. A total of 3.3 g of GL and 16.3 g of DMC were added into a 100 mL round-bottomed flask, the reaction mixture was heated to 90 °C while stirring in oil bath, then the catalyst of 3 wt% of GL was added to the reaction mixture. After the desired time, the products were separated by centrifugation and analyzed by gas chromatography using an Agilent 7890B gas chromatograph equipped with a DB-wax capillary column (30 m × 0.32 mm × 0.25 μm) and a hydrogen flame detector. The injector and detector temperatures were 250 °C and 300 °C, respectively. The yield of GC was calculated using internal standard method, in which N-butanol was the internal standard. The GL conversion, GC selectivity and yield were calculated by the following equations:

$$\text{GL conversion}(\%) = \frac{\text{mole of GL, feed} - \text{mole of GL, final}}{\text{mole of GL, feed}} \times 100 \quad (1)$$

$$\text{GC selectivity}(\%) = \frac{\text{mole of GC, produced}}{\text{mole of GL, feed} - \text{mole of GL, final}} \times 100 \quad (2)$$

$$\text{GC yield}(\%) = \frac{\text{GL conversion} \times \text{GC selectivity}}{100} \quad (3)$$

3. Results and Discussion

3.1. Effect of Preparation Method

The XRD patterns of Mg1Zr2-HT catalyst prepared by hydrothermal process and Mg1Zr2-CP catalyst prepared by co-precipitation are shown in Figure 1. It can be seen from the figure that the diffraction peaks at 30.2°, 34.9°, 50.7° and 60.2° belong to tetragonal ZrO_2 (t-ZrO_2, JCPDS No. 50-1089), and there is no monoclinic ZrO_2. t-ZrO_2 has a unique bridging hydroxyl group and strong surface basicity, which is conducive to transesterification reaction [29]. Compared with ZrO_2, the diffraction peak intensity of MgO is relatively weak, which is not due to the low content of Mg, but due to the low atomic scattering factor (atomic number) of Mg [30]. In addition, the grain sizes of Mg1Zr2-CP and Mg1Zr2-HT calculated by Scherrer formula are 13.4 nm and 13.1 nm respectively, and there is little difference between them.

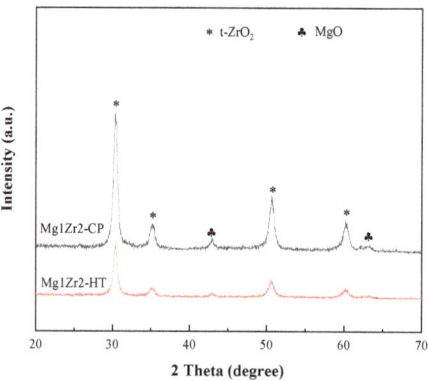

Figure 1. XRD patterns of Mg1Zr2-HT and Mg1Zr2-CP.

The textural properties and surface basicity of Mg1Zr2-HT and Mg1Zr2-CP are summarized in Table 1. It can be seen that the Mg1Zr2-HT has a larger specific surface area and pore volume than Mg1Zr2-CP. This is because the intense collision between colloidal particles promotes the secondary pore formation of composite oxides under hydrothermal conditions, whereas the condensation between colloidal particles is a very slow process at room temperature. Therefore, the hydrothermal process is conducive to the formation of a more developed pore network structure, thereby improving the specific surface area and pore volume of Mg1Zr2-HT [31]. In addition, the dissolution deposition/crystallization process also occurs in the hydrothermal process [32]. Due to the dissolution of some precursors under hydrothermal conditions, the local solubility at the junction (neck) of the two colloidal particles will be lower than that at the nearby surface. Therefore, the deposition process will occur preferentially in the neck, resulting in the reinforcement of the colloidal network structure. During the subsequent calcination, the specific surface area and pore volume of the xerogel prepared by co-precipitation decrease rapidly due to the collapse of the gel skeleton and the sintering and growth of the catalyst particles [33].

Table 1. Texture properties and surface basicity of Mg1Zr2-HT and Mg1Zr2-CP.

Catalyst	ZrO_2 Crystallite Size [a] (nm)	MgO Crystallite Size [b] (nm)	S_{BET} [c] (m^2/g)	D_P [d] (nm)	V_P [e] (cm^3/g)	Basicity [f] (µmol/g)		
						W	M + S	Total
Mg1Zr2-HT	13.1	13.2	68.8	24.0	0.41	57.1	88.2	145.3
Mg1Zr2-CP	13.4	14.9	42.8	26.9	0.28	53.2	82.6	135.8

[a] Calculated by Scherrer formula using the full width at half maximum of ZrO_2 (011) plane. [b] Calculated by Scherrer formula using the full width at half maximum of MgO (200) plane. [c] S_{BET} was measured by the multi-point BET method. [d] Average pore size was calculated from the desorption branch of the isotherm using the BJH method. [e] Total pore volume was measured at P/P_0 = 0.99. [f] Calculated by the results of CO_2-TPD.

The catalytic performance of the above two catalysts for GL transesterification was investigated, and the results are shown in Table 2. As can be seen from the data in Table 2, Mg1Zr2-HT and Mg1Zr2-CP both have good catalytic performance for GL transesterification, with GL conversion greater than 90% and GC selectivity of about 95%. Because Mg1Zr2-HT catalyst has a larger specific surface area, reactant molecules are more easily in contact with active sites, therefore have higher catalytic activity.

Table 2. Catalytic performance of Mg1Zr2-HT and Mg1Zr2-CP for transesterification of GL with DMC.

Catalyst	GL Conversion (%)	GC Selectivity (%)	GC Yield (%)
Mg1Zr2-HT	96.0	95.1	91.3
Mg1Zr2-CP	91.0	94.0	85.5

Reaction condition: GL/DMC molar ratio = 1/5, catalyst loading = 3 wt% of GL, 90 °C, 60 min.

3.2. Effect of Mg-Zr Molar Ratio

The XRD patterns of catalysts with different Mg/Zr molar ratios are shown in Figure 2. It can be seen that with the increase of the Mg/Zr ratio, the diffraction peak of t-ZrO_2 at 2θ of 30° gradually shifts to a high angle, which may be due to the doping of Mg^{2+} into the lattice of ZrO_2, and some Zr^{4+} ions are replaced by Mg^{2+}, resulting in the distortion of the crystal structure. Because the ion radius of Mg^{2+} is smaller than that of Zr^{4+} (the ion radius of Mg^{2+} and Zr^{4+} is 0.780 Å and 0.840 Å, respectively), the lattice shrinks, and the cell parameters decrease, so the corresponding 2θ shifts to high angle [34]. At a low Mg/Zr molar ratio, no diffraction peak of MgO is observed, indicating the formation of a solid solution. With the increase of Mg content, the characteristic diffraction peaks of periclase MgO (JCPDS No. 45-0946) were detected at 2θ of 43.2° (200) and 62.5° (220), and the intensity and sharpness gradually increased with the increase of Mg content, indicating that the particle size of MgO increased significantly. The lattice parameters and crystal plane spacing of Mg-Zr catalysts were analyzed by Jade, and the results are listed in Table 3. It was found that the lattice constant "a" and crystal plane spacing of (011) crystal plane of Mg-Zr catalyst decreased with the increase of Mg content, indicating that a stable and uniform Mg-Zr composite oxide structure was generated after the introduction of Mg^{2+} into t-ZrO_2.

Figure S1a displays the Mg 1s spectra of Mg-Zr composite oxides catalysts, and the XPS spectrum of a single MgO is also presented for comparison. All the catalysts exhibited a broad and intense band centered at 1360 eV related to the emission from Mg 1s of Mg^{2+} in the oxide state. More importantly, the binding energies of Mg 1s in all the mixed oxides were lower than that of pure MgO, because the Mg-Zr oxides possessed a solid solution structure. The typical Zr 3d spectra are presented in Figure S1b. For pristine ZrO_2, there appeared two peaks at 184.8 and 182.4 eV with a high intensity, which were associated with Zr $3d_{3/2}$ and Zr $3d_{5/2}$ energy states of Zr (IV) oxide species, respectively. The intensity of these two reflections gradually decreased with an increase in Mg content. Meanwhile, it is worth noting that adding Mg into ZrO_2 support could give rise to a continuous increase of Zr 3d binding energy. These observations also support that Mg^{2+} had entered into the t-ZrO_2 lattice, creating a solid solution. Dis-cussing the peak fitted O 1s spectra from Figure S1c. The peak at 531 eV, 533 eV, and 534 eV can be attributed to the presence of lattice oxygen species (O_L), oxygen vacancies (O_V), and chemisorbed oxygen species (O_C). In general, the chemical valence of Zr ion is 4, but the Mg ion has only 2 valence, thus some vacancies are generated when substitution in order to keep charge neutrality in the ionic crystal, and these vacancies are favorable for heterogeneous catalysis [35]. It is worth noting that the Mg1Zr2 catalyst has the highest concentration of oxygen vacancies.

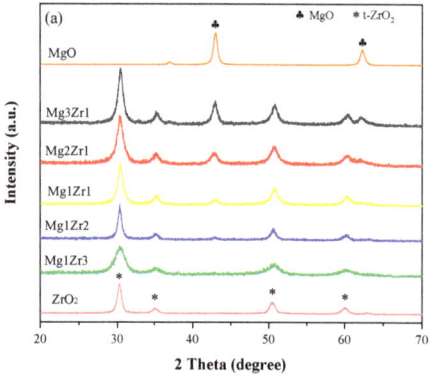
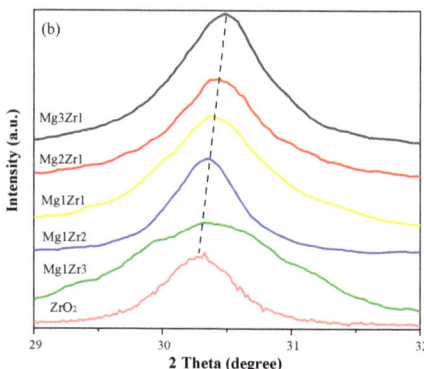

Figure 2. (a) XRD patterns of Mg-Zr composite oxides with different Mg/Zr ratios; (b) same as (a) but in region 2θ = 29–32.

Table 3. Texture properties and surface basicity of various Mg-Zr composite oxides.

Catalyst	ZrO₂ (011)		Lattice Parameter (nm)		Particle Size [a] (nm)	S_{BET} [b] (m²/g)	D_p [c] (nm)	V_p [d] (cm³/g)	Basicity [e] (μmol/g)		
	2θ (°)	d (nm)	a	c					W	M + S	Total
ZrO₂	30.25	0.2952	0.3622	0.5112	12.3	46.4	22.9	0.27	72.9	20.4	93.3
Mg1Zr3	30.30	0.2947	0.3601	0.5097	9.6	86.8	19.5	0.42	65.3	64.1	129.4
Mg1Zr2	30.33	0.2945	0.3596	0.5116	13.1	68.8	24.0	0.41	57.1	88.2	145.3
Mg1Zr1	30.36	0.2942	0.3594	0.5119	8.0	78.5	27.0	0.46	38.6	86.1	124.7
Mg2Zr1	30.38	0.2940	0.3593	0.5102	7.6	87.0	22.5	0.49	31.2	89.3	118.5
Mg3Zr1	30.45	0.2933	0.3586	0.5106	9.7	112.0	18.3	0.52	19.8	92.6	112.4
MgO	-	-	-	-	-	114.0	27.5	0.78	10.5	96.1	106.6

[a] Calculated by Scherrer formula using the full width at half maximum of ZrO₂ (011) plane. [b] S_{BET} was measured by the multi-point BET method. [c] Average pore size was calculated from the desorption branch of the isotherm using the BJH method. [d] Total pore volume was measured at $P/P_0 = 0.99$. [e] Calculated by the results of CO₂-TPD.

In order to observe the microstructure and morphology of the catalyst, the Mg-Zr composite oxides with different Mg/Zr ratios were characterized by TEM, and the results are shown in Figure 3. It can be seen from the TEM image of ZrO₂ that its particle size is relatively uniform, with an average particle size of 23 nm (based on the statistics of 91 particles in the TEM image), but its dispersion is poor, and a large number of particles agglomerate together. After adding a small amount of Mg, the uniformity of particle size of Mg1Zr3 becomes worse, indicating that the addition of Mg affects the crystallization and growth process of ZrO₂. In addition, compared with ZrO₂, there are some substances between the Mg1Zr3 particles, which may be extremely small MgO particles according to the preparation process and XRD results. With the increase of Mg content, the uniformity of ZrO₂ particle size becomes worse, and the particles with a size of about 50 nm appear in Mg1Zr2. When Mg content exceeds Zr, ZrO₂ particles gradually become smaller. Especially in Mg3Zr1, flake particle aggregates appear, and the large particle ZrO₂ disappears completely. Sádaba et al. [30] prepared the Mg-Zr catalyst by co-precipitation method. They pointed out that in the preparation process, Zr^{4+} preferentially precipitated to form $Zr(OH)_4$ or $ZrO_2(H_2O)_X$. When most of Zr^{4+} was precipitated, Mg^{2+} formed $Mg(OH)_2$ precipitation at pH 8~10. Therefore, Mg-Zr catalyst has an embedded structure with ZrO₂ as core and MgO as a shell. According to the conclusion of Sádaba et al. [30] and TEM results, it can be inferred that MgO was formed in the outer layer of ZrO₂ in the Mg-Zr catalysts prepared in this paper, which can be regarded as MgO wrapping ZrO₂. Guan et al. [36] also believed that Mg^{2+} could enter the ZrO₂ lattice to form Mg-Zr solid solution. When Mg content is large, MgO which cannot enter the lattice of ZrO₂ can appear as an independent crystal phase and attach to the surface of magnesium-zirconium solid solution. The EDX spectrum and Elemental composition (Figure S2) show the presence of Mg and Zr. Even though several random areas were selected for the EDX test, the detected Mg/Zr molar ratio was almost the same as the theoretical value. The existence of all the elements in the oxide forms can be confirmed due to the presence of the high amount of oxygen and also the presence of Mg in Mg-Zr composite oxides enhances the basicity and stability of the catalyst.

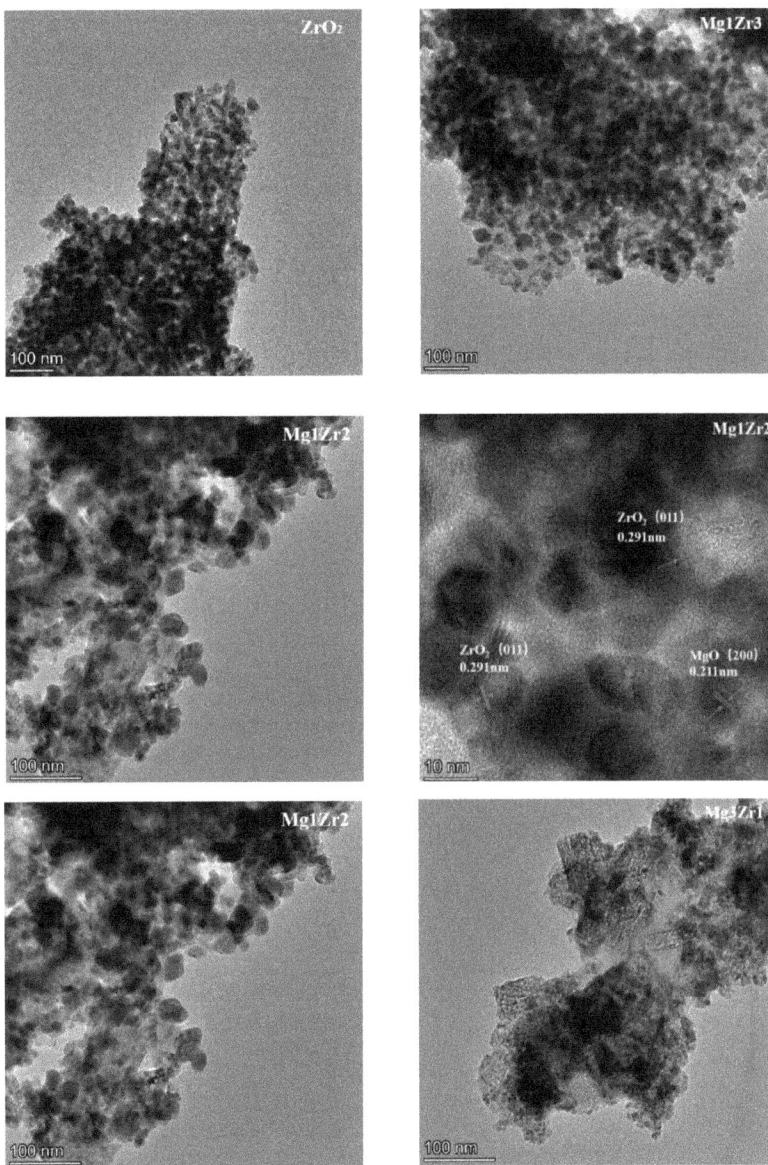

Figure 3. TEM images of Mg-Zr composite oxides with different Mg/Zr ratios.

The N_2 adsorption-desorption isotherms of Mg-Zr composite oxide catalysts are shown in Figure 4. There are obvious type IV adsorption equilibrium isotherms in the range of P/P_0 = 0.5~1.0, indicating that the catalysts had mesoporous structures. ZrO_2, Mg1Zr3, Mg1Zr2, and Mg1Zr1 catalysts all have H2 type hysteresis loops, indicating that the catalyst internal pore structure is ink bottle; The N_2 adsorption-desorption isotherms of Mg2Zr1, Mg3Zr1, and MgO catalysts have no obvious saturated adsorption platform, accompanied by H3 hysteresis loop, indicating that the pore structure of the catalyst is very irregular, combined with the TEM results, it can be seen that there is the slit hole formed by the accumulation of flake particles.

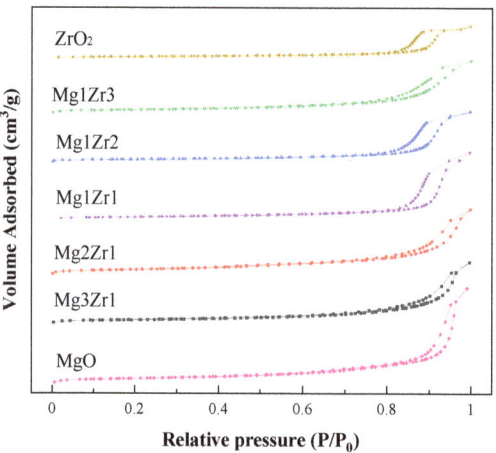

Figure 4. N_2 adsorption-desorption isotherms of Mg-Zr oxides with different Mg/Zr ratios.

The specific surface area, pore diameter, and pore volume of the catalysts are listed in Table 3. It can be seen that the specific surface area of Mg1Zr2 is 68.8 m^2/g, and then the specific surface area increases with the addition of Mg, which is consistent with the experimental results of Guan [36]. As the Mg content increased, the specific surface area of Mg-Zr oxide catalysts has an upward trend, which may be due to the multi-layer dispersion of MgO attached to the surface of magnesium-zirconium solid solution, resulting in the increase of specific surface area. MgO has the largest specific surface area, but the catalytic performance is not the best, indicating that although the structure of the catalyst has a certain impact on the catalytic performance, it is not a completely decisive factor.

To better understand the intrinsic acid-base functionalities and correlate the catalysts with their catalytic behavior, CO_2-TPD and NH_3-TPD measurements were performed to quantitatively determine the distribution of surface acidity and basicity and the number of acidic and basic sites of MgO-ZrO_2 catalysts. The CO_2-TPD characterization of Mg-Zr oxide catalysts was carried out, and the influence of the Mg/Zr molar ratio on the basicity of the catalyst was investigated. The results are shown in Figure 5 and Table 3. It can be seen from Figure 5 that the Mg/Zr molar ratio has a significant effect on the basicity of Mg-Zr oxide catalysts. When the Mg content is 0 (ZrO_2), there are mainly weak basic sites on the surface of the catalyst with a CO_2 desorption temperature lower than 200 °C. With the addition of Mg, the number of medium strong basic sites (CO_2 desorption temperature is in the range of 200–600 °C) on the catalyst surface gradually increases, while the number of weak basic sites decreases. Among them, Mg1Zr2 has the largest number of total basic sites, because it has more weak sites and medium and strong sites at the same time. With the further increase of Mg content, the number of weak basic sites decreased rapidly. The surface of the Mg3Zr1 catalyst is mainly composed of medium and strong basic sites, while the weak sites almost disappear. Its CO_2-TPD curve is similar to that of MgO. The results showed that the weak basic sites on the surface of Mg-Zr oxide catalysts were mainly provided by ZrO_2, while the medium and strong sites were mainly related to MgO. Zhang et al. [34] believed that the weak basic sites of the MgO-ZrO$_2$ catalyst were related to its surface hydroxyl group, while the medium and strong basic sites were related to metal-oxygen pairs (Mg-O and Zr-O) and low coordination oxygen atoms (O^{2-}). In addition, according to the data in Table 3, the total number of surface basic sites of ZrO$_2$ and MgO is similar, but the number of Mg-Zr oxide catalysts increases significantly. In particular, the number of total basic sites of Mg1Zr2 catalyst reaches 145.3 μmol/g, which is 55.7% and 36.3% higher than that of the two single metal oxides, respectively. It is considered that Mg^{2+} and Zr^{4+} are fully mixed during the hydrothermal preparation of the catalyst, and part of Zr^{4+} in the lattice of ZrO$_2$ is replaced by Mg^{2+} after calcination. Due to that Zr^{4+} is more positive than Mg^{2+}, the electron density of O^{2-} in Mg-Zr oxide catalysts increases, thus increasing the number of medium and strong basic sites of the catalyst [28].

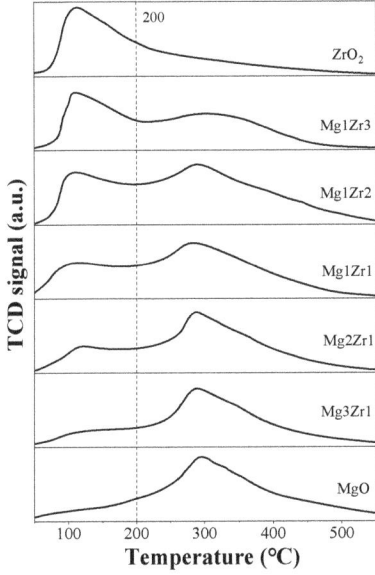

Figure 5. CO$_2$-TPD of Mg-Zr composite oxides with different Mg/Zr ratios.

As shown in Figure S3, in the NH$_3$-TPD curve of bare ZrO$_2$, there are two NH$_3$ desorption peaks at 130 °C and 530 °C, corresponding to weak acidic sites and strong acidic sites respectively. With the increase of MgO content, the medium-strength acid sites of the catalyst increased, while the weak and strong acid sites decreased. The results showed that there was no strong correlation between catalyst acidity and glycerol conversion. Although the role of acid sites in the activation of DMC cannot be completely ruled out, the effect of acid sites is less clear and predictable compared to the evident effect of basic sites [37].

The effect of the Mg/Zr ratio on the transesterification of GL and DMC to GC over Mg-Zr composite oxide was studied, and the results are shown in Figure 6. As can be seen from the figure, ZrO$_2$ and MgO alone are active for the transesterification of GL, and GL conversion is 67.2% and 73.8%, respectively. The activity of the Mg-Zr oxide catalysts was higher than that of the two single metal oxides, indicating the interaction between ZrO$_2$ and MgO and improving the performance of the catalyst. When Mg1Zr3 was used, the GL conversion was 84.0%. With the increase of Mg content, the catalyst activity increased first and then decreased. Among them, Mg1Zr2 has the highest activity for transesterification of GL, with GL conversion of 96.0% and GC selectivity of 95.3%. The Mg/Zr ratio had little effect on the selectivity of GC. The byproduct was glycidyl, and no other products were detected. According to the characterization results, Mg1Zr2 has the highest number of total basic sites. Moreover, the order of GL conversion is basically the same as that of the number of basic sites on the catalyst surface. This indicates that the influence of the Mg/Zr ratio on catalyst performance lies in the change in the number of catalyst basic sites. In this transesterification reaction, the main function of the solid catalyst is to support the abstraction of H^+ from glycerol by the basic sites so as to form glycerol anion. The higher the basicity of the catalyst, the more negative the charge of the glyceroxide anion ($C_3H_7O_3^-$), and consequently, the lower the free energy of the reaction [38]. In other words, the deprotonation of glycerol (on basic sites) is likely more important than the activation of dimethyl carbonate (on acidic sites) for the transesterification of glycerol and dimethyl carbonate [39].

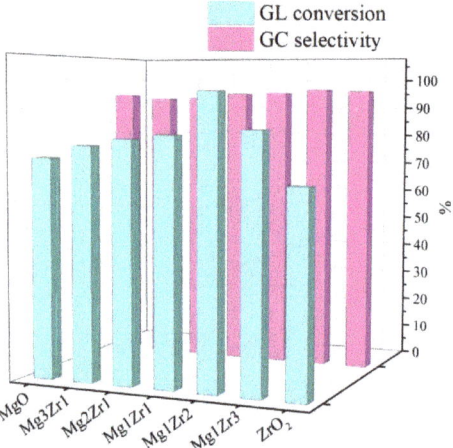

Figure 6. Catalytic performance of Mg-Zr oxides for transesterification of GL with DMC to GC (Reaction conditions: GL/DMC molar ratio = 1/5, catalyst loading = 3 wt% of GL, 90 °C, 60 min).

3.3. Effect of Reaction Conditions on Transesterification of GL over Mg1Zr2-HT

Using Mg1Zr2-HT as a catalyst, the effects of reaction time, reaction temperature, catalyst amount, and GL/DMC molar ratio on the transesterification of GL with DMC to GC were investigated.

3.3.1. Effect of Reaction Time

As shown in Figure 7a, the effect of reaction time on the transesterification of GL with DMC was investigated. It can be seen that GL conversion increased gradually with the increase in reaction time. When the reaction time was 90 min, the GL conversion was 99.0% and GC selectivity was 96.1%; With the continuous extension of reaction time, GL conversion remained unchanged and GC selectivity decreased. This was caused by the decomposition of GC into glycidyl.

3.3.2. Effect of Reaction Temperature

It can be seen from Figure 7b that increasing temperature before 90 °C is conducive to promoting the reaction. This is because the reaction equilibrium constant of this reaction increases with the increase of temperature, so heating is conducive to the reaction. At 90 °C, GL conversion was 99.0% with GC selectivity of 96.1%. When the temperature continues to rise, the decomposition of GC into glycidyl occurs more readily [40], so GC selectivity decreases.

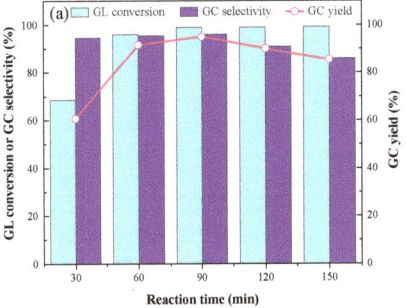
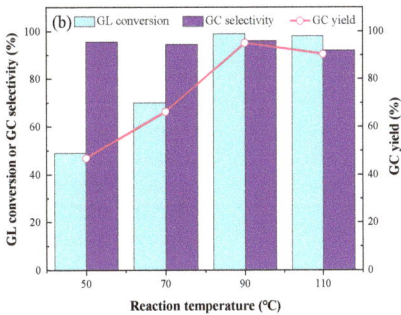
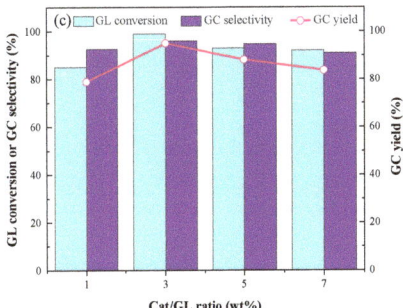
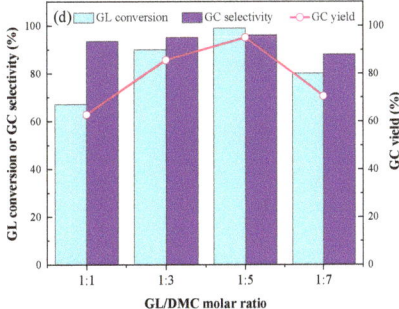

Figure 7. Effect of reaction conditions on the GL transesterification over Mg1Zr2-HT catalyst: (**a**) effect of reaction time (GL/DMC molar ratio = 1/5, catalyst amount = 3 wt% of GL, 90 °C), (**b**) effect of reaction temperature(GL/DMC molar ratio = 1/5, catalyst amount = 3 wt% of GL, 90 min), (**c**) effect of catalyst amount(GL/DMC molar ratio = 1/5, 90 °C, 90 min), (**d**) effect of GL/DMC molar ratio(catalyst amount = 3 wt% of GL, 90 °C, 90 min).

3.3.3. Effect of Catalyst Amount

The transesterification reaction of glycerol was highly influenced by the catalyst amount (wt% based on GL) and presented in Figure 7c, the increase of the amount of catalyst from 1 wt% to 3 wt%, the GL conversion and GC yield gradually increased, which was attributed to the increase in the basic sites of the transesterification catalyst. However, the amount of catalyst increased from 3 wt% to 7 wt%, and the GC yield decreased slowly, which may be due to the agglomeration of catalyst at a higher amount, which makes the reactants unable to enter the active center of the catalyst. The higher the amount of catalyst is, the greater the mass transfer resistance is, which may hinder the transesterification of GL with DMC [41].

3.3.4. Effect of the Molar Ratio of GL/DMC

The molar ratio of GL/DMC has a great influence on the GL conversion and GC yield during the transesterification. Since the transesterification reaction is essentially reversible, excessive DMC is needed to shift the reaction equilibrium to GC. From Figure 7d, it is clear that with the increase of the molar ratio of DMC/GL, the conversion of GL showed an upward trend, and when the molar ratio was 1/5 (GL/DMC), the maximum conversion was 99.0% and the GC selectivity was 96.1%. If the molar ratio of DMC/GL continues to increase, the conversion of GL and GC yield decreases. This may be due to the excessive DMC diluting the catalyst and limiting the contact between GL and the catalyst, thus reducing the reaction rate [40].

3.4. Catalyst Stability

The reusability of a catalyst is an important index to evaluate the performance of the catalyst. In this study, the reusability of Mg1Zr2-HT and Mg1Zr2-CP catalysts for transesterification of GL with DMC is compared, as shown in Figure 8. After the reaction, the catalyst was centrifuged, washed three times with methanol, dried at 100 °C, and then calcined at 600 °C in air for 3 h. As can be seen from the figure, GC selectivity was little affected by repeated use and was almost constant. However, GL conversion gradually decreased, and there were significant differences between the Mg1Zr2-HT and Mg1Zr2-CP catalysts. When a fresh catalyst was used, the GL conversion over Mg1Zr2-HT and Mg1Zr2-CP catalysts was 99.0% and 95.2%, respectively. Moreover, when repeated for the fourth time, GL conversions were 80.1% and 58.2%, respectively. The stability of Mg1Zr2-HT is much better than that of Mg1Zr2-CP.

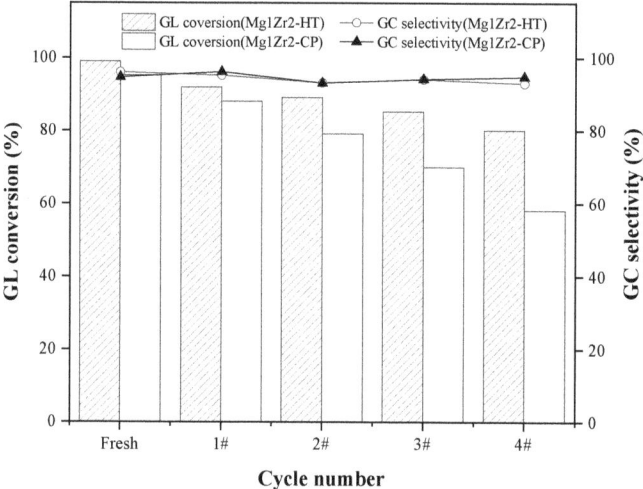

Figure 8. Reusability of Mg1Zr2-HT and Mg1Zr2-CP (Reaction condition: GL/DMC molar ratio = 1/5, catalyst loading = 3 wt%, 90 °C, 90 min).

In order to explore the reasons for the differences between the two catalysts, Mg1Zr2-HT and Mg1Zr2-CP catalysts after four times of reused were characterized by XRD, N_2 adsorption-desorption, CO_2-TPD, TEM, and XPS.

The XRD patterns of Mg1Zr2-HT-used and Mg1Zr2-CP-used catalysts after the fourth cycle are presented in Figure 9. It can be seen that there are obvious characteristic diffraction peaks at 2θ of 30.2°, 34.8°, 50.7°, 60.2°, and 62.9°, corresponding to (011), (110), (020), (121) and (202) crystal planes of tetragonal ZrO_2, respectively. The characteristic diffraction peaks appear at 2θ of 43.2° and 62.5°, corresponding to the (200) and (220) crystal planes of MgO, respectively. Based on the diffraction peaks of ZrO_2 (011) and MgO (200) crystal planes, the grain sizes of ZrO_2 and MgO in Mg1Zr2-HT-used are 15.8 nm and 15.0 nm by the Scherrer formula, respectively; The grain sizes of ZrO_2 and MgO in Mg1Zr2-CP-used are 16.6 nm and 31.4 nm, respectively. Compared with the fresh catalyst, the particle sizes of the two catalysts after repeated use both increased, but the grain size of MgO in Mg1Zr2-CP-used increased by about double, while the grain size of MgO in Mg1Zr2-HT-used increased by only 14%. Under hydrothermal conditions, ions in the solution automatically aggregate to form the most stable chemical structure that cannot be decomposed in the system during the temperature change, so they have good grain stability.

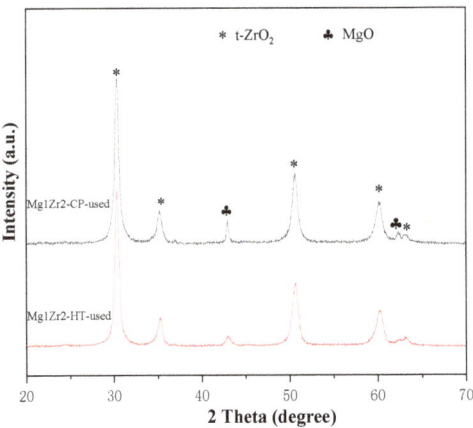

Figure 9. XRD patterns of Mg1Zr2-HT-used and Mg1Zr2-CP-used.

It can be seen from Figure 10 that after transesterification, the particle sizes of both two catalysts increased significantly and the particle sizes became uneven, indicating that the catalyst particles appeared aggregate and sintering, resulting in the gradual increase of grain size and the decrease of dispersion. The Mg1Zr2-CP-used catalyst had serious agglomeration, while the Mg1Zr2-HT-used catalyst had slight sintering but no obvious agglomeration, indicating that the catalyst prepared by the hydrothermal process had strong sintering resistance. This is because under hydrothermal conditions, the compounds in the solution may renucleate and restructure, so that the particles after hydrothermal treatment have better dispersion and grain stability than those particles only by neutralization precipitation [42].

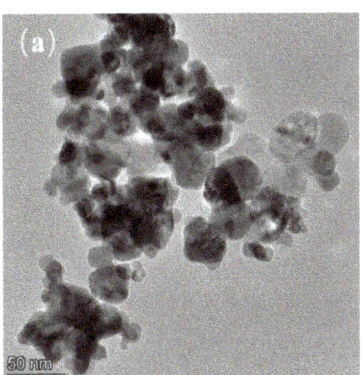 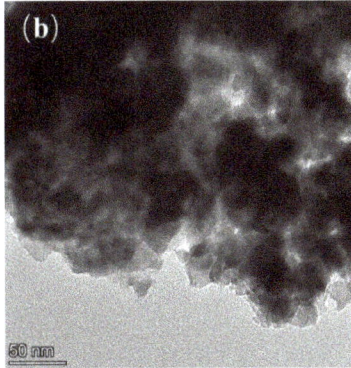

Figure 10. TEM images of Mg1Zr2-HT-used (**a**) and Mg1Zr2-CP-used (**b**).

Compares the deconvoluted Mg 1s, Zr 3d, and O 1s XPS spectra of fresh and used Mg1Zr2-HT. The relative abundances of the Mg 1s, Zr 3d, and O 1s of the samples from Figure S4 showed that the content of the oxygen vacancy decreased from 26.5% to 23.1% and the content of the chemisorbed oxygen species increased from 1.1% to 3.9%, respectively, indicating that irreversible deactivation was caused.

As can be seen from Table 4 that the Mg1Zr2-HT-used catalyst has a larger specific surface area than the Mg1Zr2-CP-used catalyst, and more active sites can be retained. This may be because the colloidal network structure formed in the hydrothermal process is more stable through dissolution-deposition, which alleviates the collapse of the structure and the sintering of particles, so that it still maintains a large specific surface area in the reaction process [43]. Since the catalytic reaction occurs on the surface of active components, the agglomeration and growth of grains lead to the decrease of active surface area, the decrease of active sites, and the reduction in catalytic activity [44]. Compared with fresh catalysts, the number of weak basic sites of Mg1Zr2-HT and Mg1Zr2-CP catalysts decreased by 12% and 13%, respectively, but the number of medium and strong basic sites decreased by 25% and 50%, respectively. Moreover, it was observed that Mg1Zr2-CP-used suffered a greater loss of basic sites than Mg1Zr2-HT-used. This is probably the reason why Mg1Zr2-HT showed better catalytic performance than Mg1Zr2-CP after four cycles. Mg1Zr2-HT possesses a more stable crystal structure to avoid the irreversible reduction of basic sites amount.

Table 4. Texture properties and surface basicity of Mg1Zr2-HT-used and Mg1Zr2-CP-used.

Catalyst	ZrO_2 Crystallite Size [a] (nm)	MgO Crystallite Size [b] (nm)	S_{BET} [c] (m^2/g)	D_P [d] (nm)	V_P [e] (cm^3/g)	Basicity [f] ($\mu mol/g$)		
						W	M + S	Total
Mg1Zr2-HT-used	15.8	15.0	40.1	30.3	0.30	50.3	66.2	116.5
Mg1Zr2-CP-used	16.6	31.4	25.5	29.7	0.19	46.2	42.0	88.2

[a] Calculated by Scherrer formula using the full width at half maximum of ZrO_2 (011) plane. [b] Calculated by Scherrer formula using the full width at half maximum of MgO (200) plane. [c] S_{BET} was measured by the multi-point BET method. [d] Average pore size was calculated from the desorption branch of the isotherm using the BJH method. [e] Total pore volume was measured at $P/P_0 = 0.99$. [f] Calculated by the results of CO_2-TPD.

4. Conclusions

In this work, Mg-Zr composite oxide catalysts with different Mg/Zr molar ratios were prepared by hydrothermal process and their activity and stability towards GC synthesis were studied. The results showed that the catalysts prepared by the hydrothermal process had larger specific surface area, smaller grain size, and higher dispersion than those prepared by the co-precipitation process. The Mg1Zr2-HT catalyst calcined at 600 °C in a nitrogen atmosphere showed the best catalytic performance, with GL conversion of 99% and GC selectivity of 96.1% under mild reaction conditions. This is attributed to the balanced strong and weak basic sites and highly dispersed MgO. Moreover, the GL conversion was demonstrated to increase in parallel with the total amount of basic sites. Compared with the Mg1Zr2-CP catalyst, the Mg1Zr2-HT catalyst has good thermal stability and reproducibility. The conversion of GL is still up to 80.1% and the selectivity of GC is 93.0% in the fourth reuse, while the regenerated Mg1Zr2-CP catalyst is 58.2% and 94.8% in the fourth reuse. The reason for the difference may be that in the cyclic reaction process, Mg1Zr2-HT has good grain stability and small growth amplitude, but the grain growth of active species in Mg1Zr2-CP is large, which will greatly reduce the effective active surface area of the catalyst, resulting in a significant decrease in the catalytic performance.

Supplementary Materials: The following supporting information can be downloaded at: https://www.mdpi.com/article/10.3390/nano12121972/s1, Figure S1: X-ray photoelectron spectra of Mg-Zr composite oxides. Regions: Mg 1s (a), Zr 3d (b) and O 1s (c); Figure S2: EDX spectrum and Elemental composition of Mg-Zr composite oxides with different Mg/Zr ratio; Figure S3: NH_3-TPD of Mg-Zr composite oxides with different Mg/Zr ratio; Figure S4: X-ray photoelectron spectra of fresh and used Mg1Zr2 composite oxides. Regions: Mg 1s (a), Zr 3d (b) and O 1s (c).

Author Contributions: Y.L.: Investigation, Data curation, Writing—Original Draft. H.Z.: Methodology, Visualization. W.X.: Formal analysis, Investigation. F.L.: Conceptualization, Writing—Review & Editing, Funding acquisition. Z.W.: Methodology, Supervision. All authors have read and agreed to the published version of the manuscript.

Funding: This work was funded by the National Natural Science Foundation of China (Nos. 21776057, U20A20152, U21A20306).

Institutional Review Board Statement: Not applicable.

Informed Consent Statement: Not applicable.

Data Availability Statement: The data presented in this study are available on request from the corresponding author.

Acknowledgments: All individuals consent to the acknowledgement.

Conflicts of Interest: The authors declare no conflict of interest.

References

1. Rozulan, N.; Halim, S.A.; Razali, N.; Lam, S.S. A Review on Direct Carboxylation of Glycerol Waste to Glycerol Carbonate and Its Applications. In *Biomass Conversion and Biorefinery*; Springer International Publishing: Cham, Switzerland, 2022; pp. 1–18.
2. Procopio, D.; Di Gioia, M.L. An Overview of the Latest Advances in the Catalytic Synthesis of Glycerol Carbonate. *Catalysts* **2022**, *12*, 50. [CrossRef]
3. Polychronopoulou, K.; Dabbawala, A.A.; Sajjad, M.; Singh, N.; Anjum, D.H.; Baker, M.A.; Charisiou, N.D.; Goula, M.A. Hydrogen Production Via Steam Reforming of Glycerol over Ce-La-Cu-O Ternary Oxide Catalyst: An Experimental and Dft Study. *Appl. Surf. Sci.* **2022**, *586*, 152798. [CrossRef]
4. Qingli, X.; Zhengdong, Z.; Kai, H.; Shanzhi, X.; Chuang, M.; Chenge, C.; Huan, Y.; Yang, Y.; Yongjie, Y. Ni Supported on MgO Modified Attapulgite as Catalysts for Hydrogen Production from Glycerol Steam Reforming. *Int. J. Hydrogen Energy* **2021**, *46*, 27380–27393. [CrossRef]
5. Dang, T.N.M.; Sahraei, O.A.; Olivier, A.; Iliuta, M.C. Effect of Impurities on Glycerol Steam Reforming over Ni-Promoted Metallurgical Waste Driven Catalyst. *Int. J. Hydrogen Energy* **2022**, *47*, 4614–4630. [CrossRef]
6. Sahraei, O.A.; Desgagnés, A.; Larachi, F.; Iliuta, M.C. A Comparative Study on the Performance of M (Rh, Ru, Ni)-Promoted Metallurgical Waste Driven Catalysts for H_2 Production by Glycerol Steam Reforming. *Int. J. Hydrogen Energy* **2021**, *46*, 32017–32035. [CrossRef]
7. Charisiou, N.D.; Siakavelas, G.; Tzounis, L.; Dou, B.; Sebastian, V.; Hinder, S.J.; Baker, M.A.; Polychronopoulou, K.; Goula, M.A. Ni/Y_2O_3–ZrO_2 Catalyst for Hydrogen Production through the Glycerol Steam Reforming Reaction. *Int. J. Hydrogen Energy* **2020**, *45*, 10442–10460. [CrossRef]
8. Keogh, J.; Deshmukh, G.; Manyar, H. Green Synthesis of Glycerol Carbonate Via Transesterification of Glycerol Using Mechanochemically Prepared Sodium Aluminate Catalysts. *Fuel* **2022**, *310*, 122484. [CrossRef]
9. Christy, S.; Noschese, A.; Lomelí-Rodriguez, M.; Greeves, N.; Lopez-Sanchez, J.A. Recent Progress in the Synthesis and Applications of Glycerol Carbonate. *Curr. Opin. Green Sustain. Chem.* **2018**, *14*, 99–107. [CrossRef]
10. Caro, S.D.; Bandres, M.; Urrutigoty, M.; Cecutti, C.; Thiebaud-Roux, S. Recent Progress in Synthesis of Glycerol Carbonate and Evaluation of Its Plasticizing Properties. *Front. Chem.* **2019**, *7*, 308. [CrossRef]
11. Lukato, S.; Kasozi, G.N.; Naziriwo, B.; Tebandeke, E. Glycerol Carbonylation with CO_2 to Form Glycerol Carbonate: A Review of Recent Developments and Challenges. *Curr. Res. Green Sustain. Chem.* **2021**, *4*, 100199. [CrossRef]
12. Zhang, J.; He, D. Lanthanum-Based Mixed Oxides for the Synthesis of Glycerol Carbonate from Glycerol and Urea. *React. Kinet. Mech. Catal.* **2014**, *113*, 375–392. [CrossRef]
13. Changmai, B.; Laskar, I.B.; Rokhum, L. Microwave-Assisted Synthesis of Glycerol Carbonate by the Transesterification of Glycerol with Dimethyl Carbonate Using Musa Acuminata Peel Ash Catalyst. *J. Taiwan Inst. Chem. Eng.* **2019**, *102*, 276–282. [CrossRef]
14. Pradhan, G.; Sharma, Y.C. Green Synthesis of Glycerol Carbonate by Transesterification of Bio Glycerol with Dimethyl Carbonate over Mg/ZnO: A Highly Efficient Heterogeneous Catalyst. *Fuel* **2021**, *284*, 118966. [CrossRef]
15. Manikandan, M.; Sangeetha, P. Optimizing the Surface Properties of MgO Nanoparticles Towards the Transesterification of Glycerol to Glycerol Carbonate. *ChemistrySelect* **2019**, *4*, 6672–6678. [CrossRef]
16. Bai, Z.; Zheng, Y.; Han, W.; Ji, Y.; Yan, T.; Tang, Y.; Chen, G.; Zhang, Z. Development of a Trapezoidal MgO Catalyst for Highly-Efficient Transesterification of Glycerol and Dimethyl Carbonate. *CrystEngComm* **2018**, *20*, 4090–4098. [CrossRef]
17. Roschat, W.; Phewphong, S.; Kaewpuang, T.; Promarak, V. Synthesis of Glycerol Carbonate from Transesterification of Glycerol with Dimethyl Carbonate Catalyzed by CaO from Natural Sources as Green and Economical Catalyst. *Mater. Today Proc.* **2018**, *5*, 13909–13915. [CrossRef]
18. Simanjuntak, F.; Kim, T.K.; Sang, D.L.; Ahn, B.S.; Kim, H.S.; Lee, H. CaO-Catalyzed Synthesis of Glycerol Carbonate from Glycerol and Dimethyl Carbonate: Isolation and Characterization of an Active Ca Species. *Appl. Catal. A Gen.* **2011**, *401*, 220–225. [CrossRef]
19. Praikaew, W.; Kiatkittipong, W.; Aiouache, F.; Najdanovic-Visak, V.; Termtanun, M.; Lim, J.W.; Lam, S.S.; Kiatkittipong, K.; Laosiripojana, N.; Boonyasuwat, S. Mechanism of CaO Catalyst Deactivation with Unconventional Monitoring Method for Glycerol Carbonate Production Via Transesterification of Glycerol with Dimethyl Carbonate. *Int. J. Energy Res.* **2022**, *46*, 1646–1658. [CrossRef]
20. Ying, T.; Yan, T.; Bo, S.; Li, H.; Jeje, A. Synthesis of No-Glycerol Biodiesel through Transesterification Catalyzed by CaO from Different Precursors. *Can. J. Chem. Eng.* **2016**, *94*, 1466–1471.
21. Liu, Y.; Xia, C.; Wang, Q.; Zhang, L.; Huang, A.; Ke, M.; Song, Z. Direct Dehydrogenation of Isobutane to Isobutene over Zn-Doped ZrO_2 Metal Oxide Heterogeneous Catalysts. *Catal. Sci. Technol.* **2018**, *8*, 4916–4924. [CrossRef]
22. Bing, W.; Wei, M. Recent Advances for Solid Basic Catalysts: Structure Design and Catalytic Performance. *J. Solid State Chem.* **2019**, *269*, 184–194. [CrossRef]

23. Zhang, X.L.; Wei, S.W.; Zhao, X.Y.; Chen, Z.; Wu, H.W.; Rong, P.; Sun, Y.; Li, Y.; Yu, H.; Wang, D.F. Preparation of Mesoporous Cao-Zro2 Catalysts without Template for the Continuous Synthesis of Glycerol Carbonate in a Fixed-Bed Reactor. *Appl. Catal. A Gen.* **2020**, *590*, 12. [CrossRef]
24. Otor, H.O.; Steiner, J.B.; García-Sancho, C.; Alba-Rubio, A.C. Encapsulation Methods for Control of Catalyst Deactivation: A Review. *ACS Catal.* **2020**, *10*, 7630–7656. [CrossRef]
25. Chandrasekar, M.; Subash, M.; Logambal, S.; Udhayakumar, G.; Uthrakumar, R.; Inmozhi, C.; Al-Onazi, W.A.; Al-Mohaimeed, A.M.; Chen, T.-W.; Kanimozhi, K. Synthesis and Characterization Studies of Pure and Ni Doped CuO Nanoparticles by Hydrothermal Method. *J. King Saud Univ.-Sci.* **2022**, *34*, 101831. [CrossRef]
26. Cui, H.; Wu, X.; Chen, Y.; Boughton, R. Synthesis and Characterization of Mesoporous MgO by Template-Free Hydrothermal Method. *Mater. Res. Bull.* **2014**, *50*, 307–311. [CrossRef]
27. Akune, T.; Morita, Y.; Shirakawa, S.; Katagiri, K.; Inumaru, K. ZrO_2 Nanocrystals as Catalyst for Synthesis of Dimethylcarbonate from Methanol and Carbon Dioxide: Catalytic Activity and Elucidation of Active Sites. *Langmuir* **2018**, *34*, 23–29. [CrossRef]
28. Yichao, W.; Yachun, C.; Guangxu, Z.; Yuntao, Y. Preparation of Mg/Sn/W Composite Oxide Catalyst by Hydrothermal Method and Its Performance Evaluation. *Petrochem. Technol.* **2019**, *48*, 1212.
29. Varkolu, M.; Burri, D.R.; Kamaraju, S.R.R.; Jonnalagadda, S.B.; Van Zyl, W.E. Transesterification of Glycerol with Dimethyl Carbonate over Nanocrystalline Ordered Mesoporous $MgO–ZrO_2$ Solid Base Catalyst. *J. Porous Mater.* **2015**, *23*, 185–193. [CrossRef]
30. Sádaba, I.; Ojeda, M.; Mariscal, R.; Fierro, J.L.G.; Granados, M.L. Catalytic and Structural Properties of Co-Precipitated Mg–Zr Mixed Oxides for Furfural Valorization Via Aqueous Aldol Condensation with Acetone. *Appl. Catal. B Environ.* **2011**, *101*, 638–648. [CrossRef]
31. Zhu, W.; Jiang, X.; Liu, F.; You, F.; Yao, C. Preparation of Chitosan—Graphene Oxide Composite Aerogel by Hydrothermal Method and Its Adsorption Property of Methyl Orange. *Polymers* **2020**, *12*, 2169. [CrossRef]
32. Pan, Q.; Xie, J.; Zhu, T.; Cao, G.; Zhao, X.; Zhang, S. Reduced Graphene Oxide-Induced Recrystallization of Nis Nanorods to Nanosheets and the Improved Na-Storage Properties. *Inorg. Chem.* **2014**, *53*, 3511–3518. [CrossRef] [PubMed]
33. Walker, R.C.; Potochniak, A.E.; Hyer, A.P.; Ferri, J.K. Zirconia Aerogels for Thermal Management: Review of Synthesis, Processing, and Properties Information Architecture. *Adv. Colloid Interface Sci.* **2021**, *295*, 102464. [CrossRef] [PubMed]
34. Zhang, X.; Wang, D.; Wu, G.; Wang, X.; Jiang, X.; Liu, S.; Zhou, D.; Xu, D.; Gao, J. One-Pot Template-Free Preparation of Mesoporous $MgO–ZrO_2$ Catalyst for the Synthesis of Dipropyl Carbonate. *Appl. Catal. A Gen.* **2018**, *555*, 130–137. [CrossRef]
35. Liu, B.; Li, C.; Zhang, G.; Yan, L.; Li, Z. Direct Synthesis of Dimethyl Carbonate from CO_2 and Methanol over $CaO–CeO_2$ Catalysts: The Role of Acid–Base Properties and Surface Oxygen Vacancies. *New J. Chem.* **2017**, *41*, 12231–12240. [CrossRef]
36. Guan, H.; Liang, J.; Zhu, Y.; Zhao, B.; Xie, Y. Structure Characterization and Monolayer Dispersion Phenomenon of $MgO-ZrO_2$ Prepared by Co-Precipitation. *Acta Phys.-Chim. Sin.* **2005**, *21*, 1011–1016.
37. Poolwong, J.; Del Gobbo, S.; D'Elia, V. Transesterification of Dimethyl Carbonate with Glycerol by Perovskite-Based Mixed Metal Oxide Nanoparticles for the Atom-Efficient Production of Glycerol Carbonate. *J. Ind. Eng. Chem.* **2021**, *104*, 43–60. [CrossRef]
38. Kumar, P.; With, P.; Srivastava, V.C.; Gläser, R.; Mishra, I.M. Glycerol Carbonate Synthesis by Hierarchically Structured Catalysts: Catalytic Activity and Characterization. *Ind. Eng. Chem. Res.* **2015**, *54*, 12543–12552. [CrossRef]
39. Chang, C.-W.; Gong, Z.-J.; Huang, N.-C.; Wang, C.-Y.; Yu, W.-Y. MgO Nanoparticles Confined in ZIF-8 as Acid-Base Bifunctional Catalysts for Enhanced Glycerol Carbonate Production from Transesterification of Glycerol and Dimethyl Carbonate. *Catal. Today* **2020**, *351*, 21–29. [CrossRef]
40. Liu, Z.; Li, B.; Qiao, F.; Zhang, Y.; Wang, X.; Niu, Z.; Wang, J.; Lu, H.; Su, S.; Pan, R. Catalytic Performance of Li/Mg Composites for the Synthesis of Glycerol Carbonate from Glycerol and Dimethyl Carbonate. *ACS Omega* **2022**, *7*, 5032–5038. [CrossRef]
41. Pradhan, G.; Sharma, Y.C. A Greener and Cheaper Approach Towards Synthesis of Glycerol Carbonate from Bio Waste Glycerol Using $CaO–TiO_2$ Nanocatalysts. *J. Clean. Prod.* **2021**, *315*, 127860. [CrossRef]
42. Zhang, Q.; Tang, T.; Wang, J.; Sun, M.; Wang, H.; Sun, H.; Ning, P. Facile Template-Free Synthesis of $Ni-SiO_2$ Catalyst with Excellent Sintering-and Coking-Resistance for Dry Reforming of Methane. *Catal. Commun.* **2019**, *131*, 105782. [CrossRef]
43. Ma, Z.; Meng, X.; Liu, N.; Yang, C.; Shi, L. Preparation, Characterization, and Isomerization Catalytic Performance of Palladium Loaded Zirconium Hydroxide/Sulfated Zirconia. *Ind. Eng. Chem. Res.* **2018**, *57*, 14377–14385. [CrossRef]
44. Grams, J.; Ruppert, A.M. Catalyst Stability—Bottleneck of Efficient Catalytic Pyrolysis. *Catalysts* **2021**, *11*, 265. [CrossRef]

Article

Alternative Controlling Agent of *Theobroma grandiflorum* Pests: Nanoscale Surface and Fractal Analysis of Gelatin/PCL Loaded Particles Containing *Lippia origanoides* Essential Oil

Ana Luisa Farias Rocha [1,2], Ronald Zico de Aguiar Nunes [1], Robert Saraiva Matos [3], Henrique Duarte da Fonseca Filho [2,4], Jaqueline de Araújo Bezerra [5], Alessandra Ramos Lima [6], Francisco Eduardo Gontijo Guimarães [6], Ana Maria Santa Rosa Pamplona [7], Cláudia Majolo [7], Maria Geralda de Souza [7], Pedro Henrique Campelo [8], Ştefan Ţălu [9,*], Vanderlei Salvador Bagnato [6,10], Natalia Mayumi Inada [6] and Edgar Aparecido Sanches [1,2]

1. Laboratory of Nanostructured Polymers (NANOPOL), Federal University of Amazonas (UFAM), Manaus 69067-005, AM, Brazil
2. Graduate Program in Materials Science and Engineering (PPGCEM), Federal University of Amazonas (UFAM), Manaus 69067-005, AM, Brazil
3. Amazonian Materials Group, Federal University of Amapá (UNIFAP), Macapá 68903-419, AP, Brazil
4. Laboratory of Nanomaterials Synthesis and Nanoscopy (LSNN), Federal University of Amazonas (UFAM), Manaus 69067-005, AM, Brazil
5. Analytical Center, Federal Institute of Education, Science and Technology of Amazonas (IFAM), Manaus 69020-120, AM, Brazil
6. São Carlos Institute of Physics (IFSC), University of São Paulo (USP), São Carlos 13563-120, SP, Brazil
7. EMBRAPA Western Amazon, Manaus AM-010 Km 29, Manaus 69010-970, AM, Brazil
8. Department of Food Technology, Federal University of Viçosa (UFV), Viçosa 36570-900, MG, Brazil
9. The Directorate of Research, Development and Innovation Management (DMCDI), Technical University of Cluj-Napoca, 15 Constantin Daicoviciu St., 400020 Cluj-Napoca, Cluj County, Romania
10. Hagler Institute for Advanced Studies, Texas A&M University, College Station, TX 77843, USA
* Correspondence: stefan.talu@auto.utcluj.ro

Citation: Rocha, A.L.F.; de Aguiar Nunes, R.Z.; Matos, R.S.; da Fonseca Filho, H.D.; de Araújo Bezerra, J.; Lima, A.R.; Guimarães, F.E.G.; Pamplona, A.M.S.R.; Majolo, C.; de Souza, M.G.; et al. Alternative Controlling Agent of *Theobroma grandiflorum* Pests: Nanoscale Surface and Fractal Analysis of Gelatin/PCL Loaded Particles Containing *Lippia origanoides* Essential Oil. *Nanomaterials* 2022, 12, 2712. https://doi.org/10.3390/nano12152712

Academic Editor: Meiwen Cao

Received: 16 July 2022
Accepted: 5 August 2022
Published: 7 August 2022

Publisher's Note: MDPI stays neutral with regard to jurisdictional claims in published maps and institutional affiliations.

Copyright: © 2022 by the authors. Licensee MDPI, Basel, Switzerland. This article is an open access article distributed under the terms and conditions of the Creative Commons Attribution (CC BY) license (https://creativecommons.org/licenses/by/4.0/).

Abstract: A new systematic structural study was performed using the Atomic Force Microscopy (AFM) reporting statistical parameters of polymeric particles based on gelatin and poly-ε-caprolactone (PCL) containing essential oil from *Lippia origanoides*. The developed biocides are efficient alternative controlling agents of *Conotrachelus humeropictus* and *Moniliophtora perniciosa*, the main pests of *Theobroma grandiflorum*. Our results showed that the particles morphology can be successfully controlled by advanced stereometric parameters, pointing to an appropriate concentration of encapsulated essential oil according to the particle surface characteristics. For this reason, the absolute concentration of 1000 µg·mL^{-1} (P_{1000} system) was encapsulated, resulting in the most suitable surface microtexture, allowing a faster and more efficient essential oil release. Loaded particles presented zeta potential around (-54.3 ± 2.3) mV at pH = 8, and particle size distribution ranging from 113 to 442 nm. The hydrodynamic diameter of 90% of the particle population was found to be up to (405 ± 31) nm in the P_{1000} system. The essential oil release was evaluated up to 80 h, with maximum release concentrations of 63% and 95% for P_{500} and P_{1000}, respectively. The best fit for the release profiles was obtained using the Korsmeyer–Peppas mathematical model. Loaded particles resulted in 100% mortality of *C. humeropictus* up to 48 h. The antifungal tests against *M. perniciosa* resulted in a minimum inhibitory concentration of 250 µg·mL^{-1}, and the P_{1000} system produced growth inhibition up to 7 days. The developed system has potential as alternative controlling agent, due to its physical stability, particle surface microtexture, as well as pronounced bioactivity of the encapsulated essential oil.

Keywords: *Lippia origanoides*; *Theobroma grandiflorum*; controlling agent; nanoscale surface; fractal analysis; controlled release; *Conotrachelus humeropictus*; *Moniliophtora perniciosa*

1. Introduction

The increasing interest in biodegradable particles has accelerated their development process for new technological applications [1–5], particularly in environmentally friendly polymeric particles containing encapsulated essential oils [6–13].

The combination of biomaterials with different physicochemical properties has allowed the development of layered particles, to protect and release secondary metabolites [14–17]. The evaluation of surface nanotexture and fractal analyses through Atomic Force Microscopy (AFM) technique has been useful to investigate the influence of texture parameters on the controlled release mechanism and concentration of encapsulated bioactive compounds [18,19].

Essential oils have long been considered as alternative natural agents for pest control [20–22]. *Lippia origanoides* Kunth [23] is popularly known as "Erva-de-Marajó" in northern Brazil. Carvacrol and thymol (the major constituents of its essential oil) present significant chemopreventive properties [24–26], antimicrobial activity against several pathogen groups [27], as well as repellency and a low toxicity [28]. The encapsulation of essential oils for controlled release formulations can improve their efficiency and reduce environmental damage [29,30].

The cupuaçu tree (*Theobroma grandiflorum* (Willd. *ex* Spreng.) K. Schum.) (Malvaceae) is one of the main fruit trees cultivated in the Brazilian Amazon. The high commercial value of the cupuaçu pulp derives from the food industry, mainly as juice, liqueur, and jelly, as well as in the manufacture of chocolate ("cupulate") from its almonds [31–33]. *Conotrachelus humeropictus* Fiedler, 1940 (Coleoptera: Curculionidae), known as "Broca-do-Cupuaçu", is the main pest of this culture in the Amazon region, especially in Rondônia and Amazonas [34]. This pest is difficult to control, as both the egg and larva are lodged in galleries inside the fruits. Infested fruits fall off before ripening or have the pulp completely destroyed [35]. Moreover, from the phytosanitary point of view, the disease caused by the fungus *Moniliophtora perniciosa* (known as "Vassoura-de-Bruxa") [36] represents the main limiting factor to the expansion of this fruit tree. This pest significantly reduces the economic production, and phytosanitary pruning is the main economic tool to control this pest [37].

The use of nanotechnology to control pests in agriculture has resulted in nanoscale materials able to enhance the stability and activity of natural controlling agents [38,39]. Reports on the encapsulation of *L. origanoides* essential oil in biodegradable particles to control *C. humeropictus* and *M. perniciosa* have not been found in the scientific literature. For this reason, particles based on gelatin and poly-ε-caprolactone (PCL) were loaded with this essential oil, aiming at the development of a controlled release formulation.

The AFM technique allowed understanding the influence of the essential oil concentration on statistical parameters (based on nanoscale surface and fractal analyses), such as roughness, peak/height distributions, and nanotexture homogeneity. Size distribution measurements and nanoparticle surface charge were evaluated, respectively, by nanoparticle tracking analysis (NTA) and zeta potential. Laser Scanning Confocal Microscopy (LSCM) and fluorescence measurements were applied to confirm the essential oil encapsulation. Encapsulation Efficiency (EE%) was measured by UV-VIS spectroscopy, and the release kinetics of the essential oil was analyzed as a log cumulative percentage of released essential oil *versus* log time by fitting the data according to the Higuchi [40] and Korsmeyer–Peppa's [41] mathematical models. Finally, the insecticidal and fungicidal efficiency of the developed formulation was assessed in vitro, respectively, against *C. humeropictus* and *M. perniciosa*.

2. Materials and Methods

2.1. Nanoparticle Development and Essential Oil Encapsulation

Colloidal system development was based on previous reports with marginal modification [14]. *L. origanoides* (SISGEN authorization code AD0C7DB) essential oil was encapsulated at absolute concentrations of 500 µg·mL^{-1} (P$_{500}$) and 1000 µg·mL^{-1} (P$_{1000}$). Unloaded particles (P$_0$) were also prepared.

Encapsulation Efficiency (EE%) was evaluated on a Epoch2 Microplate Reader Biotek (Agilent, CA, USA) [42]. From the calibration curve, the unknown concentration of essential oil was obtained by measuring the absorbance values at 278 nm. Particles were separated by centrifugation (Daiki Sciences, Seoul City, Republic of Korea) (20,000 rpm) and the supernatant absorbance allowed obtaining the percentage of free essential oil. Then, EE% was calculated using the formula: (EE%) = (amount of encapsulated essential oil/total amount of essential oil used in the formulation) × 100. Experiments were carried out in triplicate.

2.2. Zeta Potential and Nanoparticle Tracking Analysis (NTA)

Zeta potential values (in mV) were obtained from pH = 1 to 10 at 25 °C. Measurements were performed in triplicate. Size characterization was performed on a NanoSight NS300 device (Malvern Instruments, Malvern, UK), equipped with a green laser type and a SCMOS camera. Data collection and analysis were performed using the software NTA 3.0 (Malvern Instruments, Malvern, UK). Samples were diluted in MilliQ water (1:100 v/v). A standard operating procedure was created using 749 frames for 30 s. Measurements were performed in triplicate. The evaluation of the particle size distribution (PSD) was performed using the parameters Mean, Mode, SD, D_{10}, D_{50} (Median), and D_{90}, which indicate, respectively, the average, most frequent particle class size, standard deviation, and the 10%, 50%, and 90% percentiles of the analyzed particles.

2.3. Laser Scanning Confocal Microscopy (LSCM) and Fluorescence Measurements

Images were taken on a Carl Zeiss microscope (inverted model LSM 780) (ZEISS Research Microscopy Solutions, Jena, Germany), with a Ti: Sapphire LASER, a 40× objective lens, 1.2 NA, and a 0.28 mm work distance. Systems containing unloaded (P$_0$) and loaded particles (P$_{500}$ and P$_{1000}$) were centrifuged. A volume of the supernatant was discarded for visualization of the largest particles. Then, a few drops were placed on a microscopy glass slide. Measurements of fluorescence were carried out with a 63× objective and SPAD (single photon avalanche diode) detector, with a temporal resolution of 70 ps. A Coherent Chameleon tunable 690–1100 nm laser was used as the excitation source. Measurements were taken at 800 nm.

2.4. Atomic Force Microscopy (AFM)

Colloidal systems containing unloaded (P$_0$) or loaded particles ((P$_{500}$ and P$_{1000}$); 1 µL) were dripped on a glass slide and dried using liquid nitrogen. Then, the glass slides containing the formed films were fixed on a sample holder using a double-sided adhesive tape. Measurements were performed at room temperature (296 ± 1 K) and (40 ± 1)% R.H. on a Innova equipment (Bruker, MA, USA) operated in tapping mode and equipped with a silicon tip and Al coated cantilever with a spring constant of 42 N/m (Tap190AL-G) (Budget Sensors, Sofia, Bulgaria). Scans were performed using (10 × 10) µm^2 with (256 × 256) pixels at a scan rate of 1 Hz.

2.4.1. Nanoscale Surface Analysis

Topographic images obtained were processed using the commercial MountainsMap® software version 8.0 (Besançon, France) [43]. Stereometric parameters of height, Sk, and volume were obtained. In addition, quantitative parameters obtained from qualitative renderings (such as furrows and contour lines) were also obtained.

2.4.2. Fractal Analysis

Fractal analysis was carried out based on the following superficial statistical parameters: fractal dimension (FD), surface entropy (H), fractal succolarity (FS), and fractal lacunarity (FL). Fractal dimension (FD) is commonly used for quantification of surface texture homogeneity, as well as for surface complexity evaluation. However, an analysis based only on the FD parameter is not sufficient to evaluate aspects of general texture [44], because the surface irregularity usually increases as a function of FD [45]. The free software Gwyddion 2.55 [46] (Brno, Czech Republic) was used to perform calculations.

Surface entropy (H) values quantify the uniformity of the height distribution by relating pixels and heights as a function of intensity. Measurements are based on the Shannon's entropy (Equation (1)) [47]:

$$H^{(2)} = -\sum_{i=1}^{N}\sum_{j=1}^{N} p_{ij} \cdot \log p_{ij} \tag{1}$$

where p_{ij} represents the probability of finding accessible pixels on the evaluated pixel set. The AFM image (pixel matrix) was converted into a binary height matrix using the free software WSXM (Madri, Spain) [48]. Results were normalized using Equation (2) [49]:

$$H_{matralt} = \frac{H^{(2)} - H^{(2)}_{min}}{H^{(2)}_{max} - H^{(2)}_{min}} \tag{2}$$

where $H^{(2)}_{max}$ and $H^{(2)}_{min}$ represent, respectively, the uniform and non-uniform pattern surface (adopting the symbol H as the normalized value of surface entropy). A R language algorithm was programmed for H calculation using the free software RStudio 1.2.503 (Boston, MA, USA).

Additional algorithms in R language and Fortran 77 were developed to obtain fractal succolarity (FS) and fractal lacunarity (FL). Percolation can be quantified through the FS evaluation (Equation (3)) [50], while FL measures the texture homogeneity by dimensioning gaps on the fractal object surface [51].

$$FS(T(k), dir) = \frac{\sum_{k=1}^{n} P_0(T(k)) \cdot PR(T(k), p_c)}{\sum_{k=1}^{n} PR(T(k), p_c)} \tag{3}$$

where d_{ir} represents the liquid entrance direction; $T(k)$ are boxes of equal size $T(n)$; $P_0(T(k))$ is the occupancy percentage; PR represents the occupancy pressure, and p_c is the centroid position (x,y). FL was obtained from a previous report [52]. Calculations were focused on the lacunarity coefficient according to Equation (4) [53]:

$$L(r) = \alpha \cdot r^\beta \tag{4}$$

where $L(r)$ is the lacunarity, α represents an arbitrary constant, and r is the box size. The lacunarity coefficient (β) was estimated by $log(r)$ vs. $log[1 + L(r)]$. The free software Force 3.0 (Maribor, Slovenia) [54] was applied for compiling the FL algorithm. Displacement of one unit was applied, due to the small FL values.

2.5. Essential Oil Release

A colloidal system containing loaded particles (15 mL) was inserted in dialysis tubing cellulose membrane and suspended in water (85 mL) at 25 °C. The system was maintained under continuous magnetic stirring (100 rpm). A 3 mL aliquot was withdrawn from flask at regular time intervals (up to 80 h). Absorbance was measured at 278 nm on a Epoch2 Microplate Reader Biotek. The amount of released essential oil was calculated from a standard curve [55]. The cumulative release (%) of essential oil was obtained with the following equation: [Cumulative release (%) = (amount of essential oil released after time t/total amount of encapsulated essential oil) × 100]. Experiments were carried out in triplicate.

2.6. Insecticidal and Fungicidal Bioassays

2.6.1. Conotrachelus humeropictus

C. humeropictus individuals were obtained from stock colonies at the EMBRAPA Amazônia Ocidental, Manaus/AM, Brazil, without any pesticide exposure. Borers were reared on a diet of sugarcane and kept at 25 °C, with 70–85% R. H., and a 10:14 h light:dark photoperiod. Glass Petri plates (150 mm in diameter × 20 mm in height) were used as chambers.

Filter paper (150 mm in diameter) was placed in the glass Petri dishes. Each concentration (1 mL) of essential oil in natura/acetone solution (125, 250, 500, 625, 750, and 1000 µg·mL^{-1}) was uniformly applied on filter paper disk. The treated filter paper disks were air-dried for 1 min to remove solvent. Five adults were transferred from stock to the paper disk, allowing direct contact with the essential oil. Then, chambers were sealed to prevent essential oil evaporation. Acetone (1 mL) was used as negative control. Mortality was evaluated after 24 h of exposure. Individuals were considered dead if they did not move when prodded with a fine paintbrush. The experimental design was completely randomized, with three replicates. Mortality data were subjected to PROBIT analysis [56]. Then, the LD_{50} (lethal dosage that kills 50% of the exposed borers), LD_{90} (lethal dosage that kills 90% of the exposed borers), LCL (lower confidence limit) and UCL (upper confidence limit) were estimated [57], with a fiducial limit of 95%.

The toxicity of the encapsulated essential oil was also tested against *C. humeropictus*. Filter paper (150 mm in diameter) was placed in the glass Petri dishes. A volume of 1 mL of loaded particles (P_{1000}) was uniformly applied on the filter paper disk. Five adults were transferred from stock to the paper disk, allowing direct contact with the loaded particles. Then, chambers were sealed to prevent loss of essential oil. Unloaded particles (P_0) were used as negative control. The number of live borers was counted after 24 h of application. The experimental design was completely randomized based on three replicates.

2.6.2. Moniliophtora perniciosa

M. perniciosa isolates were provided by the EMBRAPA Amazônia Ocidental, Manaus/AM-Brazil. Bioassays were performed by the disk diffusion method (DDM) adapted from previous report [58]. The culture medium was prepared with potato-dextrose-agar (PDA; 15.6 g) and sucrose (8.0 g), using 400 mL of distilled water, and kept under heating until complete solubilization. Essential oil was diluted in DMSO (1:9 v/v). Different volumes (1, 0.75, 0.5, 0.25, and 0.125 mL) were added to 100 mL of culture media and then transferred to Petri dishes (90 mm in diameter × 10 cm in height). All Petri dishes were inoculated with a mycelial disc (5 mm diameter) of *M. perniciosa*. Then, the Petri-dishes were incubated for 7 days at 25 °C and the colony diameter was measured. DMSO was used in the bioassays instead of essential oil as a negative control. Four replicate plates were used for each treatment.

The Minimum Inhibitory Concentration (MIC) was interpreted as the lowest concentration that inhibited visual growth. Only plates with positive growth and quality control for purity and colony counts were considered. The mycelial growth index was obtained as the ratio of the final average growth diameter to the number of days after inoculation. The relative mycelial growth percentage (RMG%) at each tested concentration was calculated by comparing the growth on amended media (GOA) compared with the growth on the nonamended control (GOC), as follows: RMG% = (GOA/GOC) × 100. The percentage inhibition of mycelium growth at each tested concentration (I) was also calculated as the difference between the radial growth of nonamended control (C) and the radial growth of each tested concentration (T), as follows: I (%) = (1 − T/C) × 100 [59].

The efficiency of the loaded particles (P_{1000}) was tested against *M. perniciosa*, according to the same procedure as describe above. Unloaded particles (P_0) were used as negative control. Four replicate plates were considered for each treatment.

3. Results and Discussion

3.1. AFM Analysis

The morphology of gelatin/PCL particles has been extensively studied in controlled release systems for pest control [14,15,17], scaffolds [60,61], and curatives [62]. Here we focused on the particles surface morphology (unloaded and loaded with *L. origanoides* essential oil), which previously showed significant larvicidal, acaricidal, and insecticidal potential [8,14,15].

Figure 1 shows the 3D topographic images of the unloaded particles (P_0), as well as the particles loaded with 500 $\mu g \cdot mL^{-1}$ (P_{500}) and 1000 $\mu g \cdot mL^{-1}$ (P_{1000}) of essential oil. The P_0 surface presented spherical-conical grains (Figure 1a). A thinning of the rough peaks in the loaded particles (Figure 1b,c) was observed due to the encapsulation of essential oil. Furthermore, the formation of a large spherical protuberance on the P_{500} and P_{1000} surfaces was observed, probably due to the formation of air bubbles during the drying procedure. This phenomenon was also previously observed [63]. In addition, the topography qualitative analysis revealed a different surface morphology: the increasing of the essential oil concentration promoted a smoothing on the particles surface. This behavior was confirmed by the related height surface parameters analysis (Sa and Sq), as shown in Table 1.

The results were expressed as the mean value and standard deviation, where significant difference was observed (p-value < 0.05). The highest roughness value was observed in P_0 (Sq = (20.301 $\pm$ 3.030) nm). However, the Tukey test showed that both P_{500} and P_{1000} presented similar roughness values. Lower-roughness particles can present higher adhesion energy and be faster adsorbed on another surface [64]. This result indicates that the developed loaded systems represent a viable alternative to decrease particles surface roughness through the encapsulation of essential oil.

The P_0, P_{500}, and P_{1000} systems presented positive asymmetric height distributions, with Rsk values slightly greater than zero. However, the asymmetric height distribution increased in P_{1000}, showing that the height distribution was affected by the increase of the essential oil concentration (although the Tukey test also revealed no significant difference between P_{500} and P_{1000}). Greater asymmetry, whether positive or negative, suggests that a particle is more likely to be anchored or adsorbed onto another surface (probability because skewness is an index). This fact was observed because the particle created a preferential slope direction of its rough peaks (as observed in P_{1000}). In addition, all systems also showed a non-platykurtic pattern (Leptokurtic), as the Rku values were greater than 3. Consequently, the data distribution tended to deviate from the normal Gaussian behavior [65]. As shown in Table 1, the P_{1000} system presented the highest Rsk value, differing from P_{500} (p-value < 0.05). These data showed that P_{1000} presented a sharper distribution, confirming its greater tendency to be easily adsorbed on another surface.

Table 1. Surface parameters (Sa, Sq, Rsk, and Rku).

Parameters	Samples		
	P_0	P_{500}	P_{1000}
Sa (nm)	27.208 $\pm$ 3.030	8.032 $\pm$ 0.664	6.163 $\pm$ 1.352
Sq (nm)	20.301 $\pm$ 5.248	10.546 $\pm$ 1.244	8.941 $\pm$ 2.120
Rsk	0.164 $\pm$ 0.572	0.542 $\pm$ 0.064	1.406 $\pm$ 0.456
Rku	4.183 $\pm$ 0.363	4.168 $\pm$ 0.353	6.944 $\pm$ 1.009

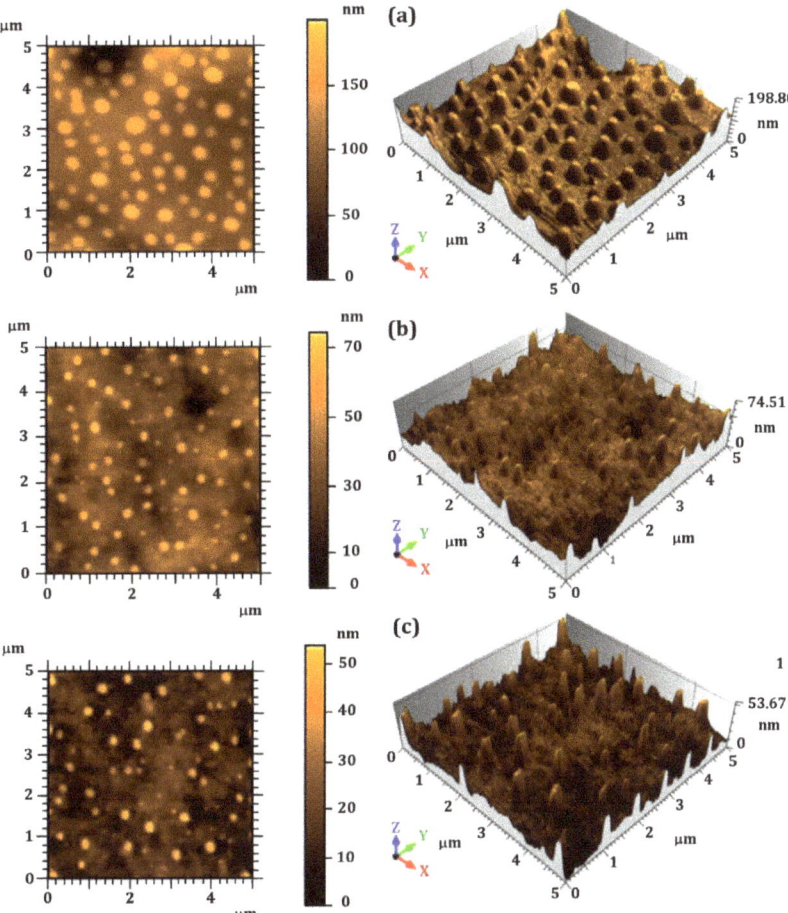

Figure 1. Two-dimensional and three-dimensional AFM micrographs: (**a**) unloaded particles (P_0), (**b**) loaded particles using 500 µg·mL^{-1} of essential oil (P_{500}), and (**c**) loaded particles using 1000 µg·mL^{-1} of essential oil (P_{1000}).

Figure 2 shows the Sk values and volume parameters concerning the height distribution of the particle surface [66,67].

Figure 2a–c indicates that the particle surface of all systems (P_0, P_{500} and P_{1000}) presented a heavy-tailed distribution (Leptokurtic), with great tapering of the height distribution (mainly in P_{1000}). On the other hand, the cumulative curve of Figure 2b (in red) showed better height distribution in P_{500}, since approximately 90% of the relative heights were found between 0 and 0.2568 nm.

Figure 2g–i shows the graphic behaviors considering the volume parameters of the particles surface. As a result of the decrease of surface roughness, especially in P_0 and P_{500}, the volume of material forming the surface topography decreased, as observed by the peak material volume (Vmp), core material volume (Vmc), dale void volume (Vvv), and core void volume (Vvc) parameters. Statistical similarity between P_{500} and P_{1000} was also identified in all parameters. This result confirms that the topography was affected by the encapsulation of essential oil. Furthermore, the particle morphology could be controlled from the observation of advanced stereometric parameters, which could be useful for quality control of the developed material, since they accurately determined the amount of material on the particle surface in different aspects [67].

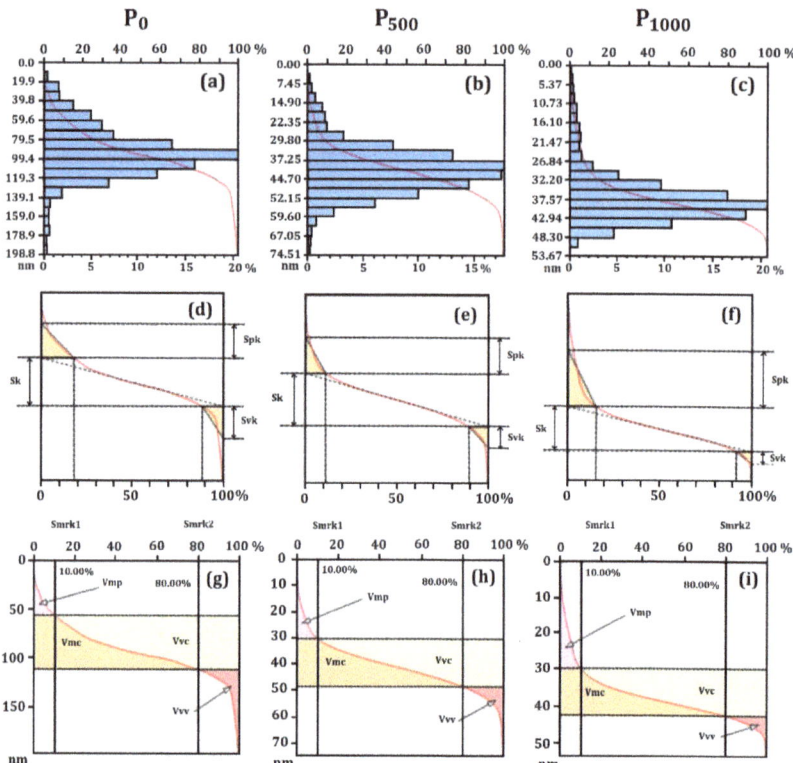

Figure 2. Sk values and volume parameters concerning the height distribution of the particle surface. (**a**–**c**) Particle surface of all systems (P_0, P_{500}, and P_{1000}) presenting a heavy-tailed distribution (Leptokurtic) with great tapering of the height distribution; (**d**–**f**) thickness of material on the particles surface, evaluated by the height distribution according to the Sk parameter family; (**e**,**f**) displacements of the Sk curve; and (**g**–**i**) graphic behaviors considering the volume parameters of the particles surface.

The thickness of material on the particles surface was evaluated by the height distribution according to the Sk parameter family (Figure 2d–f and Table 2). Most of the thickness and volume stereometric parameters exhibited a statistically significant difference (p-value < 0.05), except the valley material portion (Smr2). However, the Tukey test showed that the core thickness (Sk) values were similar to those of P_{500} and P_{1000}, while the highest S_k value was observed in P_0, whose behavior followed that of the surface roughness.

Similarly, the reduced peak height (Spk) and reduced valley depth (Svk) also exhibited similar behavior for P_{500} and P_{1000}, showing that the thickness of the material forming the particle topography did not change from P_{500} to P_{1000}. Figure 2e,f shows the displacements of the Sk curve. In addition, they also suggested that the peak material portion (Smr1) was similar in P_{500} and P_{1000}. These results indicated that the surface microtexture of the particles loaded with essential oil was similar, but still without considering the complexity of the spatial patterns.

Table 2. Sk and volume parameters of the particles surface.

Parameters	Systems		
	P0	P500	P1000
Sk (μm)	50.398 ± 10.360	23.140 ± 1.829	15.067 ± 2.938
Spk (μm)	39.308 ± 5.400	14.609 ± 0.269	19.946 ± 5.278
Svk (μm)	35.393 ± 13.872	8.009 ± 0.776	6.935 ± 3.758
Smr1 (%)	17.842 ± 1.779	12.566 ± 1.217	14.687 ± 1.106
Smr2 (%) *	89.646 ± 1.928	90.028 ± 0.799	90.623 ± 0.703
Vmp (μm/μm2)	0.001 ± 0.000	0.001 ± 0.000	0.001 ± 0.000
Vmc (μm/μm2)	0.020 ± 0.003	0.009 ± 0.001	0.006 ± 0.001
Vvc (μm/μm2)	0.036 ± 0.001	0.012 ± 0.002	0.010 ± 0.003
Vvv (μm/μm2)	0.003 ± 0.001	0.001 ± 0.000	0.001 ± 0.000

* Samples without significant difference ANOVA one-way and Tukey test (p-value > 0.05).

3.2. Surface Microtexture

Renderings of the particles surface microtexture are shown in Figure 3. Images based on furrows and contour lines were obtained for each system. This type of image has been widely used to explain the surface behavior in fluid flooding [68,69], as qualitative renderings that simulate the entrance of fluids and particle arrangement on a nanoparticle surface [52]. A significant reduction in particle size, due to the encapsulation of essential oil, was observed, which was also associated with the decreasing roughness.

Particles presented similar shapes in P_0 (Figure 3a), while P_{500} and P_{1000} (Figure 3c) acquired smaller and more randomized sizes. These results showed that the essential oil encapsulation reduced the particle size, which could result in a better and faster adsorption of the particles on their external environment.

The regions of the images presenting more intense colors are associated with rough peaks, and the darker regions are related to valleys. All parameters associated to furrows presented statistically significant differences (p-value < 0.05). However, the Tukey test showed that P_{500} and P_{1000} presented a similar behavior, exhibiting shallower furrows. These data showed the decrease of the surface roughness.

A similar configuration was also observed for the mean depth of furrows (Table 3). However, P_0 exhibited a lower mean density than those of P_{500} and P_{1000}, showing that the thinning of the rough peaks promoted a greater density of furrows, and suggesting that fluids may have a greater mobility across the particle. In addition, the contour lines of the renderings revealed that the thickness of the central part of the image affected the lines distribution, probably due to the irregular relief of those surfaces.

Table 3. Furrow parameters (maximum depth, mean depth, and mean density).

Furrow Parameters	Systems		
	P_0	P_{500}	P_{1000}
Maximum depth (μm)	78.973 ± 5.331	33.127 ± 1.762	29.623 ± 3.243
Mean depth (μm)	51.470 ± 3.118	17.722 ± 0.201	17.788 ± 1.506
Mean density (cm/cm^2)	31,933.762 ± 1044.323	42,288.498 ± 433.281	42,358.011 ± 643.838

According to these results, P_{500} e P_{1000} can be more easily penetrated by fluids, explaining the greater empty material volume in the central part of that surface.

All systems presented similar microtexture (Figure 4), because the direct texture parameters (Table 4) did not show a statistically significant difference (p-value > 0.05). Although the particles presented different morphologies, the texture distribution of the topographic patterns was similar. However, such analysis is still too qualitative to propose a specific system presenting the most uniform texture, because it does not take into account the evaluation of the spatial complexity of the surface roughness distribution, which was explored by the fractal parameters.

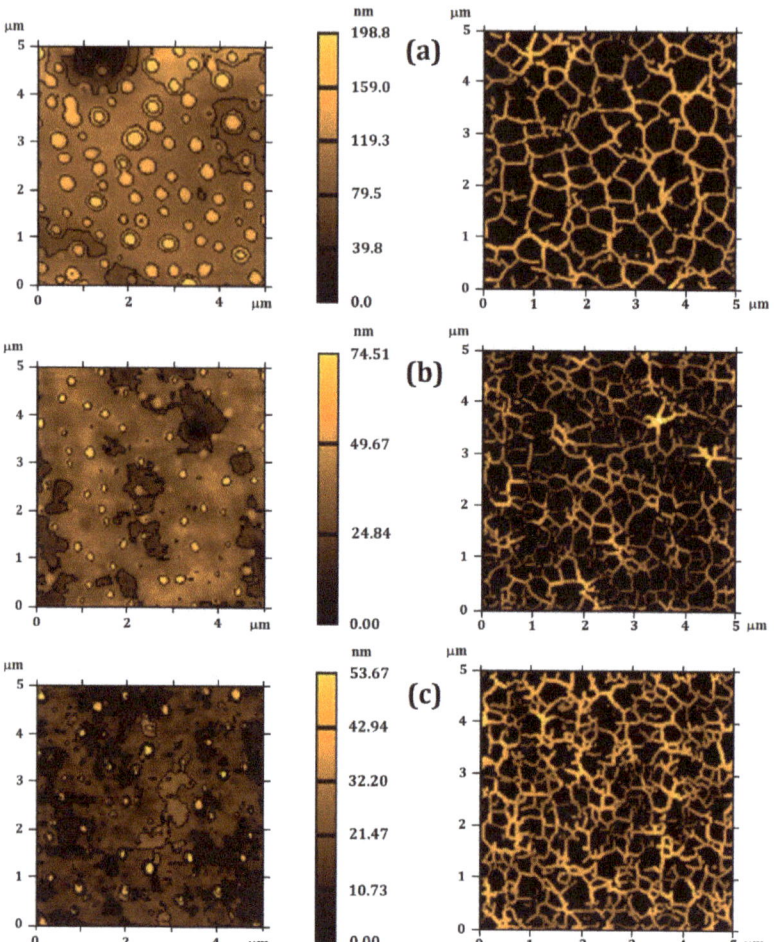

Figure 3. Renderings of the particle surface microtexture. Particles presented similar shapes in (**a**) P$_0$, while (**b**) P$_{500}$ and (**c**) P$_{1000}$ acquired smaller and more randomized sizes.

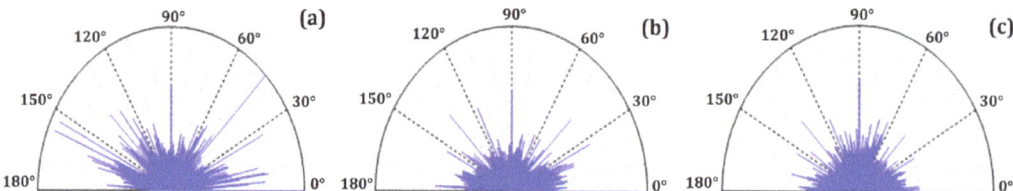

Figure 4. Surface texture directions for (**a**) P$_0$, (**b**) P$_{500}$, and (**c**) P$_{1000}$. All systems presented a similar microtexture, as the direct texture parameters did not show any statistically significant differences (p-value > 0.05).

Table 4. Surface texture isotropy (STI) and the respective directions.

Time (s)	First Direction (°) *	Second Direction (°) *	Third Direction (°) *	STI (%) *
P_0	134.995 ± 77.938	112.501 ± 38.974	88.624 ± 49.674	61.817 ± 19.551
P_{500}	165.995 ± 9.578	135.321 ± 0.453	37.626 ± 7.138	64.913 ± 7.4248
P_{1000}	67.503 ± 74.616	123.749 ± 37.310	112.511 ± 38.965	49.691 ± 17.423

* Samples without significant difference ANOVA One-Way and Tukey Test (p-value < 0.05).

3.3. Advanced Fractal Parameters

The fractal behavior of the particle surface was also evaluated, to obtain more quantitative information on the homogeneity of the microtexture. Microtexture evaluation using fractals and other related parameters has been extensively reported [70,71]. Since a fractal behavior has been attributed to objects in nature [44], several reports have focused on fractal theory to evaluate texture behavior in micro and nanoscales [72–74].

Table 5 presents the parameters fractal dimension (FD), surface entropy (H), fractal succolarity (FS), and lacunarity coefficient (β). FD is the first quantitative parameter associated with texture homogeneity. The fractal dimension presented similar values (p-value > 0.05), suggesting similar spatial complexity in all systems. For this reason, the surface microtexture was similar in P_{500} and P_{1000}, although showing different morphology. However, β was smaller in P_{1000}, suggesting more homogeneous surface microtexture. It is likely that the decrease of the surface roughness promoted the organization of surface gaps, resulting in a more homogeneous surface pattern for the system containing higher concentrations of essential oil. This homogeneity of the surface texture can allow a uniform mobility of fluids, improving its adsorption and release of essential oil.

Table 5. Fractal dimension (FD), surface entropy (H), fractal succolarity (FS), and lacunarity coefficient (β). Average results are expressed as mean values and standard deviations.

Time (s)	P_0	P_{500}	P_{1000}
FD *	2.30 ± 0.03	2.266 ± 0.006	2.29 ± 0.04
H *	0.93 ± 0.04	0.95 ± 0.03	0.90 ± 0.02
FS	0.61 ± 0.04	0.52 ± 0.01	0.59 ± 0.03
|β|	$5.74 \times 10^{-4} \pm 2.79 \times 10^{-5}$	$2.93 \times 10^{-4} \pm 6.43 \times 10^{-5}$	$1.18 \times 10^{-4} \pm 1.53 \times 10^{-5}$

* Samples without significant difference ANOVA One-Way and Tukey Test (p-value < 0.05).

On the other hand, the surface entropy analysis revealed that, although P_{500} presented more uniform height distribution (H~0.95), all particles exhibited $H \geq 0.9$ (p-value > 0.05). According to a previous report [49], surfaces with a H higher than 0.9 are significantly uniform, indicating that both P_{500} and P_{1000} can present similar adhesion and adsorption properties, although only P_{1000} presented a more homogeneous microtexture.

Although the FS values presented a significant difference (p-value < 0.05), the Tukey test revealed that P_{500} and P_{1000} were similar and could be equally penetrated by fluids. These values were close to 0.5, which is considered the ideal surface percolation value [50]. Adsorption and adhesion processes on other surfaces can also be influenced, as the entrances (allowing the interaction of ligand receptor sites between surfaces) are highly dependent on the surface texture [64]. Thus, it is important to obtain an FS value lower or close to 5, so that the encapsulated systems can release the essential oil in a controlled manner (as found in P_{500} and P_{1000}). These results revealed that the fractal parameters corroborated the results found in the stereometric parameters. However, the fractal lacunarity showed that P_{1000} presented the most suitable surface microtexture for adhesion to another surface, suggesting that this system could release the essential oil faster and more efficiently. For this reason, only the system P_{1000} was considered in further analyses.

3.4. Zeta Potential and Nanoparticle Tracking Analysis (NTA)

Zeta potential as a function of pH and NTA analysis was evaluated for the systems P_0 and P_{1000}. Zeta potential represents an important parameter for the evaluation of surface charge; besides, it is directly related to the colloidal system, influencing the particle size distribution and stability [15]. Furthermore, higher values (in module) of zeta potential are related to significant repulsion and reduction of aggregation/agglomeration [75].

A higher surface charge was found from pH $\geq$ 7 in the P_0 system (data not shown), allowing formulation stability. The surface charge ranged from (−5.0 ± 0.3) mV in pH = 7 to (−12.0 ± 0.8) mV in pH = 10. The isoelectric point was verified as close to pH = 4 and was related mainly to the type B gelatin carrier. It is known that two types of gelatin (A or B) can be produced, depending on the collagen pre-treatment [76].

The particles loaded with essential oil (P_{1000}) presented zeta potential values around (−54.3 ± 2.3) mV in pH = 8. The higher surface charge (in module) of the loaded particles can be attributed to the presence of the essential oil. The increased charges may be related to the compounds used to produce the particles and also to rearrangements among the essential oil constituents. The presence of these constituents probably resulted in an improved stabilization, due to new intermolecular interactions [15]: the surface electrostatic charge of particles can be influenced by several factors, including surface functional groups and solution ions [77]. On the other hand, electrostatic stability occurs due to the repulsion between particles, resulting from their high surface charge, never reaching the isoelectric point [78]. Thus, values equal to or greater than 30 mV (in modulus) are important for formulation stability [79]. For this reason, the surface charge of the P_{1000} system was found in a range that guarantees its stability as a colloidal system.

Unloaded (P_0) and loaded particles (P_{1000}) were characterized for number and size distribution by NTA (Figure 5). Table 6 shows the average particle size of P_0 and P_{1000}. The developed colloidal systems were compared, in terms of both size and concentration (particles/mL) as a function of encapsulated essential oil. No significant change in number of particles or in their size was observed, as registered by all the size descriptors.

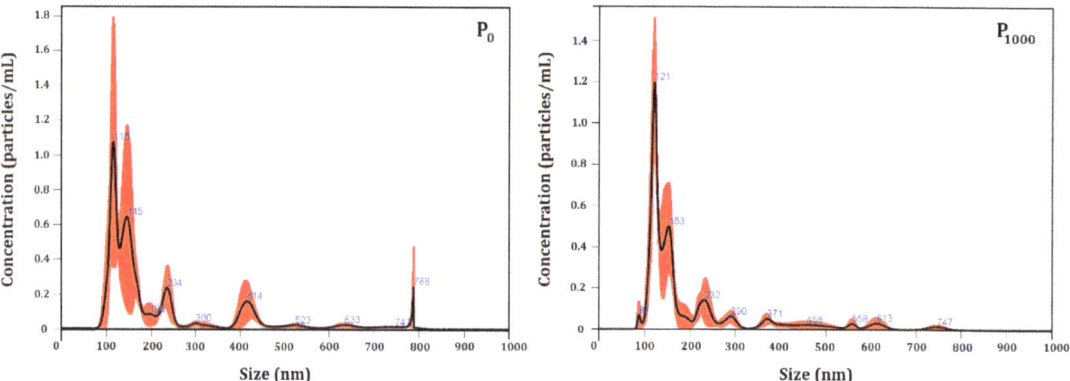

Figure 5. NTA particle size distribution analysis of P_0 and P_{1000} systems. Representative histograms of the average size distribution (black line) from three measurements of a single sample. Red areas specify the standard deviation (SD) between measurements, and blue numbers indicate the maxima of individual peaks.

The developed systems presented a polydisperse particle size distribution, ranging from 113 nm to 442 nm. Moreover, 90% of the particle population in the P_0 and P_{1000} systems presented a size up to (442 ± 12) nm and (405 ± 31) nm, respectively.

The mode parameter shows the particle size (or size range) most commonly found in the population distribution, and it is helpful to describe the midpoint for nonsymmetric distributions [80]. The value that best represents the encapsulated particle size was (128 ± 8) nm.

Table 6. Average particle size measured by NTA considering the P_0 and P_{1000} systems.

Parameters	P_0	P_{1000}
Mean (nm)	215 ± 14	202 ± 7
Mode (nm)	122 ± 12	128 ± 8
SD (nm)	161 ± 1	134 ± 15
D_{10} (nm)	113 ± 10	113 ± 3
D_{50} (nm)	135 ± 11	141 ± 8
D_{90} (nm)	442 ± 12	405 ± 31
Concentration (particles/mL)	$(6.0 \pm 0.9) \times 10^{10}$	$(5.0 \pm 0.6) \times 10^{10}$

Parameters D_{10}, D_{50}, and D_{90} indicated that 10%, 50%, or 90% of the particle's population, respectively, presented a diameter of less than or equal to the specified value.

Our results showed that the particle size distribution profile was not significantly influenced after the encapsulation of the essential oil. However, the presence of the essential oil in the P_{1000} system positively influenced its stability through the increase of the particle surface charge.

3.5. Laser Scanning Confocal Microscopy (LSCM) and Fluorescence Measurements

Figure 6 shows the particles images of the loaded particles, P_{1000}. Larger particles (μm) were selected. According to the NTA measurements, 10% of the loaded particles were larger than (405 ± 31). The essential oil was homogeneously located within the loaded particles/capsules. Moreover, an absence of essential oil was observed in the unloaded system (data not shown), as expected.

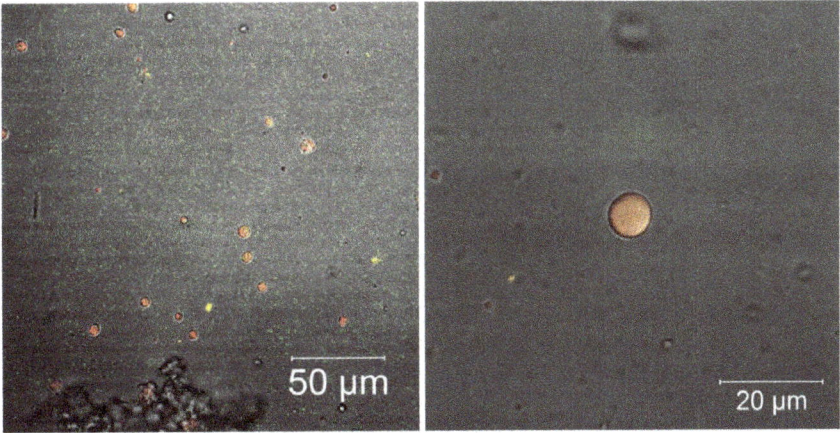

Figure 6. Confocal microscopy images of the particles from loaded system (P_{1000}).

Since the fluorescent properties of various molecules are highly dependent on the environment, this is a potentially useful method for determining material complexation [81].

Fluorescence measurements were performed on the unloaded and loaded particles. Emission spectra are presented in Figure 7 and show that the fluorescence intensity was mainly dependent on the essential oil. The luminance phenomenon of essential oil is caused by the π-electron conjugated system present in its constituents.

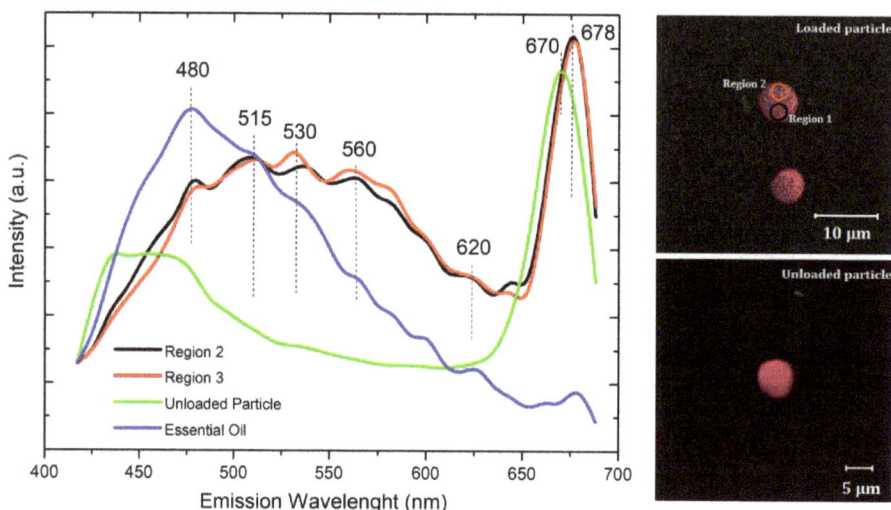

Figure 7. Fluorescence measurements of the loaded (regions 1 and 2) and unloaded particles.

The loaded particles presented a sensitive fluorescence response, under the same wavelength as the free essential oil. The emission spectrum of the P_{1000} system (regions 1 and 2) presented similar peaks, mainly at 480 nm, 515 nm, 530 nm, 560 nm, 620 nm, 670 nm, and 678 nm, confirming the essential oil encapsulation. In these cases, the fluorescence of the loaded particles was observed at a definite excitation length, owing to the fluorescent of secondary metabolites encapsulated within the polymeric particles. However, the fluorescence intensity of the loaded particles increased from 515 nm to 650 nm. In this system, well-defined and more intense emission peaks were assigned to the carriers (such as gelatin and PCL) and observed mainly at 678 nm. A blue shift of this peak was observed from 678 nm to 670 nm, due to the presence of essential oil. In conclusion, the results suggested weak interactions of an electrostatic nature that connected essential oil molecules with polymeric carriers. These interactions did not cause chemical changes in the essential oil. The emission peaks of the essential oil were not observed in the P_0 system, as expected.

Fluorescence measurements have been widely applied to evaluate chemical interactions in material complexation [82]. Similar results were observed elsewhere [83]. The composite of *bis*-eugenol/mesoporous silica presented a sensitive fluorescence response similar to that of free *bis*-eugenol obtained from clove oil. The authors suggested a weak hydrogen bond connecting the *bis*-eugenol molecules with the Si–OH groups of the silica porous wall. On the other hand, a significant enhancement of the fluorescence intensity of *Salvia sclarea* L. essential oil (SEO), due to its complexation with β-cyclodextrin (β-CD), was also investigated [82].

3.6. Controlled Release

The release kinetics were investigated, to understand the mechanisms of release of essential oil from the gelation/PCL particles as a function of the encapsulated concentration. Encapsulation efficiency (EE%) was found to be higher than 99% in both the P_{500} and P_{1000} systems.

Figure 8a shows the profile of release of essential oil. A significant difference was observed in the released concentration of essential oil in the P_{500} and P_{1000} systems. The essential oil release was evaluated up to 80 h, with maximum release concentrations of 63% and 95% for P_{500} and P_{1000}, respectively. These results agree with the fractal lacunarity values from AFM, which suggested that the P_{1000} system presented the most suitable surface microtexture for a more efficient release of essential oil. As also observed, the decrease of the surface roughness of P_{1000} resulted in a more homogeneous surface pattern. Thus, this observed homogeneity favors uniform mobility of fluids on the surface particle, as well as the solubilization of the gelatin carrier, improving its adsorption and the release of the bioactive compound. This is a possible reason for the lower concentration of essential oil released from the P_{500} system. A similar behavior of the encapsulated systems was also observed previously for gelatin/PCL particles containing essential oil from *Piper aduncum* and *Piper hispidinervum* [14].

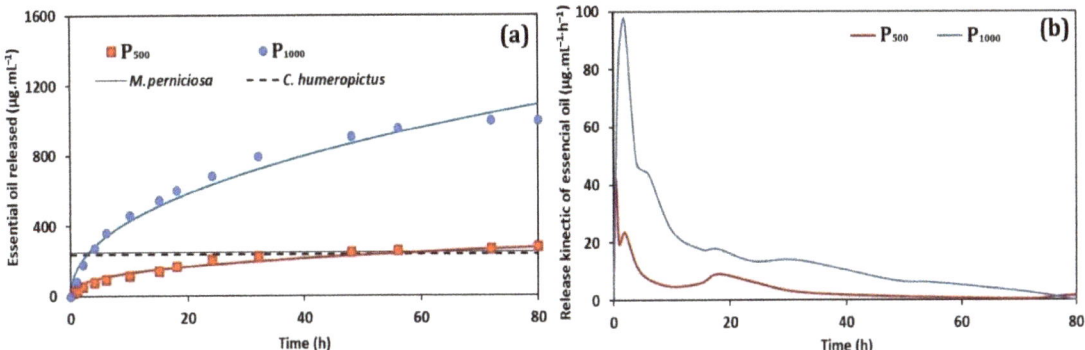

Figure 8. Controlled release curves of the P_{500} and P_{1000} systems: (**a**) concentration of released essential oil ($\mu g \cdot mL^{-1}$), and (**b**) kinetic essential oil release ($\mu g \cdot mL^{-1} h^{-1}$).

After 24 h, the P_{1000} system released $(51.5 \pm 0.3)\%$ of the total amount of encapsulated essential oil and, after 48 h, the released concentration reached $(90.2 \pm 0.4)\%$.

Figure 8b shows the derived curves from controlled release. All curves show a large release peak, representing a rapid release of essential oil in the first minutes of evaluation, resulting in the flow of essential oil into the solution. A rapid initial release followed by more sustained release was previously reported considering the essential oil of oregano in chitosan nanoparticles [84]: approximately 82% of the encapsulated essential oil was released up to 3 h. A rapid release of essential oil favors its high concentration in the medium, maintaining its effectiveness for a longer period [85].

The concentration of released essential oil observed in Figure 8a suggests that only the P_{1000} system may show effectiveness in controlling *C. humeropictus* and *M. perniciosa*, because their lethal dosages were reached (as shown in the next section). The profile of release of essential oil from the loaded particles was analyzed by applying the Higuchi [40] and Korsmeyer-Peppas [41] mathematical models. Linear regression was used to calculate the values of the release constants (k) and the correlation coefficients (R^2). The results are summarized in Table 7.

Table 7. Coefficients obtained from the controlled release according to the Higuchi and Korsmeyer–Peppas mathematical models.

Model	Coefficient	P_{500}	P_{1000}
Higuchi	K	31.1	12.46
	R^2	0.95	0.95
Korsmeyer–Peppas	K	57.0	14.4
	n	0.36	0.47
	R^2	0.99	0.99

The mathematical models presented good adjusted to the experimental curves, resulting in a R^2 from 0.95 to 0.99. The best fit to the release profiles of both P_{500} and P_{1000} was obtained using the Korsmeyer–Peppas mathematical model. Release profile curves were analyzed using a simple empirical model, $[f = kt^n]$ [86–88]. The kinetic constant k is a characteristic of a particular system considering structural and geometrical aspects; n is the release exponent representing four different mechanisms (Fickian diffusion, anomalous transport, Case-II transport, and Super Case-II transport) [89], considering spherical particles, and t is the release time.

The release mechanism by Fickian diffusion is the mechanism in which the active diffusion through the particle is exclusively determined by Fickian diffusion. In the case of anomalous transport, the active release is due both to Fickian diffusion and swelling/relaxation of the carrier. Case-II transport is controlled by the swelling and relaxation of carriers and it is independent of time. In Super Case-II transport, the release is ruled by the macromolecular relaxation of the polymeric chains [86].

In general, the n value determines the dominant release mechanism. Considering spherical particles, $n \leq 0.43$ represents a Fickian diffusion (Case I); $0.43 \leq n \leq 0.85$ represents an anomalous transport. When $n = 0.85$, the release is governed by Case-II transport, and $n > 0.85$ is related to Super Case-II transport [40].

The release assays showed that for the same period (80 h), there was a greater release of essential oil from the P_{1000} system. However, the release constant (k) values obtained for both P_{500} and P_{1000} systems (based on the Korsmeyer–Peppas mathematical model) showed that the release rate of the P_{1000} system (14.4 h^{-1}) was slower. Furthermore, the concentration of the encapsulated essential oil influenced the release mechanism. Particles containing a higher concentration of encapsulated essential oil (P_{1000}) were released according to the non-Fickian transport ($n = 47$). On the other hand, the P_{500} system presented a Fickian diffusion (Case-I) ($n = 0.36$) [90,91].

3.7. Insecticidal and Fungicidal Bioassays

The bioactivity of the *L. origanoides* essential oil against various pests can occur in different ways, causing mortality, deformation at different stages of development, as well as repellency [92]. Secondary metabolites have shown insect toxicity in the vapor phase, being reported as more toxic to microorganisms than in the contact form [14].

Our results indicated that the essential oil in natura presented an insecticidal activity against *C. humeropictus*. The LD$_{50}$ was found to be around (240 ± 25) µg·mL^{-1} after 24 h of exposure, with a lower confidence limit (LCL) and upper confidence limit (UCL), respectively, of 131 µg·mL^{-1} and 350 µg·mL^{-1}. The fiducial limit was considered as 95%.

The P_{1000} system was submitted to bioassays against *C. humeropictus*. Particles containing *L. origanoides* showed 100% mortality up to 48 h. About 80% of the borers were killed within 24 h. These results agree with the released concentration of essential oil from the controlled release curves and show that P_{1000} system was efficient against this tested pest, resulting in their control for up to 24 h. Similar works were found in the scientific literature considering other borer species [93,94].

The repellent activity of *Lippia origanoides*, *L. alba*, *Tagetes lucida*, *Rosmarinus officinalis*, *Cananga odorata*, *Eucalyptus citriodora*, and *Cymbopogon citratus* essential oils from Columbia were previously tested against the borer *Sitophilus zeamais* [95]. The authors observed that *L. origanoides* was found to be the most effective, causing (92 ± 3)% repellency at a 0.503 µL·cm^{-2} dose. The insecticidal activity of essential oils from *Thymus vulgaris* (thyme) and *Cymbopogon citratus* (lemongrass) against the devastating pest *Tuta absoluta* was also reported [96]. The resultant biological parameters for lemongrass and thyme oils were LD$_{50}$ of 1479 µL·mL^{-1} and 3046 µL·mL^{-1} for lemongrass and thyme oils, respectively, considering their fumigant toxicity.

The antifungal activity of *L. origanoides* has been extensively reported [97]. Considering the concentrations of essential oil added to the culture medium (0.125, 0.25, 0.5, 0.75, and 1 mg·mL^{-1}), the mycelial growth of *M. perniciosa* was observed only at 0.125 µg·mL^{-1}. For this reason, the tested concentration of 250 µg·mL^{-1} was considered as the MIC value.

The bioassays presented statistically significant differences (p-value < 0.05) between the essential oil and control. The treatments resulted in a percentage inhibition of mycelium growth of (57 ± 8)%, as shown in Table 8.

Table 8. Growth and inhibition parameters of *M. perniciosa*, considering the *L. origanoides* essential oil and the tested control.

	Diameter (mm)	RGM (%)	I (%)	MGI (mm/day)
L. origanoides	32 ± 6	43 ± 8	57 ± 8	4.6 ± 0.8
Control	74.8 ± 0.5	100.00	0.00	10.7 ± 0.1

RGM: relative mycelial growth percentage; I: percentage inhibition of mycelium growth; MGI: mycelial growth index. Negative control: DMSO.

The efficiency of the P_{1000} system was evaluated against *M. perniciosa*. After 7 days of incubation, no mycelial growth percentage was observed. Carvacrol has been reported as the major constituent of the essential oil from *L. origanoides* [23] and has shown efficiency in controlling bacteria [98,99], fungi [98,100], and insects [23,101]. The inhibition of *Phytophthora infestans*, a phytopathogen of potato, was reported with MIC = 150 µg mL^{-1}, confirming the efficiency of this essential oil in controlling pathogenic fungi [102].

4. Conclusions

The present study successfully developed gelatin/PCL-based particles as useful carriers of the essential oil from *L. origanoides*. The proposed colloidal system can release lethal dosage concentrations to control *C. humeropictus* and *M. perniciosa* for up to 24 h, which are the main pests of *Theobroma grandiflorum*. The AFM data also showed that the encapsulation of essential oil affected the particle's surface morphology. The surface roughness decreased as a function of the concentration of encapsulated essential oil. The homogeneity of the surface texture observed in the P_{1000} system allowed a uniform mobility of fluids on the surface, improving its adsorption and release of essential oil. These results were observed in the controlled release assays. The nanoscale surface and fractal analysis based on AFM technique represent an useful tool for quality standards in manufacturing particles containing encapsulated essential oil. For this reason, our results suggested that the developed particles containing encapsulated essential oil could be applied as a sustainable alternative controlling agent for the tested pests, combined with their biodegradability and adequate controlled release, with promising future applications.

Author Contributions: E.A.S.: project coordination and administration. A.L.F.R., R.Z.d.A.N. and E.A.S.: conceptualization, methodology and data collecting/analysis. A.L.F.R.: particles development and essential oil encapsulation. E.A.S., V.S.B., N.M.I. and S.T.: original draft preparation, data analysis and funding acquisition. J.d.A.B. and P.H.C.: controlled release assays and interpretation. A.L.F.R., E.A.S., V.S.B. and N.M.I.: Zeta potential and nanoparticle tracking analysis (NTA) measurement and data interpretation. R.S.M. and H.D.d.F.F.: AFM measurements and interpretation. A.R.L., E.A.S. and F.E.G.G.: laser scanning confocal microscopy and fluorescence measurements and interpretation. A.L.F.R., R.Z.d.A.N., A.M.S.R.P., C.M. and M.G.d.S.: insecticidal and fungicidal bioassays and interpretation. All authors have read and agreed to the published version of the manuscript.

Funding: CAPES (Coordenação de Aperfeiçoamento de Pessoal de Nível Superior—Código Financeiro 001), CNPq (Conselho Nacional de Desenvolvimento Científico e Tecnológico, grant number 403496/2013-6, Programa Sisfóton and INCT), FAPEAM (Fundação de Amparo à Pesquisa do Estado do Amazonas, Edital 004/2018—Amazonas Estratégico—062.01305/2018—Pedro H. Campelo Felix, Jaqueline de A. Bezerra, Ana Luisa F. Rocha and Ronald Z. Aguiar), FAPESP (Fundação de Amparo à Pesquisa do Estado de São Paulo/CEPOF (Centro de Pesquisas em Óptica e Fotônica) grant number 2013/07276-1 and FAPESP (EMU) grant number 09/54035-4.

Institutional Review Board Statement: Not applicable.

Informed Consent Statement: Not applicable.

Data Availability Statement: Not applicable.

Acknowledgments: The authors thank CAPES (Coordenação de Aperfeiçoamento de Pessoal de Nível Superior, CNPq (Conselho Nacional de Desenvolvimento Científico e Tecnológico), FAPEAM (Fundação de Amparo à Pesquisa do Estado do Amazonas), FAPESP (Fundação de Amparo à Pesquisa do Estado de São Paulo), CEPOF (Centro de Pesquisas em Óptica e Fotônica) for the financial support, and to the Grupo de Nanomedicina e Nanotoxicologia (GNano – IFSC/USP) for the measurements of Nanoparticle Tracking Analysis (NTA) and EMBRAPA Amazônia Ocidental (Manaus/AM) for providing the *C. humeropictus* individuals and *M. perniciosa* fungi, as well as the fungicidal bioassays infrastructure.

Conflicts of Interest: The authors declare no conflict of interest.

References

1. Abdelaziz, D.; Hefnawy, A.; Al-Wakeel, E.; El-Fallal, A.; El-Sherbiny, I.M. New biodegradable nanoparticles-in-nanofibers based membranes for guided periodontal tissue and bone regeneration with enhanced antibacterial activity. *J. Adv. Res.* **2021**, *28*, 51–62. [CrossRef] [PubMed]
2. Babaee, M.; Garavand, F.; Rehman, A.; Jafarazadeh, S.; Amini, E.; Cacciotti, I. Biodegradability, physical, mechanical and antimicrobial attributes of starch nanocomposites containing chitosan nanoparticles. *Int. J. Biol. Macromol.* **2022**, *195*, 49–58. [CrossRef]
3. Wang, C.; Gong, C.; Qin, Y.; Hu, Y.; Jiao, A.; Jin, Z.; Qiu, C.; Wang, J. Bioactive and functional biodegradable packaging films reinforced with nanoparticles. *J. Food Eng.* **2022**, *312*, 110752. [CrossRef]
4. Wang, L.; Gao, Y.; Xiong, J.; Shao, W.; Cui, C.; Sun, N.; Zhang, Y.; Chang, S.; Han, P.; Liu, F.; et al. Biodegradable and high-performance multiscale structured nanofiber membrane as mask filter media via poly(lactic acid) electrospinning. *J. Colloid Interface Sci.* **2022**, *606*, 961–970. [CrossRef]
5. Sharma, R.; Tripathi, A. Green synthesis of nanoparticles and its key applications in various sectors. *Mater. Today Proc.* **2022**, *48*, 1626–1632. [CrossRef]
6. Reddy, L.H.; Arias, J.L.; Nicolas, J.; Couvreur, P. Magnetic nanoparticles: Design and characterization, toxicity and biocompatibility, pharmaceutical and biomedical applications. *Chem. Rev.* **2012**, *112*, 5818–5878. [CrossRef] [PubMed]
7. Wang, B.; Tang, Y.; Oh, Y.; Lamb, N.W.; Xia, S.; Ding, Z.; Chen, B.; Suarez, M.J.; Meng, T.; Kulkarni, V.; et al. Controlled release of dexamethasone sodium phosphate with biodegradable nanoparticles for preventing experimental corneal neovascularization. *Nanomed. Nanotechnol. Biol. Med.* **2019**, *17*, 119–123. [CrossRef]
8. Silva, J.S.M.; Rabelo, M.S.; Lima, S.X.; Rocha, A.N.A.L.F.; Tadei, W.P.; Chaves, F.C.M.; Bezerra, J.D.E.A.; Biondo, M.M.; Campelo, P.H.; Sanches, E.A. Biodegradable nanoparticles loaded with Lippia alba essential oil: A sustainable alternative for *Aedes aegypti* larvae control. *Eur. Acad. Res.* **2020**, *VII*, 6237–6258.
9. Pandey, V.K.; Islam, R.U.; Shams, R.; Dar, A.H. A comprehensive review on the application of essential oils as bioactive compounds in nano-emulsion based edible coatings of fruits and vegetables. *Appl. Food Res.* **2022**, *2*, 100042. [CrossRef]
10. Khizar, S.; Zine, N.; Errachid, A.; Jaffrezic-Renault, N.; Elaissari, A. Microfluidic-based nanoparticle synthesis and their potential applications. *Electrophoresis* **2021**, *43*, 819–838. [CrossRef] [PubMed]
11. Corrado, I.; Di Girolamo, R.; Regalado-González, C.; Pezzella, C. Polyhydroxyalkanoates-based nanoparticles as essential oil carriers. *Polymers* **2022**, *14*, 166. [CrossRef] [PubMed]
12. Nair, A.; Mallya, R.; Suvarna, V.; Khan, T.A.; Momin, M.; Omri, A. Nanoparticles—Attractive carriers of antimicrobial essential oils. *Antibiotics* **2022**, *11*, 108. [CrossRef] [PubMed]
13. Zhang, W.; Jiang, H.; Rhim, J.-W.; Cao, J.; Jiang, W. Effective strategies of sustained release and retention enhancement of essential oils in active food packaging films/coatings. *Food Chem.* **2022**, *367*, 130671. [CrossRef] [PubMed]
14. Silva, L.S.; Mar, J.M.; Azevedo, S.G.; Rabelo, M.S.; Bezerra, J.A.; Campelo, P.H.; Machado, M.B.; Trovati, G.; Santos, L.; Fonseca, D.; et al. Encapsulation of Piper aduncum and Piper hispidinervum essential oils in gelatin nanoparticles: A possible sustainable control tool of Aedes aegypti, Tetranychus urticae and Cerataphis lataniae. *J. Sci. Food Agric.* **2018**, *99*, 685–695. [CrossRef]
15. De Oliveira, L.M.; Silva, L.S.; Mar, J.M.; Azevedo, S.G.; Rabelo, M.S.; Da Fonseca Filho, H.D.; Lima, S.X.; Bezerra, J.D.A.; Machado, M.B.; Campelo, P.H.; et al. Alternative biodefensive based on the essential oil from *Allium sativum* encapsulated in PCL/Gelatin nanoparticles. *J. Food Eng. Technol.* **2019**, *8*, 65–74. [CrossRef]
16. Kusumastuti, Y.; Istiani, A.; Rochmadi; Purnomo, C.W. Chitosan-based polyion multilayer coating on NPK fertilizer as controlled released fertilizer. *Adv. Mater. Sci. Eng.* **2019**, *2019*, 2958021. [CrossRef]
17. De Oliveira, L.M.; Lima, S.X.; Silva, L.S.; Mar, J.M.; Azevedo, S.G.; Rabelo, M.S.; Henrique, D.; Filho, F.; Campelo, P.H.; Sanches, E.A. Controlled release of *Licaria puchury-major* essential oil encapsulated in PCL/gelatin-based colloidal systems and membranes. *Am. J. Essent. Oils Nat. Prod.* **2019**, *7*, 23–29.
18. da Costa, Í.C.; Saraiva Matos, R.S.; de Azevedo, S.G.; Costa, C.A.R.; Sanches, E.A.; da Fonseca Filho, H. Microscopy-based infrared spectroscopy as a tool to evaluate the influence of essential oil on the surface of loaded bilayered-nanoparticles. *Nanotechnology* **2021**, *32*, 345703. [CrossRef]

19. de Oliveira, L.M.; Matos, R.S.; Campelo, P.H.; Sanches, E.A.; da Fonseca Filho, H.D. Evaluation of the nanoscale surface applied to biodegradable nanoparticles containing *Allium sativum* essential oil. *Mater. Lett.* **2020**, *275*, 128111. [CrossRef]
20. Nguyen, M.-H.; Nguyen, T.-H.-N.; Tran, T.-N.-M.; Vu, N.-B.-D.; Tran, T.-T. Comparison of the nematode-controlling effectiveness of 10 different essential oil-encapsulated lipid nanoemulsions. *Arch. Phytopathol. Plant Prot.* **2022**, *55*, 420–432. [CrossRef]
21. Elumalai, K.; Krishnappa, K.; Pandiyan, J.; Alharbi, N.S.; Kadaikunnan, S.; Khaled, J.M.; Barnard, D.R.; Vijayakumar, N.; Govindarajan, M. Characterization of secondary metabolites from Lamiaceae plant leaf essential oil: A novel perspective to combat medical and agricultural pests. *Physiol. Mol. Plant Pathol.* **2022**, *117*, 101752. [CrossRef]
22. Pedrotti, C.; Caro, I.M.D.D.; Franzoi, C.; Grohs, D.S.; Schwambach, J. Control of anthracnose (*Elsinoë ampelina*) in grapevines with *Eucalyptus staigeriana* essential oil. *Org. Agric.* **2022**, *12*, 81–89. [CrossRef]
23. Mar, J.M.; Silva, L.S.; Azevedo, S.G.; França, L.P.; Goes, A.F.F.; dos Santos, A.L.; Bezerra, J.d.A.; de Cássia, S.; Nunomura, R.; Machado, M.B.; et al. *Lippia origanoides* essential oil: An efficient alternative to control *Aedes aegypti*, *Tetranychus urticae* and *Cerataphis lataniae*. *Ind. Crops Prod.* **2018**, *111*, 292–297. [CrossRef]
24. Vicuña, G.C.; Stashenko, E.E.; Fuentes, J.L. Chemical composition of the *Lippia origanoides* essential oils and their antigenotoxicity against bleomycin-induced DNA damage. *Fitoterapia* **2010**, *81*, 343–349. [CrossRef]
25. Stashenko, E.E.; Martínez, J.R.; Cala, M.P.; Durán, D.C.; Caballero, D. Chromatographic and mass spectrometric characterization of essential oils and extracts from Lippia (Verbenaceae) aromatic plants. *J. Sep. Sci.* **2013**, *36*, 192–202. [CrossRef]
26. Hassan, H.F.H.; Mansour, A.M.; Salama, S.A.; El-Sayed, E.-S.M. The chemopreventive effect of thymol against dimethylhydrazine and/or high fat diet-induced colon cancer in rats: Relevance to NF-κB. *Life Sci.* **2021**, *274*, 119335. [CrossRef]
27. Borges, A.R.; de Albuquerque Aires, J.R.; Higino, T.M.M.; de Medeiros, M.d.G.F.; Citó, A.M.d.G.L.; Lopes, J.A.D.; de Figueiredo, R.C.B.Q. Trypanocidal and cytotoxic activities of essential oils from medicinal plants of Northeast of Brazil. *Exp. Parasitol.* **2012**, *132*, 123–128. [CrossRef]
28. Caballero-Gallardo, K.; Olivero-Verbel, J.; Stashenko, E.E. Repellency and toxicity of essential oils from *Cymbopogon martinii*, *Cymbopogon flexuosus* and *Lippia origanoides* cultivated in Colombia against *Tribolium castaneum*. *J. Stored Prod. Res.* **2012**, *50*, 62–65. [CrossRef]
29. Perlatti, B.; Souza Bergo, P.L.; Silva, M.F.d.G.F.d.; Batista, J.; Rossi, M. Polymeric nanoparticle-based insecticides: A controlled release purpose for agrochemicals. *Insectic. Dev. Safer More Eff. Technol.* **2013**, 521–548. [CrossRef]
30. Papanikolaou, N.E.; Kalaitzaki, A.; Karamaouna, F.; Michaelakis, A.; Papadimitriou, V.; Dourtoglou, V.; Papachristos, D.P. Nano-formulation enhances insecticidal activity of natural pyrethrins against *Aphis gossypii* (Hemiptera: Aphididae) and retains their harmless effect to non-target predators. *Environ. Sci. Pollut. Res.* **2017**, *25*, 10243–10249. [CrossRef]
31. Santos, R.A.R.S.; Silva, N.M. Primeiro registro de *Conotrachelus humeropictus* Fiedler, 1940 (Coleoptera: Curculionidae) no estado do Amapá, Brasil. *Biota Amaz.* **2020**, *10*, 69–70.
32. Pereira, A.L.F.; Abreu, V.K.G.; Rodrigues, S. Cupuassu— *Theobroma grandiflorum*. In *Exotic Fruits*; Elsevier: Amsterdam, The Netherlands, 2018; pp. 159–162.
33. Moura Rebouças, A.; Martins da Costa, D.; Priulli, E.; Teles, J.; Roberta Freitas Pires, C. Aproveitamento tecnológico das sementes de cupuaçu e de okara na obtenção do cupulate. *DESAFIOS Rev. Interdiscip. Univ. Fed. Tocantins* **2020**, *7*, 59–64. [CrossRef]
34. Lopes, C.M.D.; Silva, N.M. Impacto econômico da broca do cupuaçu, *Conotrachelus humeropictus* Field (Coleoptera: Curculionidae) nos estados do Amazonas e Rondônia. *An. Soc. Entomológica Bras.* **1998**, *27*, 481–483. [CrossRef]
35. Thomazini, M.J. Flutuação populacional e intensidade de infestação da Broca-dos-frutos em cupuaçu. *Sci. Agric.* **2002**, *59*, 463–468. [CrossRef]
36. Lopes, J.R.M.; Luz, E.D.M.N.; Bezerra, J.L. Suscetibilidade do cupuaçuzeiro e outras espécies vegetais a isolados de *Crinipellis perniciosa* obtidos de quatro hospedeiros diferentes no sul da Bahia. *Fitopatol. Bras.* **2001**, *26*, 601–605. [CrossRef]
37. Falcão, M.D.A.; De Morais, R.R.; Clement, C.R. Influência da vassoura de bruxa na fenologia do Cupuaçuzeiro. *Acta Amaz.* **1999**, *29*, 13. [CrossRef]
38. Kah, M.; Hofmann, T. Nanopesticide research: Current trends and future priorities. *Environ. Int.* **2014**, *63*, 224–235. [CrossRef]
39. Khot, L.R.; Sankaran, S.; Maja, J.M.; Ehsani, R.; Schuster, E.W. Applications of nanomaterials in agricultural production and crop protection: A review. *Crop Prot.* **2012**, *35*, 64–70. [CrossRef]
40. Boyapally, H.; Nukala, R.K.; Bhujbal, P.; Douroumis, D. Controlled release from directly compressible theophylline buccal tablets. *Colloids Surf. B Biointerfaces* **2010**, *77*, 227–233. [CrossRef]
41. Korsmeyer, R.; Peppas, N. Macromolecular and modeling aspects of swelling controlled systems. In *Controlled Release Delivery Systems*; Springer: Berlin/Heidelberg, Germany, 1983; pp. 77–90.
42. Ghasemishahrestani, Z.; Mehta, M.; Darne, P.; Yadav, A.; Ranade, S. Tunable synthesis of gelatin nanoparticles employing sophorolipid and plant extract, a promising drug carrier. *World J. Pharm. Pharm. Sci.* **2015**, *4*, 1365–1381.
43. Matos, R.S.; Pinto, E.P.; Ramos, G.Q.; da Fonseca de Albuquerque, M.D.; da Fonseca Filho, H.D. Stereometric characterization of kefir microbial films associated with *Maytenus rigida* extract. *Microsc. Res. Tech.* **2020**, *83*, 1401–1410. [CrossRef] [PubMed]
44. Mandelbrot, B.B.; Wheeler, J.A. The fractal geometry of nature. *Am. J. Phys.* **1983**, *51*, 286–287. [CrossRef]
45. de Assis, T.A.; Vivas Miranda, J.G.; de Brito Mota, F.; Andrade, R.F.S.; de Castilho, C.M.C. Geometria fractal: Propriedades e características de fractais ideais. *Rev. Bras. Ensino Fis.* **2008**, *30*, 2304.1–2304.10. [CrossRef]
46. Nečas, D.; Klapetek, P. Gwyddion: An open-source software for SPM data analysis. *Cent. Eur. J. Phys.* **2012**, *10*, 181–188. [CrossRef]

47. Nosonovsky, M. Entropy in tribology: In the search for applications. *Entropy* **2010**, *12*, 1345–1390. [CrossRef]
48. Horcas, I.; Fernández, R.; Gómez-Rodríguez, J.M.; Colchero, J.; Gómez-Herrero, J.; Baro, A.M. WSXM: A software for scanning probe microscopy and a tool for nanotechnology. *Rev. Sci. Instrum.* **2007**, *78*, 013705. [CrossRef]
49. Matos, R.S.; Lopes, G.A.C.; Ferreira, N.S.; Pinto, E.P.; Carvalho, J.C.T.; Figueiredo, S.S.; Oliveira, A.F.; Zamora, R.R.M. Superficial characterization of Kefir biofilms associated with açaí and cupuaçu extracts. *Arab. J. Sci. Eng.* **2018**, *43*, 3371–3379. [CrossRef]
50. Heitor, R.; De Melo, C.; De Melo, R.H.C.; Conci, A. Succolarity: Defining a method to calculate this fractal measure. In Proceedings of the 2008 15th International Conference on Systems, Signals and Image Processing, Bratislava, Slovakia, 25–28 June 2008; pp. 291–294. [CrossRef]
51. Henebry, G.M.; Kux, H.J. Lacunarity as a texture measure for SAR imagery. *Int. J. Remote Sens.* **1995**, *16*, 565–571. [CrossRef]
52. Țălu, Ș.; Abdolghaderi, S.; Pinto, E.P.; Matos, R.S.; Salerno, M. Advanced fractal analysis of nanoscale topography of Ag/DLC composite synthesized by RF-PECVD. *Surf. Eng.* **2020**, *36*, 713–719. [CrossRef]
53. Ricardo, L.; De Lucena, R.; Stosic, T. Utilização de lacunaridade para detecção de padrões de imagens de retinas humanas. *Rev. Estatística Univ. Fed. Ouro Preto* **2014**, *3*, 789–793.
54. Severina, O. *Programsko Okolje Force 3.0*; Fakulteta za Kemijo in Kemijsko Tehnologijo: Ljubljana, Slovenia, 2012.
55. Naidu, B.V.K.; Paulson, A.T. A new method for the preparation of gelatin nanoparticles: Encapsulation and drug release characteristics. *J. Appl. Polym. Sci.* **2011**, *121*, 3495–3500. [CrossRef]
56. Gaddum, J.H. *Probit Analysis*; Cambridge University Press: Cambridge, UK, 1948; Volume 60, ISBN 0-521-08041-X.
57. Barci, L.A.G.; de Almeida, J.E.M.; de Campos Nogueira, A.H.; do Prado, A.P. Determinação da CL90 e TL90 do isolado IBCB66 de *Beauveria bassiana* (Ascomycetes: Clavicipitaceae) para o controle de *Rhipicephalus (Boophilus) microplus* (Acari: Ixodidae). *Rev. Bras. Parasitol. Vet.* **2009**, *18*, 34–39. [CrossRef] [PubMed]
58. Pietrobelli, S.R.; Portolan, I.B.; Moura, G.S.; Franzener, G. Preparados de plantas bioativas na indução de fitoalexinas e no controle in vitro de fitopatógenos do tomateiro/Preparations of bioative plants in phytoalexins induction and in vitro control of tomato phytopathogens. *Brazilian J. Dev.* **2020**, *6*, 102316–102331. [CrossRef]
59. Kim, C.H.; Hassan, O.; Chang, T. Diversity, pathogenicity, and fungicide sensitivity of Colletotrichum species associated with apple anthracnose in South Korea. *Plant Dis.* **2020**, *104*, 2866–2874. [CrossRef]
60. Gautam, S.; Dinda, A.K.; Mishra, N.C. Fabrication and characterization of PCL/gelatin composite nanofibrous scaffold for tissue engineering applications by electrospinning method. *Mater. Sci. Eng. C* **2013**, *33*, 1228–1235. [CrossRef] [PubMed]
61. Gautam, S.; Chou, C.F.; Dinda, A.K.; Potdar, P.D.; Mishra, N.C. Surface modification of nanofibrous polycaprolactone/gelatin composite scaffold by collagen type I grafting for skin tissue engineering. *Mater. Sci. Eng. C* **2014**, *34*, 402–409. [CrossRef] [PubMed]
62. Shokrollahi, M.; Bahrami, S.H.; Nazarpak, M.H.; Solouk, A. Multilayer nanofibrous patch comprising chamomile loaded carboxyethyl chitosan/poly(vinyl alcohol) and polycaprolactone as a potential wound dressing. *Int. J. Biol. Macromol.* **2020**, *147*, 547–559. [CrossRef]
63. Tyrrell, J.W.G.; Attard, P. Images of nanobubbles on hydrophobic surfaces and their interactions. *Phys. Rev. Lett.* **2001**, *87*, 176104. [CrossRef]
64. Israelachvili, J.N. *Intermolecular and Surface Forces*, 3rd ed.; Elsevier: Amsterdam, The Netherlands, 2011; pp. 1–676. [CrossRef]
65. Yu, N.; Polycarpou, A.A. Contact of rough surfaces with asymmetric distribution of asperity heights. *J. Tribol.* **2002**, *124*, 367–376. [CrossRef]
66. Leach, R. *Characterisation of Areal Surface Texture*; Springer: Berlin/Heidelberg, Germany, 2013; ISBN 9783642364587.
67. Franco, L.A.; Sinatora, A. 3D surface parameters (ISO 25178-2): Actual meaning of Spk and its relationship to Vmp. *Precis. Eng.* **2015**, *40*, 106–111. [CrossRef]
68. Solaymani, S.; Țălu, Ș.; Nezafat, N.B.; Rezaee, S.; Kenari, M.F. Diamond nanocrystal thin films: Case study on surface texture and power spectral density properties. *AIP Adv.* **2020**, *10*, 045206. [CrossRef]
69. Țălu, Ș.; Ghaderi, A.; Stępień, K.; Mwema, F.M. Advanced Micromorphology Analysis of Cu/Fe NPs Thin Films. *IOP Conf. Ser. Mater. Sci. Eng.* **2019**, *611*, 012016. [CrossRef]
70. Mahboob Kanafi, M.; Kuosmanen, A.; Pellinen, T.K.; Tuononen, A.J. Macro-and micro-texture evolution of road pavements and correlation with friction. *Int. J. Pavement Eng.* **2015**, *16*, 168–179. [CrossRef]
71. Țălu, Ș.; Matos, R.S.; Pinto, E.P.; Rezaee, S.; Mardani, M. Stereometric and fractal analysis of sputtered Ag-Cu thin films. *Surf. Interfaces* **2020**, *21*, 100650. [CrossRef]
72. Gonçalves, E.C.M.; Pinto, E.P.; Ferreira, N.; De Sergipe, U.F.; Matos, R.S. Fractal study of kefir biofilms. In Proceedings of the XVIII Brazil MRS Meeting, Balneário Camboriú, Brazil, 25–26 September 2019.
73. Omar, M.; Salcedo, C.; Ronald, R.; Zamora, M.; Tavares, C. Study fractal leaf surface of the plant species Copaifera sp. using the Microscope Atomic-Force-AFM. *Rev. ECIPerú* **2016**, *13*, 10–16. [CrossRef]
74. Szerakowska, S.; Woronko, B.; Sulewska, M.J.; Oczeretko, E. Spectral method as a tool to examine microtextures of quartz sand-sized grains. *Micron* **2018**, *110*, 36–45. [CrossRef]
75. Honary, S.; Zahir, F. Effect of zeta potential on the properties of nano-drug delivery systems—A review (Part 2). *Trop. J. Pharm. Res.* **2013**, *12*, 265–273. [CrossRef]
76. Young, S.; Wong, M.; Tabata, Y.; Mikos, A.G. Gelatin as a delivery vehicle for the controlled release of bioactive molecules. *J. Control. Release* **2005**, *109*, 256–274. [CrossRef]

77. Mahmoudi, M.; Lynch, I.; Ejtehadi, M.R.; Monopoli, M.P.; Bombelli, F.B.; Laurent, S. Protein-nanoparticle interactions: Opportunities and challenges. *Chem. Rev.* **2011**, *111*, 5610–5637. [CrossRef]
78. Campelo, P.H.; Junqueira, L.A.; de Resende, J.V.; Zacarias, R.D.; de Fernandes, R.V.B.; Botrel, D.A.; Borges, S.V. Stability of lime essential oil emulsion prepared using biopolymers and ultrasound treatment. *Int. J. Food Prop.* **2017**, *20*, S564–S579. [CrossRef]
79. Roland, I.; Piel, G.; Delattre, L.; Evrard, B. Systematic characterization of oil-in-water emulsions for formulation design. *Int. J. Pharm.* **2003**, *263*, 85–94. [CrossRef]
80. Seibert, J.B.; Viegas, J.S.R.; Almeida, T.C.; Amparo, T.R.; Rodrigues, I.V.; Lanza, J.S.; Frézard, F.J.G.; Soares, R.D.O.A.; Teixeira, L.F.M.; de Souza, G.H.B.; et al. Nanostructured systems improve the antimicrobial potential of the essential oil from Cymbopogon densiflorus leaves. *J. Nat. Prod.* **2019**, *82*, 3208–3220. [CrossRef] [PubMed]
81. Marques, H.M.C. A review on cyclodextrin encapsulation of essential oils and volatiles. *Flavour Fragr. J.* **2010**, *25*, 313–326. [CrossRef]
82. Tian, X.-N.; Jiang, Z.-T.; Li, R. Inclusion interactions and molecular microcapsule of *Salvia sclarea* L. essential oil with β-cyclodextrin derivatives. *Eur. Food Res. Technol.* **2008**, *227*, 1001–1007. [CrossRef]
83. Guntero, V.; Ferretti, C.; Mancini, P.; Kneeteman, M. Synthesis and encapsulation of bis-eugenol in a mesoporous solid material: Enhancement of the antioxidant activity of a natural compound from clove oil. *Chem. Sci. Int. J.* **2018**, *22*, 1–10. [CrossRef]
84. Hosseini, S.F.; Zandi, M.; Rezaei, M.; Farahmandghavi, F. Two-step method for encapsulation of oregano essential oil in chitosan nanoparticles: Preparation, characterization and in vitro release study. *Carbohydr. Polym.* **2013**, *95*, 50–56. [CrossRef]
85. Peng, H.; Xiong, H.; Li, J.; Xie, M.; Liu, Y.; Bai, C.; Chen, L. Vanillin cross-linked chitosan microspheres for controlled release of resveratrol. *Food Chem.* **2010**, *121*, 23–28. [CrossRef]
86. Korsmeyer, R.W.; Gurny, R.; Doelker, E.; Buri, P.; Peppas, N.A. Mechanisms of solute release from porous hydrophilic polymers. *Int. J. Pharm.* **1983**, *15*, 25–35. [CrossRef]
87. Maderuelo, C.; Zarzuelo, A.; Lanao, J.M. Critical factors in the release of drugs from sustained release hydrophilic matrices. *J. Control. Release* **2011**, *154*, 2–19. [CrossRef]
88. Siepmann, J.; Siepmann, F. Mathematical modeling of drug dissolution. *Int. J. Pharm.* **2013**, *453*, 12–24. [CrossRef] [PubMed]
89. Vincekovi ́c, M.; Juri ́c, S.; Đermi ́c, E.; Topolovec-Pintari ́c, S. Kinetics and mechanisms of chemical and biological agents release from biopolymeric microcapsules. *J. Agric. Food Chem.* **2017**, *65*, 9608–9617. [CrossRef]
90. de Oliveira, E.F.; Paula, H.C.B.; Paula, R.C.M. de Alginate/cashew gum nanoparticles for essential oil encapsulation. *Colloids Surf. B Biointerfaces* **2014**, *113*, 146–151. [CrossRef]
91. Paula, H.C.B.; Oliveira, E.F.; Carneiro, M.J.M.; De Paula, R.C.M. Matrix Effect on the spray drying nanoencapsulation of *Lippia sidoides* essential oil in chitosan-native gum blends. *Planta Med.* **2017**, *83*, 392–397. [CrossRef]
92. Isman, M.B. Botanical insecticides, deterrents, and repellents in modern agriculture and an increasingly regulated world. *Annu. Rev. Entomol.* **2006**, *51*, 45–66. [CrossRef]
93. Babarinde, S.A.; Olaniran, O.A.; Ottun, A.T.; Oderinde, A.E.; Adeleye, A.D.; Ajiboye, O.; Dawodu, E.O. Chemical composition and repellent potentials of two essential oils against larger grain borer, *Prostephanus truncatus* (Horn.) (Coleoptera: Bostrichidae). *Biocatal. Agric. Biotechnol.* **2021**, *32*, 101937. [CrossRef]
94. Campolo, O.; Cherif, A.; Ricupero, M.; Siscaro, G.; Grissa-Lebdi, K.; Russo, A.; Cucci, L.M.; Di Pietro, P.; Satriano, C.; Desneux, N.; et al. Citrus peel essential oil nanoformulations to control the tomato borer, *Tuta absoluta*: Chemical properties and biological activity. *Sci. Rep.* **2017**, *7*, 13036. [CrossRef]
95. Chaudhari, A.K.; Singh, V.K.; Kedia, A.; Das, S.; Dubey, N.K. Essential oils and their bioactive compounds as eco-friendly novel green pesticides for management of storage insect pests: Prospects and retrospects. *Environ. Sci. Pollut. Res.* **2021**, *28*, 18918–18940. [CrossRef]
96. Ngongang, M.D.T.; Eke, P.; Sameza, M.L.; Mback, M.N.L.N.; Lordon, C.D.-; Boyom, F.F. Chemical constituents of essential oils from *Thymus vulgaris* and *Cymbopogon citratus* and their insecticidal potential against the tomato borer, *Tuta absoluta* (Lepidoptera: Gelechiidae). *Int. J. Trop. Insect Sci.* **2022**, *42*, 31–43. [CrossRef]
97. Brandão, R.M.; Ferreira, V.R.F.; Batista, L.R.; Alves, E.; Santiago, W.D.; Barbosa, R.B.; Caetano, A.R.S.; Nelson, D.L.; das G. Cardoso, M. Antifungal activity and the effect of the essential oil of *Lippia origanoides* Kunth on *Aspergillus* mycotoxins production. *Aust. J. Crop Sci.* **2021**, *15*, 1005–1012. [CrossRef]
98. Jesus, F.P.K.; Ferreiro, L.; Bizzi, K.S.; Loreto, É.S.; Pilotto, M.B.; Ludwig, A.; Alves, S.H.; Zanette, R.A.; Santurio, J.M. In vitro activity of carvacrol and thymol combined with antifungals or antibacterials against Pythium insidiosum. *J. Mycol. Med.* **2015**, *25*, e89–e93. [CrossRef]
99. Takahashi, H.; Nakamura, A.; Fujino, N.; Sawaguchi, Y.; Sato, M.; Kuda, T.; Kimura, B. Evaluation of the antibacterial activity of allyl isothiocyanate, clove oil, eugenol and carvacrol against spoilage lactic acid bacteria. *LWT* **2021**, *145*, 111263. [CrossRef]
100. Yang, R.; Miao, J.; Shen, Y.; Cai, N.; Wan, C.; Zou, L.; Chen, C.; Chen, J. Antifungal effect of cinnamaldehyde, eugenol and carvacrol nanoemulsion against *Penicillium digitatum* and application in postharvest preservation of citrus fruit. *LWT* **2021**, *141*, 110924. [CrossRef]

101. Kordali, S.; Cakir, A.; Ozer, H.; Cakmakci, R.; Kesdek, M.; Mete, E. Antifungal, phytotoxic and insecticidal properties of essential oil isolated from Turkish Origanum acutidens and its three components, carvacrol, thymol and *p*-cymene. *Bioresour. Technol.* **2008**, *99*, 8788–8795. [CrossRef]
102. Arango Bedoya, O.; Hurtado Benavides, A.M.; Pantoja Daza, D.; Santacruz Chazatar, L. Actividad inhibitoria del aceite esencial de *Lippia origanoides* H.B.K sobre el crecimiento de *Phytophthora infestans*. *Acta Agronómica* **2014**, *64*, 116–124. [CrossRef]

Review

Application of Peptides in Construction of Nonviral Vectors for Gene Delivery

Yujie Yang, Zhen Liu, Hongchao Ma * and Meiwen Cao *

State Key Laboratory of Heavy Oil Processing, Department of Biological and Energy Chemical Engineering, College of Chemical Engineering, China University of Petroleum (East China), 66 Changjiang West Road, Qingdao 266580, China
* Correspondence: mahc@upc.edu.cn (H.M.); mwcao@upc.edu.cn (M.C.)

Abstract: Gene therapy, which aims to cure diseases by knocking out, editing, correcting or compensating abnormal genes, provides new strategies for the treatment of tumors, genetic diseases and other diseases that are closely related to human gene abnormalities. In order to deliver genes efficiently to abnormal sites in vivo to achieve therapeutic effects, a variety of gene vectors have been designed. Among them, peptide-based vectors show superior advantages because of their ease of design, perfect biocompatibility and safety. Rationally designed peptides can carry nucleic acids into cells to perform therapeutic effects by overcoming a series of biological barriers including cellular uptake, endosomal escape, nuclear entrance and so on. Moreover, peptides can also be incorporated into other delivery systems as functional segments. In this review, we referred to the biological barriers for gene delivery in vivo and discussed several kinds of peptide-based nonviral gene vectors developed for overcoming these barriers. These vectors can deliver different types of genetic materials into targeted cells/tissues individually or in combination by having specific structure–function relationships. Based on the general review of peptide-based gene delivery systems, the current challenges and future perspectives in development of peptidic nonviral vectors for clinical applications were also put forward, with the aim of providing guidance towards the rational design and development of such systems.

Keywords: peptide; gene delivery; nonviral vector; self-assembly; gene therapy

Citation: Yang, Y.; Liu, Z.; Ma, H.; Cao, M. Application of Peptides in Construction of Nonviral Vectors for Gene Delivery. *Nanomaterials* 2022, 12, 4076. https://doi.org/10.3390/nano12224076

Academic Editor: Alicia Rodríguez-Gascón

Received: 21 October 2022
Accepted: 16 November 2022
Published: 19 November 2022

Publisher's Note: MDPI stays neutral with regard to jurisdictional claims in published maps and institutional affiliations.

Copyright: © 2022 by the authors. Licensee MDPI, Basel, Switzerland. This article is an open access article distributed under the terms and conditions of the Creative Commons Attribution (CC BY) license (https://creativecommons.org/licenses/by/4.0/).

1. Introduction

Gene therapy refers to the introduction of exogenous genes into target cells to correct defective and abnormal genes for the purpose of treating diseases. With the development of modern molecular biology and progress of human genome project, gene therapy has become a promising strategy to treat cancer, gene diseases, infectious diseases, cardiovascular diseases and nervous system diseases [1]. The therapeutic nucleic acids used in gene therapy include plasmid DNA, siRNA and other free nucleic acids [2]. However, it is difficult for these nucleic acids to reach the target tissue due to their large molecular weight and huge number of negative charges [3]. Therefore, developing safe and effective gene delivery vectors is essential for gene therapy.

Generally, gene delivery vectors are categorized into two types, that is, viral vectors and nonviral ones. Typically, viral vectors use modified viruses including retroviruses, lentiviruses, adenoviruses and adeno-associated viruses to carry genes into cells due to their advantages of high infection level of host cells [4]. Their accurate programmed infection characteristics and efficient delivery ability of exogenous gene into host cells make them the most widely used gene vectors in clinical trials. However, viral vectors have inherent disadvantages such as potential carcinogenic effects, limited DNA encapsulation ability, lack of targeting ability and difficulty in production [5]. Moreover, they may also activate the host's immune system and reduce the effectiveness of subsequent gene delivery [6]. These defects greatly limit the usage of virus vectors in clinical treatment and further promote the development of nonviral gene delivery systems [7]. Compared with viral vectors, nonviral vectors are usually easier to synthesize and operate, having lower immune response, larger loading capacity of genetic material and better targeting ability. Recently, a large number of efficient and safe nonviral vectors have been designed for gene therapy. When using nonviral vectors to deliver nucleic acids such as DNA [8], messenger (m)RNA [9], short interfering (si)RNA [10] and micro (mi)RNA into cells [11], they need to overcome several biological barriers (Figure 1). First, the vectors should protect the nucleic acids from degradation by endonucleases and exonucleases and help them evade immune detection [12–14]. Second, the vectors need to contain specific groups and ligands both to make nucleic acid molecules exude from the bloodstream to the target tissue and to mediate cell entry. Third, siRNA and miRNA mimics must be loaded into the RNA-induced silencing complex, while mRNA must bind to the translational machinery and DNA must be further transported to the nucleus to play its function (Figure 2) [15]. The commonly used nonviral vectors include cationic liposomes, cationic polymers, dendrimers, peptides and so on [16]. Among them, peptides have been considered as unique tools for delivering nucleic acid drugs due to their excellent biocompatibility and biodegradability, ease of production and modification as well as being able to respond to external stimuli [17–19].

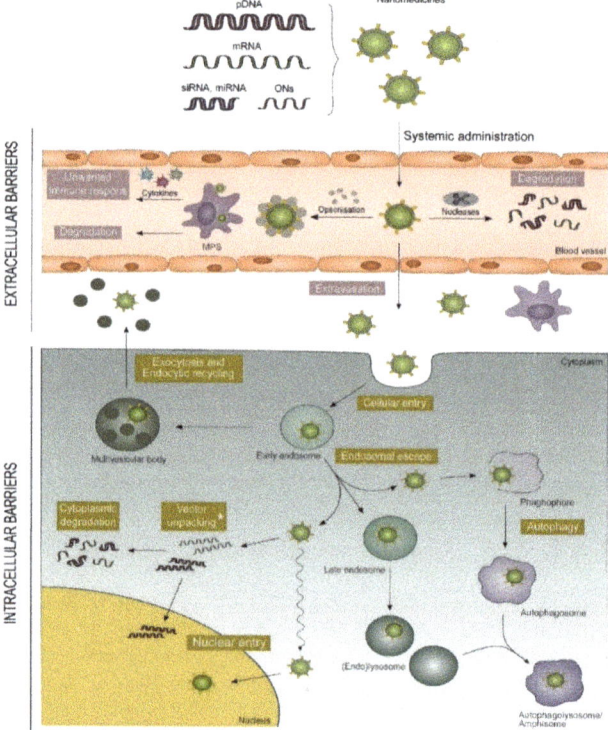

Figure 1. Biological barriers to overcome when using nonviral vectors to deliver nucleic acids in vivo. Nucleic acids bind to peptides through electrostatic interactions, transferring them across the cell membrane via endocytosis, providing endosome escape, and ultimately releasing the associated nucleic acids in the cytoplasm or nucleus. Reprinted with permission from Ref. [20]. Copyright 2018, Elsevier.

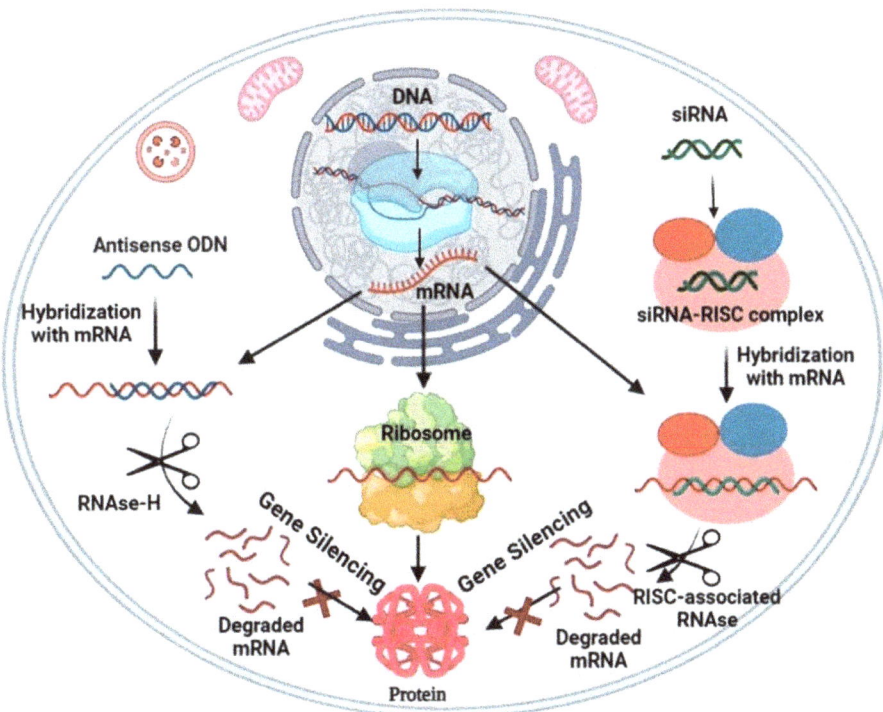

Figure 2. Schematic illustration of different cellular pathways involved in gene silencing. Reprinted with permission from Ref. [21]. Copyright 2022, Elsevier.

Nowadays, many peptides have been incorporated as functional components into nonviral gene delivery systems to overcome various biological obstacles and deliver nucleic acid drugs to target sites with high efficiency. Peptides used as non-viral gene vectors can be divided into the following types according to their functions: cell penetrating peptides (CPPs), membrane active peptides, targeting peptides, and nuclear localization signal (NLS) peptides (Table 1). In this review, we first talk about the strategies for constructing peptide–nucleic acid complexes, and then summarize the applications of these peptides in gene delivery, as well as how to combine these peptides with other nonviral vectors to achieve the purpose of improving transfection efficiency.

Table 1. Types of peptides designed for use in gene delivery.

Peptide Type	Name	Sequence [a]	Reference
CPPs	CHAT	CHHHRRRWRRRHHHC	[22]
	LH2	Ac, T, C-LHHLCHLLHHLCHLAG Ac-GALHCLHHLLHCLHHL Ac -LHHLCHLLHHLCHLGA Ac -LHHLCHLLHHLCHLGA	[23,24]
	SRCRP2-11	GRVEVLYRGSW	[25]
	SRCRP2-11-R	GRVRVLYRGSW	
	R$_8$	RRRRRRRR	[26–29]
	Penetratin	RQIKIWFQNRRMKWKK	[30]
	WTAS	PLKTPGKKKKGKPGKRKEQEKKKRRTR	[31]
	PF14	Stearyl-AGYLLGKLLOOLAAAALOOLL-NH2	[32]
	CPP	CGRRMKWKK	[33]

Table 1. Cont.

Peptide Type	Name	Sequence [a]	Reference
Targeted peptides	circular NGR	CNGRCG	[28]
	NGR	NGR	[33,34]
	RGD	RGD	[29,35–37]
	Trivalent cRGD	HCACAE[cyclo(RGD-D-FK)]E[cyclo(RGD-D-FK)]$_2$	[38]
	cRGD	cyclo(RGD-D-FK)	[39,40]
	cyclic iRGD	cyclo (CRGDKGPDC)	[41]
Membrane active peptides	RALA	WEARLARALARALARHLARALAHALHACEA	[42–44]
	HALA2	WEARLARALARALARHLARALAHALHACEA	[45]
	(LLHH)3	CLLHHLLHHLLHH	[46]
	(LLKK)3-H6	LLKKLLKKLLKKCHHHHHH	[46]
	LAH4	KKALLALALHHLAHLALHLALALKKA	[47]
	KH27K	KHHHHHHHHHHHHHHHHHHHHHHHHHHHK	[48,49]
	G3	GIIKKIIKKIIKKI	[50]
	Melittin	GIGAVLEVLTTGLPALISWIEEEEQQ	[51]
	CMA-1	EEGIGAVLKVLTTGLPALISWIKRKRQQC	[52]
	CMA-2	GIGAVLKVLTTGLPALISWIHHHHEEC	[53,54]
	CMA-3	GIGAVLKVLTTG LPALISWIKRKREEC	[54]
	CMA-4	EEGIGAVLKVLTTG LPALISWIHHHHQQC	[52]
	NMA-3	CGIGAVLKVLTTGLPALISWI KRKREE	[52,53]
	acid-Melittin	GIGAVLKVLTTGLPALISWIKRKRQQ	[51]
	Mel-L6A10	GIGAIEKVLETGLPTLISWIKNKRKQ	[55]
	RV-23	RIGVLLARLPKLFSLFKLMGKKV	[53]
NLS peptides	SV40 T antigen	PKKKRKV	[56–60]
	Mouse FGF3	RLRRDAGGRGGVYEHLGGAPRRRK	[61]
	NLSV404	PKKKRKVGPKKKRKVGPKKKVGPKKKRKVGC	[62]
	Ku7O$_2$	CKVTKRKHGAAGAASKRPKGKVTKRKHGAAGAASKRPK	[63]
Other peptides	Smart peptide	Nap-FFGPLGLAG-(CK$_m$)$_n$C	[64]
	24-mer β-annulus peptide	INHVGGTGGAIMAPVAVTRQLVGS	[65,66]
	β-annulus-GGGCG peptide	INHVGGTGGAIMAPVAVTRQLVGSGGGCG	[67]
	H4K5HC$_{BZI}$C$_{BZI}$H	HHHHKKKKKC12LLHC$_{BZI}$C$_{BZI}$HLLGSPD	[68]
	K3C6SPD	KKKC6WLVFFAQQGSPD	[69,70]
	CC	REGVAKALRAVANALHYNASALEEVADALQKVKM	[71]
	Surfactant-like peptide	IIIVVVAAAGGGKKK	[72]

[a] All peptide sequences are given in the one-letter code amino acid name (Table A1, Appendix A).

2. Construction of Peptide–Nucleic Acid Complexes for Gene Delivery

To achieve the purpose of gene delivery, the functional peptides should be first fused with nucleic acids to form complexes so as to play the roles of gene condensing, protection, and delivery. Three main strategies can be adopted to achieve peptide/nucleic acid fusion. The first is to link the peptide segment covalently with nucleic acid to produce a conjugated molecule. For this strategy, the functional peptide segments are conjugated to the to-be-delivered nucleic acid via chemical bonds (e.g., ester bond, disulfide bridge, thiol-maleimide linkage) [73]. The superior advantage of this strategy is that the peptide–nucleic acid conjugated molecule has defined structure and stoichiometry as well as high stability, which can lead to repeatable delivery performance. This approach is particularly suitable for charge-neutral nucleic acid analogs such as phosphonodiamidate morpholino oligomer (PMO) and peptide nucleic acid (PNA) [74,75]. The peptide–nucleic acid conjugate can easily cross the cell membranes and enter the nucleus and fulfill its biological functions. Currently, this strategy has exhibited promise in clinical trials. For example, peptide-PNA conjugates have been utilized in preclinical studies targeting c-myc for severe combined immunodeficiency, while peptide-PMO conjugates have been employed for Duchenne muscular dystrophy [76,77]. However, for this strategy, the covalent bond formation may reduce the biological activity of nucleic acids or inhibit their release and expression in cells, which may hinder their application. The second is the noncovalent complexation strategy, which is to complex peptides with nucleic acids directly via noncovalent forces. For this strategy, the peptides are usually designed to have various positive charges, which can fist bind with negatively charged nucleic acids to result in charge neutralization and then induce hydrophobic collapse of the nucleic acid molecules into condensed nanoparticles [78]. This strategy has superior advantages including ease of vector construction, high loading amount of gene drug, and controllable genome release by introducing stimuli responsibility. It is suitable for delivery of most nucleic acids involving plasmid DNA, siRNA, mRNA and so on. Peptide–nucleic acid nanocomposites obtained by this method are easy to prepare and have been attempted to treat a series of diseases including cancer and cardiovascular diseases [79,80]. However, it should be noted that the peptide should be well designed to endow the peptide carrier with high functionality and avoid loss of peptide function because of its electrostatic binding with nucleic acids. The third strategy is to modify functional peptide segments on the surface of specific nanoparticles to produce composite nanoplatforms, which can further be used to complex with nucleic acids for delivery purposes. This strategy can take advantage of the nanoparticles to facilitate cellular uptake as well as to give multifunctionalities [81], which is especially suitable for development of systems for combined therapy. In summary, the above three strategies, each having specific features in peptide/nucleic acid fusion, have been extensively used in gene delivery.

3. Application of CPPs in Gene Delivery

Composed of 10–20 amino acids, CPPs are one class of peptides which have the potential to penetrate bio-membrane and transport bioactive substances into cells [82]. In recent years, a variety of substances such as hydrophilic proteins, nucleic acids and even nanoparticles have been carried by CPPs across cell membrane into the cytoplasm to serve specific functions. This rapid intracellular transport is not destructive to cell membranes, and the active substances can be delivered into a variety of cells regardless of the cell type. Use of CPPs to deliver nucleic acids and drugs for gene therapy and disease treatment has therefore attracted extensive attention. For example, Emma et al. designed a new 15-amino acid linear peptide CHAT that contains six arginine residues, the minimum number of residues required for cell uptake [22]. The cysteine residues located at both ends can enhance the stability of the delivery system and achieve cargo release in cells. Experiments demonstrated that CHAT peptide can transfect plasmid (p)DNA into various cell lines, resulting in successful reporter-gene expression in vivo in 4T1 and MDA-MB-231 breast xenograft models (Figure 3a). The transfection efficiency in tumor sites is comparable to that of commercial transfectants, making it a low-cost, easily formulated delivery system for the administration of nucleic acid therapeutics. However, some inherent properties of CPPs limit their clinical application. First, when CPPs are administered in vivo, they are penetrable only at concentrations above micromoles, which will cause many systemic side effects. In this case, designing new CPPs and improving their ability to penetrate cell membranes are of great importance for enhancing the safety of CPP application. Recently, a pH-active CPP called dimer LH2 was designed by Dougherty and co-workers because they found that amphiphilic CPPs in dimeric form showed higher cell-penetrating activity compared with the monomeric ones [23]. As expected, dimer LH2 can effectively deliver nucleic acid drugs to triple-negative breast cancer cell MDA-MB-231 with only tens of nanomolar concentration, showing strong membrane penetrating ability and antitumor effects [24]. In addition to using CPPs as carriers to deliver pDNA into cells, naked siRNA must be protected and delivered by carriers to enter the cell, because it is unstable, and readily degraded by nucleases in the serum environment and absorbed by tissues [83]. To solve this problem, Martina et al. used DMBT1-derived peptides with membrane penetrating ability as carriers to prepare siRNA delivery nanoparticles, which can complex with siRNA and transfect human breast metastatic adenocarcinoma MCF7 cells [25]. The delivered siRNA exhibited effective gene silencing in MCF7-recombinant cells. The study laid the foundation for developing a new vector for therapeutic siRNA delivery.

Second, most CPPs can be internalized by all cell types and lack the ability to target specific tissues as particular objectives. This imprecise feature will lead to their low stability in blood, poor tissue penetration and limited cell uptake, thus greatly reducing their targeting efficiency towards specific tissues. To solve this problem, several strategies have been developed to improve the specificity of CPPs to pathological tissues. Among them, combing targeting molecules such as RGD (Arginine-Glycine-Aspartic acid), NGR (Asparagine-Glycine-Arginine) peptide, folic acid (FA) and hyaluronic acid with CPPs is a very effective strategy [84–86]. These targeting molecules are usually overexpressed in tumor types, but not in normal cells. Therefore, they can improve the targeting effect for pathological tissues, whilst healthy tissues are not affected by drug delivery. For example, Qi-ying Jiang conjugated the target ligand of FA and the CPP segment of octaarginine (R_8) to an existing vector (PEI600-CD) composed of β-cyclodextrin and low-molecular-weight polyethylenimine (PEI) to produce a new gene vector FA-PC/R_8-PC [26]. This vector can form ternary nanocomplexes with pDNA, and further deliver it to tumor sites in vivo with excellent gene transfection efficiency (Figure 3b). Moreover, hyaluronic acid coupled with CPPs can effectively deliver siRNA to macrophages within the atherosclerotic plaques and enhance gene delivery to macrophages in antiatherosclerotic therapy [30], which is a promising nanocarrier for efficient macrophage-targeted gene delivery and antiatherogen (Figure 3c).

In addition to being used as vectors for gene delivery alone, CPPs can also be combined with other non-viral vectors such as liposomes and cationic polymers to achieve high gene transfection efficiency. Integrating different types of functional vectors into one gene delivery system can exert a synergistic effect between the components, improving the low permeability and poor selectivity of CPPs, and so enhance the gene delivery efficiency. Ikramy et al. developed an efficient gene delivery system by combining a CPP segment (R_8) and pH-sensitive cationic lipid (YSK05) [27]. Positive nanoparticles can be formed by attaching high density R8 to the surface of YSK05 nanoparticles. The particles can further encapsulate pDNA to produce complexes that can lead to high gene transfection efficiency due to the synergistic effect between R_8 and YSK05. Obdulia and co-workers also developed a gene delivery vector by co-assembly of CPP (WTAS) and a poly β-amino ester (PBAE) polymer [31]. The WTAS-PBAE vector showed high transfection rate, and the results of cell transfection experiments with GL26 cells revealed that WTAS-PBAE vector loaded with GFP pDNA led to virtually complete transfection (> 90%). This excellent transfection efficiency makes it a very promising gene delivery vector for delivering a variety of genetic materials. In addition, the combination of CPPs and inorganic nanoparticles also shows great potential in the application of delivering nucleic acid drugs. For example, Dowaidar et al. found that the conjugation of CPPs-oligonucleotides with magnetic iron oxide nanoparticles can promote cellular uptake of the plasmid and improve the transfection efficiency, which opens up a new way for selective and efficient gene therapy [32].

4. Application of Targeted Peptides in Gene Delivery

During gene delivery, an off-target effect may occur when the therapeutic nucleic acids bind to non-specific cells, which is undesirable and will decrease the therapeutic effect of gene therapy. Therefore, selectively delivering vector-nucleic acid complexes to the target cells and exerting the therapeutic effect at specific sites are critical to improve the transfection efficiency of gene therapy [87]. Conjugating targeting ligands such as FA, hyaluronic acid and biomolecules including peptides and proteins can greatly increase the targeting of the gene delivery systems because they can specifically bind to the receptors on cells. Among them, peptides are excellent gene delivery targeting ligands due to their good biocompatibility, ease of synthesis and modification as well as their high response to stimuli. Thus far, more than 700 targeted peptides have been discovered for targeting different cells. The most widely used target peptides among them are NGR and RGD which can specifically recognize tumor angiogenic markers and provide new venues for exploring tumor targeting agents [84].

The NGR motif, whose tumor-targeting ability relies on its specific interaction with CD13 (aminopeptidase N), was identified from a tumor homing peptide. It is often selectively overexpressed in neovascular and some tumor cells, but seldom expressed in quiet vascular endothelial cells. NGR peptides have now been used to promote the targeted delivery of therapeutic agents and enhance antitumor effects [88]. A bi-functional peptide, NGR-10R, which consists of an N-terminal circular NGR motif (CNGRCG) and a C-terminal R_8 sequence was designed for gene therapy. The R_8 sequence at the end of NGR-10R can bind to siRNA through electrostatic interaction to form NGR-10R/siRNA nanoparticles. Thanks to the NGR motif, NGR-10R/siRNA nanoparticles can be specifically delivered to MDA-MB-231 cells and localized around the nucleus, thus robustly repressing gene expression in MDA-MB-231 and HUVEC (a $CD13^+/\alpha_V\beta_3^+$ cell) (Figure 4a) [28]. In the study of Yang, as a targeted peptide, NGR plays a navigational effect, enabling the pcCPP/NGR-LP dual-modified liposomes vector to accumulate at the tumor site. Finally, with the aid of CPPs, the siRNA-loaded vector enters target cells efficiently [33]. In addition to targeting siRNA to MDA-MB-231 cells, the NGR motif can effectively deliver siRNA to HT-1080 cells and downregulate target genes with the synergistic effect of other vectors. Chen et al. designed the LPD-poly(ethylene glycol) (PEG)-NGR vector by modifying PEGylated LPD using the NGR motif. It can target CD13 expressed in the tumor cells or tumor vascular endothelium, effectively delivering siRNA to the cytoplasm of HT-1080 cells and silence the target gene [34].

Different from NRG, the RGD peptide can specifically bind to integrin in tumor endothelial cells and act as ligand to target tumor cells that overexpress $\alpha_V\beta_3$ integrin [89,90]. As an attractive tumor cell receptor, integrin plays a major role in promoting the proliferation, migration, invasion and survival of tumor cells. Therefore, gene vectors modified by RGD peptide can block cell–cell and cell-matrix adhesions by competing with adhesion proteins for cell surface integrins, thus achieving targeted selectivity to tumor cells and improving the efficiency of gene transfection. In view of this, a large number of RGD peptide-based gene vectors have been developed. Recently, lung cancer and bronchial cancer have become the most deadly cancers due to the aggravation of air pollution. In order to develop new targeted, effective and less painful therapies, Yang et al. synthesized the RRPH (RGD-R_8-PEG-HA) which is composed of peptide (RGD-R_8) and PEGylation on HA to coat PFC (plasmid complex). The obtained RRPHC nanoparticles (RRPH coated PFC complex) achieve long-term circulation and tumor tissue-penetration while maintaining the high transfection efficiency of PFC [29]. Kim et al. designed a targeted gene vector, RGD/PEI/WSC, which can combine the RGD to chitosan and PEI, for $\alpha_V\beta_3$ integrin-overexpressing tumor cells [35]. In vivo experiments show that the vector can suppress the growth of PC3 prostate tumor cell xenograft model by silencing BCL2 mRNA, which is expected to be a good candidate for a specific targeted gene vector without cytotoxicity (Figure 4b).

Oncolytic adenovirus has been widely used in clinical trials of cancer gene therapy [91,92]. Moreover, tumor targeted gene virus therapy (CTGVT) may be an effective strategy for the treatment of advanced or metastatic cancer [93]. In a previous study, Luo et al. found that replicating adenovirus (AD-ZD55-miR-143) showed specific anti-rectal cancer efficacy in vitro. However, its anti-tumor effect in vivo is not ideal, because the vector does not increase the chance of reaching target cells. To solve this problem, they developed AD-RGD-survivin-ZD55-miR-143, a novel triple regulatory oncolytic adenovirus which significantly enhanced the anti-tumor effect and directly broadened the treatment options for colorectal cancer [36]. RGD peptides with a circular structure, i.e., cyclic (c)RGDs can also be used for tumor targeting studies—being more active due to their conformation-less assembly than linear RGD oligopeptides. Moreover, (c)RCDs are resistant to proteolysis and have higher affinity to integrin receptors [94]. Therefore, many five membered ring RGDs containing pentapeptides have been used to endow gene vectors with tumor targeting [95]. Alam et al. reported that cRGDs can selectively enter cancer cells overexpressing $\alpha_V\beta_3$ integrin carrying siRNA for gene silencing [38]. A further study indicated that cRGDs can specifically guide siRNA to cells expressing $\alpha_V\beta_3$, resulting in effective knocking out of selected genes and significantly reducing tumor growth [39]. In addition, cRGDs were employed to promote cellular internalization of polyplex micelles encapsulating anti-angiogenic pDNA by tumor vascular endothelial cells, which abundantly express RGD-specific $\alpha_v\beta_3$ and $\alpha_v\beta_5$ integrin receptors and thereby exhibit anti-tumor activity against pancreatic adenocarcinoma upon systemic injection [96,97]. Moreover, liposomes modified with cRGD peptide can be used to deliver drugs to targeted cancer cells [40].

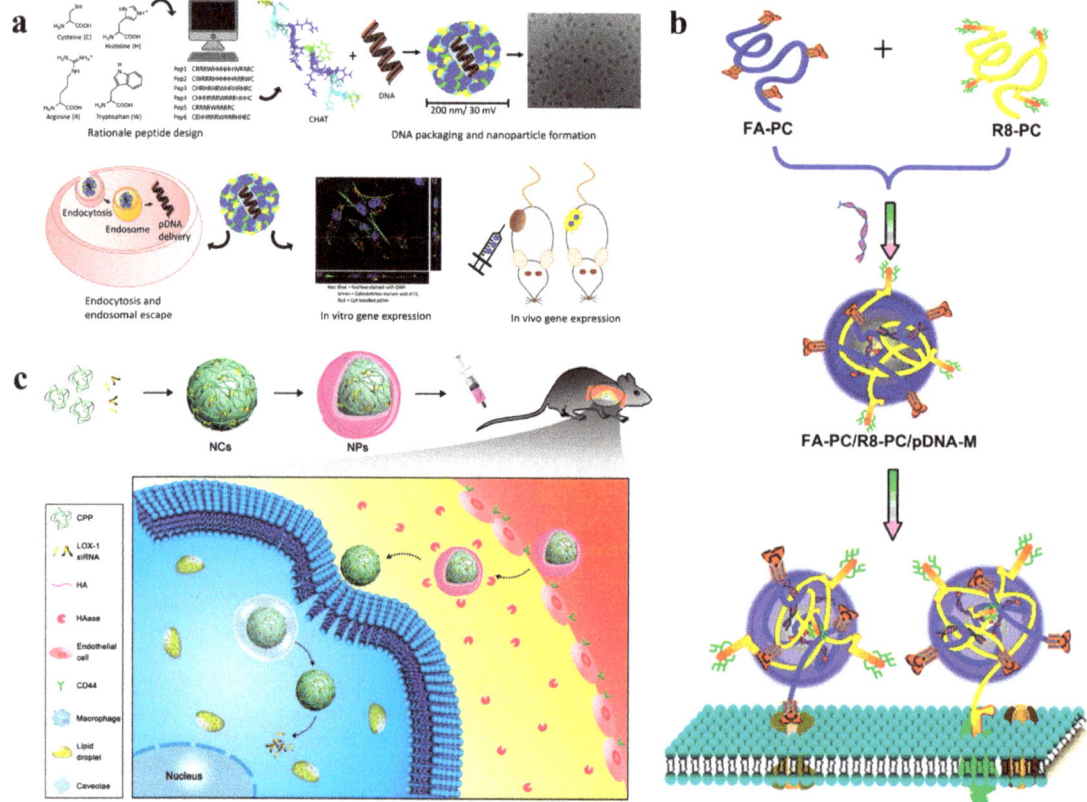

Figure 3. (a) CHAT peptide condenses pDNA to produce cationic nanoparticles less than 200 nm in diameter. The complex can cross the cell membrane through endocytosis and successfully escape from the endosomes, obtaining high transfection efficiency. Reprinted with permission from Ref. [22]. Copyright 2020, Elsevier. (b) The process of preparation of nanoparticles formed from FA-PC/R8-PC/pDNA complex. Reprinted with permission from Ref. [26]. Copyright 2011, Elsevier. (c) CPPs condense siRNA and deliver it to macrophages. Reprinted with permission from Ref. [30]. Copyright 2018, American Chemical Society.

Our group is also devoted to designing peptide carriers with targeting functions. Recently, we have designed an amphiphilic peptide Ac-RGDGPLGLAGI$_3$GR$_8$-NH$_2$ with two charged chain segments distributed at the end and a hydrophobic chain segment in the middle [37]. It can selectively kill cancer cells through the specific recognition and binding of RGD fragments to cancer cell membranes and cleavage of PLGLA fragments by tumor-overexpressed matrix metalloproteinase-7 enzymes. The R$_8$ sequence can induce efficient condensation of DNA into dense nanoparticles, resist enzymatic degradation of DNA, ensure successful delivery of DNA into cells, and improve the expression level as well as transfection rate of target genes [87]. Moreover, we also combined the cRGD peptide to gold nanoparticles (AuNPs) which has been widely used in the delivery of nucleic acid molecules due to its good biocompatibility and easy surface functionalization [98,99]. We designed the peptide of sequence (CRGDKGPDC)GPLGLAGIIIGRRRRRRR-NH$_2$ (CPIR28) which was grafted onto the surface of AuNPs by the one-pot synthesis method [41]. The CPIR28-AuNPs nanocomposite can effectively condense DNA and improve the intracellular transport of genes (Figure 4c).

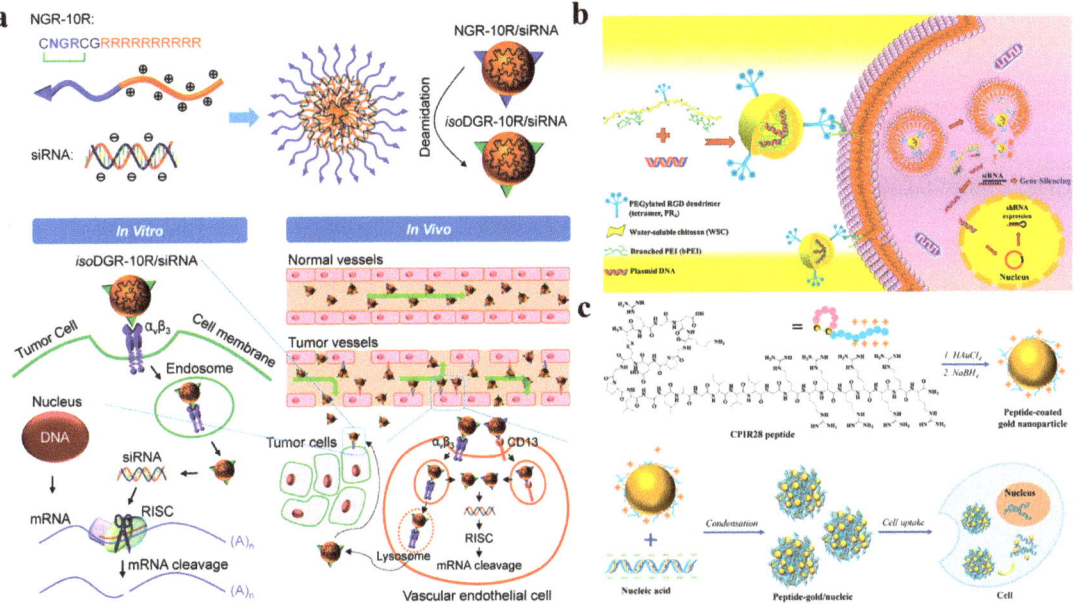

Figure 4. (a) Bi-functional NGR-10R peptide condenses siRNA to form spherical nanostructures which can enter cells by receptor $\alpha_v\beta_3$ and CD13 mediated endocytosis. After escaping from the endosomes/lysosomes, siRNA is released into the cytoplasm and loaded by the RISC. Reprinted with permission from Ref. [28]. Copyright 2015, Biomaterials Science. (b) RPgWSC-pDNA complexes can suppress solid tumor growth by silencing BCL2 mRNA. Reprinted with permission from Ref. [35]. Copyright 2017, Elsevier. (c) CRIP28-AuNPs form nanocomplexes with nucleic acids by electrostatic interaction for cellular delivery. Reprinted with permission from Ref. [41]. Copyright 2022, Elsevier.

5. Application of Membrane Active Peptides in Gene Delivery

After cell uptake, successful release of vector/nucleic acid complexes from endosomes is a major obstacle for effective gene therapy. After the vector/nucleic acid complexes cross the membrane barrier into the cell through endocytosis, vesicles will enclose them and develop into early endosomes, which then mature to form late endosomes and then fuse with lysosomes. In order to exert the therapeutic effect of nucleic acid drugs, the complexes need to escape from the endosomes and enter into the cytoplasm. Otherwise, the nucleic acid drugs will be degraded by hydrolases [46]. Therefore, developing vectors with endosomal escape ability is essential for efficient gene delivery. There are two ways to achieve endosomal escape. First, considering the acidic environment inside the endosomes, materials with a buffer effect in the acidic environment, such as chlorine and calcium, can be added to assist endosomal escape. These buffer agents can prevent endosomes from binding to lysosomes, vacuolate endosomes and then decrease the membrane stability. However, these chemicals are generally only used in vitro and not suitable for clinical applications due to their potential cytotoxicity. Nevertheless, the acidic endosomal environment suggests that we can introduce amino acids with a acidic buffering effect into the carrier to destroy the endosome membrane by proton pump for the purpose of endosomal escape. Since only histidine has a buffering effect among the 20 common amino acids due to its imidazole group, it is often embedded into the carrier to improve endosomal escape during delivery of nucleic acids. RALA, which is a 30-mer cationic amphipathic helical peptide, contains seven hydrophilic arginine residues on one side of the helix, and hydrophobic leucine residues on the other side. When the pH drops, the α-helicity of RALA increases to achieve endosomal escape and release of the cargo [42]. Therefore, Vimal K et al. used RALA peptides to condense mRNA and effectively deliver them to dendritic cells [43]. Subsequently, the RALA-mRNA nanocomplexes successfully escaped from endosomes and expressed mRNA in the cell cytosol to promote antigen specific T cell proliferation as well as evoking T cell immunity in vivo (Figure 5a). In addition to delivering mRNA, RALA can also deliver siRNA with high efficiency. Eoghan J. Mulholland et al. reported that RALA is an effective siRNA carrier targeting the FK506-binding protein and has great potential in promoting angiogenesis for advanced wound healing applications (Figure 5b) [44]. Recent studies have found that the introduction of histidine into RALA peptide can further improve the endosomal escape ability of the vectors, thereby increasing the transfection efficiency. For example, Liu et al. designed a new peptide-based vector HALA2 with ability of endosomal escape and high cell transfection efficiency by adjusting the ratio of histidine and arginine in the RALA peptide [45]. HALA2 replaced two arginines close to the C-terminal of RALA with histidine, which reduced the number of positively charged amino acids in HALA2 from 7 to 5, resulting in a better transfection rate than RALA. In addition, introducing histidine fragments into other kinds of vectors can also improve their endosomal escape ability. Chitosan has the advantages of non-toxicity, non-immunogenicity, biodegradability and good biocompatibility as a gene vector. However, chitosan cannot mediate the escape of endosome due to its low endosomal escape rate and poor buffer capacity. For this reason, Liu et al. introduced histidine into chitosan and obtained a new vector with good solubility, strong binding ability to siRNA and excellent endosomal escape performance [100].

Secondly, using membrane active peptides with membrane destruction capability to destroy the endosomal membrane can also realize endosomal escape and release the vector/nucleic acid complex into the cytoplasm. Recently, a series of membrane active peptides have been designed. For example, $(LLHH)_3$ and $(LLKK)_3$-H_6 are two typical amphiphilic membrane active peptides that can destroy endosomal membranes and regulate the "proton sponge effect". Introducing them into vectors containing rigid acyl and polyarginine, Yang et al. designed two multifunctional peptide vectors, C_{18}-C$(LLKK)_3$-H_6-R_8 and C_{18}-C$(LLHH)_3$C-R_8. They found that each functional fragment showed a synergistic effect, and the presence of membrane active peptide significantly improved the endosomal escape efficiency and transfection rate, which greatly promotes the application of peptide-based vectors in the treatment of genetic diseases [46]. In the past few years, Bechinger and co-workers have been devoted to developing pH-responsive cationic amphiphilic membrane active peptides rich in histidine residues for gene delivery. They have designed a variety of LAH4-based peptides which have been proven to be able to bind to plasmid DNA and facilitate its cellular uptake and endosomal escape [47,101–103]. Among them, some derivative peptides of LAH4 not only have the ability to bind to plasmid DNA, but also have strong siRNA and mRNA delivery capabilities [47]. To date, the interactions of LAH4-based peptides and bio-membrane have been studied in detail by biophysical methods, and the results indicate that these peptides show strong delivery capacity for a variety of cargoes, including nucleic acids, peptides and proteins [104]. The histidine-rich amphiphilic peptide KH27K has also been developed as a "proton sponge" escape endosomal agent. Unlike LAH4, KH27K is currently mainly used to deliver virus particles into the cell to achieve the intracellular release of the virus, and this "membrane release" activity is consistent with its pH dependent hemolysis activity. However, there is no clear study on the intracellular delivery of nucleic acid molecules [48,49].

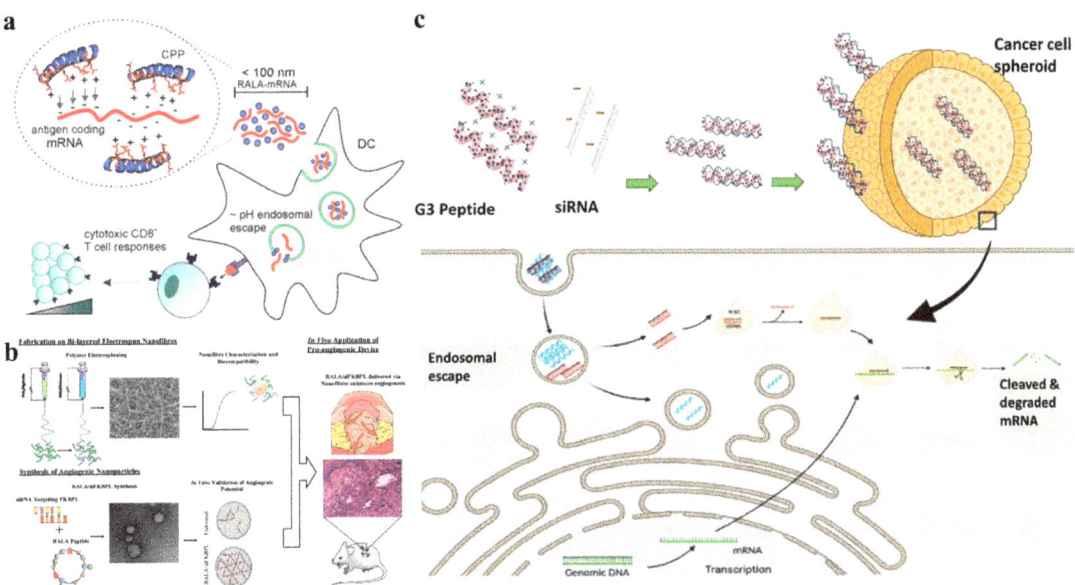

Figure 5. (**a**) RALA peptides condense mRNA into nanoparticles, releasing mRNA in dendritic cell cytosol to promote antigen specific T cell proliferation. Reprinted with permission from Ref. [43]. Copyright 2017, Wiley Online Library. (**b**) RALA peptides form a complex with siRNA to deliver siRNA into the cell and promote the regeneration of blood vessels. Reprinted with permission from Ref. [44]. Copyright 2019, Elsevier. (**c**) G3 peptide was assembled with siRNA and delivered to cancer cells, where siRNA was released to regulate gene expression in cancer cells. Reprinted with permission from Ref. [50]. Copyright 2021, American Chemical Society.

Antibacterial peptides (AMPs) with an α-helical amphiphilic structure can also effectively promote endosomal escape. They are primarily found in bacteria and have activities against a variety of microorganisms. Most of them are composed of nearly 50% hydrophobic residues and are usually positively charged due to the presence of lysine and arginine fragments. The spatially separated hydrophobic and charged regions endow them with membrane interaction activity. In view of the characteristics of AMPs, Cirillo et al. designed a short cationic amphiphilic α-helical peptide $G(IIKK)_3I-NH_2$ with endosomal escape ability and high affinity towards colon cancer cells [50]. They report that when interacting with negatively charged DPPG small unilamellar vesicles, the peptides fold into α-helical structure helping to carry nucleic acids across the cell membrane and achieving endosomal escape, thus enabling the protection and selective delivery of siRNA to cancer cells (Figure 5c). Melittin is a multifunctional AMP that can inhibit many Gram-negative and Gram-positive bacteria. It is widely used to facilitate the endosomal escape of nanoparticles because of its significant cleavage activity in mammals both in vivo and in vitro. However, this amphiphilic peptide from bee venom has obvious toxicity to mammalian cells. If it is directly used to deliver nucleic acids, the transfection efficiency will be reduced due to the increase of cytotoxicity [105]. Therefore, melittin analogues have been designed in order to reduce the toxicity while promoting the ability to promote endosomal escape [106]. Glutamic acid and histidine residues on peptides are negatively charged due to deprotonation in the extracellular medium; however, in endosomes with a pH of about 5, the two amino acids are protonated, which reduces the hydrophilicity of the peptide and exposes its cleavage activity. Therefore, the method of replacing the basic amino acids in melittin with glutamic acid or histidine can be used to enhance the cleavage ability of the pH sensitive peptide. In views of this principle, a series of novel pH-sensitive peptides have been developed. Melittin analogues such as CMA-1, CMA-2, CMA-3, CMA-4, NMA-3 [52] and acid-melittin [51] have been obtained and used to conjugate with PEI to improve the intracellular endosomal escape of the PEI/DNA complex. Compared with CMA-1-PEI and CMA-4-PEI that covalently linked PEI to the N-terminal of peptide, C-terminal modified CMA-2-PEI, CMA-3-PEI and acid-melittin-PEI complexes showed strong cleavage activity at pH 5. The transfection experiments also showed that CMA-2-PEI and CMA3-PEI complexes induced significant gene expression [53,54]. Not all N-terminal modified melittin analogues have poor cleavage ability. For example, Kloeckner et al. proved that the transfection efficiency can be significantly improved by introducing N-terminal PEI-coupled melittin analogue NMA-3 into the EGF/OEI-HD-1 complex gene vector [52]. In addition, considering the effect of glutamate replacement location on peptide cleavage activity, Tamemoto et al. designed four melittin analogues and studied the optimal position of glutamate substitution. The results showed that a novel attenuated cationic cleavage peptide MEL-L6A10 with higher delivery activity, relatively lower cytotoxicity and higher endolytic activity can be designed by placing Glu on the boundary of the hydrophobic/hydrophilic region [55]. RV-23 is a pH-sensitive endolytic peptide extracted from Rana Linnaeus. Zhang et al. obtained a pH-sensitive endolytic peptide by replacing the positive charge residues in RV with glutamate. This substituted RV-23 peptide can promote the obvious destruction of cell intima and promote the entry of the carrier/nucleic acid complex into the cytoplasm. Thus, the gene transfection rate was significantly increased and the PEI-mediated cell transfection rate promoted [53].

6. Application of NLS Peptides in Gene Delivery

In gene delivery, some nucleic acid drugs, such as siRNA and mRNA, can directly play a therapeutic role in the cytoplasm after endosomal escape. However, for pDNA, DNA needs to be further transferred into the nucleus to realize its therapeutic effect. In such cases, whether DNA can be assisted to enter the nucleus is a key factor to evaluate the delivery capacity of non-viral gene vectors [56]. Macromolecules such as proteins cannot directly enter the nucleus due to the strong impedance from the nuclear envelope, and their transport into the nucleus must be regulated by the nuclear pore complex (NPC) [107,108]. When the protein enters the nucleus, the NLS (a short cationic peptide sequence) on the proteins can be recognized by the corresponding nuclear transporter, which helps them reach the nucleus through NPC with the assistance of transporter and nucleoporin [109,110]. Based on this, introducing NLS peptide sequences into non-viral vectors may achieve efficient delivery of the therapeutic DNA into the nucleus. Generally, NLS peptides can be divided into two categories, termed monopartite NLS (MP NLS) and bipartite NLS (BP NLS). The MP NLS is a single cluster composed of 4–8 basic amino acids, and the most common MP NLS is the basic heptad-peptide derived from SV40 virus large T antigen. Since this NLS is only related to nuclear transport and has no effect on improving cell uptake, it needs to enter the cytoplasm first to assist gene drugs to enter the nucleus [57]. The MP NLS peptides are often combined with CPPs to fabricate vectors which can promote transmembrane transport, nuclear localization and further realize targeting delivery of pDNA. For example, Yan et al. constructed a new nucleus-targeted NLS (KALA-SA) vector by combining MP NLS, KALA (a cationic CPP) and stearic acid (SA). Besides enhancing cytoplasmic transport, this vector realized targeting localization and provided a promising strategy for the treatment of lung cancer [56]. Moreover, conjugating MP NLS peptide with targeted peptide RGD can also achieve an excellent therapeutic effect. Following this strategy, Ozcelik modified MP NLS peptide and RGD peptide onto AuNPs with radio-sensitizer ability to initiate X-ray radiation-induced cell death and achieve the effect of killing or inhibiting cancer cells while retaining the normal cells. Interestingly, the results indicated that AuNPs with both cancer cell targeting and nuclear targeting capabilities are far more specific and lethal than AuNPs modified by NLS or RGD alone [58]. In order to significantly improve the delivery capacity, Hao et al. integrated NLS with CPPs (TAT) and RGD (REDV) with a selectively targeting function for endothelial cells to obtain the REDV-TAT-NLS triple tandem peptides [59].

By inserting glycine sequences with different repeats into the triple tandem peptides, the functions of each peptide were synergistically performed. The peptide complexes can be used as vector to deliver pZNF580 plasmid in endothelial cells, which can significantly improve the revascularization ability of human umbilical vein endothelial cells in vitro and in vivo, thus providing a promising and effective delivery option for angiogenesis treatment of vascular diseases (Figure 6a). Recent studies revealed that Mice Fibroblast Growth Factor 3 (FGF3) is a peptide containing multiple NLS peptides. RLRR and RRRK are two peptide sequences that can induce nuclear localization in this NLS. Introducing the RRRK peptide fragment into PAMAM non-viral vectors can significantly improve the transfection efficiency and gene expression of the vectors [61]. In addition, using four NLS derived from SV40 virus with glycine residues as spacers, Ritter synthesized the NLS tetramer of SV40 large T antigen. This lysine-rich peptide solves the past problem of NLS interfering with gene expression by covalent binding to nucleic acid molecules: it binds and concentrates nucleic acid molecules by electrostatic interaction to form stable polymers with nuclear transport properties [62]. More importantly, NLS has also been used in clustered regularly interspersed short palindromic repeats (CRISPR)/CRISPR-associated protein 9 (Cas9) gene editing technology which is widely studied nowadays. As a nuclear targeting peptide, NLS can specifically transport the vector into the nucleus, so that the Cas9/sgRNA plasmids can be accurately delivered to the tumor sites. Studies have shown that combination of NLS peptides with other non-viral vectors can significantly improve the gene editing ability of Cas9/sgRNA. For example, using NLS peptide and AS1411 aptamer as delivery vector, Cas9/sgRNA can achieve effective genome editing in targeted tumor cells [60], down-regulate the expression of FAK protein in tumor cells, and thus lead to tumor cell apoptosis (Figure 6b).

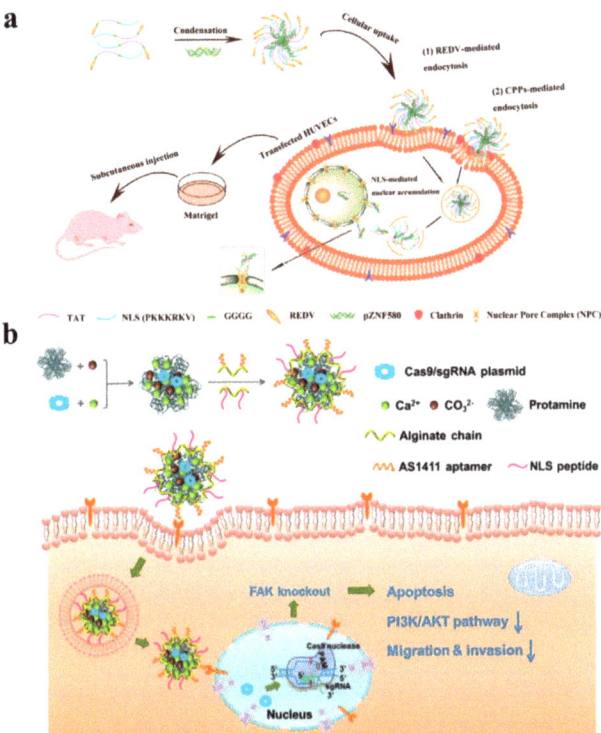

Figure 6. (a) The REDV-G4-TAT-G4-NLS peptide assembles with pZNF580 plasmid to form nanocomplexes, which are transported to endothelial cells by the targeting effect of REDV. After the transmembrane

and endosomal escape, the complexes enter the nucleus by the action of NLS to promote the expression of pZNF580 plasmid and enhance the revascularization ability of cells. Reprinted with permission from Ref. [59]. Copyright 2017, American Chemical Society. (**b**) The Cas9/sgRNA plasmid gene delivery system was prepared by the self-assembly method, which can specifically deliver the plasmid to the nuclei of tumor cells by the targeting of NLS, and knock down the protein tyrosine kinase 2 (PTK2) gene to the down-regulated local adhesion kinase (FAK). Reprinted with permission from Ref. [60]. Copyright 2019, American Chemical Society.

In addition to adding MP NLS to various nonviral vectors to achieve efficient nuclear delivery of therapeutic DNA, BP NLS composed of two or more positively charged amino acid clusters have also been developed and used for gene delivery. Matschke synthesized a modified NLS dimer structure, NLS-Ku7O$_2$. Highly efficient nuclear transport and transgenic expression were realized by co-assembling this BP NLS-Ku7O$_2$ with PEI and DNA into a ternary gene carrier complex [63].

7. Application of Other Peptides in Gene Delivery

To date, great success has been achieved in developing nonviral vectors using materials including peptides, proteins, dendrimer and liposomes. Although the gene transduction efficiency has been improved, the gene expression level is still far lower than that of viral vectors and cannot meet the clinical requirements. However, the inherent toxicity, immunogenicity and complex preparation process of viral vectors greatly limit their clinical application [64]. Therefore, great efforts have been devoted to building supramolecular assemblies that can simulate both the viral structure and function. The therapeutic nucleic acids are encapsulated into these supramolecular assemblies and delivered into cells, in the hope of obtaining efficient gene delivery vectors while reducing the inherent risk of viruses [111–113]. Recently, because of the good biocompatibility and low cytotoxicity of peptides, more and more research has been focused on imitating the virus structure through the co-assembly of peptide and nucleic acid [114,115]. Spherical viral capsids have discrete nanospace, good cell transfection ability and biodegradability, and can therefore be used as nanocarriers for nucleic acid drug delivery [116–118]. Inspired by the spherical virus, Matsuura found that the 24-mer β-annulus peptide involved in dodecahedral skeleton formation of tomato bushy stunt virus can spontaneously assemble into a "spherical artificial virus-like capsid" with a size of 30–50 nm. The cationic interior of the artificial viral capsid is hollow, allowing DNA molecules to be effectively encapsulated [65,66]. Based on the above, Matsuura K. used β-cyclic GGGCG peptide as the binding site of AuNPs, which finally self-assembled into nanocapsules with a diameter of 50 nm. This strategy extends the design of artificial viral capsids and can be further used for the delivery of nucleic acid molecules [67]. The short peptide H4K5HC$_{BZl}$C$_{BZl}$H obtained by rational design is also a spherical viral capsid. Compared with the past research on spherical artificial viruses, this spherical viral capsid has a low aspect ratio because of adding the cysteine in the center of the short peptide H4K5HC$_{BZl}$C$_{BZl}$H. This nanostructure can not only mimic the sequential decomposition of spherical viruses in response to stimuli, but also simulate the complex morphology and intracellular transformation of spherical viruses, making it an effective DNA delivery vector [68]. In addition to spherical artificial virus particles, filamentous, rod-shaped and cocoon-like virus particles have also been developed as artificial viruses. For example, the short peptide K3C6SPD which contains three fragments including N-terminal cationic fragment, β-sheet forming fragment and C-terminal hydrophilic fragment can be co-assembled to obtain cocoon-like artificial virus particles (Figure 7a) [69,70]. Ruff designed triblock molecules SP-CC-PEG which can self-assemble into mushroom nanostructures [71]. Using self-assembled non-centrosymmetric nanostructures similar to supramolecular mushrooms as caps, virus-like particles with a certain length are created and then wrapped on DNA to generate filamentous particles (Figure 7b). Marchetti designed a triblock peptide C−S10−B containing a segment of artificial lysine capsid using a de novo design method. Through electrostatic interaction, it interacted with the phosphate of single stranded or double stranded DNA and co-assembled into coronavirus-like particles, mimicking the corresponding function of viral capsid proteins [119]. These theoretical studies provide new ideas for current nucleic acid delivery.

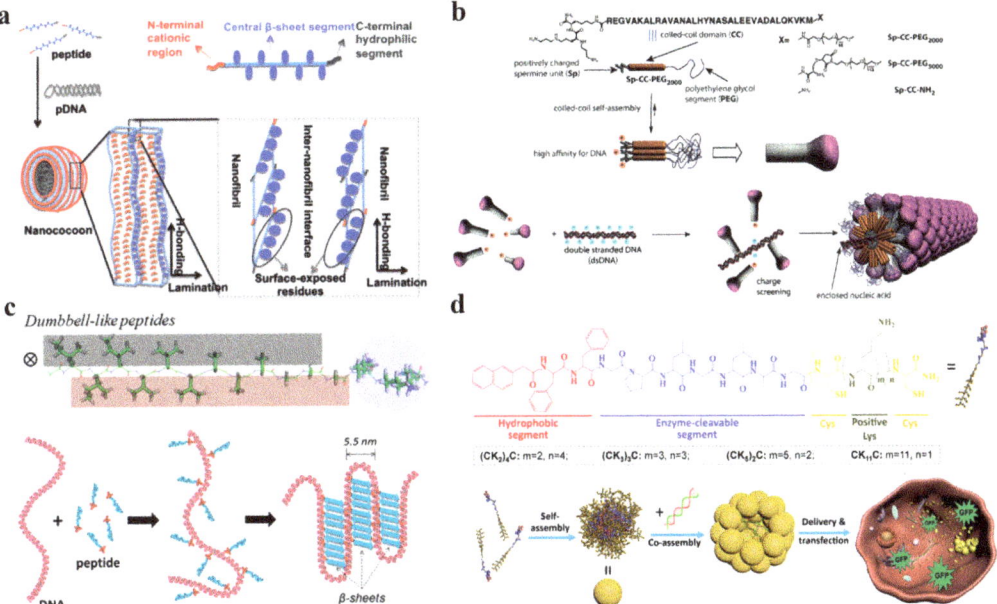

Figure 7. (a) Co-assembly of the K3C6SPD short peptide with plasmid DNA develops cocoon-like viral mimics. Reprinted with permission from Ref. [70]. Copyright 2017, German Chemical Society. (b) The mushroom shaped nanostructures SP-CC-PEG created by the synergistic self-assembly of three functional fragments, which has high affinity with DNA by electrostatic interaction, is used to prepare synthetic filamentous viruses. Reprinted with permission from Ref. [71]. Copyright 2013, American Chemical Society. (c) The dumbbell-like peptide, $I_3V_3A_3G_3K_3$, binds onto the DNA chain through electrostatic interactions, and then self-associates into β-sheets under hydrophobic interactions and hydrogen bonding, the resulting final formed structure being able to imitate the essence of viral capsid to condense and wrap DNA. Reprinted with permission from Ref. [72]. Copyright 2018, American Chemical Society. (d) NapFFGPLGLAG$(CK_m)_n$C peptides, containing the multifunctional segment, self-assemble into stable nanospheres which can encapsulate DNA by interacting with DNA in the interior, and finally realize intracellular delivery and release of genome. Reprinted with permission from Ref. [64]. Copyright 2022, Elsevier.

The efficient delivery of nucleic acids has been achieved by constructing new nonviral delivery systems using single or several lysines as functional fragments. Furthermore, many studies have shown that cationic poly(L-lysine) (PLL) can also be used to achieve efficient nucleic acid transport in vivo. PLL can mediate condensation of anionic nucleic acids to form smaller nanoparticles and protect them from enzymatic and physical degradation [120]. Yugyeong Kim et al. synthesized a new cationic AB2 miktoarm block copolymer consisting of two cationic PLL blocks and one PEG block, which can form effective nanocomplexes with pDNA. The nanocomplexes can release pDNA effectively under reducing conditions and show high level of gene expression [121]. However, for PLL, its in vitro transfection efficiency is poor in the absence of any covalently attached functional moieties to promote gene targeting or uptake [120]. To solve this problem, researchers have discovered a new cationic poly-amino acid, that is, poly(L-ornithine) (PLO). Compared to PLL that contains a tetramethylene spacer, PLO possesses a trimethylene unit in the side chain. It can complex with pDNA or mRNA and enhance transfection efficiency [122]. One big issue of nonviral gene delivery is unnecessary uptake by the reticuloendothelial system, mainly the liver. In general, 60–70% of nucleic acid molecules are taken up by scavenger receptors on liver Kupffer cells when being injected into the body without the protection of carrier molecules. This nonspecific scavenging behavior results in a significant reduction in the efficiency of drug entry into target tissues [123]. Lysine polymer exhibits excellent potential in solving this problem by avoiding unwanted uptake by the reticuloendothelial system. Recently, Anjaneyulu Dirisala et al. found that oligo(L-lysine) conjugated linear or two-armed PEG can transiently and selectively mask liver scavenger cells, effectively inhibiting sinusoidal clearance of nonviral gene carriers, thereby increasing their gene transduction efficiency in target tissues [124].

The formation of artificial viruses is based on the non-covalent interaction of peptide/peptide or peptide/DNA. By rational design of the peptide structure, the morphology, stability and transfection efficiency of the peptide/DNA hybrid structure can be regulated to construct artificial viruses [125]. In recent years, our group has been focusing on the design and study of different surfactant-like peptides to induce effective DNA condensation and so produce artificial viruses for protecting DNA from enzymatic degradation. For example, we designed six surfactant-like peptides with the same amino acid composition but different primary sequences. Because the peptide residues have different side chain size and hydrophobicity, this can lead to different self-assembled structures [126]. Among them, $I_3V_3A_3G_3K_3$ is a dumbbell-like peptide which can effectively induce DNA condensation into a virus-like structure through non-covalent interactions such as electrostatic interaction, hydrophobic interaction and hydrogen bonding [72]. The final formed structure can imitate the essence of a viral capsid to condense and wrap DNA, which is conducive to effective gene delivery in the later stage (Figure 7c). AKAEAKAE, another peptide segment we designed, has strong β-sheet forming capability and can co-assemble with PNA to obtain peptide nucleic acid-peptide conjugate, $T'_3(AKAE)_2$. It can condense DNA at low micromole concentrations, which suggests it can be a gene delivery vector [112,127]. NapFFGPLGLAG$(CK_m)_nC$ peptides have been developed by introducing several functional segments, that is, an aromatic segment of Nap-FF to promote peptide assembly by providing hydrophobic interaction, an enzyme-cleavable segment of GPLGLA to target cancer cells, and several positively charged K residues for DNA binding. These peptides can self-assemble into homogenous capsid-like nanospheres with high stability under the synergy of functional segments [64]. Moreover, they can further co-assemble with DNA to protect the genome from enzymatic digestion and greatly improve the efficiency of gene delivery (Figure 7d).

8. Concluding Remarks and Future Perspectives

Developing versatile vectors to deliver therapeutic nucleic acids into target cells/tissues is critical for gene therapy. As promising candidates, peptide-based vectors have been widely used for delivering therapeutic nucleic acids. In addition to condensing nucleic acids to form nanoparticles for protecting them from being degraded by enzymes, the rationally designed functional peptides can also help to overcome a series of biological barriers including crossing cell membrane, escaping from endosome, entering the nucleus, etc., and finally release the therapeutic nucleic acids at the target sites. These functional peptides can not only be used alone to overcome such biological barriers in gene delivery, but also can be combined to form multifunctional peptide vectors. Moreover, they can also be introduced into other nonviral gene delivery systems as functional elements to enhance the delivery capacity, which greatly expands the application of peptides in gene therapy. However, it is worth noting that although there have been a large number of reports on peptide-based gene delivery systems, most of them are still in the stage of theoretical research and animal experiments, and there are still many challenges before peptide vectors being considered for clinical use. First, the peptide-based vectors often suffer from short circulating half-time and poor chemical/physical stability, which greatly hinder the use of peptide–nucleic acid complexes in clinical trials. Effective strategies such as modifying the peptides with unnatural amino acids should be developed to improve the structural stability of the peptide-based gene delivery systems. Secondly, although peptide sequences with different functions can be combined to overcome various barriers for efficient gene delivery, this approach carries the risk of reducing individual functions. Therefore, the combination of peptide with other components without affecting the function of each part is still a problem to be solved. Thirdly, how to precisely control the microstructures of the peptide–nucleic acids complexes so as to achieve effective cellular uptake and gene transfection at targeted sites is another important issue. Modifying the peptidic vectors with stimulus-responsive fragments to design smart delivery systems so that they can perceive changes in the disease microenvironment and trigger gene release may be an effective way to solve this problem. In summary, although there has been much study and great success in the field of peptide-based gene vectors, researchers still need to move forward to find solutions for promoting peptidic gene delivery systems for them to become a gene therapy product that can be approved for clinical applications. Such research would not only promote the rapid development of peptide-based gene delivery systems, but also enable some emerging gene therapy strategies, such as CRISPR/CAS9 technology and mRNA vaccines to be applied in the human body at an early date.

Author Contributions: Conceptualization, M.C.; software, Y.Y., Z.L. and H.M.; resources, M.C.; writing—original draft preparation, Y.Y., Z.L. and H.M.; writing—review and editing, H.M. and M.C.; supervision, M.C.; project administration, M.C.; funding acquisition, M.C. All authors have read and agreed to the published version of the manuscript.

Funding: This work was supported by the National Natural Science Foundation of China (22172194, 21872173, 21972167).

Conflicts of Interest: The authors declare no conflict of interest.

Appendix A

Table A1. Full names and corresponding one-letter codes of the amino acids.

Full Amino Acid Names	One-Letter Codes
Alanine	A
Arginine	R
Asparagine	N
Aspartic acid	D
Cysteine	C
Glutamine	Q
Glutamic acid	E
Glycine	G
Histidine	H
Isoleucine	I
Leucine	L
Lysine	K
Methionine	M
Phenylalanine	F
Proline	P
Serine	S
Threonine	T
Tryptophan	W
Tyrosine	Y
Valine	V

References

1. Luo, M.; Lee, L.K.C.; Peng, B.; Choi, C.H.J.; Tong, W.Y.; Voelcker, N.H. Delivering the Promise of Gene Therapy with Nanomedicines in Treating Central Nervous System Diseases. *Adv. Sci.* **2022**, *9*, 2201740. [CrossRef] [PubMed]
2. Zhu, Y.; Shen, R.; Vuong, I.; Reynolds, R.A.; Shears, M.J.; Yao, Z.-C.; Hu, Y.; Cho, W.J.; Kong, J.; Reddy, S.K.; et al. Multistep screening of DNA/lipid nanoparticles and co-delivery with siRNA to enhance and prolong gene expression. *Nat. Commun.* **2022**, *13*, 4282. [CrossRef] [PubMed]
3. Kuriyama, N.; Yoshioka, Y.; Kikuchi, S.; Okamura, A.; Azuma, N.; Ochiya, T. Challenges for the Development of Extracellular Vesicle-Based Nucleic Acid Medicines. *Cancers* **2021**, *13*, 6137. [CrossRef] [PubMed]
4. Isgrig, K.; McDougald, D.S.; Zhu, J.; Wang, H.J.; Bennett, J.; Chien, W.W. AAV2.7m8 is a powerful viral vector for inner ear gene therapy. *Nat. Commun.* **2019**, *10*, 427. [CrossRef] [PubMed]
5. Shirley, J.L.; de Jong, Y.P.; Terhorst, C.; Herzog, R.W. Immune Responses to Viral Gene Therapy Vectors. *Mol. Ther.* **2020**, *28*, 709–722. [CrossRef] [PubMed]
6. Bouard, D.; Alazard-Dany, D.; Cosset, F.L. Viral vectors: From virology to transgene expression. *Br. J. Pharmacol.* **2009**, *157*, 153–165. [CrossRef] [PubMed]
7. Kang, Z.; Meng, Q.; Liu, K. Peptide-based gene delivery vectors. *J. Mater. Chem. B* **2019**, *7*, 1824–1841. [CrossRef]
8. Munagala, R.; Aqil, F.; Jeyabalan, J.; Kandimalla, R.; Wallen, M.; Tyagi, N.; Wilcher, S.; Yan, J.; Schultz, D.J.; Spencer, W.; et al. Exosome-mediated delivery of RNA and DNA for gene therapy. *Cancer Lett.* **2021**, *505*, 58–72. [CrossRef]
9. Gao, Y.; Men, K.; Pan, C.; Li, J.; Wu, J.; Chen, X.; Lei, S.; Gao, X.; Duan, X. Functionalized DMP-039 Hybrid Nanoparticle as a Novel mRNA Vector for Efficient Cancer Suicide Gene Therapy. *Int. J. Nanomed.* **2021**, *16*, 5211–5232. [CrossRef]
10. Sun, W.; Liu, X.Y.; Ma, L.L.; Lu, Z.L. Tumor Targeting Gene Vector for Visual Tracking of Bcl-2 siRNA Transfection and Anti-Tumor Therapy. *ACS Appl. Mater. Interfaces* **2020**, *12*, 10193–10201. [CrossRef]
11. Peng, X.; Ma, X.; Lu, S.; Li, Z. A Versatile Plant Rhabdovirus-Based Vector for Gene Silencing, miRNA Expression and Depletion, and Antibody Production. *Front. Plant Sci.* **2020**, *11*, 627880. [CrossRef] [PubMed]
12. Dirisala, A.; Uchida, S.; Tockary, T.A.; Yoshinaga, N.; Li, J.; Osawa, S.; Gorantla, L.; Fukushima, S.; Osada, K.; Kataoka, K. Precise tuning of disulphide crosslinking in mRNA polyplex micelles for optimising extracellular and intracellular nuclease tolerability. *J. Drug Target.* **2019**, *27*, 670–680. [CrossRef]
13. Devoldere, J.; Dewitte, H.; De Smedt, S.C.; Remaut, K. Evading innate immunity in nonviral mRNA delivery: Don't shoot the messenger. *Drug Discov. Today* **2016**, *21*, 11–25. [CrossRef] [PubMed]
14. Shaikh, S.; Nazam, N.; Rizvi, S.M.D.; Ahmad, K.; Baig, M.H.; Lee, E.J.; Choi, I. Mechanistic Insights into the Antimicrobial Actions of Metallic Nanoparticles and Their Implications for Multidrug Resistance. *Int. J. Mol. Sci.* **2019**, *20*, 2468. [CrossRef] [PubMed]
15. Vickers, T.A.; Crooke, S.T. siRNAs targeted to certain polyadenylation sites promote specific, RISC-independent degradation of messenger RNAs. *Nucleic Acids Res.* **2012**, *40*, 6223–6234. [CrossRef]
16. Wang, D.; Fan, Z.; Zhang, X.; Li, H.; Sun, Y.; Cao, M.; Wei, G.; Wang, J. pH-Responsive Self-Assemblies from the Designed Folic Acid-Modified Peptide Drug for Dual-Targeting Delivery. *Langmuir* **2021**, *37*, 339–347. [CrossRef]
17. Wang, H.; Feng, Z.; Xu, B. Supramolecular Assemblies of Peptides or Nucleopeptides for Gene Delivery. *Theranostics* **2019**, *9*, 3213–3222. [CrossRef]
18. Balbino, T.A.; Serafin, J.M.; Malfatti-Gasperini, A.A.; de Oliveira, C.L.; Cavalcanti, L.P.; de Jesus, M.B.; de La Torre, L.G. Microfluidic Assembly of pDNA/Cationic Liposome Lipoplexes with High pDNA Loading for Gene Delivery. *Langmuir* **2016**, *32*, 1799–1807. [CrossRef]
19. Wang, D.; Sun, Y.; Cao, M.; Wang, J.; Hao, J. Amphiphilic short peptide modulated wormlike micelle formation with pH and metal ion dual-responsive properties. *RSC Adv.* **2015**, *5*, 95604–95612. [CrossRef]
20. Vermeulen, L.M.P.; Brans, T.; De Smedt, S.C.; Remaut, K.; Braeckmans, K. Methodologies to investigate intracellular barriers for nucleic acid delivery in nonviral gene therapy. *Nano Today* **2018**, *21*, 74–90. [CrossRef]
21. Hadianamrei, R.; Zhao, X. Current state of the art in peptide-based gene delivery. *J. Control. Release* **2022**, *343*, 600–619. [CrossRef] [PubMed]
22. McErlean, E.M.; Ziminska, M.; McCrudden, C.M.; McBride, J.W.; Loughran, S.P.; Cole, G.; Mulholland, E.J.; Kett, V.; Buckley, N.E.; Robson, T.; et al. Rational design and characterisation of a linear cell penetrating peptide for nonviral gene delivery. *J. Control. Release* **2021**, *330*, 1288–1299. [CrossRef] [PubMed]
23. Dougherty, P.G.; Wen, J.; Pan, X.; Koley, A.; Ren, J.G.; Sahni, A.; Basu, R.; Salim, H.; Kubi, G.A.; Qian, Z.; et al. Enhancing the Cell Permeability of Stapled Peptides with a Cyclic Cell-Penetrating Peptide. *J. Med. Chem.* **2019**, *62*, 10098–10107. [CrossRef] [PubMed]
24. Nam, S.H.; Jang, J.; Cheon, D.H.; Chong, S.-E.; Ahn, J.H.; Hyun, S.; Yu, J.; Lee, Y. pH-Activatable cell penetrating peptide dimers for potent delivery of anticancer drug to triple-negative breast cancer. *J. Control. Release* **2021**, *330*, 898–906. [CrossRef]
25. Tuttolomondo, M.; Casella, C.; Hansen, P.L.; Polo, E.; Herda, L.M.; Dawson, K.A.; Ditzel, H.J.; Mollenhauer, J. Human DMBT1-Derived Cell-Penetrating Peptides for Intracellular siRNA Delivery. *Mol. Ther. Nucleic Acids* **2017**, *8*, 264–276. [CrossRef] [PubMed]
26. Jiang, Q.-Y.; Lai, L.-H.; Shen, J.; Wang, Q.-Q.; Xu, F.-J.; Tang, G.-P. Gene delivery to tumor cells by cationic polymeric nanovectors coupled to folic acid and the cell-penetrating peptide octaarginine. *Biomaterials* **2011**, *32*, 7253–7262. [CrossRef] [PubMed]

27. Khalil, I.A.; Kimura, S.; Sato, Y.; Harashima, H. Synergism between a cell penetrating peptide and a pH-sensitive cationic lipid in efficient gene delivery based on double-coated nanoparticles. *J. Control. Release* **2018**, *275*, 107–116. [CrossRef]
28. Huang, Y.; Cheng, Q.; Jin, X.; Ji, J.L.; Guo, S.; Zheng, S.; Wang, X.; Cao, H.; Gao, S.; Liang, X.J.; et al. Systemic and tumor-targeted delivery of siRNA by cyclic NGR and isoDGR motif-containing peptides. *Biomater. Sci.* **2016**, *4*, 494–510. [CrossRef]
29. Yang, S.; Ou, C.; Wang, L.; Liu, X.; Yang, J.; Wang, X.; Wang, M.; Shen, M.; Wu, Q.; Gong, C. Virus-esque nucleus-targeting nanoparticles deliver trojan plasmid for release of anti-tumor shuttle protein. *J. Control. Release* **2020**, *320*, 253–264. [CrossRef]
30. Zhao, Y.; He, Z.; Gao, Y.; Tang, H.; He, J.; Guo, Q.; Zhang, W.; Liu, J. Fine Tuning of Core-Shell Structure of Hyaluronic Acid/Cell-Penetrating Peptides/siRNA Nanoparticles for Enhanced Gene Delivery to Macrophages in Antiatherosclerotic Therapy. *Biomacromolecules* **2018**, *19*, 2944–2956. [CrossRef]
31. Covarrubias-Zambrano, O.; Shrestha, T.B.; Pyle, M.; Montes-Gonzalez, M.; Troyer, D.L.; Bossmann, S.H. Development of a Gene Delivery System Composed of a Cell-Penetrating Peptide and a Nontoxic Polymer. *ACS Appl. Bio Mater.* **2020**, *3*, 7418–7427. [CrossRef] [PubMed]
32. Dowaidar, M.; Abdelhamid, H.N.; Hallbrink, M.; Freimann, K.; Kurrikoff, K.; Zou, X.; Langel, U. Magnetic Nanoparticle Assisted Self-assembly of Cell Penetrating Peptides-Oligonucleotides Complexes for Gene Delivery. *Sci. Rep.* **2017**, *7*, 9159. [CrossRef] [PubMed]
33. Yang, Y.; Yang, Y.; Xie, X.; Wang, Z.; Gong, W.; Zhang, H.; Li, Y.; Yu, F.; Li, Z.; Mei, X. Dual-modified liposomes with a two-photon-sensitive cell penetrating peptide and NGR ligand for siRNA targeting delivery. *Biomaterials* **2015**, *48*, 84–96. [CrossRef] [PubMed]
34. Chen, Y.; Wu, J.J.; Huang, L. Nanoparticles targeted with NGR motif deliver c-myc siRNA and doxorubicin for anticancer therapy. *Mol. Ther.* **2010**, *18*, 828–834. [CrossRef] [PubMed]
35. Kim, Y.M.; Park, S.C.; Jang, M.K. Targeted gene delivery of polyethyleneimine-grafted chitosan with RGD dendrimer peptide in αvβ3 integrin-overexpressing tumor cells. *Carbohydr. Polym.* **2017**, *174*, 1059–1068. [CrossRef]
36. Luo, Q.; Song, H.; Deng, X.; Li, J.; Jian, W.; Zhao, J.; Zheng, X.; Basnet, S.; Ge, H.; Daniel, T.; et al. A Triple-Regulated Oncolytic Adenovirus Carrying MicroRNA-143 Exhibits Potent Antitumor Efficacy in Colorectal Cancer. *Mol. Ther. Oncolytics* **2020**, *16*, 219–229. [CrossRef]
37. Wang, Y.; Nie, Y.; Ding, Z.; Yao, M.; Du, R.; Zhang, L.; Wang, S.; Li, D.; Wang, Y.; Cao, M. An amphiphilic peptide with cell penetrating sequence for highly efficient gene transfection. *Colloids Surf. A Physicochem. Eng. Asp.* **2020**, *590*, 124529. [CrossRef]
38. Alam, M.R.; Ming, X.; Fisher, M.; Lackey, J.G.; Rajeev, K.G.; Manoharan, M.; Juliano, R.L. Multivalent Cyclic RGD Conjugates for Targeted Delivery of Small Interfering RNA. *Bioconjugate Chem.* **2011**, *22*, 1673–1681. [CrossRef]
39. Liu, X.; Wang, W.; Samarsky, D.; Liu, L.; Xu, Q.; Zhang, W.; Zhu, G.; Wu, P.; Zuo, X.; Deng, H.; et al. Tumor-targeted in vivo gene silencing via systemic delivery of cRGD-conjugated siRNA. *Nucleic Acids Res.* **2014**, *42*, 11805–11817. [CrossRef]
40. Khatri, N.; Baradia, D.; Vhora, I.; Rathi, M.; Misra, A. cRGD grafted liposomes containing inorganic nano-precipitate complexed siRNA for intracellular delivery in cancer cells. *J. Control. Release* **2014**, *182*, 45–57. [CrossRef]
41. Xu, X.; Liu, Y.; Yang, Y.; Wu, J.; Cao, M.; Sun, L. One-pot synthesis of functional peptide-modified gold nanoparticles for gene delivery. *Colloids Surf. A Physicochem. Eng. Asp.* **2022**, *640*, 128491. [CrossRef]
42. Jena, L.N.; Bennie, L.A.; McErlean, E.M.; Pentlavalli, S.; Glass, K.; Burrows, J.F.; Kett, V.L.; Buckley, N.E.; Coulter, J.A.; Dunne, N.J.; et al. Exploiting the anticancer effects of a nitrogen bisphosphonate nanomedicine for glioblastoma multiforme. *J. Nanobiotechnology* **2021**, *19*, 127. [CrossRef] [PubMed]
43. Udhayakumar, V.K.; De Beuckelaer, A.; McCaffrey, J.; McCrudden, C.M.; Kirschman, J.L.; Vanover, D.; Van Hoecke, L.; Roose, K.; Deswarte, K.; De Geest, B.G.; et al. Arginine-Rich Peptide-Based mRNA Nanocomplexes Efficiently Instigate Cytotoxic T Cell Immunity Dependent on the Amphipathic Organization of the Peptide. *Adv. Healthc Mater.* **2017**, *6*, 1601412. [CrossRef]
44. Mulholland, E.J.; Ali, A.; Robson, T.; Dunne, N.J.; McCarthy, H.O. Delivery of RALA/siFKBPL nanoparticles via electrospun bilayer nanofibres: An innovative angiogenic therapy for wound repair. *J. Control. Release* **2019**, *316*, 53–65. [CrossRef]
45. Liu, Y.; Wan, H.-H.; Tian, D.-M.; Xu, X.-J.; Bi, C.-L.; Zhan, X.-Y.; Huang, B.-H.; Xu, Y.-S.; Yan, L.-P. Development and Characterization of High Efficacy Cell-Penetrating Peptide via Modulation of the Histidine and Arginine Ratio for Gene Therapy. *Materials* **2021**, *14*, 4674. [CrossRef]
46. Yang, S.; Meng, Z.; Kang, Z.; Sun, C.; Wang, T.; Feng, S.; Meng, Q.; Liu, K. The structure and configuration changes of multifunctional peptide vectors enhance gene delivery efficiency. *RSC Adv.* **2018**, *8*, 28356–28366. [CrossRef] [PubMed]
47. Ali, S.; Dussouillez, C.; Padilla, B.; Frisch, B.; Mason, A.J.; Kichler, A. Design of a new cell penetrating peptide for DNA, siRNA and mRNA delivery. *J. Gene Med.* **2022**, *24*, e3401. [CrossRef]
48. Vanova, J.; Hejtmankova, A.; Zackova Suchanova, J.; Sauerova, P.; Forstova, J.; Hubalek Kalbacova, M.; Spanielova, H. Influence of cell-penetrating peptides on the activity and stability of virus-based nanoparticles. *Int. J. Pharm.* **2020**, *576*, 119008. [CrossRef]
49. Ferrer-Miralles, N.; Corchero, J.L.; Kumar, P.; Cedano, J.A.; Gupta, K.C.; Villaverde, A.; Vazquez, E. Biological activities of histidine-rich peptides; merging biotechnology and nanomedicine. *Microb. Cell Factories* **2011**, *10*, 101. [CrossRef]
50. Cirillo, S.; Tomeh, M.A.; Wilkinson, R.N.; Hill, C.; Brown, S.; Zhao, X. Designed Antitumor Peptide f or Targeted siRNA Delivery into Cancer Spheroids. *ACS Appl. Mater. Interfaces* **2021**, *13*, 49713–49728. [CrossRef]
51. Blenke, E.O.; Sleszynska, M.; Evers, M.J.; Storm, G.; Martin, N.I.; Mastrobattista, E. Strategies for the Activation and Release of the Membranolytic Peptide Melittin from Liposomes Using Endosomal pH as a Trigger. *Bioconjugate Chem.* **2017**, *28*, 574–582. [CrossRef] [PubMed]

52. Kloeckner, J.; Boeckle, S.; Persson, D.; Roedl, W.; Ogris, M.; Berg, K.; Wagner, E. DNA polyplexes based on degradable oligoethylenimine-derivatives: Combination with EGF receptor targeting and endosomal release functions. *J. Control. Release* **2006**, *116*, 115–122. [CrossRef]
53. Zhang, S.K.; Song, J.W.; Li, S.B.; Gao, H.W.; Chang, H.Y.; Jia, L.L.; Gong, F.; Tan, Y.X.; Ji, S.P. Design of pH-sensitive peptides from natural antimicrobial peptides for enhancing polyethylenimine-mediated gene transfection. *J. Gene Med.* **2017**, *19*, e2955. [CrossRef] [PubMed]
54. Boeckle, S.; Fahrmeir, J.; Roedl, W.; Ogris, M.; Wagner, E. Melittin analogs with high lytic activity at endosomal pH enhance transfection with purified targeted PEI polyplexes. *J. Control. Release* **2006**, *112*, 240–248. [CrossRef] [PubMed]
55. Tamemoto, N.; Akishiba, M.; Sakamoto, K.; Kawano, K.; Noguchi, H.; Futaki, S. Rational Design Principles of Attenuated Cationic Lytic Peptides for Intracellular Delivery of Biomacromolecules. *Mol. Pharm.* **2020**, *17*, 2175–2185. [CrossRef]
56. Yan, C.; Shi, W.; Gu, J.; Lee, R.J.; Zhang, Y. Design of a Novel Nucleus-Targeted NLS-KALA-SA Nanocarrier to Delivery Poorly Water-Soluble Anti-Tumor Drug for Lung Cancer Treatment. *J. Pharm. Sci.* **2021**, *110*, 2432–2441. [CrossRef]
57. Li, Q.; Hao, X.; Wang, H.; Guo, J.; Ren, X.K.; Xia, S.; Zhang, W.; Feng, Y. Multifunctional REDV-G-TAT-G-NLS-Cys peptide sequence conjugated gene carriers to enhance gene transfection efficiency in endothelial cells. *Colloids Surf. B Biointerfaces* **2019**, *184*, 110510. [CrossRef]
58. Ozcelik, S.; Pratx, G. Nuclear-targeted gold nanoparticles enhance cancer cell radiosensitization. *Nanotechnology* **2020**, *31*, 415102. [CrossRef]
59. Hao, X.; Li, Q.; Guo, J.; Ren, X.; Feng, Y.; Shi, C.; Zhang, W. Multifunctional Gene Carriers with Enhanced Specific Penetration and Nucleus Accumulation to Promote Neovascularization of HUVECs in Vivo. *ACS Appl. Mater. Interfaces* **2017**, *9*, 35613–35627. [CrossRef]
60. Liu, B.Y.; He, X.Y.; Xu, C.; Ren, X.H.; Zhuo, R.X.; Cheng, S.X. Peptide and Aptamer Decorated Delivery System for Targeting Delivery of Cas9/sgRNA Plasmid To Mediate Antitumor Genome Editing. *ACS Appl. Mater. Interfaces* **2019**, *11*, 23870–23879. [CrossRef]
61. Lee, J.; Jung, J.; Kim, Y.J.; Lee, E.; Choi, J.S. Gene delivery of PAMAM dendrimer conjugated with the nuclear localization signal peptide originated from fibroblast growth factor 3. *Int. J. Pharm.* **2014**, *459*, 10–18. [CrossRef] [PubMed]
62. Ritter, W.; Plank, C.; Lausier, J.; Rudolph, C.; Zink, D.; Reinhardt, D.; Rosenecker, J. A novel transfecting peptide comprising a tetrameric nuclear localization sequence. *J. Mol. Med. (Berl.)* **2003**, *81*, 708–717. [CrossRef] [PubMed]
63. Matschke, J.; Bohla, A.; Maucksch, C.; Mittal, R.; Rudolph, C.; Rosenecker, J. Characterization of Ku70(2)-NLS as bipartite nuclear localization sequence for nonviral gene delivery. *PLoS ONE* **2012**, *7*, e24615. [CrossRef] [PubMed]
64. Cao, M.; Zhang, Z.; Zhang, X.; Wang, Y.; Wu, J.; Liu, Z.; Sun, L.; Wang, D.; Yue, T.; Han, Y.; et al. Peptide Self-assembly into stable Capsid-Like nanospheres and Co-assembly with DNA to produce smart artificial viruses. *J. Colloid Interface Sci.* **2022**, *615*, 395–407. [CrossRef] [PubMed]
65. Matsuura, K.; Watanabe, K.; Matsuzaki, T.; Sakurai, K.; Kimizuka, N. Self-assembled synthetic viral capsids from a 24-mer viral peptide fragment. *Angew Chem. Int. Ed. Engl.* **2010**, *49*, 9662–9665. [CrossRef] [PubMed]
66. Fujita, S.; Matsuura, K. Encapsulation of CdTe Quantum Dots into Synthetic Viral Capsids. *Chem. Lett.* **2016**, *45*, 922–924. [CrossRef]
67. Matsuura, K.; Ueno, G.; Fujita, S. Self-assembled artificial viral capsid decorated with gold nanoparticles. *Polym. J.* **2014**, *47*, 146–151. [CrossRef]
68. Ni, R.; Chau, Y. Nanoassembly of Oligopeptides and DNA Mimics the Sequential Disassembly of a Spherical Virus. *Angew. Chem. -Int. Ed.* **2020**, *59*, 3578–3584. [CrossRef]
69. Ni, R.; Liu, J.; Chau, Y. Ultrasound-facilitated assembly and disassembly of a pH-sensitive self-assembly peptide. *RSC Adv.* **2018**, *8*, 29482–29487. [CrossRef]
70. Ni, R.; Chau, Y. Tuning the Inter-nanofibril Interaction To Regulate the Morphology and Function of Peptide/DNA Co-assembled Viral Mimics. *Angew. Chem. Int. Ed. Engl.* **2017**, *56*, 9356–9360. [CrossRef]
71. Ruff, Y.; Moyer, T.; Newcomb, C.J.; Demeler, B.; Stupp, S.I. Precision templating with DNA of a virus-like particle with peptide nanostructures. *J. Am. Chem. Soc.* **2013**, *135*, 6211–6219. [CrossRef] [PubMed]
72. Cao, M.; Wang, Y.; Zhao, W.; Qi, R.; Han, Y.; Wu, R.; Wang, Y.; Xu, H. Peptide-Induced DNA Condensation into Virus-Mimicking Nanostructures. *ACS Appl. Mater. Interfaces* **2018**, *10*, 24349–24360. [CrossRef] [PubMed]
73. Järver, P.; Coursindel, T.; Andaloussi, S.E.; Godfrey, C.; Wood, M.J.; Gait, M.J. Peptide-mediated Cell and In Vivo Delivery of Antisense Oligonucleotides and siRNA. *Mol. Ther. -Nucleic Acids* **2012**, *1*, e27. [CrossRef] [PubMed]
74. Vázquez, O.; Seitz, O. Cytotoxic peptide–PNA conjugates obtained by RNA-programmed peptidyl transfer with turnover. *Chem. Sci.* **2014**, *5*, 2850–2854. [CrossRef]
75. Shabanpoor, F.; Gait, M.J. Development of a general methodology for labelling peptide-morpholino oligonucleotide conjugates using alkyne-azide click chemistry. *Chem. Commun.* **2013**, *49*, 10260–10262. [CrossRef] [PubMed]
76. López-Vidal, E.M.; Schissel, C.K.; Mohapatra, S.; Bellovoda, K.; Wu, C.-L.; Wood, J.A.; Malmberg, A.B.; Loas, A.; Gómez-Bombarelli, R.; Pentelute, B.L. Deep Learning Enables Discovery of a Short Nuclear Targeting Peptide for Efficient Delivery of Antisense Oligomers. *JACS Au* **2021**, *1*, 2009–2020. [CrossRef]
77. Eilers, W.; Gadd, A.; Foster, H.; Foster, K. Dmd Treatment: Animal Models. *Neuromuscul. Disord.* **2018**, *28*, S92–S93. [CrossRef]

78. Urello, M.; Hsu, W.-H.; Christie, R.J. Peptides as a Material Platform for Gene Delivery: Emerging Concepts and Converging Technologies. *Acta Biomater.* **2020**, *117*, 40–59. [CrossRef]
79. Samec, T.; Boulos, J.; Gilmore, S.; Hazelton, A.; Alexander-Bryant, A. Peptide-based delivery of therapeutics in cancer treatment. *Mater. Today Bio* **2022**, *14*, 100248. [CrossRef]
80. Khan, M.M.; Filipczak, N.; Torchilin, V.P. Cell penetrating peptides: A versatile vector for co-delivery of drug and genes in cancer. *J. Control. Release* **2021**, *330*, 1220–1228. [CrossRef]
81. Liu, Y.; An, S.; Li, J.; Kuang, Y.; He, X.; Guo, Y.; Ma, H.; Zhang, Y.; Ji, B.; Jiang, C. Brain-targeted co-delivery of therapeutic gene and peptide by multifunctional nanoparticles in Alzheimer's disease mice. *Biomaterials* **2016**, *80*, 33–45. [CrossRef] [PubMed]
82. Lehto, T.; Simonson, O.E.; Mäger, I.; Ezzat, K.; Sork, H.; Copolovici, D.-M.; Viola, J.R.; Zaghloul, E.M.; Lundin, P.; Moreno, P.M.D.; et al. A peptide-based vector for efficient gene transfer in vitro and in vivo. *Mol. Ther.* **2011**, *19*, 1457–1467. [CrossRef] [PubMed]
83. Shajari, N.; Mansoori, B.; Davudian, S.; Mohammadi, A.; Baradaran, B. Overcoming the Challenges of siRNA Delivery: Nanoparticle Strategies. *Curr. Drug Deliv.* **2017**, *14*, 36–46. [CrossRef] [PubMed]
84. Dissanayake, S.; Denny, W.A.; Gamage, S.; Sarojini, V. Recent developments in anticancer drug delivery using cell penetrating and tumor targeting peptides. *J. Control. Release* **2017**, *250*, 62–76. [CrossRef]
85. Wang, H.; Lin, S.; Wang, S.; Jiang, Z.; Ding, T.; Wei, X.; Lu, Y.; Yang, F.; Zhan, C. Folic Acid Enables Targeting Delivery of Lipodiscs by Circumventing IgM-Mediated Opsonization. *Nano Lett.* **2022**, *22*, 6516–6522. [CrossRef]
86. Shin, J.M.; Oh, S.J.; Kwon, S.; Deepagan, V.G.; Lee, M.; Song, S.H.; Lee, H.-J.; Kim, S.; Song, K.-H.; Kim, T.W.; et al. A PEGylated hyaluronic acid conjugate for targeted cancer immunotherapy. *J. Control. Release* **2017**, *267*, 181–190. [CrossRef]
87. Cao, M.; Lu, S.; Wang, N.; Xu, H.; Cox, H.; Li, R.; Waigh, T.; Han, Y.; Wang, Y.; Lu, J.R. Enzyme-Triggered Morphological Transition of Peptide Nanostructures for Tumor-Targeted Drug Delivery and Enhanced Cancer Therapy. *ACS Appl. Mater. Interfaces* **2019**, *11*, 16357–16366. [CrossRef]
88. Kapoor, P.; Singh, H.; Gautam, A.; Chaudhary, K.; Kumar, R.; Raghava, G.P. TumorHoPe: A database of tumor homing peptides. *PLoS ONE* **2012**, *7*, e35187. [CrossRef]
89. Ahmad, K.; Lee, E.J.; Shaikh, S.; Kumar, A.; Rao, K.M.; Park, S.Y.; Jin, J.O.; Han, S.S.; Choi, I. Targeting integrins for cancer management using nanotherapeutic approaches: Recent advances and challenges. *Semin. Cancer Biol.* **2021**, *69*, 325–336. [CrossRef]
90. Liu, F.; Yan, J.R.; Chen, S.; Yan, G.P.; Pan, B.Q.; Zhang, Q.; Wang, Y.F.; Gu, Y.T. Polypeptide-rhodamine B probes containing laminin/fibronectin receptor-targeting sequence (YIGSR/RGD) for fluorescent imaging in cancers. *Talanta* **2020**, *212*, 120718. [CrossRef]
91. Koch, J.; Schober, S.J.; Hindupur, S.V.; Schoning, C.; Klein, F.G.; Mantwill, K.; Ehrenfeld, M.; Schillinger, U.; Hohnecker, T.; Qi, P.; et al. Targeting the Retinoblastoma/E2F repressive complex by CDK4/6 inhibitors amplifies oncolytic potency of an oncolytic adenovirus. *Nat. Commun.* **2022**, *13*, 4689. [CrossRef] [PubMed]
92. Kasala, D.; Lee, S.H.; Hong, J.W.; Choi, J.W.; Nam, K.; Chung, Y.H.; Kim, S.W.; Yun, C.O. Synergistic antitumor effect mediated by a paclitaxel-conjugated polymeric micelle-coated oncolytic adenovirus. *Biomaterials* **2017**, *145*, 207–222. [CrossRef] [PubMed]
93. Yamamoto, Y.; Hiraoka, N.; Goto, N.; Rin, Y.; Miura, K.; Narumi, K.; Uchida, H.; Tagawa, M.; Aoki, K. A targeting ligand enhances infectivity and cytotoxicity of an oncolytic adenovirus in human pancreatic cancer tissues. *J. Control. Release* **2014**, *192*, 284–293. [CrossRef] [PubMed]
94. Yu, J.; Xie, X.; Xu, X.; Zhang, L.; Zhou, X.; Yu, H.; Wu, P.; Wang, T.; Che, X.; Hu, Z. Development of dual ligand-targeted polymeric micelles as drug carriers for cancer therapy in vitro and in vivo. *J. Mater. Chem. B* **2014**, *2*, 2114–2126. [CrossRef] [PubMed]
95. Zhou, L.Y.; Zhu, Y.H.; Wang, X.Y.; Shen, C.; Wei, X.W.; Xu, T.; He, Z.Y. Novel zwitterionic vectors: Multi-functional delivery systems for therapeutic genes and drugs. *Comput. Struct. Biotechnol. J.* **2020**, *18*, 1980–1999. [CrossRef]
96. Vachutinsky, Y.; Oba, M.; Miyata, K.; Hiki, S.; Kano, M.R.; Nishiyama, N.; Koyama, H.; Miyazono, K.; Kataoka, K. Antiangiogenic gene therapy of experimental pancreatic tumor by sFlt-1 plasmid DNA carried by RGD-modified crosslinked polyplex micelles. *J. Control. Release* **2011**, *149*, 51–57. [CrossRef]
97. Dirisala, A.; Osada, K.; Chen, Q.; Tockary, T.A.; Machitani, K.; Osawa, S.; Liu, X.; Ishii, T.; Miyata, K.; Oba, M.; et al. Optimized rod length of polyplex micelles for maximizing transfection efficiency and their performance in systemic gene therapy against stroma-rich pancreatic tumors. *Biomaterials* **2014**, *35*, 5359–5368. [CrossRef]
98. Ghosh, P.; Han, G.; De, M.; Kim, C.K.; Rotello, V.M. Gold nanoparticles in delivery applications. *Adv. Drug Deliv. Rev.* **2008**, *60*, 1307–1315. [CrossRef]
99. Nativo, P.; Prior, I.A.; Brust, M. Uptake and Intracellular Fate of Surface-Modified Gold Nanoparticles. *ACS Nano* **2008**, *2*, 1639–1644. [CrossRef]
100. Liu, T.; Lin, M.; Wu, F.; Lin, A.; Luo, D.; Zhang, Z. Development of a nontoxic and efficient gene delivery vector based on histidine grafted chitosan. *Int. J. Polym. Mater. Polym. Biomater.* **2022**, *71*, 717–727. [CrossRef]
101. Kichler, A.; Leborgne, C.; Danos, O.; Bechinger, B. Characterization of the gene transfer process mediated by histidine-rich peptides. *J. Mol. Med. (Berl)* **2007**, *85*, 191–201. [CrossRef] [PubMed]
102. Lointier, M.; Aisenbrey, C.; Marquette, A.; Tan, J.H.; Kichler, A.; Bechinger, B. Membrane pore-formation correlates with the hydrophilic angle of histidine-rich amphipathic peptides with multiple biological activities. *Biochim. Biophys. Acta Biomembr.* **2020**, *1862*, 183212. [CrossRef] [PubMed]

103. Kichler, A.; Leborgne, C.; März, J.; Danos, O.; Bechinger, B. Histidine-rich amphipathic peptide antibiotics promote efficient delivery of DNA into mammalian cells. *Proc. Natl. Acad. Sci. USA* **2003**, *100*, 1564–1568. [CrossRef] [PubMed]
104. Vanova, J.; Ciharova, B.; Hejtmankova, A.; Epperla, C.P.; Skvara, P.; Forstova, J.; Hubalek Kalbacova, M.; Spanielova, H. VirPorters: Insights into the action of cationic and histidine-rich cell-penetrating peptides. *Int. J. Pharm.* **2022**, *611*, 121308. [CrossRef]
105. Zhang, S.K.; Gong, L.; Zhang, X.; Yun, Z.M.; Li, S.B.; Gao, H.W.; Dai, C.J.; Yuan, J.J.; Chen, J.M.; Gong, F.; et al. Antimicrobial peptide AR-23 derivatives with high endosomal disrupting ability enhance poly(l-lysine)-mediated gene transfer. *J. Gene Med.* **2020**, *22*, e3259. [CrossRef]
106. Peeler, D.J.; Thai, S.N.; Cheng, Y.; Horner, P.J.; Sellers, D.L.; Pun, S.H. pH-sensitive polymer micelles provide selective and potentiated lytic capacity to venom peptides for effective intracellular delivery. *Biomaterials* **2019**, *192*, 235–244. [CrossRef]
107. Cao, X.; Shang, X.; Guo, Y.; Zheng, X.; Li, W.; Wu, D.; Sun, L.; Mu, S.; Guo, C. Lysosomal escaped protein nanocarriers for nuclear-targeted siRNA delivery. *Anal. Bioanal. Chem.* **2021**, *413*, 3493–3499. [CrossRef]
108. Huang, G.; Zhang, Y.; Zhu, X.; Zeng, C.; Wang, Q.; Zhou, Q.; Tao, Q.; Liu, M.; Lei, J.; Yan, C.; et al. Structure of the cytoplasmic ring of the Xenopus laevis nuclear pore complex by cryo-electron microscopy single particle analysis. *Cell Res.* **2020**, *30*, 520–531. [CrossRef]
109. Mangipudi, S.S.; Canine, B.F.; Wang, Y.; Hatefi, A. Development of a Genetically Engineered Biomimetic Vector for Targeted Gene Transfer to Breast Cancer Cells. *Mol. Pharm.* **2009**, *6*, 1100–1109. [CrossRef]
110. Lu, J.; Wu, T.; Zhang, B.; Liu, S.; Song, W.; Qiao, J.; Ruan, H. Types of nuclear localization signals and mechanisms of protein import into the nucleus. *Cell Commun. Signal.* **2021**, *19*, 60. [CrossRef]
111. Noble, J.E.; De Santis, E.; Ravi, J.; Lamarre, B.; Castelletto, V.; Mantell, J.; Ray, S.; Ryadnov, M.G. A De Novo Virus-Like Topology for Synthetic Virions. *J. Am. Chem. Soc.* **2016**, *138*, 12202–12210. [CrossRef] [PubMed]
112. Cao, M.; Wang, N.; Zhou, P.; Sun, Y.; Wang, J.; Wang, S.; Xu, H. Virus-like supramolecular assemblies formed by cooperation of base pairing interaction and peptidic association. *Sci. China Chem.* **2015**, *59*, 310–315. [CrossRef]
113. van der Aa, M.A.; Mastrobattista, E.; Oosting, R.S.; Hennink, W.E.; Koning, G.A.; Crommelin, D.J. The nuclear pore complex: The gateway to successful nonviral gene delivery. *Pharm. Res.* **2006**, *23*, 447–459. [CrossRef] [PubMed]
114. Matsuura, K.; Ota, J.; Fujita, S.; Shiomi, Y.; Inaba, H. Construction of Ribonuclease-Decorated Artificial Virus-like Capsid by Peptide Self-assembly. *J. Org. Chem.* **2020**, *85*, 1668–1673. [CrossRef] [PubMed]
115. Kong, J.; Wang, Y.; Zhang, J.; Qi, W.; Su, R.; He, Z. Rationally Designed Peptidyl Virus-Like Particles Enable Targeted Delivery of Genetic Cargo. *Angew. Chem. Int. Ed. Engl.* **2018**, *57*, 14032–14036. [CrossRef]
116. Nakamura, Y.; Inaba, H.; Matsuura, K. Construction of Artificial Viral Capsids Encapsulating Short DNAs via Disulfide Bonds and Controlled Release of DNAs by Reduction. *Chem. Lett.* **2019**, *48*, 544–546. [CrossRef]
117. Wen, A.M.; Steinmetz, N.F. Design of virus-based nanomaterials for medicine, biotechnology, and energy. *Chem. Soc. Rev.* **2016**, *45*, 4074–4126. [CrossRef]
118. Cao, M.; Shen, Y.; Wang, Y.; Wang, X.; Li, D. Self-Assembly of Short Elastin-like Amphiphilic Peptides: Effects of Temperature, Molecular Hydrophobicity and Charge Distribution. *Molecules* **2019**, *24*, 202. [CrossRef]
119. Marchetti, M.; Kamsma, D.; Vargas, E.C.; Garcia, A.H.; van der Schoot, P.; de Vries, R.; Wuite, G.J.L.; Roos, W.H. Real-Time Assembly of Viruslike Nucleocapsids Elucidated at the Single-Particle Level. *Nano Lett.* **2019**, *19*, 5746–5753. [CrossRef]
120. Walter, E.; Merkle, H.P. Microparticle-mediated transfection of non-phagocytic cells in vitro. *J. Drug Target.* **2002**, *10*, 11–21. [CrossRef]
121. Kim, Y.; Uthaman, S.; Nurunnabi, M.; Mallick, S.; Oh, K.S.; Kang, S.W.; Cho, S.; Kang, H.C.; Lee, Y.K.; Huh, K.M. Synthesis and characterization of bioreducible cationic biarm polymer for efficient gene delivery. *Int. J. Biol. Macromol.* **2018**, *110*, 366–374. [CrossRef]
122. Dirisala, A.; Uchida, S.; Li, J.; Van Guyse, J.F.R.; Hayashi, K.; Vummaleti, S.V.C.; Kaur, S.; Mochida, Y.; Fukushima, S.; Kataoka, K. Effective mRNA Protection by Poly(l-ornithine) Synergizes with Endosomal Escape Functionality of a Charge-Conversion Polymer toward Maximizing mRNA Introduction Efficiency. *Macromol. Rapid Commun.* **2022**, *43*, e2100754. [CrossRef] [PubMed]
123. Collard, W.T.; Yang, Y.; Kwok, K.Y.; Park, Y.; Rice, K.G. Biodistribution, metabolism, and in vivo gene expression of low molecular weight glycopeptide polyethylene glycol peptide DNA co-condensates. *J. Pharm. Sci.* **2000**, *89*, 499–512. [CrossRef]
124. Dirisala, A.; Uchida, S.; Toh, K.; Li, J.; Osawa, S.; Tockary, T.A.; Liu, X.; Abbasi, S.; Hayashi, K.; Mochida, Y.; et al. Transient stealth coating of liver sinusoidal wall by anchoring two-armed PEG for retargeting nanomedicines. *Sci. Adv.* **2020**, *6*, eabb8133. [CrossRef] [PubMed]
125. Cao, M.; Wang, Y.; Hu, X.; Gong, H.; Li, R.; Cox, H.; Zhang, J.; Waigh, T.A.; Xu, H.; Lu, J.R. Reversible Thermoresponsive Peptide-PNIPAM Hydrogels for Controlled Drug Delivery. *Biomacromolecules* **2019**, *20*, 3601–3610. [CrossRef]
126. Cao, M.; Lu, S.; Zhao, W.; Deng, L.; Wang, M.; Wang, J.; Zhou, P.; Wang, D.; Xu, H.; Lu, J.R. Peptide Self-Assembled Nanostructures with Distinct Morphologies and Properties Fabricated by Molecular Design. *ACS Appl. Mater. Interfaces* **2017**, *9*, 39174–39184. [CrossRef]
127. Cao, M.; Zhao, W.; Zhou, P.; Xie, Z.; Sun, Y.; Xu, H. Peptide nucleic acid-ionic self-complementary peptide conjugates: Highly efficient DNA condensers with specific condensing mechanism. *RSC Adv.* **2017**, *7*, 3796–3803. [CrossRef]

Article

Effect of Inlet Flow Strategies on the Dynamics of Pulsed Fluidized Bed of Nanopowder

Syed Sadiq Ali [1], Agus Arsad [2], Kenneth L. Roberts [3] and Mohammad Asif [4,*]

1 School of Chemical and Energy Engineering, Faculty of Engineering, Universiti Teknologi Malaysia, Johor Bahru 81310, Johor, Malaysia
2 UTM-MPRC Institute for Oil and Gas, School of Chemical and Energy Engineering, Faculty of Engineering, Universiti Teknologi Malaysia, Johor Bahru 81310, Johor, Malaysia
3 SmartState Center for Strategic Approaches to the Generation of Electricity (SAGE), College of Engineering and Computing, University of South Carolina, Columbia, SC 29208, USA
4 Department of Chemical Engineering, King Saud University, P.O. Box 800, Riyadh 11421, Saudi Arabia
* Correspondence: masif@ksu.edu.sa; Tel.: +966-114-676-849; Fax: +966-114-678-770

Abstract: The use of fluidization assistance can greatly enhance the fluidization hydrodynamics of powders that exhibit poor fluidization behavior. Compared to other assistance techniques, pulsed flow assistance is a promising technique for improving conventional fluidization because of its energy efficiency and ease of process implementation. However, the inlet flow configuration of pulsed flow can significantly affect the bed hydrodynamics. In this study, the conventional single drainage (SD) flow strategy was modified to purge the primary flow during the non-flow period of the pulse to eliminate pressure buildup in the inlet flow line while providing a second drainage path to the residual gas. The bed dynamics for both cases, namely, single drainage (SD) and modified double drainage (MDD), were carefully monitored by recording the overall and local pressure drop transients in different bed regions at two widely different pulsation frequencies of 0.05 and 0.25 Hz. The MDD strategy led to substantially faster bed dynamics and greater frictional pressure drop in lower bed regions with significantly mitigated segregation behavior. The spectral analysis of the local and global pressure transient data in the frequency domain revealed a pronounced difference between the two flow strategies. The application of the MDD inlet flow strategy eliminated the disturbances from the pulsed fluidized bed irrespective of the pulsation frequency.

Keywords: fluidization; pulsation; frequency; flow strategy; flow spike; disturbances

1. Introduction

Fluidized bed technology holds great promise for improving the process efficiency in the petrochemical and chemical industries. Its key strength lies in efficient solid dispersion that ensures effective utilization of the solids' surface area and significantly enhances the surface-based rate processes. In the fluidized bed mode of contact, lower pressure drop, greater interfacial contact, efficient gas–solid mixing, and higher heat and mass transfer rates provide a substantial advantage over fixed bed or packed bed processes [1–5].

Particle properties have a strong bearing on fluidization behavior, as pointed out by Geldart, who classified the powders into different groups based on physical properties [6]. The fluidization of fine and ultrafine particles of group C classification is difficult because of their cohesiveness caused by strong inter-particle forces (IPF) [7,8], which often result in poor interphase phase mixing and severe bed non-homogeneities [9–12]. In particular, ultrafine powders display agglomerate bubbling fluidization (ABF) due to the formation of large multi-level agglomerates, leading to low bed expansion and high minimum fluidization velocity. As a result, the high surface area characteristics of ultrafine particles become severely compromised. Moreover, size-based segregation often occurs along the height of the bed during ABF [13–17]. The size of hydrophobic silica agglomerates in the lower bed region was 5–10 times larger than the ones in the upper layer [18]. Zhao et al. [19] reported more severe size segregation, where an order of magnitude agglomerate size difference between the upper and lower layers occurred, such that agglomerates as large as 2000 μm were found in the lower region. Although higher gas flow tends to mitigate the effect of IPFs, other problems, such as elutriation and entrainment, occur at high velocities. Therefore, various assistance strategies have been suggested in the literature to provide additional energy to overcome cohesive IPFs and improve the hydrodynamics of the fluidized bed. One such strategy is vibration assistance, which can be employed either internally or externally. External vibrations involve oscillating, shaking, or vibrating the complete test section by using a vibrator or an electric motor [19,20]. Despite their proven capability, the implementation of external vibration-based fluidization assistance techniques, whether at a laboratory scale or large-scale units, is challenging and expensive. On the other hand, internal vibrations directly transfer energy to the solid phase in the bed through techniques such as acoustic perturbation and high-shear mixer [21–27]. Similar to external vibrations, these techniques require additional equipment, leading to higher costs. In some cases, the premixing of the resident solid phase with inert or magnetic particles has been suggested to alter the interparticle force equilibrium of ultrafine particles in the bed [15,28–31]. The compatibility and post-processing issues, however, limit the application of particle premixing.

An important prerequisite for large-scale applications of any assisted technique is its amenability to scale up and easy implementation without being energy intensive and any major process modification requirement. One such technique is the pulsation of inlet fluid flow to the fluidized bed [32–34]. The flow pulsation shortened the drying time and improved the bed homogeneity in drying porous pharmaceutical granules [35]. The constant and falling drying rates of the fluidized biomass particles were enhanced by using optimal pulsation frequency [36]. The flow pulsation also promoted the density-based segregation of coal particles [37,38]. During the fluidization of ultrafine nanoparticles using the square wave pulsation strategy, the channeling and plug formation were suppressed, leading to improved bed homogeneity [16,33,39–41]. Moreover, the minimum fluidization velocity significantly decreased, indicating the deagglomeration of large-sized nanoagglomerates [16,39,40]. Besides promoting uniform bed expansion, pulsed flow also helped decrease the bubble velocity and size [40].

The efficacy of pulsed flow strongly depends on inlet flow and deaeration configurations. The effect of the deaeration strategy has been extensively studied in the context of bed collapse [42–44]. Once the inlet flow is stopped, the deaeration of the residual air critically affects the bed collapse process, which is clearly reflected in the evolution of pressure transient profiles in different regions of the bed [43,44]. The ratio of the distributor to bed pressure drop showed a pronounced effect on the collapse process. As this ratio was increased from 0.005 to 0.03, the difference between the two deaeration strategies was substantially mitigated [43]. When the residual air escapes only from the top of the bed, known as single drainage (SD) deaeration, the collapse process is slow. Providing the residual air dual pathways (i.e., from the top as well as from the bottom of the bed through the plenum), also called dual drainage (DD) deaeration, leads to faster bed transients [42]. Of equal importance is the design of the inlet flow configuration because the line pressure inevitably builds up during the no-flow phase of the pulse when the bed collapses. Therefore, once the valve opens to allow the inlet flow, the line pressure leads to a flow spike. This phenomenon leads to intense size segregation of nanoagglomerates along the height of the fluidized bed, thereby affecting the collapse dynamics monitored in different regions of the bed. To suppress the initial flow spike by eliminating the line pressure buildup, Ali et al. suggested to vent the inlet flow to the atmosphere during the no-flow phase of the pulse while allowing dual deaeration routes to residual gas. This modified dual drainage (MDD) strategy significantly suppressed the size segregation of agglomerates and improved the bed homogeneity [18].

The foregoing discussion, although mainly in the context of bed collapse, is of great importance for pulsed fluidized beds. The flow spike resulting from the pressure buildup during the no-flow phase of the square pulse could severely compromise the efficacy of pulsation assistance. This phenomenon could be complicated by pulsation frequency. At a shorter pulsation frequency, the line pressure builds up because of the longer duration of the no-flow phase of the pulse. Therefore, a careful investigation has been undertaken using the MDD strategy for the fluidized bed pulsed with two widely different square wave frequencies, namely, 0.05 and 0.25 Hz. The local and global bed dynamics for both frequencies were monitored using highly sensitive pressure transducers with a response time of 1 ms. The results were compared with SD pulsed bed to obtain a greater understanding of the pulsed fluidized bed. The present study substantially extends the scope of the conventional bed collapse to examine how the regular intermittency of bed collapse affects the evolution of local and global bed transients in the presence of two different deaeration strategies. Modifying the inlet configuration to suppress the initial peak adds another dimension to the problem, that is, how the occurrence of a short-term event affects the subsequent development of fluidization hydrodynamics.

2. Experimental

The experimental setup consisted of a test section that was a 1.6 m long transparent perplex column with a 0.07 m internal diameter (Figure 1). A calming section with a length of 0.3 m was attached beneath the test section to eliminate the effects of the entry of the fluidizing air. A distributor with 0.025 fractional open area and 2 mm perforations on a circular pitch was used to ensure a uniform distribution of the fluidizing gas across the cross-sectional area of the test section. The perforations of the distributor were covered by a fine nylon mesh filter of 20 μm to prevent the falling of particles through the distributor. A disengagement section with a length of 0.5 m and a diameter of 0.14 m was attached above the test section to suppress particle entrainment with the exiting gas. The test section was washed with an anti-static fluid prior to the experiments.

The overall and local pressure drop transients in four different regions of the bed were measured by positioning the pressure taps along the bed height (Table 1). Five highly sensitive bidirectional differential-pressure transducers (Omega PX163-005BD5V; 1 ms response time; range: ±2.5″ H_2O) recorded the pressure transients at a rate of 100 Hz by using data acquisition system (DAQ) and Labview software. The lower and upper ports of pressure transducers were located diametrically opposite sides of the column to ensure reliable monitoring of cross-sectionally averaged local bed dynamics in different bed regions (Figure 1).

Table 1. Pressure tap positions used to record region-wise pressure transients.

Pressure Drop	Bed Region	Pressure Tap Positions (from the Distributor)
ΔP_1	Lower	0.5–0.1 m
ΔP_2	Lower middle	0.1–0.2 m
ΔP_3	Upper middle	0.2–0.3 m
ΔP_4	Upper	0.3 m–open
ΔP_g	Overall	0.05 m–open

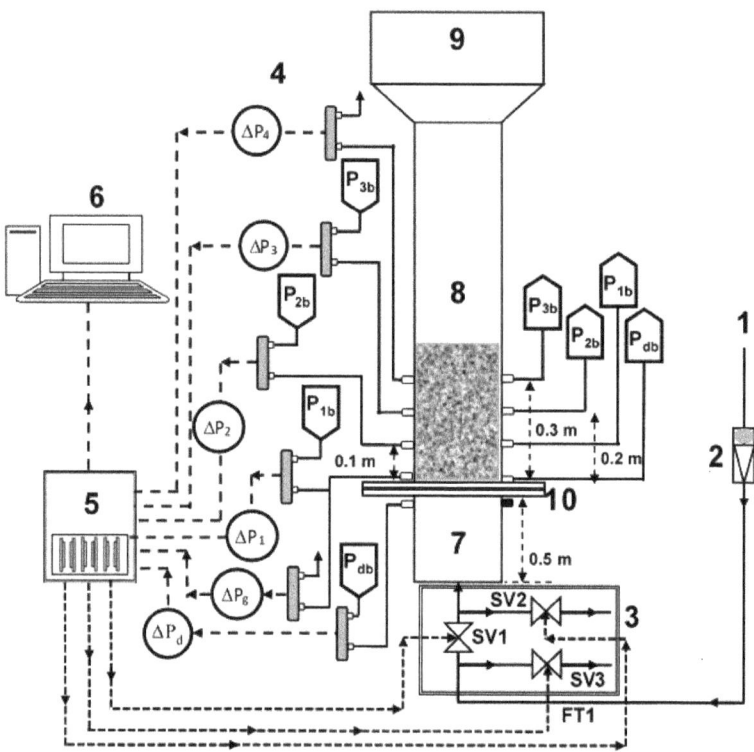

Figure 1. Experimental set-up; (1) Compressed air; (2) Flowmeter; (3) 2-way solenoid valves (for SD and MDD flow strategy); (4) Pressure transducers; (5) Data acquisition system; (6) Computer; (7) Calming section; (8) Test section; (9) Disengagement section; (10) Distributor.

The primary dimension of hydrophilic nanosilica (Aerosil 200) reported by the manufacturer (Evonik, GmBH) was 12 nm with a density of 2200 kg/m^3 [45]. However, the dry particle size analysis (Malvern Panalytical Mastersizer 2000) yielded an average size of 12.5 µm due to the multi-level agglomeration of particles under the effect of IPFs [1,24]. This behavior was clearly evident in the morphological characterization of the nanopowder sample by SEM [13,16]. The specific surface area of the powder was 0.62 m^2/g, which was several orders of magnitude smaller than the reported value of 200 ± 25 m^2/g [45].

Three different two-way solenoid valves (Model: Omega SV 3310) were used (Figure 1). These valves were controlled using digital IO signals from the DAQ to provide two different inlet flow strategies during pulsation [18].

2.1. Single Drainage (SD) Configuration

In this configuration, only the primary inlet valve, marked as SV1 in Figure 1, was employed. When energized, SV1 allowed the inlet airflow and stopped it when de-energized. The other two valves, namely, SV2 and SV3, remained closed throughout the experiment. This strategy provided only one passage, that is, the top of the bed, for the escape of the trapped residual air during the collapse process. Since the square wave flow pulsations were implemented in our experiments, the closing of the valve during the no-flow phase of the square wave led to the buildup of the line pressure across SV1. This pressure buildup led to an initial airflow spike when the valve opened to allow the inlet flow to the test section.

2.2. Modified Dual Drainage (MDD) Configuration

In this configuration, SV1 was kept open during inlet flow to the test section, while the two other valves (SV2 and SV3) remained closed. However, when SV1 closed, cutting off the airflow, SV2 and SV3 were energized to remain fully open. SV2 provided an alternate passage for the residual air to escape through the plenum, while SV3 vented the primary airflow to the atmosphere, thereby preventing the buildup of the pressure drop. This strategy completely eliminated the initial airflow spike when SV1 opened to allow the inlet flow to the test section.

The opening and closing frequency of the solenoid valves was controlled using a digital IO signal from DAQ and Labview. High-pressure air under ambient conditions was used as the fluidizing gas. Gilmont flowmeters were used to set the initial airflow in the column. The particle bed was allowed to achieve a steady state before the start of the pulsation experiments using two different frequencies, namely 0.05 and 0.25 Hz. Whereas 0.05 Hz pulsations with a time period of 20 s allowed the complete collapse of the bed between two successive pulsations [13], 0.25 Hz pulsations with a much shorter time period of 4 s allowed only partial bed collapse before the occurrence of another pulsation event [1]. Local and global bed dynamics were monitored for four identical pulses of 0.05 Hz and six pulses of 0.25 Hz pulsation.

3. Results and Discussion

Evolution of Local Pressure Drop Transients

Figure 2 shows the local bed dynamics of the 0.05 Hz pulsed fluidized bed under different inlet and deaeration strategies (i.e., SD and MDD). The broken red vertical line indicates the start of the flow pulse, while the collapse process theoretically initiates as soon as the inlet flow is stopped during the no-flow phase of the pulse. At 0.05 Hz, the complete pulse cycle lasted for 20 s that comprised 10 s of inlet flow at a fixed velocity, followed by 10 s of complete flow interruption. A wide spectrum of velocities within 13–87 mm/s were considered in our experiments. The peaks at the onset of the flow pulse, often pronounced for SD configuration, were due to the pressure buildup across the solenoid valve, resulting in the initial flow spike. This phenomenon can promote size-based segregation along the height of the bed [18]. By contrast, the venting of the primary flow in the MDD strategy did not allow the line pressure buildup, thereby suppressing the initial flow spike.

The effect of velocity variation on the SD fluidized bed was not always notable, except in the lower middle region that is represented by ΔP_2 in Figure 2e. Since the bed was not fluidized at 13 mm/s, its dynamics inevitably differed from others. Unlike SD, the effect of velocity on the local dynamics of the MDD fluidized bed was significantly pronounced with substantially faster dynamics. Once the pulsed flow began, the pressure drop almost attained a steady state value within a span of 1 s. The collapse process also finished in one second, with the pressure drop attaining a zero value. However, some exceptions were seen for the upper region (ΔP_4) and the upper middle region (ΔP_3) of the bed. The faster dynamics of the MDD strategy could be due to the availability of second escape routes for the residual air through the plenum in addition to the top of the bed. Being closer to the lower drainage pathway, the difference between the two strategies in behavior was inevitably more pronounced in the lower region (Figure 2g,h). The evolution of pressure transients and their dependence on velocity in Figure 2h clearly highlight the improved bed hydrodynamics obtained using the MDD strategy. However insignificant, the initial pressure drop spike in the MDD pulsed bed could have resulted from the time lag in the closing of the vent valve and the opening of the flow valve.

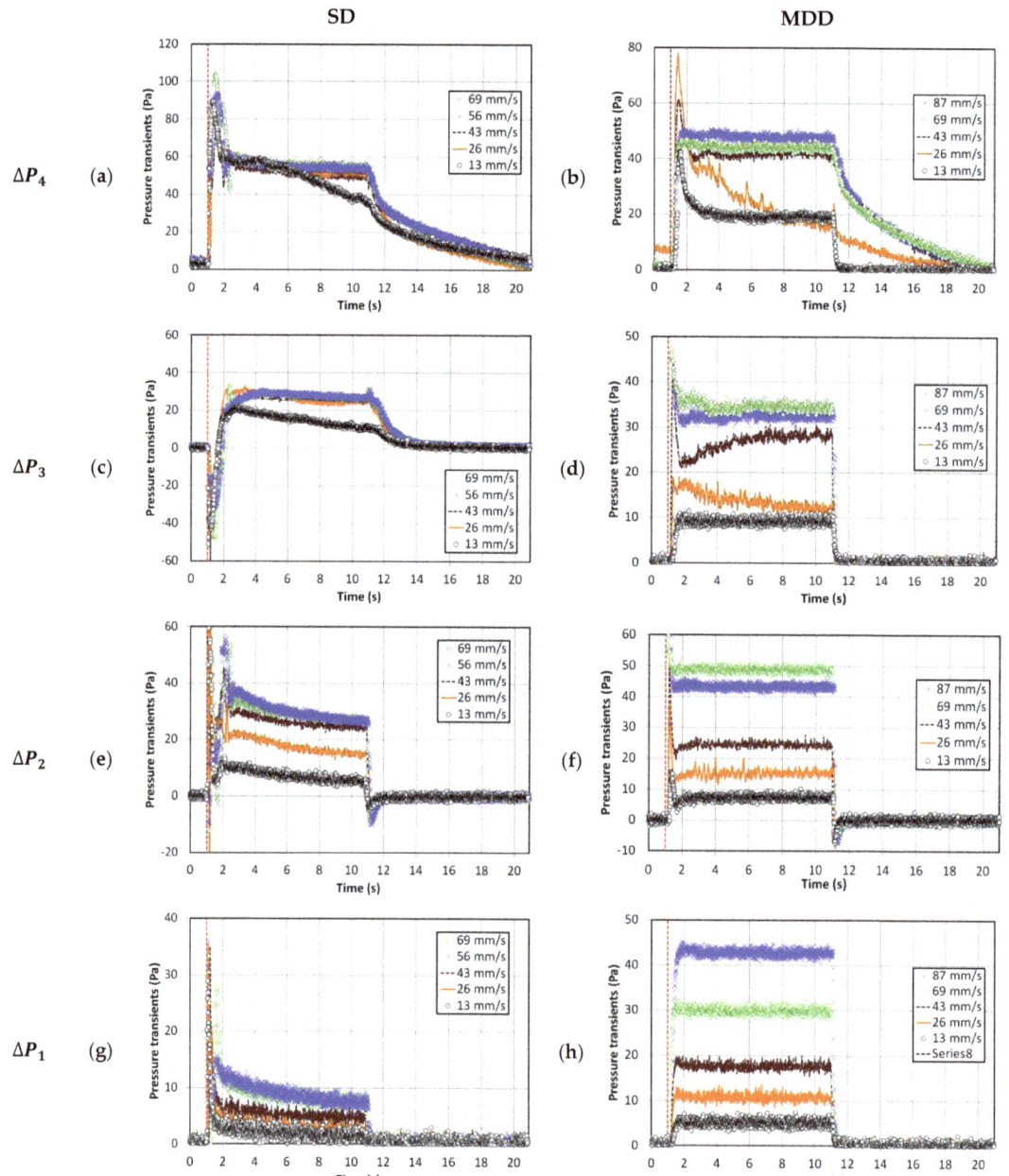

Figure 2. Local pressure drop transients with pulsation frequency 0.05 Hz; (**a**) Upper region (SD); (**b**) Upper region (MDD); (**c**) Upper middle region (SD); (**d**) Upper middle region (MDD); (**e**) Lower middle region (SD); (**f**) Lower middle region (MDD); (**g**) Lower region (SD); (**h**) Lower region (MDD).

In the lower bed region (Figure 2g), the accumulation of rigid and large agglomerates due to segregation resulted in a lower pressure drop than that in the regions above, that is, ΔP_2 to ΔP_4. However, in Figure 2h, the pressure drop was comparable with that in the middle region with a strong dependence on the velocity, which indicated the presence of smaller agglomerates in the lower region due to feeble segregation tendencies.

The behavior of the higher frequency pulsed fluidized bed is shown in Figure 3. The inlet flow occurred for 2 s only, followed by 2 s of complete interruption. Therefore, neither the expansion nor the collapse process could reach a steady condition in most cases, irrespective of whether SD or MDD strategy was implemented. The evolution of MDD pressure transients were rather predictable, whereas a great deal of disturbances was evident for the SD transients. These disturbances were strongly affected by the change in velocity. In the case of MDD, ΔP_1 and ΔP_2 showed faster transients than ΔP_3 and ΔP_4 owing to the availability of the lower deaeration route through the plenum. Away from the distributor, the transients were slower, especially at higher velocities, because of the presence of a greater amount of residual gas. On the other hand, the complex, unpredictable pressure drop transients in the SD pulsed fluidized bed developed due to the interaction between the solid particles, whether rising or falling and the upward rising gas flow through the bed. The difference between the local bed dynamics in the two cases of different frequencies was also observed in the initial pressure drop spike, which appeared to be significantly mitigated for the higher frequency pulsed flow owing to a lower line pressure buildup due to the shorter duration of the no-flow phase of the pulsed flow.

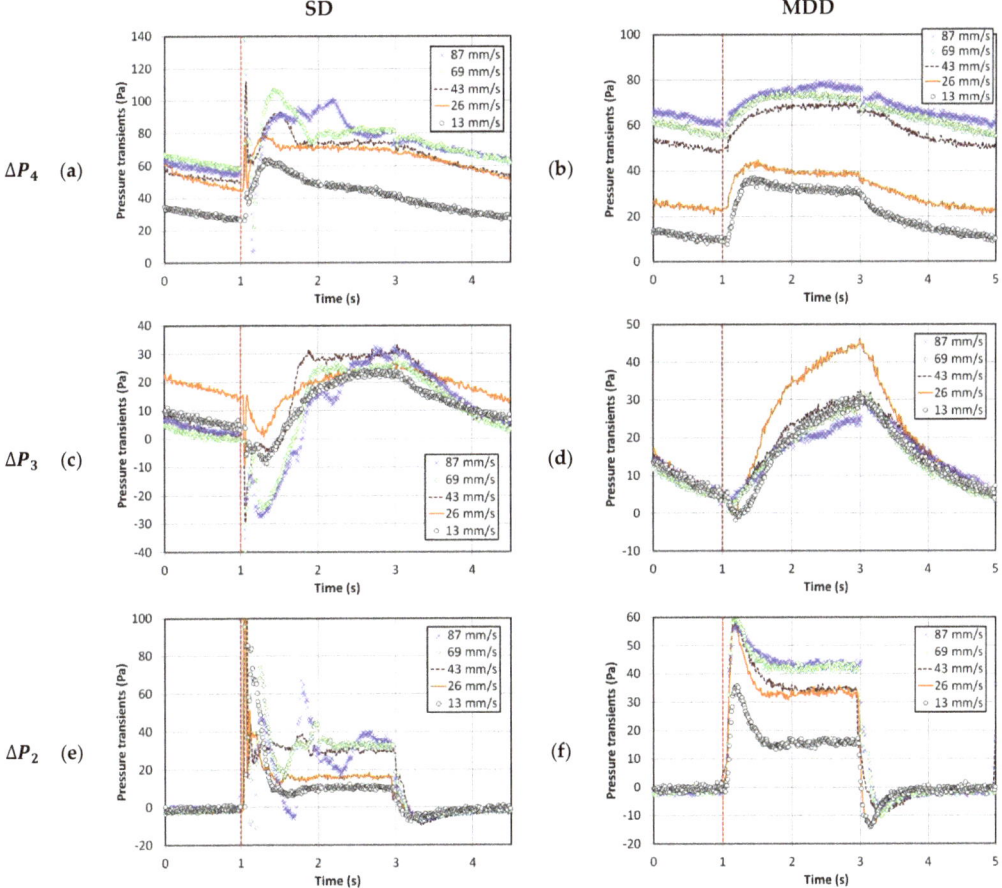

Figure 3. *Cont.*

ΔP_1 (g)

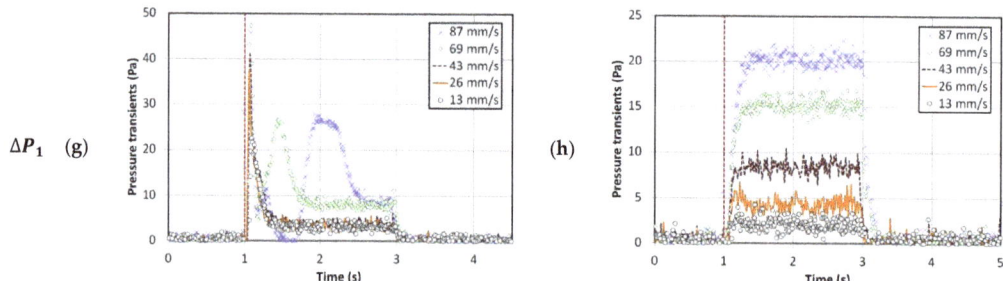

Figure 3. Local pressure drop transients with pulsation frequency 0.25 Hz; (**a**) Upper region (SD); (**b**) Upper region (MDD); (**c**) Upper middle region (SD); (**d**) Upper middle region (MDD); (**e**) Lower middle region (SD); (**f**) Lower middle region (MDD); (**g**) Lower region (SD); (**h**) Lower region (MDD).

The pressure drop ranges were of comparable magnitude in the lower region (ΔP_1) for both strategies at higher frequency pulsation in Figure 3g,h. Moreover, $\Delta P_1 < \Delta P_2$, suggesting similar segregation behavior for both flow strategies. At higher frequencies, although the pulsed bed with SD configuration showed a smaller initial flow spike, their hydrodynamics were more susceptible to intense disturbances.

The global transients are shown in Figure 4 for both cases of 0.05 and 0.25 Hz pulsed fluidized bed. The trends in Figures 2 and 3 were witnessed again. The smooth pressure transients for MDD, irrespective of the pulsation frequency, are a clear reflection of the improved hydrodynamics. At lower frequencies, another aspect of the bed hydrodynamics that was apparently not evident before occurred. Higher pressure drop values were noticed at higher velocities, such as 69 and 87 mm/s, indicating better contact between the solid and fluid phases causing greater frictional losses.

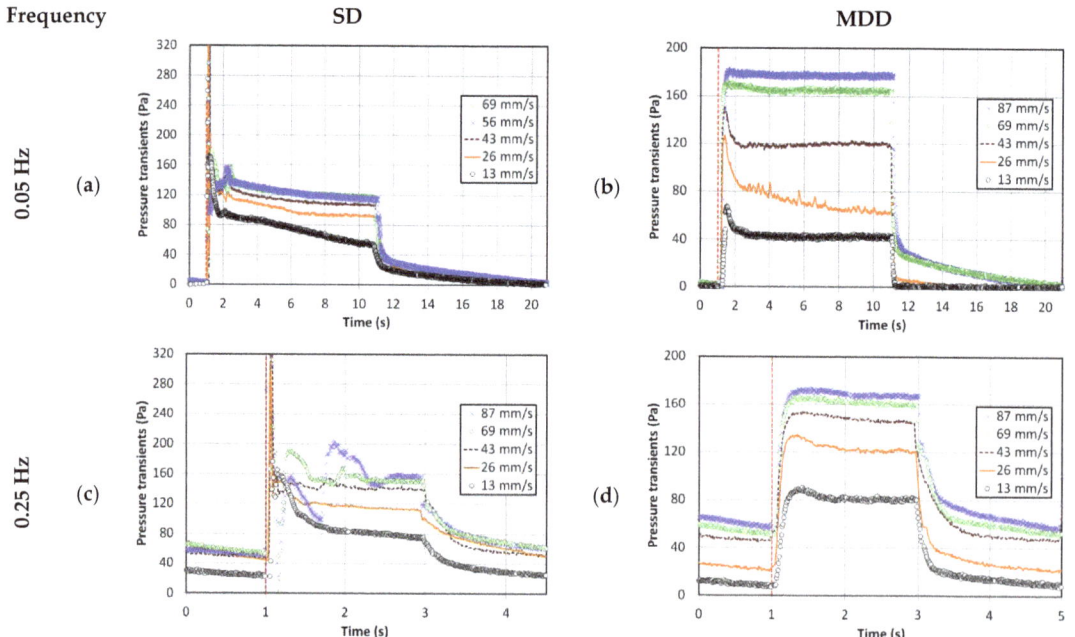

Figure 4. Global pressure drop transients for different pulsation frequencies; (**a**) 0.05 Hz frequency SD; (**b**) 0.05 Hz frequency MDD; (**c**) 0.25 Hz frequency SD; (**d**) 0.25 Hz frequency MDD.

The bed hydrodynamics were further investigated by computing the mean value of the pressure drop from the pressure transients. The latter portion of the flow pulse, immediately before the flow cutoff, was used to evaluate the mean. The case of 0.05 Hz pulsed fluidized bed is considered in Figure 5. Only the defluidization part of the experimental run was compared for SD and MDD, owing to its repeatability. The difference was significantly pronounced in the lower region of the bed (ΔP_1), which was monitored in the bed region from 5–100 mm above the distributor (Figure 5d). At higher flowrates, the difference between the two pressure drop values reached several folds with a steeper rise for the MDD, indicating the presence of smaller agglomerates. A similar difference, albeit less pronounced, was again observed for ΔP_2 (Figure 5c). A smoother pressure drop profile was seen with the MDD in the upper middle region (Figure 5b). The upper region, represented by ΔP_4, fully fluidized at approximately 20 mm/s, with a higher pressure drop for SD. Given that the total weight of solid particles in both cases were the same, the total pressure drops across the bed for the fully fluidized bed in both cases should be comparable. The lower pressure drop values in the lower region obtained with SD were therefore compensated in the upper region of the bed. Owing to the size segregation of nanoagglomerates, the bed showed partial fluidization. The upper region with smaller agglomerates fluidized at 20 mm/s, whereas the lower region with large agglomerates remained un-fluidized even at higher gas velocities.

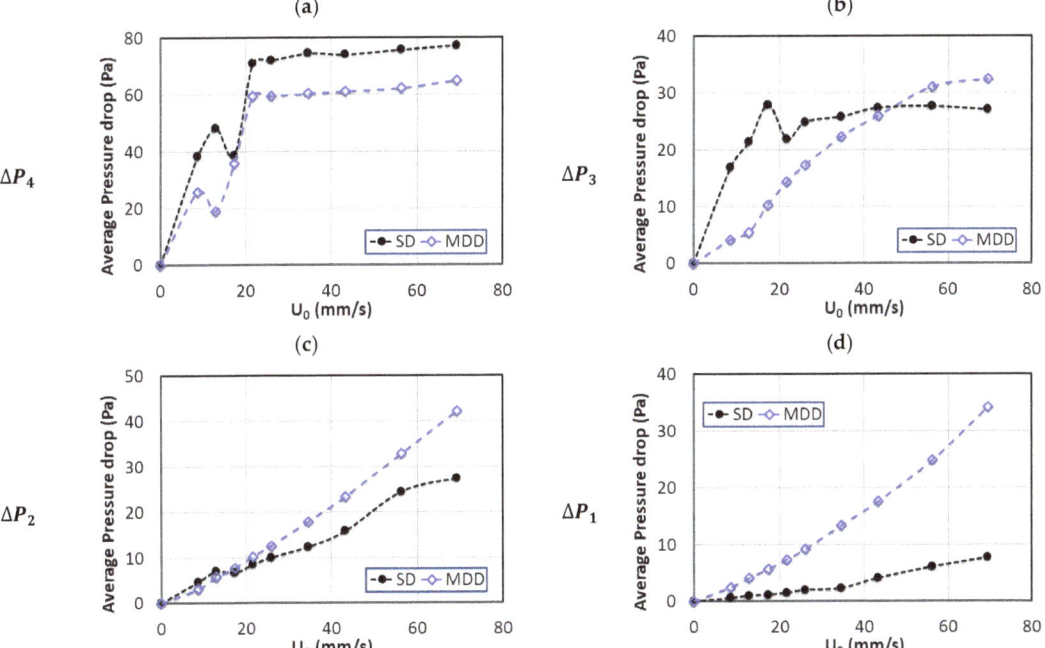

Figure 5. Variations in local pressure drop with air superficial velocity for defluidization runs of 0.05 Hz pulsed fluidized bed; (**a**) Upper region (ΔP_4); (**b**) Upper middle region (ΔP_3); (**c**) Lower region (ΔP_2); (**d**) Lower region (ΔP_1).

Similar behavior was observed for the case of 0.25 Hz pulsed bed in Figure 6. A clear difference between SD and MDD was detected in the lower region, where a higher pressure drop was obtained with MDD, and the difference between the two strategies was more pronounced at higher velocities. However, the difference was not as prominent as it was in the lower frequency case. The trend was reversed for the upper middle region (ΔP_3)

and the upper region (ΔP_4) due to the material balance consideration, as explained in the preceding paragraph. The MDD bed hydrodynamics appeared to be more sensitive to the frequency change. The upper middle region showed a decrease in the pressure drop at higher velocities due to the bed expansion that caused the migration of solids to the upper region, where this phenomenon was reflected in the increase in the pressure drop.

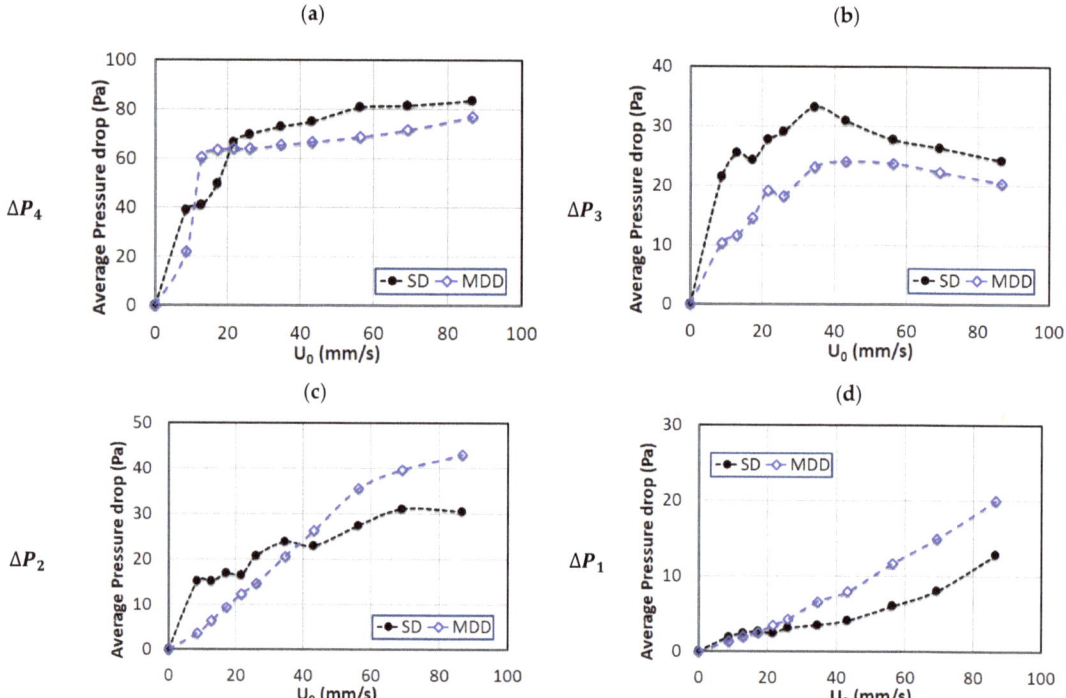

Figure 6. Variations in local pressure drop with air superficial velocity for defluidization runs of 0.25 Hz pulsed fluidized bed; (**a**) Upper region (ΔP_4); (**b**) Upper middle region (ΔP_3); (**c**) Lower region (ΔP_2); (**d**) Lower region (ΔP_1).

A more revealing insight into the bed dynamics is shown in Figure 7, which presents the amplitude spectra of the signals in the frequency domain. The "fft" function of MATLAB was used for computing the spectra. The figure considers the global pressure drop signals that include the disturbances in the whole bed. The comparison for both cases of SD and MDD is also shown. The case of the 0.05 Hz pulsed bed is considered in Figure 7a for different velocities. The difference between the SD and MDD was significant. The amplitude spectra of the pressure drop transients of the SD within the 10–50 Hz range were dominated by the small amplitude events, which decreased in intensity as the frequency increased. This finding clearly indicated the disturbances occurring in the pulsed fluidized bed with SD configuration. The amplitude profile showed the existence of a wide spectrum of pressure fluctuations beginning from below 10 Hz and extending up to 30 Hz. It was caused by the solid particles falling under gravity and were obstructed by the upward flow of the residual air exiting from the top of the bed, thereby generating pressure drop fluctuations. This phenomenon was completely absent in the case of MDD. The case of 0.25 Hz pulsed bed is considered in Figure 7b. The behavior was similar to the one observed earlier for 0.05 Hz. The amplitude of fluctuations for SD at this frequency was higher than that obtained for the case of the lower frequency. This fact was already pointed out while discussing the real-time bed dynamics of SD configuration in Figure 3.

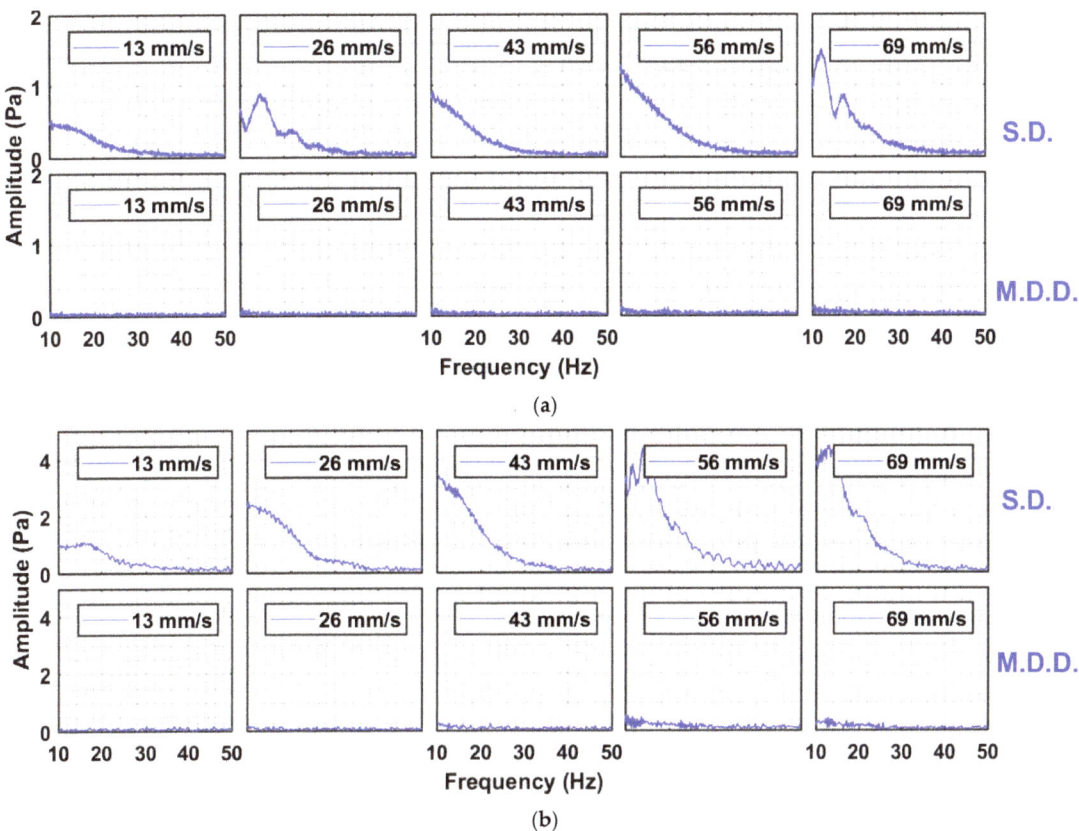

Figure 7. Amplitude spectral analysis of global pressure drop transients of SD and MDD pulsed fluidized bed at (**a**) 0.05 Hz; (**b**) 0.25 Hz.

To obtain further insight into the local bed dynamics, the local amplitude spectra of pressure transients in different regions of the bed at 0.05 Hz flow pulsation are shown in Figure 8 for two different velocities (i.e., 26 and 69 mm/s). Interestingly, the difference between the local dynamics was notable. The case of lower velocity, that is, 26 mm/s, is considered in Figure 8a. The lower region represented by ΔP_1 showed no fluctuations because large and hard agglomerates in the lower region were still stationary at 26 mm/s; therefore, no visible change occurred whether the flow was started or cutoff. The lower middle region pressure transients (ΔP_2) were greatly affected as the falling particles achieved higher kinetic energy by traversing a greater distance in the bed in reaching this region when obstructed by the upwards moving residual air, indicating a great deal of fluctuations. For the same reason, this effect was substantially mitigated in the upper middle region (ΔP_3) because of the lower kinetic energy of falling solids. In the upper region, we observed fluctuations with a lot of distinct frequencies as the smaller particles interacted with the residual air exiting the bed as small bubbles. As the velocity was increased to 69 mm/s, the phenomenon described became more intense (Figure 8b).

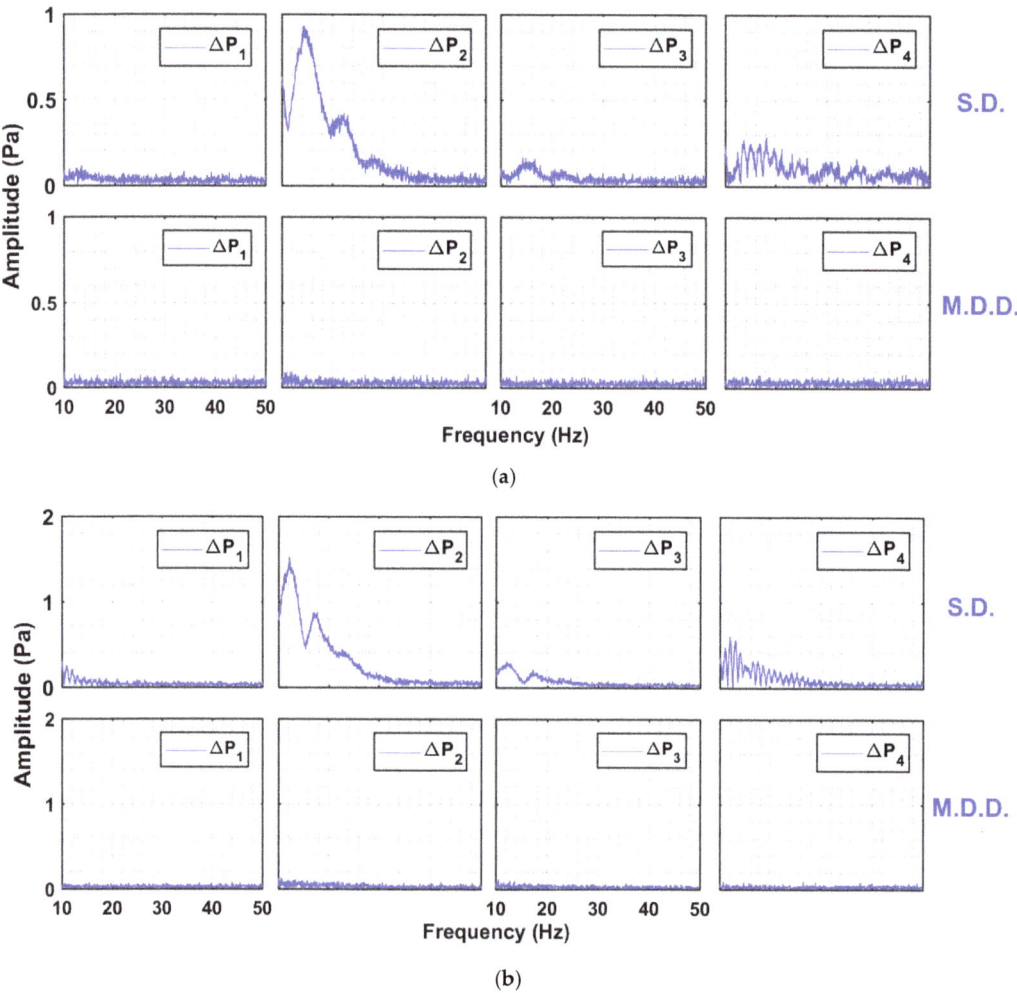

Figure 8. Amplitude spectral analysis of local pressure drop transients of 0.05 Hz pulsed fluidized bed for one complete pulse at (**a**) $U_0 = 26$ mm/s; (**b**) $U_0 = 69$ mm/s.

Similar observations were persistent at high-frequency pulsations (Figure 9a,b). The MDD configuration completely eliminated the disturbances throughout the bed, irrespective of gas velocity. For the SD configuration, the local bed hydrodynamics presented a completely different picture. The amplitude was the highest in the ΔP_2 region, where the swarm of falling particles possessed the highest kinetic energy. In the region above, that is, upper middle, ΔP_3 showed clear periodic events of small amplitude in the range of 10–20 Hz. Increasing the fluid velocity increased the amplitude by more than threefold due to the increased disturbances in the bed. The continuous spectra of ΔP_2 at 26 mm/s changed into discrete high amplitude events occurring at 12 Hz followed by those at 20 Hz, probably due to the development of flow structures.

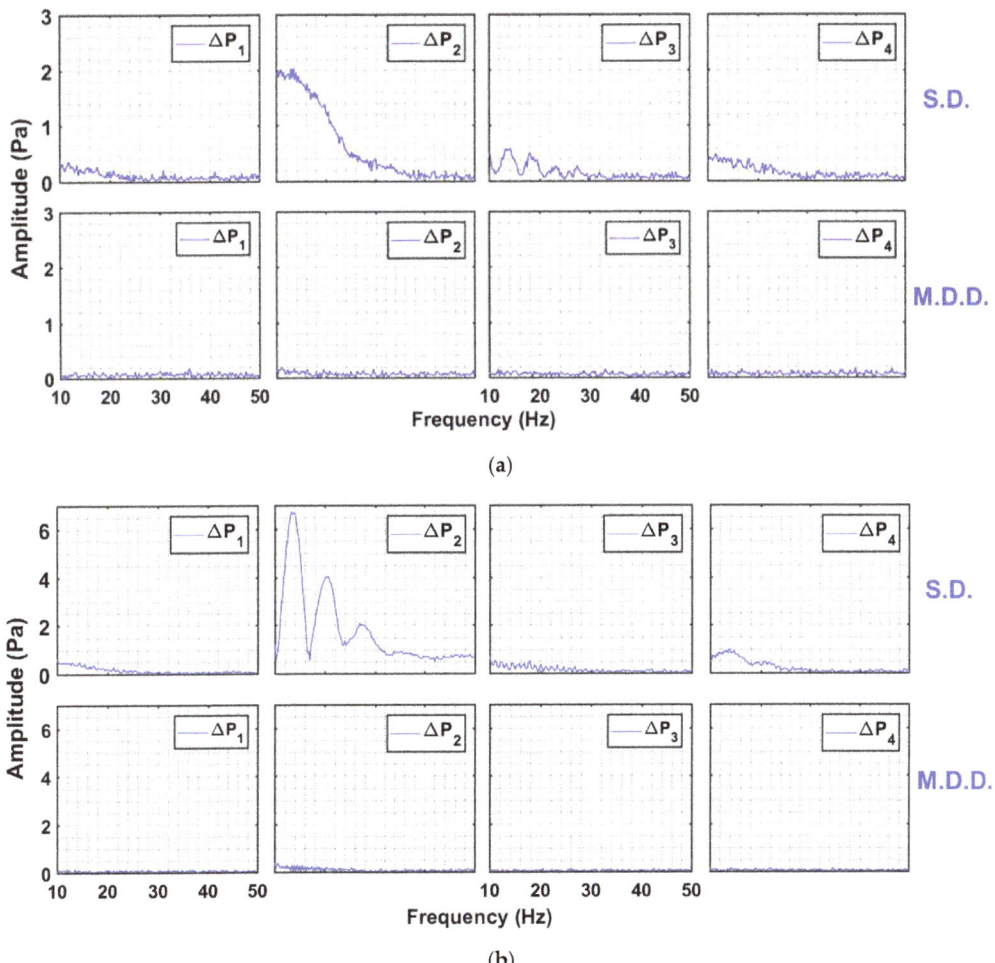

Figure 9. Amplitude spectral analysis of local pressure drop transients of 0.05 Hz pulsed fluidized bed for one complete pulse at (**a**) $U_0 = 26$ mm/s; (**b**) $U_0 = 69$ mm/s.

The average diameters from local dynamics were calculated during the collapse of the bed. The mathematical model based on mass balance proposed by Nie and Liu was used [46]. The diameters of the agglomerates are reported in Figure 10. For 0.05 Hz in Figure 10a–c, the segregation was clearly visible for SD configuration. The range of diameters for upper, middle, and lower regions were 15–22, 30–90, and 140–260 μm, respectively. The agglomerate size in the upper region was almost constant with airflow rate, while agglomerates in the middle and lower regions showed a consistent increase with airflow rate because finer particles moved to the upper region due to segregation. For the case of MDD configuration, the size ranges were 20–30, 90–140, and 100–200 μm in the upper, middle, and lower regions, respectively. The size difference between the agglomerates in the lower and middle regions was not as pronounced as those in the case of the SD configuration. The MDD configuration suppressed the segregation tendencies in the pulsed fluidized beds. At higher frequencies (Figure 10d–f), the size ranges were similar for both flow strategies with strong segregation behavior that was seen for the case of 0.05 Hz SD configuration. This means that the disturbances generated due to high-frequency pulsation developed a similar impact on the hydrodynamics as the initial airflow spike in the lower frequency. Moreover, the slope of the curve in Figure 10d decreased with increasing airflow, signifying the addition of finer particles in the region. Moreover, the slope of curves in Figure 10e,f was lower than that in Figure 10b,c. This finding could be due to the deagglomeration phenomenon, wherein the size of larger particles decreased, and finer particles moved to the upper region.

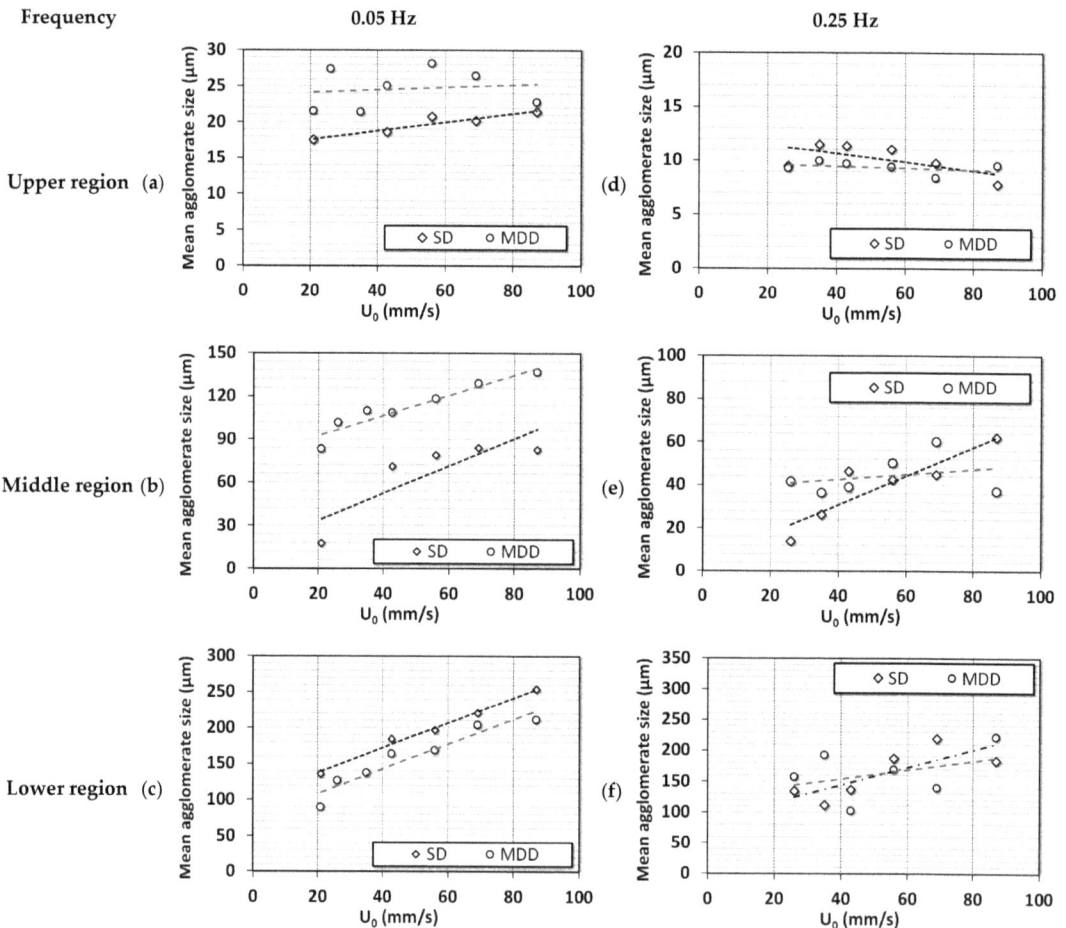

Figure 10. Changes in the average agglomerate diameter of ultrafine particles with the change in fluid velocity in different regions of the bed of nanosilica; (a) Upper region (0.05 Hz); (b) Middle region (0.05 Hz); (c) Lower region (0.05 Hz); (d) Upper region (0.25 Hz); (e) Middle region (0.25 Hz); (f) Lower region (0.25 Hz).

4. Conclusions

The inlet flow strategies greatly affected the hydrodynamics of pulsed fluidized beds. At a low pulsation frequency of 0.05 Hz, large size-based segregation was observed when the SD flow strategy was used. This was caused by the initial airflow spike resulting from pressure built up across the solenoid valve in the collapse process. This phenomenon was eliminated using the MDD airflow strategy, leading to a subdued segregation behavior. The region-wise pressure drop during defluidization displayed the difference in the degree of segregation, especially in the lower region due to the different flow strategies. However, at a higher pulsation frequency of 0.25 Hz, the flow spike was feeble due to the lower time period of pressure built up across the solenoid valve. The difference in stratification was less between both flow strategies when high-frequency pulsation was used. The particle diameter calculated from the local bed dynamics signified that the segregation was prominent for both flow strategies at high-frequency flow pulsation. The disturbances developed due to the frequent expansion and collapse of the bed promoted segregation and deagglomeration.

Author Contributions: Conceptualization, S.S.A. and M.A.; Methodology, S.S.A. and M.A.; Software, S.S.A.; Validation, S.S.A.; Formal Analysis, S.S.A.; Resources, M.A.; Data Curation, S.S.A.; Writing—Original Draft Preparation, S.S.A.; Writing—Review & Editing, A.A., K.L.R. and M.A.; Supervision, M.A.; Funding Acquisition, M.A. All authors have read and agreed to the published version of the manuscript.

Funding: This research was funded by Researchers Supporting Project, King Saud University, Riyadh, Saudi Arabia, RSP2023R42.

Acknowledgments: The authors acknowledge the Researchers Supporting Project, RSP2023R42, King Saud University, Riyadh, Saudi Arabia, for the financial support.

Conflicts of Interest: The authors declare no conflict of interest.

References

1. Ali, S.S.; Hossain, S.K.S.; Asif, M. Dynamics of partially collapsing pulsed fluidized bed. *Can. J. Chem. Eng.* **2021**, *99*, 2333–2344. [CrossRef]
2. Lehmann, S.; Hartge, E.-U.; Jongsma, A.; Deleeuw, I.-M.; Innings, F.; Heinrich, S. Fluidization characteristics of cohesive powders in vibrated fluidized bed drying at low vibration frequencies. *Powder Technol.* **2019**, *357*, 54–63. [CrossRef]
3. Bakhurji, A.; Bi, X.; Grace, J.R. Hydrodynamics and solids mixing in fluidized beds with inclined-hole distributors. *Particuology* **2019**, *43*, 19–28. [CrossRef]
4. Raganati, F.; Chirone, R.; Ammendola, P. Gas–solid fluidization of cohesive powders. *Chem. Eng. Res. Des.* **2018**, *133*, 347–387. [CrossRef]
5. Fedorov, A.V.; Yazykov, N.A.; Bulavchenko, O.A.; Saraev, A.A.; Kaichev, V.V.; Yakovlev, V. CuFeAl Nanocomposite Catalysts for Coal Combustion in Fluidized Bed. *Nanomaterials* **2020**, *10*, 1002. [CrossRef] [PubMed]
6. Geldart, D. Types of gas fluidization. *Powder Technol.* **1973**, *7*, 285–292. [CrossRef]
7. Lettieri, P.; Macrì, D. Effect of Process Conditions on Fluidization. *KONA Powder Part. J.* **2016**, *33*, 86–108. [CrossRef]
8. Asif, M.; Ali, S.S. Bed collapse dynamics of fluidized beds of nano-powder. *Adv. Powder Technol.* **2013**, *24*, 939–946. [CrossRef]
9. Ali, S.S.; Arsad, A.; Roberts, K.; Asif, M. Effect of Voidage on the Collapsing Bed Dynamics of Fine Particles: A Detailed Region-Wise Study. *Nanomaterials* **2022**, *12*, 2019. [CrossRef]
10. van Ommen, J.R.; Valverde, J.; Pfeffer, R. Fluidization of nanopowders: A review. *J. Nanoparticle Res.* **2012**, *14*, 737. [CrossRef]
11. Shabanian, J.; Jafari, R.; Chaouki, J. Fluidization of Ultrafine Powders. *Int. Rev. Chem. Eng.* **2012**, *4*, 16–50.
12. Shabanian, J.; Chaouki, J. Influence of interparticle forces on solids motion in a bubbling gas-solid fluidized bed. *Powder Technol.* **2016**, *299* (Suppl. C), 98–106. [CrossRef]
13. Asif, M.; Al-Ghurabi, E.H.; Ajbar, A.; Kumar, N.S. Hydrodynamics of Pulsed Fluidized Bed of Ultrafine Powder: Fully Collapsing Fluidized Bed. *Processes* **2020**, *8*, 807. [CrossRef]
14. Ali, S.S.; Asif, M.; Ajbar, A. Bed collapse behavior of pulsed fluidized beds of nano-powder. *Adv. Powder Technol.* **2014**, *25*, 331–337. [CrossRef]
15. Ali, S.S.; Asif, M. Effect of particle mixing on the hydrodynamics of fluidized bed of nanoparticles. *Powder Technol.* **2017**, *310*, 234–240. [CrossRef]
16. Al-Ghurabi, E.H.; Shahabuddin, M.; Kumar, N.S.; Asif, M. Deagglomeration of Ultrafine Hydrophilic Nanopowder Using Low-Frequency Pulsed Fluidization. *Nanomaterials* **2020**, *10*, 388. [CrossRef]
17. Tamadondar, M.R.; Zarghami, R.; Tahmasebpoor, M.; Mostoufi, N. Characterization of the bubbling fluidization of nanoparticles. *Particuology* **2014**, *16*, 75–83. [CrossRef]
18. Ali, S.S.; Arsad, A.; Asif, M. Effect of modified inlet flow strategy on the segregation phenomenon in pulsed fluidized bed of ultrafine particles: A collapse bed study. *Chem. Eng. Process. Process Intensif.* **2021**, *159*, 108243. [CrossRef]
19. Zhao, Z.; Liu, D.; Ma, J.; Chen, X. Fluidization of nanoparticle agglomerates assisted by combining vibration and stirring methods. *Chem. Eng. J.* **2020**, *388*, 124213. [CrossRef]
20. Lee, J.-R.; Lee, K.S.; Park, Y.O.; Lee, K.Y. Fluidization characteristics of fine cohesive particles assisted by vertical vibration in a fluidized bed reactor. *Chem. Eng. J.* **2020**, *380*, 122454. [CrossRef]

21. Ding, P.; Pacek, A. De-agglomeration of goethite nano-particles using ultrasonic comminution device. *Powder Technol.* **2008**, *187*, 1–10. [CrossRef]
22. Ding, P.; Orwa, M.; Pacek, A. De-agglomeration of hydrophobic and hydrophilic silica nano-powders in a high shear mixer. *Powder Technol.* **2009**, *195*, 221–226. [CrossRef]
23. Al-Ghurabi, E.H.; Ali, S.S.; Alfadul, S.M.; Shahabuddin, M.; Asif, M. Experimental investigation of fluidized bed dynamics under resonant frequency of sound waves. *Adv. Powder Technol.* **2019**, *30*, 2812–2822. [CrossRef]
24. Ajbar, A.; Bakhbakhi, Y.; Ali, S.; Asif, M. Fluidization of nano-powders: Effect of sound vibration and pre-mixing with group A particles. *Powder Technol.* **2011**, *206*, 327–337. [CrossRef]
25. Raganati, F.; Ammendolab, P.; Chirone, R. CO_2 capture by adsorption on fine activated carbon in a sound assisted fluidized bed. *Chem. Eng. Trans.* **2015**, *43*, 1033–1038.
26. Ali, S.S.; Arsad, A.; Hossain, S.S.; Asif, M. A Detailed Insight into Acoustic Attenuation in a Static Bed of Hydrophilic Nanosilica. *Nanomaterials* **2022**, *12*, 1509. [CrossRef]
27. Lee, J.-R.; Lee, K.S.; Hasolli, N.; Park, Y.O.; Lee, K.Y.; Kim, Y.H. Fluidization and mixing behaviors of Geldart groups A, B and C particles assisted by vertical vibration in fluidized bed. *Chem. Eng. Process. Process Intensif.* **2020**, *149*, 107856. [CrossRef]
28. Zhou, L.; Diao, R.; Zhou, T.; Wang, H.; Kage, H.; Mawatari, Y. Behavior of magnetic Fe3O4 nano-particles in magnetically assisted gas-fluidized beds. *Adv. Powder Technol.* **2011**, *22*, 427–432. [CrossRef]
29. Emiola-Sadiq, T.; Wang, J.; Zhang, L.; Dalai, A. Mixing and segregation of binary mixtures of biomass and silica sand in a fluidized bed. *Particuology* **2021**, *58*, 58–73. [CrossRef]
30. Ali, S.S.; Basu, A.; Alfadul, S.M.; Asif, M. Nanopowder Fluidization Using the Combined Assisted Fluidization Techniques of Particle Mixing and Flow Pulsation. *Appl. Sci.* **2019**, *9*, 572. [CrossRef]
31. Ali, S.S.; Al-Ghurabi, E.H.; Ibrahim, A.A.; Asif, M. Effect of adding Geldart group A particles on the collapse of fluidized bed of hydrophilic nanoparticles. *Powder Technol.* **2018**, *330*, 50–57. [CrossRef]
32. Ireland, E.; Pitt, K.; Smith, R. A review of pulsed flow fluidisation; the effects of intermittent gas flow on fluidised gas–solid bed behaviour. *Powder Technol.* **2016**, *292*, 108–121. [CrossRef]
33. Ali, S.S.; Asif, M. Fluidization of nano-powders: Effect of flow pulsation. *Powder Technol.* **2012**, *225*, 86–92. [CrossRef]
34. Sung, W.C.; Jung, H.S.; Bae, J.W.; Kim, J.Y.; Lee, D.H. Segregation phenomena of binary solids in a pulsed fluidized bed. *Powder Technol.* **2022**, *410*, 117881. [CrossRef]
35. Akhavan, A.; van Ommen, J.R.; Nijenhuis, J.; Wang, X.S.; Coppens, M.-O.; Rhodes, M.J. Improved Drying in a Pulsation-Assisted Fluidized Bed. *Ind. Eng. Chem. Res.* **2009**, *48*, 302–309. [CrossRef]
36. Jia, D.; Cathary, O.; Peng, J.; Bi, X.; Lim, C.J.; Sokhansanj, S.; Liu, Y.; Wang, R.; Tsutsumi, A. Fluidization and drying of biomass particles in a vibrating fluidized bed with pulsed gas flow. *Fuel Process. Technol.* **2015**, *138*, 471–482. [CrossRef]
37. Dong, L.; Zhang, Y.; Zhao, Y.; Peng, L.; Zhou, E.; Cai, L.; Zhang, B.; Duan, C. Effect of active pulsing air flow on gas-vibro fluidized bed for fine coal separation. *Adv. Powder Technol.* **2016**, *27*, 2257–2264. [CrossRef]
38. Saidi, M.; Tabrizi, H.B.; Chaichi, S.; Dehghani, M. Pulsating flow effect on the segregation of binary particles in a gas–solid fluidized bed. *Powder Technol.* **2014**, *264*, 570–576. [CrossRef]
39. Liu, Y.; Ohara, H.; Tsutsumi, A. Pulsation-assisted fluidized bed for the fluidization of easily agglomerated particles with wide size distributions. *Powder Technol.* **2017**, *316*, 388–399. [CrossRef]
40. Akhavan, A.; Rahman, F.; Wang, S.; Rhodes, M. Enhanced fluidization of nanoparticles with gas phase pulsation assistance. *Powder Technol.* **2015**, *284*, 521–529. [CrossRef]
41. Ali, S.S.; Al-Ghurabi, E.H.; Ajbar, A.; Mohammed, Y.A.; Boumaza, M.; Asif, M. Effect of Frequency on Pulsed Fluidized Beds of Ultrafine Powders. *J. Nanomater.* **2016**, *2016*, 23. [CrossRef]
42. Lorences, M.J.; Patience, G.S.; Díez, F.V.; Coca, J. Fines effects on collapsing fluidized beds. *Powder Technol.* **2003**, *131*, 234–240. [CrossRef]
43. Cherntongchai, P.; Innan, T.; Brandani, S. Mathematical description of pressure drop profile for the 1-valve and 2-valve bed collapse experiment. *Chem. Eng. Sci.* **2011**, *66*, 973–981. [CrossRef]
44. Cherntongchai, P.; Brandani, S. A model for the interpretation of the bed collapse experiment. *Powder Technol.* **2005**, *151*, 37–43. [CrossRef]
45. Available online: https://www.l-i.co.uk/contentfiles/270.pdf (accessed on 10 December 2022).
46. Nie, Y.; Liu, D. Dynamics of collapsing fluidized beds and its application in the simulation of pulsed fluidized beds. *Powder Technol.* **1998**, *99*, 132–139. [CrossRef]

Disclaimer/Publisher's Note: The statements, opinions and data contained in all publications are solely those of the individual author(s) and contributor(s) and not of MDPI and/or the editor(s). MDPI and/or the editor(s) disclaim responsibility for any injury to people or property resulting from any ideas, methods, instructions or products referred to in the content.

Article

Functional Silane-Based Nanohybrid Materials for the Development of Hydrophobic and Water-Based Stain Resistant Cotton Fabrics Coatings

Silvia Sfameni [1,2], Tim Lawnick [3], Giulia Rando [2,4], Annamaria Visco [1,5], Torsten Textor [3,*] and Maria Rosaria Plutino [2,*]

1. Department of Engineering, University of Messina, Contrada di Dio, S. Agata, 98166 Messina, Italy
2. Institute for the Study of Nanostructured Materials, ISMN–CNR, Palermo, c/o Department ChiBioFarAm, University of Messina, Viale F. Stagno d'Alcontres 31, 98166 Messina, Italy
3. TEXOVERSUM School of Textiles, Reutlingen University, 72762 Reutlingen, Germany
4. Department of ChiBioFarAm, University of Messina, Viale F. Stagno d'Alcontres 31, Vill. S. Agata, 98166 Messina, Italy
5. Institute for Polymers, Composites and Biomaterials CNR IPCB, Via Paolo Gaifami 18, 95126 Catania, Italy
* Correspondence: torsten.textor@reutlingen-university.de (T.T.); mariarosaria.plutino@cnr.it (M.R.P.)

Citation: Sfameni, S.; Lawnick, T.; Rando, G.; Visco, A.; Textor, T.; Plutino, M.R. Functional Silane-Based Nanohybrid Materials for the Development of Hydrophobic and Water-Based Stain Resistant Cotton Fabrics Coatings. *Nanomaterials* 2022, 12, 3404. https://doi.org/10.3390/nano12193404

Academic Editor: Meiwen Cao

Received: 2 September 2022
Accepted: 23 September 2022
Published: 28 September 2022

Publisher's Note: MDPI stays neutral with regard to jurisdictional claims in published maps and institutional affiliations.

Copyright: © 2022 by the authors. Licensee MDPI, Basel, Switzerland. This article is an open access article distributed under the terms and conditions of the Creative Commons Attribution (CC BY) license (https://creativecommons.org/licenses/by/4.0/).

Abstract: The textile-finishing industry, is one of the main sources of persistent organic pollutants in water; in this regard, it is necessary to develop and employ new sustainable approaches for fabric finishing and treatment. This research study shows the development of an efficient and eco-friendly procedure to form highly hydrophobic surfaces on cotton fabrics using different modified silica sols. In particular, the formation of highly hydrophobic surfaces on cotton fabrics was studied by using a two-step treatment procedure, i.e., first applying a hybrid silica sol obtained by hydrolysis and subsequent condensation of (3-Glycidyloxypropyl)trimethoxy silane with different alkyl(trialkoxy)silane under acid conditions, and then applying hydrolyzed hexadecyltrimethoxysilane on the treated fabrics to further improve the fabrics' hydrophobicity. The treated cotton fabrics showed excellent water repellency with a water contact angle above 150° under optimum treatment conditions. The cooperative action of rough surface structure due to the silica sol nanoparticles and the low surface energy caused by long-chain alkyl(trialkoxy)silane in the nanocomposite coating, combined with the expected roughness on microscale due to the fabrics and fiber structure, provided the treated cotton fabrics with excellent, almost super, hydrophobicity and water-based stain resistance in an eco-sustainable way.

Keywords: sol–gel; (3-Glycidyloxypropyl)trimethoxy silane; functional cotton fabrics; hydrophobicity; nanohybrid coatings

1. Introduction

Textiles are critical to a country's growth and industrialization. In recent decades, many efforts have been made to develop innovative and nanostructured surface treatments in order to modify the mechanical and surface properties of natural and synthetic fabrics [1], thus replacing commonly used hazardous chemicals with products that are respectful of the environment and of health, while maintaining functional characteristics [2,3]. New multifunctional protective and smart textiles have been developed in response to growing technical breakthroughs, new standards, and a customer demand for textiles that are not only attractive but also practical [4–6]. In this regard, silica-based organic-inorganic nanostructured finishes could be considered an interesting alternative [7,8].

In recent years, the sol–gel approach has shown to be a creative and efficient method of improving the characteristics of fibers [9–13]. This approach comprises a diverse synthetic pathway that may be used to create novel materials with high molecular homogeneity and excellent physical and chemical characteristics. Due to its biocompatibility and non-toxicity, the sol–gel technique has been used to confer several functional properties to different textiles materials [14–17], such as antimicrobial [18–22], self-cleaning [23], water repellency [24–26], flame retardancy [27–29], and sensing [30–33], as well as improving the dye ability of fabric samples (see Figure 1) [34].

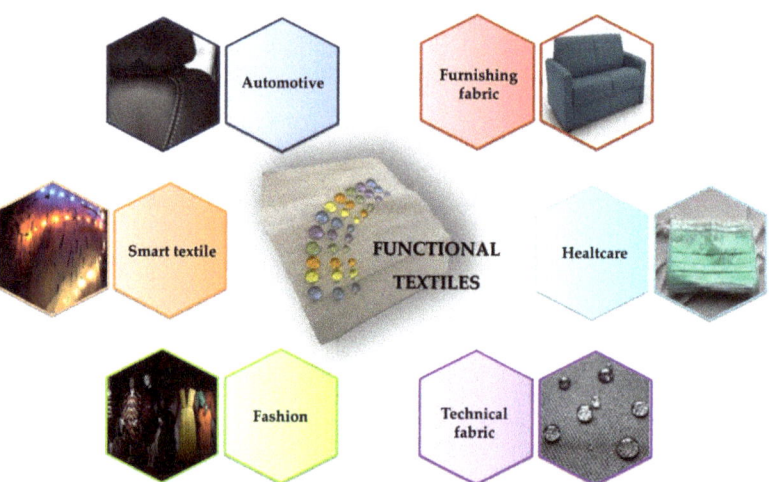

Figure 1. Functional and technical textiles for improved performance, protection, and health: types, application, and market.

Sol–gel synthesis and applications follow a two-step procedure based on the hydrolysis and condensation of metal or semi-metal alkoxides: after forming a hydrolyzed metal alkoxide solution at room temperature, textile materials are impregnated with the latter, and the samples are cured at a specific temperature to obtain a porous 3D fully inorganic or hybrid organic–inorganic nanostructured coating. Consequently, there are many alternatives for the formulation and application of sol–gel coatings in the field of textile functionalization, as choosing the correct and opportune functional silane precursor, which allows for the desired chemical and physical properties improvement of the fabric. Because of the moderate processing conditions required and the use of ordinary commercial textile finishing machines, in recent years, there has been a surge in interest in the application of the sol–gel approach to produce functional coated textiles [35–39], for example water-repellent fabrics. In general, surfaces that exhibit water contact angles > 150° (on which water drops remain almost spherical and easily roll off, also able to remove dirt particles in their path), are usually called superhydrophobic surfaces [40,41]. Superhydrophobic surfaces have recently attracted significant attention within the scientific community because of their unique water-repellent, anti-icing, anti-contamination, anti-sticking, and self-cleaning properties and their potential for practical applications [42,43].

Much of this research has been inspired by lotus leaves and has demonstrated that superhydrophobic surfaces may be produced by combining the right surface roughness and low surface free-energy [44–46]. The surface of lotus leaf was first examined by Barthlott in 1970 using scanning electron microscopy and it was found that the surface has small micro-protrusions covered with nano-hairs which are covered with low surface free-energy wax substances [46].

Surface roughness and surface free-energy were used to create superhydrophobic surfaces on cotton textiles. Different nanoparticles, including zinc oxide, titanium dioxide, silica nanoparticles (SNP), or alkoxysilane-based nano-sols [47] were added to cotton fabric to provide surface roughness. Fluorocarbons and silicones are examples of substances with low surface free-energy that might change the surface energy of cotton substrate [48].

In particular, fluoroalkylsilanes were used to further increase the surface water-repellency. Most recently, the ECHA's committee (Committee for Risk Assessment—ECHA—European Union) recommended restriction for some perfluoroalkyl substances (PFAS) regarding some application fields. In particular, fluoro-chemical finishing products are banned for textile applications in all EU states, while only some exemptions (i.e., in medical, technical, and workwear textiles) are accepted, but a complete restriction is expected in EU shortly, with a movement towards more widespread use of hydrophobic alkyl silanes. Currently, there are reports of the creation of rough surface micro/nanostructures using silane nanoparticles or nano-sols [49–52] and the subsequent modification with hydrophobic materials (e.g., fluoroalkylsilane, long-chain alkyl(trialkoxy)silane) to create superhydrophobic surfaces through a multi-step procedure [53,54].

Lakshmi et al. [55] produced superhydrophobic sol–gel nanocomposite coatings by adding silica nanoparticles to an acid-catalyzed ethanol–water solution of methyltriethoxysilane (MTEOS), while Huang et al. [56] created superhydrophobic surfaces by covering a silane-based coating in ethanol with a low surface-energy material 1H, 1H, 2H, 2H-perfluorooctyltrichlorosilane. By spraying an ethanol suspension of silica sol and silica microspheres, Shang et al.'s method [57] produced superhydrophobic silica coatings that were then hydrophobically treated with a solution of 1H, 1H, 2H, 2H-perfluorodecyltriethoxysilane (PFDTS). In order to create superhydrophobic silica films, Ramezani et al. [58] examined the two-step dip coating method using a sol–gel procedure. They coated a silica-based solution, and then modified it with isooctyltrimethoxysilane as a hydrophobic agent. According to studies [59,60], fluorine-based hybrid materials are the most successful in reducing the free-energy surface. However, some of the molecules are carcinogenic, highly costly, and not environmentally friendly.

In this work, co-condensation of (3-Glycidyloxypropyl)trimethoxysilane (hereafter, GPTMS or G) and different non-fluoro compounds, i.e., Hexadecyltrimethoxysilane C16, Triethoxy(octyl)silane C8 and Triethoxy(ethyl)silaneC2, as showed in Figure 2, was conducted in the presence of an acid catalyst to obtain functional nanohybrids via a one-step process.

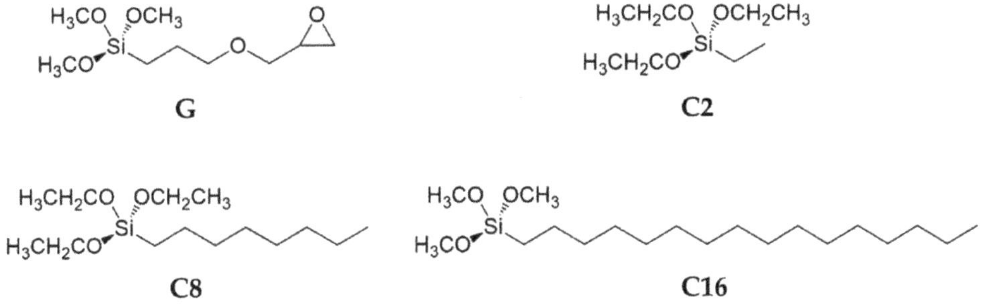

Figure 2. Alkyl(trialkoxy)silanes employed in this work.

By varying the length of the chain of the alky(trialkoxy)silane, R-Si(OR')$_3$, modified silane-based nanocomposite hydrosols, R-Si(O-)$_3$, were obtained with high dispersion stability. By applying R-Si(O-)$_3$ nanocomposite hydrosols to cotton fabrics, almost superhydrophobic cotton surfaces were obtained, as well as surface roughness and low surface energy. This study aimed to employ a multicoating eco-friendly technique in sol–gel textile finishing by examining the impact of various alkyl(trialkoxy) silane precursors on the silica-based mesh and, finally, to study the implemented mechanical characteristics of the treated cotton fabric. GPTMS is a useful molecule capable of forming extensive cross-links between the silanol groups of the polyoxysilane matrix and promoting adhesion through the opening of the epoxy ring on the treated polymers. It is a silica precursor that is frequently used for silica-based hybrid textile finishing [61,62].

The characteristics and the bi-functionality of the GPTMS, as well as its potential as a new textile finishing agent, should be investigated because there has not yet been much research on the impact of GPTMS synthetic parameters on the mechanical properties of fabrics made with both natural and synthetic polymers [63,64]. Indeed, because the chemical structure of fabric substrates is significant for the stability of the applied coatings, which is dependent on the thermodynamic affinity between the silica precursor and the selected textile samples, natural cotton textiles were employed in the current investigation.

Cotton fabrics were chosen as model substrates owing to their unique properties such as high hydroxyl group content, hydrophilic nature, and broad use, which allows them to be used not only in fabrics and garments but also in technical or smart textiles. Moreover, the use of alkyl(trialkoxy)silane has numerous advantages; it is low-cost and once polymerized is a non-toxic material [65–68], and a promising alternative for achieving durable hydrophobic fabrics. The final goal of this work was to illustrate an easy, environmentally friendly, and adaptable technique for generating hybrid coatings that are compatible with cellulose fabrics and their physical intrinsic features so that they can find applications in different sectors such as textiles [69], biomedical [70], furnishings [71], environmental remediation [72] and sensing [73]. The hydrophobicity was evaluated by WCA and WSA measurements.

By the characterization methods, the morphological qualities, surface chemistry, and durability of the sol–gel coatings were mainly evaluated using optical microscopy and SEM, comparing treated and untreated cotton textiles as a reference. Moreover, the water based anti-stain performances of the treated fabrics and, qualitatively, their oil–water separation ability towards paraffin oil were evaluated. In fact, functionalizing textiles with coatings based on the use of GPTMS in conjunction with functional alkyl(trialkoxy)silane could result in useful multifunctional nanocomposites for potential applications in the field of advanced, environmentally friendly nanohybrid materials, which would then find use in numerous nanotechnology fields.

2. Materials and Methods

2.1. Fabric

Knitted pure cotton fabric 100% (scoured and bleached, 1.4 g/cm^2 or 0.014 g/cm^2 and 0.2 mm thick) was used as natural fabric and it was provided by the School of Textile and Design (University of Reutlingen, Germany).

2.2. Chemicals

The (3-Glycidyloxypropyl)trimethoxysilane (G),Triethoxy(ethyl)silane (C2), Triethoxy (octyl)silane (C8) and Hexadecyltrimethoxysilane (C16), were all purchased at the highest purity level and used as received from Sigma Aldrich (Merk GaA, Darmstadt, Germany), without any further purification. Hydrochloric acid HCl 37% was used as sol–gel catalyst. Ethanol 96% vol. was purchased from Sigma Aldrich and used as solvent.

2.3. Preparation of the Nanosol Solution

The sol–gel solution was prepared by mixing the G precursor in combination with an equimolar amount of each of the three different alkoxysilanes featuring increasing length of the hydrocarbon chain (namely, C2, C8, C16). The obtained mixture was stirred while ethanol was added slowly at room temperature. Ethanol was used as dilution medium while HCl was added dropwise to induce the hydrolysis–condensation reaction. The resulting mixture was vigorously stirred at room temperature for 24 h.

2.4. Sol–Gel Treatment of Cotton Fabrics

Cotton fabrics were cut into square pieces (10 × 15 cm), weighted and then impregnated with the solution using the dip-pad-dry-cure method (Figure 3).

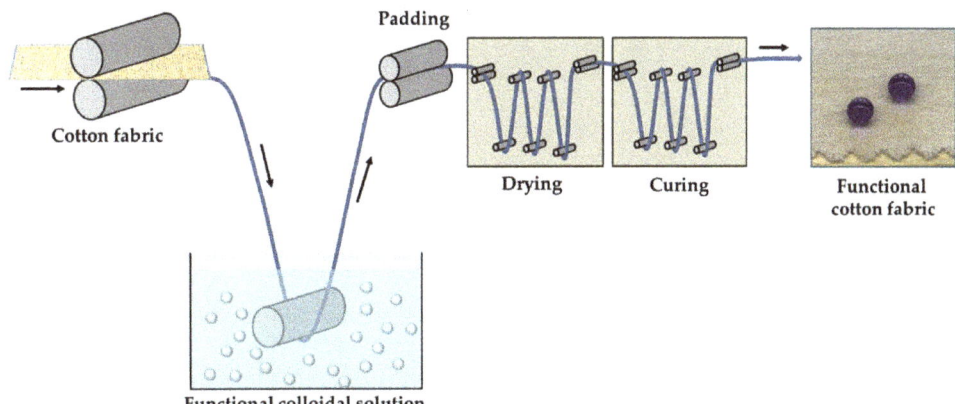

Figure 3. Pad-dry-cure process employed for finishing cotton fabrics.

First, the cotton fabric samples were immersed in the solutions for 5 min at room temperature before being washed with water. Second, an automated padder (simple two roller lab-padder of Mathis, Oberhasli, Switzerland) with a nip pressure of 2 kg/cm^2, was used to pad the cotton fabric samples. They were then dried at 80 °C for 6 min.

The process was repeated three times. In addition, samples of cotton were dipped in the alkyl(trialkoxy)silane-based ethanol solution (1.0 g, 30 mL) for 5 min, giving rise to a double-coating deposition (Figure 4).

Figure 4. Double sol–gel-based coating application for the development of the treated functional cotton fabrics.

The impregnated fabrics were finally put in the oven support and dried to a constant weight in the oven at 130 °C for 6 min: during this time the evaporation of water and ethanol and the sol–gel reactions took place.

This was confirmed by the color change of the fabrics as shown in Figure 5 and then modified cotton fabric was weighted, after being climatized for 24 h in a standard climate chamber. The composition of the functional nanohybrid sols employed for the double deposition process is shown in Table 1.

Table 1. Composition of the functional nanohybrid sols of each deposition.

Sample Code	First Deposition	Second Deposition
G	G	G
G/C2_C′2	G and C2	C2
G/C8 _C′8	G and C8	C8
G/C16 _C′16	G and C16	C16
G/C2_C′16	G and C2	C16
G/C8_C′16	G and C8	C16

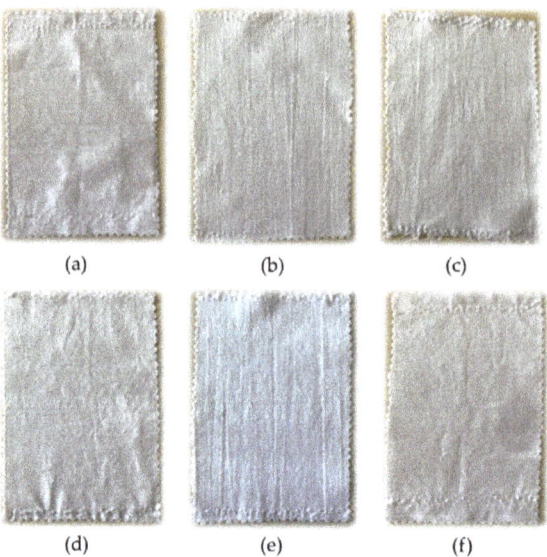

Figure 5. Cotton modified with sol–gel solution. (**a**) COT + G, (**b**) COT + G/C2_C′2, (**c**) COT + G/C8_C′8, (**d**) COT + G/C16_C′16, (**e**) COT + G/C2_C′16, and (**f**) COT + G/C8_C′16.

Subsequently, total dry-solid add-ons on the cotton samples (the weight gain, A wt.%) was determined by weighing each sample before (W_i) and after the impregnation with the solution and the subsequent thermal treatment (W_f) (Table 2).

Table 2. Composition (wt. %) of the investigated treated cotton fabrics (COT).

Sample Code	W_i	W_f	Total Add-on wt.% (A)
COT + G	2.140 g	2.163 g	1.06%
COT + G/C2_C′2	2.237 g	2.292 g	2.39%
COT + G/C8 _C′8	2.121 g	2.201 g	3.63%
COT + G/C16 _C′16	2.157 g	2.217 g	2.70%
COT + G/C2_C′16	2.210 g	2.308 g	4.24%
COT + G/C8_C′16	2.188 g	2.196 g	0.36%

The weight gain of the treated fabric was calculated using the following formula:

$$A = \frac{W_f - W_i}{W_f} \times 100 \quad (1)$$

2.5. Characterization and Functional Properties of Treated Fabrics

Wettability. Aqueous liquid repellency: water/alcohol solution tests were performed using a test reagent formulated using the AATCC test method 193-2007 Aqueous Liquid Repellency: Water/Alcohol Solution Resistance Test. Using a 5 µL water droplet at room temperature, the sessile drop technique (according to the international standard ASTM D7334) was used to measure the static water contact angles (WCA). One representative WCA was generated by averaging ten readings. The AATCC Test Method 22-2005, which is applicable to any textile fabric, was used to conduct the spray testing in order to examine the dynamic wettability of the treated samples. Three fabric samples, measuring 150 mm × 150 mm, are required to obtain one representative value for the spray testing. The tester's funnel is filled with 250 mL of distilled water, which is then sprayed onto a sample of cotton at a 45° angle. Three knocks are applied before the sample is removed.

The water repellency rating (WRR) is used to examine the extent of the wetting. Valuation is carried out by comparing the wetted sample's appearance with the wetted pristine cotton sample used as standard. Better hydrophobicity is indicated by a higher rating. The maximum and minimum ratings are 100 and 0, respectively.

Optical microscopy. Optical images were recorded by means of a Hirox digital microscope, model KH8700 (Hirox, Tokyo, Japan) by mounting a MX(G)-5040Z lens at room temperature.

Scanning Electron Microscopy (SEM) Analysis. The two-dimensional morphology and structure of the surface fibers of the of the original and treated cotton fabrics were observed at 2.0 kV using scanning electron microscope (SEM, SU-70, Hitachi, Chiyoda, Tokyo, Japan) and a magnification of 1000 × and 4000 × for the insets. All the samples were sputter-coated with Aurum prior to testing.

Self-cleaning ability. To evaluate the wetting behavior, several liquids including coffee, milk, tea, methylene-blue-dyed water, pH = 1 acid (HCl), pH = 14 alkali (NaOH), and salt solution (NaCl) were individually placed onto the GC16_C'16-modified cotton fibers. In order to test the self-cleaning abilities, soil was applied to the surface of the modified cotton fibers and washed with blue-dyed water.

Oil/water separation ability. The oil/water separation capabilities of the modified cotton textiles were tested using paraffin oil. The paraffin oil was colored using the coloring agent oil red before to the oil/water separation experiments.

Moisture analysis. The moisture-transfer properties of all the cotton fabrics samples were evaluated by using the KERN DBS moisture meter (KERN & SOHN GmbH-TYPE DBS60-3) that often replaces others drying processes, such as the laboratory dryer, because it allows for shorter measurement times. The moisture-transfer properties of all the cotton fabrics samples were evaluated through the principle of thermogravimetry. In this method, to determine the difference in moisture in a material, the sample is weighed before and after drying. In the case of the KERN DBS moisture meter, the radiation penetrates the sample and is transformed into thermal energy, heating up from the inside out. A small amount of radiation is reflected by the sample and this reflection is larger in dark samples than in light ones. Therefore, light samples, such as cotton in this case, reflect more thermal radiation than dark ones and therefore require a higher drying temperature, which is why a drying temperature of 130 °C is used for the analysis. Moisture measurement protocol (unit indicating the result: M/W, drying mode: TIME, drying temperature: 130 °C). The hygroscopicity ratio was calculated by Equation (2), which was used to as an indicator for evaluating the hygroscopicity of these cotton fabrics.

$$\text{Hygroscopicity ratio}(\%) = \frac{m_2 - m_1}{m_1} \times 100\% \qquad (2)$$

where, m_2 is the weight of the conditioned sample and m_1 is the initial weight of original samples.

Air-permeability test. The air permeability of treated fabrics, which serves as an indication of their breathability, was investigated. The permeability of the samples was measured by the use of an apparatus (FX3300, Tex Test AG, Schwerzenbach, Switzland) under the air pressure of 125 Pa, according to the ASTM D737-96 standard test method.

3. Results

3.1. Nanosol Synthesis and Application on Cotton Fabrics

The sol–gel technique is a very versatile method leading to the formation of different kind of interesting functional nano- and micro-structured materials, with a fine control and tuning of their surface chemistry and of the bulk nanocomposite/nanohybrid properties of the end-products. In this study, the sol–gel synthesis and application followed a two-step pathway (Figure 6a,b) to finally yield the desired hydrophobic cotton fabrics.

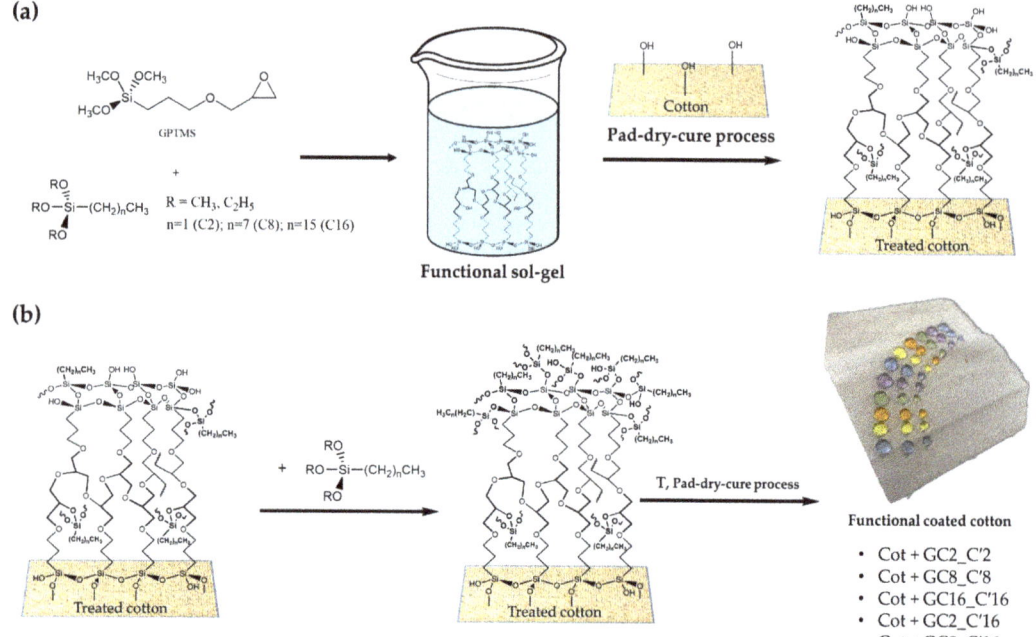

Figure 6. Two-step synthetic pathways for the development of the final functional coated cotton fabrics involving the condensation reaction between the cotton cellulose and alkoxysilane ends (**a**), and subsequent anchorage of alkyl(trialkoxy)silanes (**b**).

In the first step, the functional sol was prepared by reaction of the bifunctional GPTMS alkoxysilane and either the C2, or C8, or C16 alkyl(trialkoxy)silane, respectively, featuring different length alkyl chains. As previously reported [63,64], the functional nanosol solution is obtained by subsequent hydrolysis and condensation reaction, thus producing colloidal particles or dissolved pre-condensed polymeric hybrid polymers. Once applied on cotton fabrics and with additional heat treatment at higher temperatures by a pad-dry-cure process, the gel will give rise to a functional xerogel.

In order to form a more efficient hydrophobic coated cotton, it seemed to be worthwhile to use a second pad-dry-cure step by employing C2, or C8, or C16 alkyl(trialkoxy)silanes, respectively, thus giving rise to five functional treated cotton samples, namely Cot + GC2_ C′2, Cot + GC8_ C′8, Cot + GC16_ C′16, Cot + GC2_ C′16, Cot + GC8_ C′16; Cot + G that was used as cotton fabric reference in all experimental measurements.

3.2. Wettability Measurement

Water contact angle (WCA) measurement was used to explore the static hydrophobicity of the treated cotton samples, and the spraying test was used to evaluate the dynamic water repellence.

3.2.1. Aqueous Liquid Repellency: Water/Alcohol Solution Test

To evaluate the level of anti-wettability or repellency, the contact angles of liquids with different surface tensions was measured, using a test reagent formulated using the AATCC test method 193-2007 Aqueous Liquid Repellency [74]. The aqueous-liquid repellency test (also known as the water-rating method WRA) describes a procedure whereby 20 µL drops of solution with increasing concentration of isopropyl alcohol is placed onto the fabric. If the drops of a solution do not wet the textile within 10 s, the next solution with a higher share of isopropanol is applied. The rating number is assigned based on the solution with the highest isopropanol share that does not wet the textile within the 10 s.

Table 3 outlines which concentration of isopropyl alcohol solution equates to which rating number.

Table 3. Aqueous-liquid repellency test.

Sample Code	Aqueous-Solution Repellency-Grade Number	Composition (by Volume)
Cot + G	0	100% Water
Cot + GC2_C′2	1	98:2/water:isopropyl alcohol
Cot + GC8_C′8	3	90:10/water:isopropyl alcohol
Cot + GC16_C′16	3	90:10/water:isopropyl alcohol
Cot + GC2_C′16	3	90:10/water:isopropyl alcohol
Cot + GC8_C′16	4	80:20/water:isopropyl alcohol

The fabric must be able to repel the solution for 10 s to be deemed successful. As a static test, this method could be considered to be more stringent than other water-drop methods as it makes use of solutions with surface tensions lower than that of water.

3.2.2. Sessile Drop Method

As a way to examine the effect of the length of the hydrocarbon chains and their distribution/orientation into the nano composite sol on the hydrophobicity of the treated cotton fabric, the cotton fabrics samples were coated by using three different sols, as prepared by changing the length of the functional alkyl(trialkoxy)silane, from 2 to 8 to 16 methylene groups. In a typical process, a deionized water droplet (ca. 5 µL) was dropped carefully onto the surface at ambient temperature and the images were captured using the accessory digital camera. All the water contact-angle values reported herein were obtained as averages of five measurements performed on different points of the sample surface so as to improve the accuracy [75–77]. The results were shown in Figures 7 and 8.

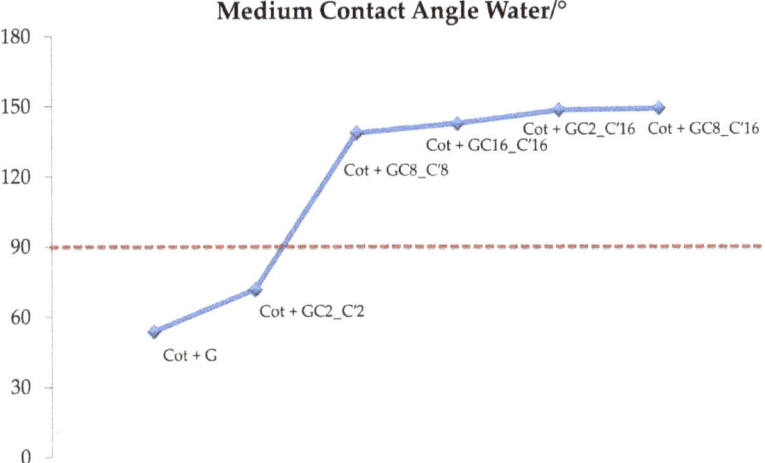

Figure 7. Schematic graph representative of the static water-contact angles of Wenzel θw values. The line between the data points is only a guide for the eye.

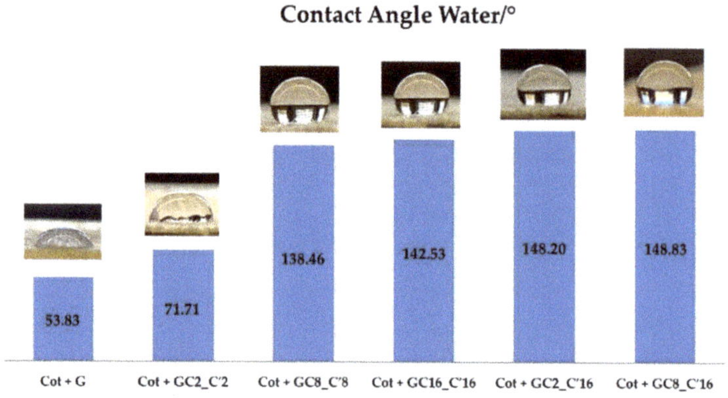

Figure 8. Histograms of the contact angle θw of the cotton fabrics treated by the nanocomposite sol samples G, GC2_C′2, GC8_C′8, GC16_C′16, GC2_C′16, and GC8_C′16 with photos of the representative drops.

WCAs for the treated cotton fabrics ranged from 71.71° to 142.53°. In particular, the cotton fabric coated by the nanocomposite hydrosol with the shortest hydrocarbon chain (GC2_C′2) displayed the poor hydrophobicity, showing the lowest WCA of 71.71°. However, the cotton fabric that had been treated with the nanocomposite sol bearing the longest hydrocarbon chain (GC16_C′16) demonstrated exceptional hydrophobicity with the maximum WCA of 142.53°. In this regard, it was shown that nanocomposite sols, treated with mixed alkyl(trialkoxy)silane (GC2_C′16 and GC8_C′16), are beneficial for the formation of a brush effect surface topography [67] on the coated fabrics, thus resulting in an improvement of the hydrophobicity of treated fabrics of 148.20° and 148.83°, respectively.

Table 4 and Figure 8 indicate the hydrophobicity of the treated cotton samples.

Table 4. Static water contact angles of Wenzel $θ_w$ values.

Sample Code	Static Water Contact Angle $θ_w$ [°]
Cot + G	53.83 ± 0.82
Cot + GC2_C′2	71.71 ± 0.33
Cot + GC8_C′8	138.46 ± 0.40
Cot + GC16_C′16	142.53 ± 0.34
Cot + GC2_C′16	148.20 ± 0.80
Cot + GC8_C′16	148.83 ± 0.29

According to Figure 7, the WCA of the treated cotton increased as well as the length of the alkyl chain. The final treatment with HDTMS-based sol led to a lowering of the surface energy of the cotton fabric, with a consequent improvement in hydrophobicity. As indicated in Figure 8, the WCA of all cotton fabric samples was less than 150°. Although a water droplet can sit on the surface of the coated cotton cloth, it does not achieve super-hydrophobicity.

Furthermore, as shown in Figure 9, when we exposed the coated cotton textile to water droplets, the fabric successfully showed hydrophobicity similar to that present in the rose petals. The double-coating synthetic method used to improve the cotton surface hydrophobicity by pad-dry-cure deposition of the prepared nanohybrid coatings, as demonstrated using water contact angle, was shown to be successful. As a matter of fact, while the control sample had a water contact angle of 53.8°, with the incorporation of hydrophobic long-alkyl-chain silane coupling agent onto surface of cotton fabrics, the water contact angle for the Cot + GC8_C′16 nanohybrid increased to 148.83°. These values are generally higher (ca. 30–40°) than those obtained with a "grafting to" chemisorption method of the corresponding GC* functional sol–gel, as recorded on a glass slide coated after an opportune commercial primer and tie-coat layer [67].

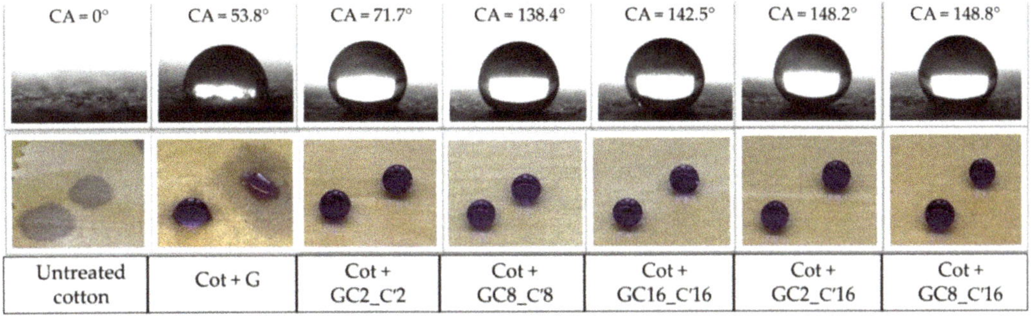

CA = 0°	CA = 53.8°	CA = 71.7°	CA = 138.4°	CA = 142.5°	CA = 148.2°	CA = 148.8°
Untreated cotton	Cot + G	Cot + GC2_C'2	Cot + GC8_C'8	Cot + GC16_C'16	Cot + GC2_C'16	Cot + GC8_C'16

Figure 9. The images show the change in water contact angle and colored water droplets sitting on coated cotton fabrics and untreated cotton.

In this study the key step was shown to be the second coating application of the bulk functional C2, C8, and C16 silanes, especially those featuring increase of the length of the alkyl(trialkoxy) chain (i.e., C8 and C16), by the "grafting from" chemisorption covalent technique [78–80] (Figure 10a). The treated cotton fabrics exploited the observed surface hydrophobicity, as shown in Figure 10b.

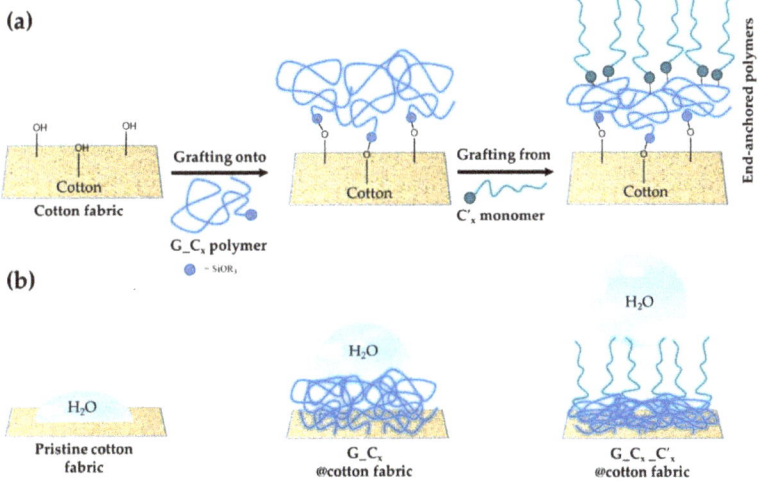

Figure 10. Double surface functionalization of cotton fabrics with alkyl(trialkoxy)silane polymer shell by "grafting to" or "grafting from" covalent grafting techniques (**a**) and corresponding observed surface hydrophobicity of the coated cotton fabrics (**b**).

In this process, on the first coated cotton surface, characterized by a silane-based 3D polymeric matrix obtained through a "grafting onto" procedure, a second functional coating was deposited, and at the end a polymerization process under conventional dry-cure conditions occurred. This latter "grafting from" process could bring a better control of the in-situ grafting density, composition, and molar ratio of the polymer brush shells. It is also not limited by the steric hindrance of the incoming functional alkyl chains, thus resulting significantly in lower surface energy and higher hydrophobicity [81] (Figure 10b).

Moreover, it has been reported that improvement of the properties, morphological characteristics, and fiber roughness of different functional coatings on textile fabrics, leading to enhanced mechanical and hydrophobic behaviors, can be achieved by employing a double-layer deposition approach [82–84].

3.2.3. Spray Test

Spray testing (AATCC 22-2005) was also used to study the dynamic wettability of the treated cotton fabric [85]. Spray testing quantifies the degree of wetness when the fabric is sprayed with water. The water-stain characteristics at different wetting degrees (in ISO standard ratings) are listed in Table 5. The wettability level of the uncoated fabric was 0, referring to no water repellency. After one coating of the sample COT + G, the wettability level was found to be 0, which was similar to uncoated fabric. However, the wettability level of the coated fabrics increased up to 50 after two coatings and remained at the same value after three coatings. This result indicates that the water repellent property of the fabrics could be enhanced by increasing the number of depositions.

Table 5. Wettability levels specified in the ATTCC 22 standard for spray tests.

Sample Code	Wettability Level	Water-Stain Characteristics
Cot + G	50 (ISO 1)	Complete wetting of the entire specimen face
Cot + GC2_C'2	50 (ISO 1)	Complete wetting of the entire specimen face
Cot + GC8_C'8	100 (ISO 5)	No wetting of the specimen face
Cot + GC16_C'16	100 (ISO 5)	No wetting of the specimen face
Cot + GC2_C'16	100 (ISO 5)	No wetting of the specimen face
Cot + GC8_C'16	100 (ISO 5)	No wetting of the specimen face

According to the spray test, the treated cotton fabric samples COT + GC8_C'8 and COT + GC16_C'16 had a rating number corresponding to 100. The 250 mL of water in contact with the treated cotton immediately slipped away from the fabric, leaving only a few drops of water attached. In the light of the obtained results, cotton fabrics possess excellent hydrophobicity with a low surface energy. In contrast, when the cotton fabric was coated by the GPTMS-based sol without alkysilane modification, its surface shows higher surface energy. In addition, when a water droplet was applied on the pristine cotton fabric, it quickly spread.

3.2.4. Self-Cleaning Ability Measurement

The self-cleaning capabilities of the GC16_C'16-modified cotton textiles are displayed in Figure 11. The GC16_C'16-modified cotton fabrics, as shown in Figure 11a, exhibited excellent almost super hydrophobicity against blue-dyed water, strong acid (HCl, pH = 1), strong alkali (NaOH, pH = 14), and salt solution (NaCl 0.9 wt.%, pH = 7). They also exhibited excellent, almost super, hydrophobicity against milk, coffee, and tea.

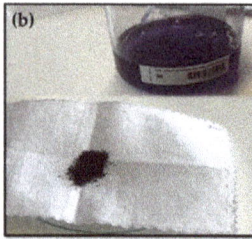
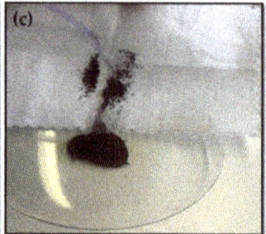
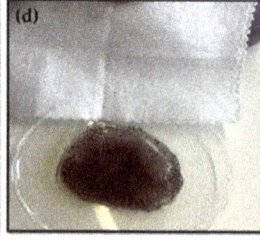

Figure 11. Photograph demonstrating the almost super-hydrophobicity of GC16_C'16-modified cotton textiles against various liquids (**a**) and real-time photographs demonstrating their water-based stain resistance performance (**b–d**).

It can be also seen that different liquids all exist in a similar spherical shape and this indicates that the coating has a wide range of adaptability in practical applications and implying that they may be flexible to self-cleaning under a variety of settings. Moreover, it can be clearly observed that the dust on the surface of the coating was completely removed after washing with water. As a matter of fact, water could readily flow over the surface of the sample and remove dust, as illustrated in Figure 11b–d, indicating that GC16_C'16 coating, in particular, has a very good water-based stain resistance.

It is assumed that the hydrophobic properties of the coating may be principally due to the air trapped in the nanoscale gaps of the almost super hydrophobic surface, which decreases the area of interaction between the soil and the coating and assists soil to roll-off from its surface [86,87].

3.3. Oil/Water Separation Ability

It is well known that cotton fabric is capable of absorbing large amounts of water due to its high hydrophilicity [88]. The above-mentioned, GC8_C'8, GC16_C'16, GC2_C'16, and GC8_C'16 coatings endowed the cotton fabrics with almost super hydrophobicity. In contrast, the modified cotton fabrics also have reduced surface energy that allows a repelling of water, but wetting with oils is still possible—that means the surfaces can be able to separate water from oil [89].

Figure 12 shows a sequence of images in which a cotton fabric treated with the GC16_C'16 sol was used to qualitatively study its oil/water separation ability by using paraffin oil as a model oil [90,91]. Paraffin oil was further dyed red for better observation [92]. Therefore, a droplet of red-tinted paraffin oil (with a density lower than that of water) was located in the watch glass, as illustrated in Figure 12a–d. A piece of treated fabric was soaked in the oil/water mixture to make it completely contact with the oil. It can be clearly observed that oil droplets quickly spread and even permeated into fabric, indicating the lipophilic nature of the coating. This selective adsorption is indicative of remarkable hydrophobicity or oleophilicity of the sample [93].

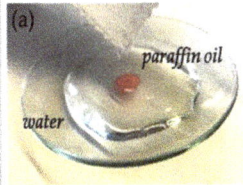

Figure 12. Real-time photos demonstrating the removal of paraffin oil (**a**–**d**) droplets from water using GC16_ C'16 modified cotton fibers.

In particular, the GC16_C'16 coating used in the current experiment can selectively absorb paraffin oil from water, further indicating that the fabric was hydrophobically modified by the coating. This demonstrates, qualitatively, that the as-prepared modified cotton textiles had good separation capacity against the tested oil/water system liquids, indicating that they might have interesting uses in industry for very efficient and long-term oil/water separation processes [94].

3.4. Morphological Characterization
3.4.1. Optical Microscopy (MO)

In order to evaluate at the microstructural level some modification in the roughness of the surface, low-magnification micrographs of the raw cotton fabric were performed and are shown in Figure 13a [95].

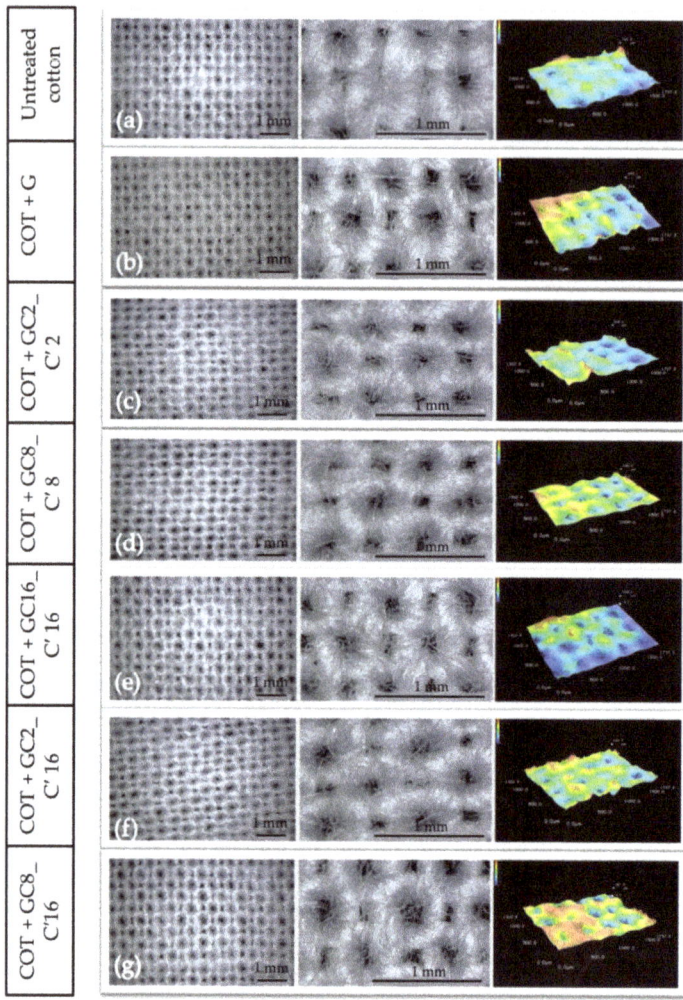

Figure 13. Optical images of untreated (**a**) and treated (**b**–**g**) cotton fabrics at different magnification and 3D image of the roughness of the analyzed samples.

It is difficult to differentiate, at this scale, between the treatment processes applied to the cotton fabric, regardless of their weaving density, mesh size, or physical appearance. Therefore, it is possible that the flux and permeability of the cotton fabric itself are not greatly affected by a microscale porosity occlusion that can lead to a lowering of the breathability of the textile [96]. Figure 13b–g shows the activated cotton fabric with the different sols at high magnifications as obtained by MO. Unfortunately, no distinct structural/morphological changes are shown, therefore a SEM analysis at the nanoscale level was performed in order to investigate how the nanohybrid functional agents can affect the fiber roughness.

3.4.2. Morphological Characterization by SEM Analysis

To evaluate the surface roughness changes of the cotton fabric at the nanoscale level after coating with the functional nanohybrids, a SEM analysis was performed. In Figure 14a the fiber surface of the raw cotton fabric had relatively smooth morphology. In Figure 14b, after deposition of three layers of GPTMS coatings, the fiber surface of the activated cotton fabrics showed a different surface morphology. Moreover, the fabric surface shows some coarse particles but the natural structure of the single cotton fiber looks flattened when compared the inserts (a) and (b).

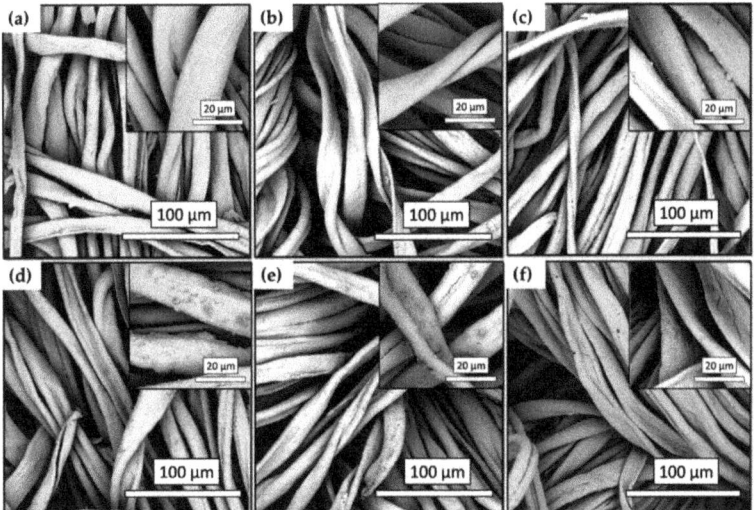

Figure 14. SEM images of untreated cotton (**a**) and GC2_C′2-coated, (**b**) GC8_C′8-coated, (**c**) GC16_C′16-coated, (**d**) GC2_C′16-coated, (**e**) GC8_C′16-coated, and (**f**) modified cotton fabrics (the inserts are partially enlarged images).

Surface roughness is a common indicator of product quality and occasionally even included as a technical requirement for obtaining the required fabric surface functionality [97]. In this regard, by comparing Figure 14a–f, it can be concluded that the addition of an alkyl(trialkoxy)silane with a long hydrocarbon chain to the coating films transforms, to some extent, the surface morphology from flat texture to rough surface. In short, the activation process promotes the alkyl(trialkoxy)silane-graft copolymerization reaction and contributes to the almost super hydrophobicity and hydrophobic stability of the cotton fabric. By the addition of a rougher surface nanoarchitecture on the micro-scaled fabric, the structural basis for transforming the extreme hydrophilicity of the cotton fabric into stable almost super hydrophobicity was therefore achieved [98,99].

3.5. Moisture-Adsorption Analysis

The determination of humidity is always of enormous importance when in the production process there is absorption or lack of humidity to and from the products. In numerous quantities of products and finishings, moisture content is both a quality characteristic and an important cost factor [100].

As shown in Table 6, the untreated cotton sample had a water absorption of 4.31%. The thin hydrophilic fabrics can easily absorb water vapor and water can pass to the other face. It is well established that original cotton fabrics exhibit high breathability as well as hygroscopicity. These outstanding properties of such cotton fabrics are attributed to their abundant hydrophilic groups (hydroxyl). This tendency and sensitivity of cellulose fabrics towards moisture, leads to a limit in their use [101]. Therefore, a proper surface modification can greatly affect the moisture adsorption of textiles [102,103].

Table 6. Moisture-adsorption standard test.

Sample Code	Weight (g)	Drying Temperature (°C)	Drying Time (min) [1]	Humidity (%)
Untreated cotton	2.156	130	5–6	4.31
COT + G	2.181	130	2	3.62
COT + GC2_C′2	2.230	130	3	4.44
COT+ GC8_C′8	2.198	130	2	3.55
COT + GC16_C′16	2.209	130	3	4.07
COT + GC2_C′16	2.280	130	2–3	3.90
COT + GC8_C′16	2.206	130	3–4	4.12

[1] Drying time until constant weight.

As a matter of fact, the GPTMS-modified cotton fabrics samples, compared with original ones, had lower WVT values. By applying pure GPTMS-sol, this water absorption was the same because the modified GPTMS-based coating was potentially still hydrophilic. The molecular chain similarly owns abundant hydrophilic groups (i.e., carboxyl, hydroxyl). Thus, the surface of GPTMS-modified cotton fibers still possessed a hydrophilic nature and abundant highly active hydrophilic groups on the fiber surface.

To achieve a significant change in the response to water absorption, these GPTMS-sols must be modified with a strong hydrophobic additive such as hexadecyltriethoxysilane (16 carbon alkyl(trialkoxy) chain). This decrease is consequently accentuated by the addition of the alkyl(trialkoxy)silane monomers, proportionally with the increase in the length of the alkyl(trialkoxy) chain.

3.6. Air-Permeability Measurement

Air permeability is one of the most important properties of a fabric, mainly intended for technical or smart textile applications and it is highly related on its porosity [104]. Additionally, air permeability is a crucial property for fabric applications in order to assess the chances of reducing the physiological strain on the human body and the hazards of heat stress [105]. It is obvious that non-coated fabrics, due to their low thicknesses and high porosities, show higher permeability in comparison to the coated fabrics and therefore, finishings can affect this behavior [106,107].

The breathability and physical properties of pristine cotton fabric and coated cotton fabrics were measured in terms of the air permeability and the obtained results are summarized in Figure 15.

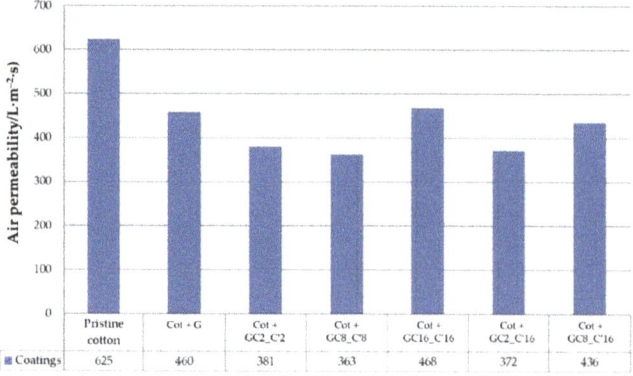

Figure 15. Histogram of the air permeability of the cotton fabrics treated by the nanocomposite sol samples G, GC2_C′2, GC8_C′8, GC16_C′16, GC2_C′16, and GC8_C′16.

Herein, it can be seen that the treatment of fabrics with the hydrophobic alkyl(trialkoxy) silane solutions moderately decreases the air permeability of fabric by a maximum of about 40%. Especially for the Cot + GC16_C'16 and Cot + GC8_C'16 samples, the coating did not highly influence the air permeability of the cotton fabric, assessing an overall good breathability of the textile support, and therefore, making them suitable for applications in a wide range of industrial-related sectors.

4. Conclusions

In this research, functional alkyl(trialkoxy)silane-modified hybrid nanostructured materials were developed and successfully employed as eco-friendly hydrophobic and water-based stain resistant coatings for cotton fabrics the via sol–gel technique and cure/pad applications. In particular, the aim of the present work was to investigate different functional alkyl(trialkoxy)silanes as precursors to obtain efficient and stable hybrid sol–gel GPTMS-based coatings and to further reduce the cotton surface energy, thus improving hydrophobicity and water-based stain resistance properties on textiles in an eco-sustainable way. This method reveals a promising application for the future finishing and functionalization of ordinary fabrics since it is straightforward, affordable, and ecologically friendly.

Morphological characterizations were performed on all the samples by optical microscopy and SEM. This last revealed an improvement on the surface roughness of the treated fabrics.

The investigation of the fabrics hydrophobicity via water contact angle (WCA) measurements showed that the treated fabrics exhibited high static contact angles (up to ca. 150°). Moreover, this was confirmed by a spray test, performed according to the AATCC 22 standard, in order to evaluate the dynamic surface-wettability of the coated samples. The water-based stain resistance of the treated fabric, was also demonstrated towards different tested liquids, solutions, and soil. Therefore, an oil/water separation experiment, was performed revealing, qualitatively, good ability of the GC16_C'16-modified samples, in particular, to retain paraffin oil, representing a valuable approach for possible efficient industrial and long-term oil/water separation approaches.

The quality characteristics of the fabrics were additionally evaluated by moisture-adsorption analysis and air-permeability test, observing with the latter an overall good breathability of the coated cotton fabrics compared to the pristine one.

All experimental findings, indicated that the synergic action of the rough surface structures and their low surface energy, caused by the chosen functional alkyl(trialkoxy)silane in the sol–gel nanohybrid coatings, provided treated cotton fabrics with excellent hydrophobicity and therefore water repellency by an eco-friendly approach.

Thus the results demonstrate the effectiveness of this nanohybrid sol–gel based functional double-coating treatment for cotton fabrics, for the preparation of hydrophobic surfaces that may have applications in different sectors ranging from textile and biomedical to water separation, providing a valuable contribution to eco-friendly hydrophobic surface treatments, with the possibility of being scaled to other types of fabrics.

Author Contributions: Conceptualization, M.R.P., T.T. and S.S.; methodology, T.T. and M.R.P.; validation, S.S.; investigation, M.R.P., T.T., S.S. and G.R.; resources, M.R.P.; data curation, M.R.P., T.L., T.T., A.V., S.S. and G.R.; writing—original draft preparation, M.R.P., T.T., S.S. and G.R.; writing—review and editing M.R.P., T.T., A.V. and S.S.; supervision, M.R.P. and T.T. All authors have read and agreed to the published version of the manuscript.

Funding: The research is supported by "Technology and materials for safe low consumption and low life-cycle cost vessels and crafts" project (PON MUR 2014–2020; CUP ARS01_00293). Area of specialization "BLUE GROWTH".

Institutional Review Board Statement: Not applicable.

Informed Consent Statement: Not applicable.

Data Availability Statement: Not applicable.

Acknowledgments: All authors wish to thank S. Romeo, G. Napoli, and F. Giordano for the informatic and technical assistance in all instrumentation set-up and subsequent data-fitting. This work was also performed within the framework of the PhD program of S.S., as financed by Confindustria (Noxosorkem Group S.r.l.) and CNR, and G.R., as financed by PON-MUR "Ricerca e Innovazione 2014–2020" RESTART project; MUR, Confindustria and CNR are gratefully acknowledged.

Conflicts of Interest: The authors declare no conflict of interest.

References

1. Toprak, T.; Anis, P. Textile industry's environmental effects and approaching cleaner production and sustainability, an overview. *J. Text. Eng. Fash. Technol.* **2017**, *2*, 429–442. [CrossRef]
2. Schindler, W.D.; Hauser, P.J. *Chemical Finishing of Textiles*; Woodhead Publishing: Cambridge, UK, 2004.
3. Hasanbeigi, A.; Price, L. A technical review of emerging technologies for energy and water efficiency and pollution reduction in the textile industry. *J. Clean. Prod.* **2015**, *95*, 30–44. [CrossRef]
4. Trovato, V.; Sfameni, S.; Rando, G.; Rosace, G.; Libertino, S.; Ferri, A.; Plutino, M.R. A review on stimuli-responsive smart materials for wearable health technology: Retrospective, perspective and prospective. *Molecules* **2022**, *27*, 5709. [CrossRef]
5. Sobha, K.; Surendranath, K.; Meena, V.; Jwala, T.K.; Swetha, N.; Latha, K.S.M. Emerging trends in nanobiotechnology. *Biotechnol. Mol. Biol. Rev.* **2010**, *5*, 1–12.
6. Ahmad, S.; Fatma, A.; Manal, E.; Ghada, A.M. Applications of Nanotechnology and Advancements in Smart Wearable Textiles: An Overview. *Egypt. J. Chem.* **2020**, *63*, 2177–2184.
7. Sanchez, C.; Julián, B.; Belleville, P.; Popall, M. Applications of hybrid organic–inorganic nanocomposites. *J. Mater. Chem.* **2005**, *15*, 3559. [CrossRef]
8. Wang, Z.; Lan, T.; Pinnavaia, T. Hybrid organic-inorganic nanocomposites formed from an epoxy polymer and a layered silicic acid (magadiite). *Chem. Mater.* **1996**, *8*, 2200–2204.
9. Mahltig, B.; Textor, T. *Nanosols and Textiles*; World Scientific: Singapore, 2008; ISBN 978-981-283-350-1.
10. Mahltig, B.; Haufe, H.; Böttcher, H. Functionalisation of textiles by inorganic sol–gel coatings. *J. Mater. Chem.* **2005**, *15*, 4385. [CrossRef]
11. Ismail, W.N.W. Sol–gel technology for innovative fabric finishing—A Review. *J. Sol-Gel Sci. Technol.* **2016**, *78*, 698–707. [CrossRef]
12. Ielo, I.; Giacobello, F.; Sfameni, S.; Rando, G.; Galletta, M.; Trovato, V.; Rosace, G.; Plutino, M.R. Nanostructured Surface Finishing and Coatings: Functional Properties and Applications. *Materials* **2021**, *14*, 2733. [CrossRef]
13. Trovato, V.; Rosace, G.; Colleoni, C.; Sfameni, S.; Migani, V.; Plutino, M.R. Sol-gel based coatings for the protection of cultural heritage textiles. *IOP Conf. Ser. Mater. Sci. Eng.* **2020**, *777*, 12007. [CrossRef]
14. Giacobello, F.; Ielo, I.; Belhamdi, H.; Plutino, M.R. Geopolymers and Functionalization Strategies for the Development of Sustainable Materials in Construction Industry and Cultural Heritage Applications: A Review. *Materials* **2022**, *15*, 1725. [CrossRef]
15. Trovato, V.; Mezzi, A.; Brucale, M.; Abdeh, H.; Drommi, D.; Rosace, G.; Plutino, M.R. Sol-Gel Assisted Immobilization of Alizarin Red S on Polyester Fabrics for Developing Stimuli-Responsive Wearable Sensors. *Polymers* **2022**, *14*, 2788. [CrossRef]
16. Filipic, J.; Glazar, D.; Jerebic, S.; Kenda, D.; Modic, A.; Roskar, B.; Vrhovski, I.; Stular, D.; Golja, B.; Smolej, S.; et al. Tailoring of antibacterial and UV protective cotton fabric by an in-situ synthesis of silver particles in the presence of a sol–gel matrix and sumac leaf extract. *Tekstilec* **2020**, *63*, 4–13. [CrossRef]
17. Puoci, F.; Saturnino, C.; Trovato, V.; Iacopetta, D.; Piperopoulos, E.; Triolo, C.; Bonomo, M.G.; Drommi, D.; Parisi, O.I.; Milone, C.; et al. Sol–Gel Treatment of Textiles for the Entrapping of an Antioxidant/Anti-Inflammatory Molecule: Functional Coating Morphological Characterization and Drug Release Evaluation. *Appl. Sci.* **2020**, *10*, 2287. [CrossRef]
18. Zhang, Y.Y.; Xu, Q.B.; Fu, F.Y.; Liu, X.D. Durable antimicrobial cotton textiles modified with inorganic nanoparticles. *Cellulose* **2016**, *23*, 2791–2808. [CrossRef]
19. Cuk, N.; Sala, M.; Gorjanc, M. Development of antibacterial and UV protective cotton fabrics using plant food waste and alien invasive plant extracts as reducing agents for the in-situ synthesis of silver nanoparticles. *Cellulose* **2021**, *28*, 3215–3233. [CrossRef]
20. Ielo, I.; Giacobello, F.; Castellano, A.; Sfameni, S.; Rando, G.; Plutino, M.R. Development of Antibacterial and Antifouling Innovative and Eco-Sustainable Sol–Gel Based Materials: From Marine Areas Protection to Healthcare Applications. *Gels* **2022**, *8*, 26. [CrossRef]
21. Saleemi, S.; Naveed, T.; Riaz, T.; Memon, H.; Awan, J.A.; Siyal, M.I.; Xu, F.; Bae, J. Surface Functionalization of Cotton and PC Fabrics Using SiO_2 and ZnO Nanoparticles for Durable Flame Retardant Properties. *Coatings* **2020**, *10*, 124. [CrossRef]
22. Ibrahim, H.M.; Zaghloul, S.; Hashem, M.; El-Shafei, A. A green approach to improve the antibacterial properties of cellulosebased fabrics using Moringa oleifera extract presence of silver nanoparticles. *Cellulose* **2021**, *28*, 549–564. [CrossRef]
23. Colleoni, C.; Massafra, M.R.; Rosace, G. Photocatalytic properties and optical characterization of cotton fabric coated via sol–gel with non-crystalline TiO_2 modified with poly(ethylene glycol). *Surf. Coat. Technol.* **2012**, *207*, 79–88. [CrossRef]
24. Colleoni, C.; Guido, E.; Migani, V.; Rosace, G. Hydrophobic behaviour of non fluorinated sol-gel based cotton and polyester fabric coatings. *J. Ind. Text.* **2015**, *44*, 815–834. [CrossRef]

25. Sfameni, S.; Rando, G.; Galletta, M.; Ielo, I.; Brucale, M.; De Leo, F.; Cardiano, P.; Cappello, S.; Visco, A.; Trovato, V.; et al. Design and Development of Fluorinated and Biocide-Free Sol–Gel Based Hybrid Functional Coatings for Anti-Biofouling/Foul-Release Activity. *Gels* **2022**, *8*, 538. [CrossRef]
26. Textor, T.; Mahltig, B. A sol–gel based surface treatment for preparation of water repellent antistatic textiles. *Appl. Surf. Sci.* **2010**, *256*, 1668–1674. [CrossRef]
27. Alongi, J.; Colleoni, C.; Rosace, G.; Malucelli, G. The role of pre-hydrolysis on multi step sol-gel processes for enhancing the flame retardancy of cotton. *Cellulose* **2013**, *20*, 525–535. [CrossRef]
28. Ahmed, M.; Morshed, M.; Farjana, S.; Ana, S. Fabrication of new multifunctional cotton–modal–recycled aramid blended protective textiles through deposition of a 3D-polymer coating: High fire retardant, water repellent and antibacterial properties. *New J. Chem.* **2020**, *44*, 12122–12133. [CrossRef]
29. Castellano, A.; Colleoni, C.; Iacono, G.; Mezzi, A.; Plutino, M.R.; Malucelli, G.; Rosace, G. Synthesis and characterization of a phosphorous/nitrogen based sol-gel coating as a novel halogen- and formaldehyde-free flame retardant finishing for cotton fabric. *Polym. Degrad. Stab.* **2019**, *162*, 148–159. [CrossRef]
30. Caldara, M.; Colleoni, C.; Guido, E.; Rosace, G. Optical sensor development for smart textiles. In Proceedings of the 12th World Textile Conference AUTEX, Zadar, Croatia, 13–15 June 2012; pp. 1149–1452.
31. Caldara, M.; Colleoni, C.; Guido, E.; Re, V.; Rosace, G. Optical monitoring of sweat pH by a textile fabric wearable sensor based on covalently bonded litmus-3-glycidoxypropyltrimethoxysilane coating. *Sens. Actuators B Chem.* **2016**, *222*, 213–220. [CrossRef]
32. Trovato, V.; Rosace, G.; Colleoni, C.; Plutino, M.R. Synthesis and characterization of halochromic hybrid sol-gel for the development of a pH sensor fabric. *IOP Conf. Ser. Mater. Sci. Eng.* **2017**, *254*, 72027. [CrossRef]
33. Ielo, I.; Rando, G.; Giacobello, F.; Sfameni, S.; Castellano, A.; Galletta, M.; Drommi, D.; Rosace, G.; Plutino, M.R. Synthesis, Chemical–Physical Characterization, and Biomedical Applications of Functional Gold Nanoparticles: A Review. *Molecules* **2021**, *26*, 5823. [CrossRef]
34. Mahltig, B.; Textor, T. Combination of silica sol and dyes on textiles. *J. Sol-Gel Sci. Technol.* **2006**, *39*, 111–118. [CrossRef]
35. AbouElmaaty, T.; Abdeldayem, S.; Ramadan, S.; Sayed-Ahmed, K.; Plutino, M. Coloration and Multi-Functionalization of Polypropylene Fabrics with Selenium Nanoparticles. *Polymers* **2021**, *13*, 2483. [CrossRef]
36. Ibrahim, N.; Eid, B.; Abd El-Aziz, E.; AbouElmaaty, T.; Ramadan, S. Loading of chitosan Nano metal oxide hybrids ontocotton/polyester fabrics to impart permanent and effective multifunctions. *Int. J. Biol. Macromol.* **2017**, *105*, 769–776. [CrossRef]
37. Aya, R.; Abdel Rahim, A.; Amr, H.; El-Amir, M. Improving Performance and Functional Properties of Different Cotton Fabrics by Silicon Dioxide Nanoparticles. *J. Eng. Sci.* **2019**, *4*, 1–17.
38. AbouElmaaty, T.; Elsisi, H.G.; Elsayad, G.M.; Elhadad, H.H.; Sayed-Ahmed, K.; Plutino, M.R. Fabrication of New Multifunctional Cotton/Lycra Composites Protective Textiles through Deposition of Nano Silica Coating. *Polymers* **2021**, *13*, 2888. [CrossRef]
39. Trovato, V.; Vitale, A.; Bongiovanni, R.; Ferri, A.; Rosace, G.; Plutino, M.R. Development of a Nitrazine Yellow-glycidyl methacrylate coating onto cotton fabric through thermal-induced radical polymerization reactions: A simple approach towards wearable pH sensors applications. *Cellulose* **2021**, *28*, 3847–3868. [CrossRef]
40. Li, X.M.; Reinhoudt, D.; Crego, C.M. What do we need for a superhydrophobic surface? A review on the recent progress in the preparation of superhydrophobic surfaces. *Chem. Soc. Rev.* **2007**, *36*, 1350–1368. [CrossRef]
41. Richard, E.; Lakshmi, R.V.; Aruna, S.T.; Basu, B.J. A simple cost-effective and eco-friendly wet chemical process for the fabrication of superhydrophobic cotton fabrics. *Appl. Surf. Sci.* **2013**, *277*, 302–309. [CrossRef]
42. Xu, B.; Cai, Z.; Wang, W.; Ge, F. Preparation of superhydrophobic cotton fabrics based on SiO2 nanoparticles and ZnO nanorod arrays with subsequent hydrophobic modification. *Surf. Coat. Technol.* **2010**, *204*, 1556–1561. [CrossRef]
43. Zhang, X.; Shi, F.; Niu, J.; Jiang, Y.; Wang, Z. Superhydrophobic surfaces: From structural control to functional application. *J. Mater. Chem* **2008**, *18*, 621. [CrossRef]
44. Patankar, N.A. Mimicking the Lotus Effect: Influence of Double Roughness Structures and Slender Pillars. *Langmuir* **2004**, *20*, 8209–8213. [CrossRef]
45. Cassie, A.B.D.; Baxter, S. Wettability of porous surfaces. *Trans. Faraday Soc.* **1944**, *40*, 546–551. [CrossRef]
46. Balani, K.; Batista, R.G.; Lahiri, D.; Agarwal, A. The hydrophobicity of a lotus leaf: A nanomechanical and computational approach. *Nanotechnology* **2009**, *20*, 305707. [CrossRef]
47. Heiman-Burstein, D.; Dotan, A.; Dodiuk, H.; Kenig, S. Hybrid Sol-Gel Superhydrophobic Coatings Based on Alkyl Silane-Modified Nanosilica. *Polymers* **2021**, *13*, 539. [CrossRef]
48. Xue, C.H.; Jia, S.T.; Zhang, J.; Ma, J.Z. Large-area fabrication of superhydrophobic surfaces for practical applications: An overview. *Sci. Technol. Adv. Mater.* **2010**, *11*, 033002. [CrossRef]
49. Zhang, M.; Wang, S.; Wang, C.; Li, J. A facile method to fabric at superhydrophobic cotton fabrics. *Appl. Surf. Sci.* **2012**, *261*, 561. [CrossRef]
50. Huang, W.Q.; Xing, Y.J.; Yu, Y.; Shang, S.M.; Dai, J.J. Enhanced washing durability of hydrophobic coating on cellulose fabric using polycarboxylic acids. *Appl. Surf. Sci.* **2011**, *257*, 4443. [CrossRef]
51. Rando, G.; Sfameni, S.; Galletta, M.; Drommi, D.; Cappello, S.; Plutino, M.R. Functional Nanohybrids and Nanocomposites Development for the Removal of Environmental Pollutants and Bioremediation. *Molecules* **2022**, *27*, 4856. [CrossRef]

52. Bae, G.Y.; Jeong, Y.G.; Min, B.G. Superhydrophobic PET fabrics achieved by silica nanoparticles and water-repellent agent. *Fibers Polym.* **2010**, *11*, 976. [CrossRef]
53. Bae, G.Y.; Geun, Y.; Min, B.G.; Jeong, Y.G.; Lee, S.C.; Jang, J.H.; Koo, G.H. Superhydrophobicity of cotton fabrics treated with silica nanoparticles and water-repellent agent. *J. Colloid Interface Sci.* **2009**, *337*, 170–175. [CrossRef]
54. Lehocky, M.; Amaral, P.F.F.; St'ahel, P.; Coelho, M.A.Z.; Timmons, A.M.; Coutinho, J.A.P. Preparation and characterization of organosilicon thin films for selective adhesion of yarrowialipolytica yeast cells. *J. Chem. Technol. Biotechnol.* **2007**, *366*, 360–366. [CrossRef]
55. Lakshmi, R.V.; Bera, P.; Anandan, C.; Basu, B. Effect of the size of silica nanoparticles on wettability and surface chemistry of sol–gel superhydrophobic and oleophobic nanocomposite coatings. *J. Appl. Surf. Sci.* **2014**, *320*, 780. [CrossRef]
56. Huang, W.H.; Lin, C.S. Robust superhydrophobic transparent coatings fabricated by a low-temperature sol–gel process. *Appl. Surf. Sci.* **2014**, *305*, 702. [CrossRef]
57. Shang, Q.; Zhou, Y.; Xiao, G. A simple method for the fabrication of silica-based superhydrophobic surfaces. *Coat. Technol. Res.* **2014**, *11*, 509. [CrossRef]
58. Ramezani, M.; Vaezi, M.R.; Kazemzadeh, A. Preparation of silane-functionalized silica films via two-step dip coating sol–gel and evaluation of their superhydrophobic properties. *Appl. Surf. Sci.* **2014**, *317*, 147. [CrossRef]
59. Zou, H.; Lin, S.; Tu, Y.; Liu, G.; Hu, J.; Li, F.; Miao, L.; Zhang, G.; Luo, H.; Liu, F.; et al. Simple approach towards fabrication of highly durable and robust superhydrophobic cotton fabric from functional di block copolymer. *J. Mat. Chem. A* **2013**, *1*, 11246. [CrossRef]
60. Gao, Y.; Huang, Y.G.; Feng, S.J.; Gu, G.T.; Qing, F.L. Novel superhydrophobic and highly oleophobic PFPE-modified silica nanocomposite. *J. Mater. Sci.* **2010**, *45*, 460. [CrossRef]
61. Guido, E.; Colleoni, C.; De Clerck, K.; Plutino, M.R.; Rosace, G. Influence of catalyst in the synthesis of a cellulose-based sensor: Kinetic study of 3-glycidoxypropyltrimethoxysilane epoxy ring opening by Lewis acid. *Sens. Actuators B Chem.* **2014**, *203*, 213–222. [CrossRef]
62. Trovato, V.; Colleoni, C.; Castellano, A.; Plutino, M.R. The key role of 3-glycidoxypropyltrimethoxysilane sol–gel precursor in the development of wearable sensors for health monitoring. *J. Sol-Gel Sci. Technol.* **2018**, *87*, 27–40. [CrossRef]
63. Plutino, M.R.; Guido, E.; Colleoni, C.; Rosace, G. Effect of GPTMS functionalization on the improvement of the pH-sensitive methyl red photostability. *Sens. Actuators B* **2017**, *238*, 281–291. [CrossRef]
64. Rosace, G.; Guido, E.; Colleoni, C.; Brucale, M.; Piperopoulos, E.; Milone, C.; Plutino, M.R. Halochromic resorufin-GPTMS hybrid sol-gel: Chemical-physical properties and use as pH sensor fabric coating. *Sens. Actuators B Chem.* **2017**, *241*, 85–95. [CrossRef]
65. Plutino, M.R.; Colleoni, C.; Donelli, I.; Freddi, G.; Guido, E.; Maschi, O.; Mezzi, A.; Rosace, G. Sol-gel 3-glycidoxypropyltriethoxysilane finishing on different fabrics: The role of precursor concentration and catalyst on the textile performances and cytotoxic activity. *J. Colloid Interface Sci.* **2017**, *506*, 504–517. [CrossRef]
66. Ielo, I.; Galletta, M.; Rando, G.; Sfameni, S.; Cardiano, P.; Sabatino, G.; Drommi, D.; Rosace, G.; Plutino, M.R. Design, synthesis and characterization of hybrid coatings suitable for geopolymeric-based supports for the restoration of cultural heritage. *IOP Conf. Ser. Mater. Sci. Eng.* **2020**, *777*, 012003. [CrossRef]
67. Sfameni, S.; Rando, G.; Marchetta, A.; Scolaro, C.; Cappello, S.; Urzì, C.; Visco, A.; Plutino, M.R. Development of eco-friendly hydrophobic and fouling-release coatings for blue-growth environmental applications: Synthesis, mechanical characterization and biological activity. *Gels* **2022**, *8*, 528. [CrossRef]
68. Zhu, D.; Hu, N.; Schaefer, D.W. *Chapter 1—Water-Based Sol–Gel Coatings for Military Coating Applications*; Elsevier: Amsterdam, The Netherlands, 2020; pp. 1–27.
69. Khatri, M.; Qureshi, U.A.; Ahmed, F.; Khatri, Z.; Kim, I.S. Dyeing of Electrospun Nanofibers BT. In *Handbook of Nanofibers*; Barhoum, A., Bechelany, M., Makhlouf, A.S.H., Eds.; Springer International Publishing: Cham, Switzerland, 2019; pp. 373–388, ISBN 978-3-319-53655-2.
70. El-Ghazali, S.; Khatri, M.; Kobayashi, S.; Kim, I.S. 1—An overview of medical textile materials. In *Medical Textiles from Natural Resources*; The Textile Institute Book, Series; Mondal, M.I.H., Ed.; Woodhead Publishing: Sawston, UK, 2022; pp. 3–42, ISBN 978-0-323-90479-7.
71. Basak, S.; Samanta, K.K.; Chattopadhyay, S.K. Fire retardant property of cotton fabric treated with herbal extract. *J. Text. Inst.* **2015**, *106*, 1338–1347. [CrossRef]
72. Lin, T.-C.; Lee, D.-J. Cotton fabrics modified for use in oil/water separation: A perspective review. *Cellulose* **2021**, *28*, 4575–4594. [CrossRef]
73. Hu, B.; Pu, H.; Sun, D.-W. Multifunctional cellulose based substrates for SERS smart sensing: Principles, applications and emerging trends for food safety detection. *Trends Food Sci. Technol.* **2021**, *110*, 304–320. [CrossRef]
74. Malshe, P.; Mazloumpour, M.; El-Shafei, A.; Hauser, P. Multi-functional military textile: Plasma-induced graft polymerization of a C6 fluorocarbon for repellent treatment on nylon–cotton blend fabric. *Surf. Coat. Technol.* **2013**, *217*, 112–118. [CrossRef]
75. Marmur, A.A. Guide to the Equilibrium Contact Angles Maze. In *Contact angle, Wettability and Adhesion*; CRC Press: Boca Raton, FL, USA, 2009; pp. 1–18.
76. Ubuo, E.E.; Udoetok, I.A.; Tyowua, A.T.; Ekwere, I.O.; Al-Shehri, H.S. The Direct Cause of Amplified Wettability: Roughness or Surface Chemistry? *J. Compos. Sci.* **2021**, *5*, 213. [CrossRef]

77. Wenzel, R.N. Resistance of solid surfaces to wetting by water. *Ind. Eng. Chem.* **1936**, *28*, 988–994. [CrossRef]
78. Glosz, K.; Stolarczyk, A.; Jarosz, T. Siloxanes—Versatile Materials for Surface Functionalisation and Graft Copolymers. *Int. J. Mol. Sci.* **2020**, *21*, 6387. [CrossRef]
79. Han, Y.; Hu, J.; Xin, Z. In-Situ Incorporation of Alkyl-Grafted Silica into Waterborne Polyurethane with High Solid Content for Enhanced Physical Properties of Coatings. *Polymers* **2018**, *10*, 514. [CrossRef]
80. Sánchez-Milla, M.; Gómez, R.; Pérez-Serrano, J.; Sánchez-Nieves, J.; de la Mata, F.J. Functionalization of silica with amine and ammonium alkyl chains, dendrons and dendrimers: Synthesis and antibacterial properties. *Mater. Sci. Eng. C* **2020**, *109*, 110526. [CrossRef]
81. Radhakrishnan, B.; Ranjan, R.; Brittain, W.J. Surface initiated polymerizations from silica nanoparticles. *Soft Matter* **2006**, *2*, 386–396. [CrossRef]
82. Donnini, J.; Corinaldesi, V.; Nanni, A. Mechanical properties of FRCM using carbon fabrics with different coating treatments. *Compos. Part B Eng.* **2016**, *88*, 220–228. [CrossRef]
83. Koo, K.; Park, Y. Characteristics of double-layer coated fabrics with and without phase change materials and nano-particles. *Fibers Polym.* **2014**, *15*, 1641–1647. [CrossRef]
84. Mazzon, G.; Zahid, M.; Heredia-Guerrero, J.A.; Balliana, E.; Zendri, E.; Athanassiou, A.; Bayer, I.S. Hydrophobic treatment of woven cotton fabrics with polyurethane modified aminosilicone emulsions. *Appl. Surf. Sci.* **2019**, *490*, 331–342. [CrossRef]
85. Lin, H.; Rosu, C.; Jiang, L.; Sundar, V.A.; Breedveld, V.; Hess, D.W. Nonfluorinated Superhydrophobic Chemical Coatings on Polyester Fabric Prepared with Kinetically Controlled Hydrolyzed Methyltrimethoxysilane. *Ind. Eng. Chem. Res.* **2019**, *58*, 15368–15378. [CrossRef]
86. Pavlidou, S.; Paul, R. 3—Soil repellency and stain resistance through hydrophobic and oleophobic treatments. In *Waterproof and Water Repellent Textiles and Clothing*; The Textile Institute Book, Series; Williams, J., Ed.; Woodhead Publishing: Sawston, UK, 2018; pp. 73–88. ISBN 978-0-08-101212-3.
87. Peran, J.; Ercegović Ražić, S. Application of atmospheric pressure plasma technology for textile surface modification. *Text. Res. J.* **2019**, *90*, 1174–1197. [CrossRef]
88. Liu, Y.; Xin, J.H.; Choi, C.-H. Cotton Fabrics with Single-Faced Superhydrophobicity. *Langmuir* **2012**, *28*, 17426–17434. [CrossRef]
89. Li, Y.; Yu, Q.; Yin, X.; Xu, J.; Cai, Y.; Han, L.; Huang, H.; Zhou, Y.; Tan, Y.; Wang, L.; et al. Fabrication of superhydrophobic and superoleophilic polybenzoxazine-based cotton fabric for oil–water separation. *Cellulose* **2018**, *25*, 6691–6704. [CrossRef]
90. He, Y.; Wan, M.; Wang, Z.; Zhang, X.; Zhao, Y.; Sun, L. Fabrication and characterization of degradable and durable fluoride-free super-hydrophobic cotton fabrics for oil/water separation. *Surf. Coat. Technol.* **2019**, *378*, 125079. [CrossRef]
91. Tang, H.; Fu, Y.; Yang, C.; Zhu, D.; Yang, J. A UV-driven superhydrophilic/superoleophobic polyelectrolyte multilayer film on fabric and its application in oil/water separation. *RSC Adv.* **2016**, *6*, 91301–91307. [CrossRef]
92. Wang, J.; Chen, Y. Oil–water separation capability of superhydrophobic fabrics fabricated via combining polydopamine adhesion with lotus-leaf-like structure. *J. Appl. Polym. Sci.* **2015**, *132*. [CrossRef]
93. Zhao, Y.; Liu, E.; Fan, J.; Chen, B.; Hu, X.; He, Y.; He, C. Superhydrophobic PDMS/wax coated polyester textiles with self-healing ability via inlaying method. *Prog. Org. Coat.* **2019**, *132*, 100–107. [CrossRef]
94. Huang, G.; Huo, L.; Jin, Y.; Yuan, S.; Zhao, R.; Zhao, J.; Li, Z.; Li, Y. Fluorine-free superhydrophobic PET fabric with high oil flux for oil–water separation. *Prog. Org. Coat.* **2022**, *163*, 106671. [CrossRef]
95. Canal, L.P.; González, C.; Molina-Aldareguía, J.M.; Segurado, J.; LLorca, J. Application of digital image correlation at the microscale in fiber-reinforced composites. *Compos. Part A Appl. Sci. Manuf.* **2012**, *43*, 1630–1638. [CrossRef]
96. Ghaffari, S.; Yousefzadeh, M.; Mousazadegan, F. Investigation of thermal comfort in nanofibrous three-layer fabric for cold weather protective clothing. *Polym. Eng. Sci.* **2019**, *59*, 2032–2040. [CrossRef]
97. Wan, T.; Stylios, G.K. Effects of coating process on the surface roughness of coated fabrics. *J. Text. Inst.* **2017**, *108*, 712–719. [CrossRef]
98. Hsieh, C.-T.; Chen, W.-Y.; Wu, F.-L. Fabrication and superhydrophobicity of fluorinated carbon fabrics with micro/nanoscaled two-tier roughness. *Carbon* **2008**, *46*, 1218–1224. [CrossRef]
99. Park, S.; Kim, J.; Park, C.H. Influence of micro and nano-scale roughness on hydrophobicity of a plasma-treated woven fabric. *Text. Res. J.* **2016**, *87*, 193–207. [CrossRef]
100. Zarrinabadi, E.; Abghari, R.; Nazari, A.; Mirjalili, M. Environmental effects of enhancement of mechanical and hydrophobic properties of polyester fabrics using silica/kaolinite/silver nanocomposite: A facile technique for synthesis and RSM optimization. *Eurasian J. Biosci.* **2018**, *12*, 1–14.
101. Forsman, N.; Lozhechnikova, A.; Khakalo, A.; Johansson, L.-S.; Vartiainen, J.; Österberg, M. Layer-by-layer assembled hydrophobic coatings for cellulose nanofibril films and textiles, made of polylysine and natural wax particles. *Carbohydr. Polym.* **2017**, *173*, 392–402. [CrossRef]
102. Glampedaki, P.; Jocic, D.; Warmoeskerken, M.M.C.G. Moisture absorption capacity of polyamide 6,6 fabrics surface functionalised by chitosan-based hydrogel finishes. *Prog. Org. Coat.* **2011**, *72*, 562–571. [CrossRef]
103. Tissera, N.D.; Wijesena, R.N.; Perera, J.R.; de Silva, K.M.N.; Amaratunge, G.A.J. Hydrophobic cotton textile surfaces using an amphiphilic graphene oxide (GO) coating. *Appl. Surf. Sci.* **2015**, *324*, 455–463. [CrossRef]

104. Zupin, Ž.; Hladnik, A.; Dimitrovski, K. Prediction of one-layer woven fabrics air permeability using porosity parameters. *Text. Res. J.* **2011**, *82*, 117–128. [CrossRef]
105. Liu, R.; Li, J.; Li, M.; Zhang, Q.; Shi, G.; Li, Y.; Hou, C.; Wang, H. MXene-Coated Air-Permeable Pressure-Sensing Fabric for Smart Wear. *ACS Appl. Mater. Interfaces* **2020**, *12*, 46446–46454. [CrossRef]
106. Jiang, J.; Shen, Y.; Yu, D.; Yang, T.; Wu, M.; Yang, L.; Petru, M. Porous Film Coating Enabled by Polyvinyl Pyrrolidone (PVP) for Enhanced Air Permeability of Fabrics: The Effect of PVP Molecule Weight and Dosage. *Polymers* **2020**, *12*, 2961. [CrossRef]
107. Oh, J.-H.; Ko, T.-J.; Moon, M.-W.; Park, C.H. Nanostructured fabric with robust superhydrophobicity induced by a thermal hydrophobic ageing process. *RSC Adv.* **2017**, *7*, 25597–25604. [CrossRef]

MDPI
St. Alban-Anlage 66
4052 Basel
Switzerland
Tel. +41 61 683 77 34
Fax +41 61 302 89 18
www.mdpi.com

Nanomaterials Editorial Office
E-mail: nanomaterials@mdpi.com
www.mdpi.com/journal/nanomaterials